Springer INdAM Series

Volume 53

This series will publish textbooks, multi-authors books, thesis and monographs in English language resulting from workshops, conferences, courses, schools, seminars, doctoral thesis, and research activities carried out at INDAM - Istituto Nazionale di Alta Matematica, http://www.altamatematica.it/en. The books in the series will discuss recent results and analyze new trends in mathematics and its applications.

THE SERIES IS INDEXED IN SCOPUS

Gilberto Bini
Editor

Algebraic Geometry between Tradition and Future

An Italian Perspective

Springer

Editor
Gilberto Bini
Department of Mathematics and Computer Science
University of Palermo
Palermo, Italy

ISSN 2281-518X ISSN 2281-5198 (electronic)
Springer INdAM Series
ISBN 978-981-19-8280-4 ISBN 978-981-19-8281-1 (eBook)
https://doi.org/10.1007/978-981-19-8281-1

This Springer imprint is published by the registered company Springer Nature Singapore Pte Ltd.
The registered company address is: 152 Beach Road, #21-01/04 Gateway East, Singapore 189721, Singapore

Contents

About the Editor

Gilberto Bini received his Ph.D. from Scuola Normale Superiore in July 2000. After that, he spent the following 4 years in the USA (University of Michigan) and in the Netherlands (Universiteit van Amsterdam). He became a "Ricercatore di Geometria" at the Università degli Studi di Milano in 2004. In 2015, he started his job as an associate professor at Università degli Studi di Milano where he worked until 2019. The year afterwards, he moved to the Università degli Studi di Palermo where he started his job as a full professor of geometry. His main research interests focus on projective geometry, especially the classification of complex projective varieties, as well as its application to Grassmann tensors and computer vision. He authored various peer-reviewed publications (proceedings, papers, books) in international journals. Through the years, he has organised different outreach activities (exhibitions, laboratories, public lectures, etc.). From 2015 until 2021, he was an editor of *Matematica, Cultura e Società*, the journal of the Unione Matematica Italiana. Last year, he co-organised with Claudio Fontanari the INdAM Workshop "Italian Algebraic Geometry between Tradition and Future".

Introduction

Gilberto Bini

Learn from yesterday, live for today, hope for tomorrow.
The important thing is not to stop questioning.

Albert Einstein

This volume collects the proceedings of the *Italian Algebraic Geometry between Tradition and Future*, an INdAM Workshop that was held in Rome on December 6–8, 2021. The *Istituto Nazionale di Alta Matematica* (INdAM) was founded by Francesco Severi who died on December 8, 1961. The workshop was also meant as a celebration of this anniversary. For these purposes, the contribution by Claudio Fontanari and Stefano Gattei is a homage to Severi's mathematical library at INdAM, with a list of 1400 items, which reflects Severi's life and works. Indeed, as recalled by Leonard Roth:[1]

Severi's scientific work presents several features that, taken together, must make his career a rarity. To begin with, there is the uniformly high level of his very considerable scientific production: as a rule Severi attacks only important questions of general character and is usually of great difficulty. (...) In the second place, one cannot fail to observe an essential unity of outlook. (...) Severi passes from one topic to another only to turn back at some future time. His production resembles a vast network linking many nodal points, and his thought may pass from one such point to another by seemingly devious paths. (...) Third, there is the topicality of Severi's work. Few of his papers have been relegated to the museum; the rest remain part of the living body of geometrical discipline. Aside from their theoretical

[1] *Francesco Severi. Obituary note. Journal of the London Mathematical Society, Volume 38, Issue 1, 1963, pp. 282–307*, https://londmathsoc.onlinelibrary.wiley.com/doi/10.1112/jlms/s1-38.1.282, p. 282–283.

G. Bini (✉)
Università di Palermo, Dipartimento di Matematica e Informatica, Palermo, Italy
e-mail: gilberto.bini@unipa.it

G. Bini (ed.), *Algebraic Geometry between Tradition and Future*,
Springer INdAM Series 53, https://doi.org/10.1007/978-981-19-8281-1_1

importance, they follow the best of all recipes for scientific longevity: embedded in many of them is a most interesting problem which still defies all attempts at solution. Finally, there is the sheer length of Severi's career: it extends to 60 years of continuous activity in which again and again he returns to the problems of his youth, sometimes after an interval of 40 or 50 years (...)

Two years before his death, Francesco Severi published the last volume of the *Treatise on Algebraic Systems Over a Surface and an Algebraic Variety*. Very likely, it was published at the twilight of an incredible season for algebraic geometry in Italy, which began almost a century beforehand when Luigi Cremona was assigned the chair of Geometria Superiore in Bologna in 1860. This prolific period has had a prominent influence on the evolution of complex algebraic geometry and still inspires modern research in the area, both at the national and international levels.

In this volume, we opt to move a century backwards from Francesco Severi's times which are depicted in Angelo Guerraggio's contribution. A different viewpoint is presented on the one hand by Maria Giulia Lugaresi, who reports the biography of Fabio Conforto, a contemporary of Francesco Severi, and on the other hand by Livia Giacardi, who reviews the life and works of Alessandro Terracini before and after the racial laws were issued in 1938. Persecution is also one of the main focuses of the contributions by Livia Giacardi, Erika Luciano and Elena Scalambro, which tells about the biography of Gino Fano.

A new approach to classification of projective varieties is introduced in the 1980s from seminal work by Shigefumi Mori, who laid the foundations of the Minimal Model Program: see the contribution by Marco Andreatta for a review article on these topics from the time of Gino Fano to date. In higher dimensions, the classification problem of projective varieties is quite difficult yet very challenging. Rationality questions for threefolds were addressed by Gino Fano and Ugo Morin, who were the first to study rationality problems for algebraic varieties of dimension greater than or equal to 3. Moreover, they had an influence on more recent solutions of some of these problems, including the proof of the nonrationality of the cubic threefold by Herbert Clemens and Phillip Griffiths. In the same year, Artin and Mumford found a counterexample to the nonrationality of a complex unirational threefold, which is revisited in the contribution written by Alessandro Verra.

At the turn of the twentieth century, the 'fundamental problem' consisted of finding an algebraic proof of the Fundamental Theorem by Henri Poincaré, which stated the equality of the irregularity and the dimension of the Picard variety of a complex nonsingular projective surface. Edoardo Sernesi writes about this problem from the point of view of Guido Zappa who noted a mistake in one of Francesco Severi's proofs. The paper by Ciro Ciliberto illustrates Enriques' contribution to the fundamental problem.

Needless to say, Federigo Enriques and Guido Castelnuovo are the two main characters in the classification of complex nonsingular projective surfaces. In this volume, there are various papers on the life and works of Guido Castelnuovo, more precisely one by Ciro Ciliberto and Claudio Fontanari, another one by Claudio Fontanari and, last but not the least, one by Enrico Rogora. Currently, the investigation of algebraic surfaces is still a very active research area. For instance, the geometry of algebraic cycles on most surfaces is considerably mysterious, as claimed by Bloch's conjecture, which Francesco Severi attempted to solve but failed, alongside many other claimed issues that still need proving.

Birational maps appeared in a series of papers by Luigi Cremona, who can be acknowledged as the father of algebraic geometry in Italy. He had many connections with foreigner scholars and active students: to mention a few, Guido Ascoli, Eugenio Bertini and Giovanni Battista Guccia. Many more are detailed in the contribution by Nicla Palladino and Maria Alessandra Vaccaro. Some aspects of Cremona transformations, and their relations to cubic surfaces, are brought up to the reader's attention in the contribution by Aldo Brigaglia, which concludes our journey back to the origins of algebraic geometry in Italy.

Francesco Severi's Mathematical Library

Claudio Fontanari and Stefano Gattei

When writers die they become books—
which is, after all, not too bad an incarnation.

Jorge Luis Borges

Abstract We introduce and offer the inventory of the mathematical library of Francesco Severi (1879–1961), which is held at the Istituto Nazionale di Alta Matematica (INdAM) in Rome.

Keywords Mathematical library · Francesco Severi · INdAM

An investigation into the scientific and philosophical literature found in the private libraries of modern scientists, philosophers, and scholars allows us an unprecedented vantage point in an author's world. A long-neglected field of research, the subject has become one of increasing interest for historians—as shown, to mention but one example, by Antonio Favaro's pioneering studies on the private library of Galileo.[1]

[1] Antonio Favaro, "La libreria di Galileo Galilei," *Bullettino di Bibliografia e di Storia delle Scienze Matematiche e Fisiche*, XIX, 1886, pp. 219–293; Id., "Appendice prima alla libreria di

Claudio Fontanari's research work was partially supported by GNSAGA of INdAM and by PRIN 2017 "Moduli Theory and Birational Classification."

C. Fontanari (✉)
Dipartimento di Matematica, Università degli Studi di Trento, Trento, Italy
e-mail: claudio.fontanari@unitn.it

S. Gattei
Dipartimento di Sociologia e Ricerca Sociale, Università degli Studi di Trento, Trento, Italy
e-mail: stefano.gattei@unitn.it

G. Bini (ed.), *Algebraic Geometry between Tradition and Future*,
Springer INdAM Series 53, https://doi.org/10.1007/978-981-19-8281-1_2

1 The Relevance of Private Libraries in Historical Perspective

Library catalogues and inventories, once they have been organized and made available to the scientific community, may be profitably used not only in order to reconstruct the "material" history of institutions, but also, more broadly, to provide an in-depth account of individual authors or entire periods of intellectual history.

Private library catalogues and inventories, gathered across the past four or five centuries, become a powerful and effective tool for scholars aiming to

(a) Account for the course of studies, intellectual competences, and interests of an author, by working on the list of his or her books
(b) Inquire about the circulation and fortune of the author's own writings, widening the scope of such inquiry to the libraries of subsequent generations
(c) Become familiar with the intellectual features of a given time
(d) Study the history of the transformations of private libraries through time, thereby contributing to a sort of "sociology of library structures"
(e) Take notice of the changes that take place in the "system of knowledge," of the mutual relationships among different disciplines (it may be interesting to investigate, for instance, the way in which the set of scientific or philosophical texts evolves in the private libraries of jurists, linguists, naturalists, etc.)
(f) Follow the circulation of texts, or an array of texts, of specific relevance or rarity

Most importantly, getting to know private libraries might allow, in many cases, to rectify widespread but threadbare interpretations of the thought and production of a given author, whose legacy is all too often left to the sort of vulgate found in

Galileo Galilei," *Bullettino di Bibliografia e di Storia delle Scienze Matematiche e Fisiche*, XX, 1887, pp. 372–376; and Id., "Appendice seconda alla 'Libreria di Galileo'," *Atti e memorie della R. Accademia di Scienze, lettere ed arti in Padova*, XII, 1895–1896, pp. 44–50. Favaro's attempted reconstruction of Galileo's private library is especially noteworthy, as he did not have a specific inventory of Galileo's own books available. Favaro appealed to several resources: the inventory of the inheritance of Galileo's son, Vincenzo, who died in 1649; the inventory of the books found in the house of his widow, Sestilia Bocchineri Galilei, in January 1663; and the inventory of the rich library of Vincenzo Viviani, who inherited a large part of the books that had once belonged to Galileo, and which are now held at the National Library of Florence and in some private collections. See also Pierluigi Pizzamiglio, "Le biblioteche di Copernico e di Galileo. Il ruolo della stampa nella nascita della scienza moderna," in Carlo Vinti (ed.), *Galileo e Copernico. Alle origini del pensiero scientifico moderno*, Assisi: Porziuncola, 1990, pp. 115–140; Maurizio Torrini, "La biblioteca di Galileo e dei galileiani," *Intersezioni*, XXI, 3, 2001, pp. 545–558; Michele Camerota, "La biblioteca di Galileo. Alcune integrazioni desunte dal carteggio," in Francesca Maria Crasta (ed.), *Biblioteche filosofiche private in età moderna e contemporanea*, Florence: Le Lettere, 2010, pp. 81–95; and Crystal Hall, "Galileo's Library Reconsidered," *Galilæana*, XII, 2015, pp. 29–82. For a very interesting case of a book long-thought to have belonged to Galileo, but which did not turn up as such, yet may still shed new light on his early intellectual development, see Stefano Gattei, "Amicus Galilæus sed Magis Amica Veritas: The True Story of 'Galileo's Philoponus'," *History of Universities*, 31, 2, 2018, pp. 82–129.

textbooks or offered by more or less charitable commentator—not to mention the fact that unknown and unexpected elements may surface, shedding new light on the author's personality and interests and suggesting richer readings of his or her work.

The document summary of the Greek library of Franciscus Patricius (1529–1597), for example, reveals his overall project to bring to light the encyclopedia of Platonic knowledge—in which music, mathematics, poetry, rhetoric, and art of memory occupy central positions beside theology and philosophy—and contrast it to that Aristotle's own. Analogously, skimming through the inventory of Wilhelm Dilthey's (1833–1911) library, one's attention is caught by the breadth and articulation of a philosophical reflection assiduously dedicated—notwithstanding the author's own proclamation of the autonomy of the "humanities"—to keeping dialogue with the natural science of the period alive. And again, Friedrich Nietzsche's (1844–1900) library testifies, in the variety of the collected texts and of the documented interdisciplinary interests, to an approach characterized by the "patience of the philologist," foreign to the "cult of genius" in all its forms, closely connected to the evolution both of positive knowledge and of different scientific disciplines. Or, to offer but a last example, the study of the private library of Karl Popper (1902–1994) provides key new elements to the understanding of his problematic relations with the members of the Vienna Circle, his extensive study of the works of Karl Marx, as well as his literary and musical preferences that implicitly find their way into Popper's published works.[2]

Indeed, the catalogues of private libraries represent an orienting basis, a sort of "frame of reference" for the definition and the assessment of the library collections themselves. Thanks to their multifaceted features, these instruments offer scholars the opportunity to study individual authors in depth and hands-on, so to say, bringing to light implicit but at times fairly important aspects in the creation of philosophical traditions and fields of research. But the chief intention, in the reconstruction of important philosophical and scientific libraries—usually through notary deeds or sale catalogues, or by way of explicit or implicit references in published works, manuscripts, and correspondences, or else through bequests or donations—is not that of privileging intellectual biography, as in historiographical research, nor of offering material of mere erudition. The aim is, rather, to expand the use of private libraries in order to assist and enrich historical and philosophical investigations, taking into account the growing relevance of research fields like the history of culture or the history of ideas, as well as the German Begriffsgeschichte.

[2] See Maria Muccillo, "La biblioteca greca di Francesco Patrizi," in Eugenio Canone (ed.), *Bibliotecae selectae. Da Cusano a Leopardi*, Florence: Olschki, 1993, pp. 73–118; Wilhelm Dilthey, *Archive der Literatur in ihrer Bedeutung für das Studium der Geschichte der Philosophie* [1889], in *Gesammelte Schriften*, Vol. 4: *Die Jugendgeschichte Hegels und andere Abhandlungen zur Geschichte des deutschen Idealismus*, Leipzig-Berlin: B. G. Teubner, 1921, pp. 555–575; Giuliano Campioni, Paolo D'Iorio, Maria Cristina Fornari, Francesco Fronterotta, Andrea Orsucci (eds.), *Nietzsches persönliche Bibliothek*, Berlin-New York: Walter De Gruyter, 2003; Stefano Gattei, "La biblioteca di Karl Popper," in *L'oggetto libro '96: Arte della stampa, mercato e collezionismo*, Milan: Sylvestre Bonnard, 1996, pp. 72–93.

In this perspective, the study of private library collections has acquired an ever-increasing relevance in recent years, especially as part of the renewal in the vast range of book studies. More specifically, the exploration of private library collections is a fundamental aspect of the methodological debate on the history of libraries, whether a historical-institutional approach is adopted or else a more bibliographic one. It is no accident that private collections have become the subject of much discussion in the most recent occasions of collective debate about these issues.[3]

Private collections presuppose a considerably vaguer definition of a library than the one that derives from a study of the political-administrative events of individual library institutions. In fact, it implies a completely different level of analysis and requires a focus on the cultural profile of the bibliographic identity of each collection. This is necessary both as a fundamental element to analyze the stratification of the individual collections and as an indication of research interests, disciplinary horizons, and bibliophilic orientations that characterize historical periods, groups of readers, and individual personalities. In other words, private collections of books, whether they merge with public libraries at some point or are auctioned and end up on the shelves of other private libraries, are always privileged places of memory, material traces of a world which is otherwise destined to deteriorate and eventually vanish.

2 Francesco Severi and His Private Library

The importance of the study of private library collections holds equally for modern and contemporary authors, for scholars of the humanities, and for men of science.[4] As far as we know, no in-depth study of private libraries of mathematicians has been produced so far. A symposium on Francesco Severi is a welcome opportunity to offer the very starting point of such study, namely, a list of his books, as found in the library that is now held at the Istituto Nazionale di Alta

[3] See, for example, Angela Nuovo (ed.), *Biblioteche private in età moderna*, Milan: Sylvestre Bonnard, 2005; Fiammetta Sabba (ed.), *Le biblioteche private come paradigma bibliografico*, Rome: Bulzoni, 2008; Francesca Maria Crasta (ed.), *Biblioteche filosofiche private in età moderna e contemporanea*, Florence: Le Lettere, 2010; and Renzo Ragghianti and Alessandro Savorelli (eds.), *Biblioteche filosofiche private. Strumenti e prospettive di ricerca*, Pisa: Edizioni della Normale, 2014.

[4] In the field of history of mathematics, see, for example, Erika Luciano and Clara Silvia Roero, "Corrado Segre's library at the Department of Mathematics, Florence University," in Gianfranco Casnati *et al.* (eds.), *From Classical to Modern Algebraic Geometry*, Basel: Birkhäuser, 2016, pp. 211–230; Erika Luciano and Elena Scalambro, "Sul ruolo euristico dei patrimoni matematici: il *case-study* delle collezioni di A. Terracini," *Rivista di Storia dell'Università di Torino*, IX, 2, 2020, pp. 273–332; Erika Luciano, "G. Fubini: une expérience de patrimonialisation en théorie des nombres," in *Patrimonialisation des mathématiques*, special issue of *Philosophia Scientiae*, 26, 2, 2022, pp. 123–144.

Matematica (INdAM), of which Severi was both founding member and president for life.

Francesco Severi died on December 8, 1961, after a long and debilitating illness. As pointed out by Leonard Roth in his obituary, "In 1957 the struggle changed his character; it became bitter and tragic. He fought his adversary to the very end. During the last four years of his life he continued to keep all his affairs in a progressively enfeebled grasp."[5] In particular, a few years before his death, Severi decided to donate his private library to INdAM, which by the Decreto del Presidente della Repubblica (no. 126, 27 May 1959) was allowed to accept the donation. During the meeting board of directors of INdAM held in Rome on June 10, 1959, Francesco Severi claimed that, at the moment of the gift, he was already ill, and hence he was not able to personally check the number of volumes. The board accepted the donation and gave Beniamino Segre the task of managing the Severi Library, by coordinating a suitable executive staff.[6]

Presumably under the direction of Beniamino Segre, each volume belonging to the Severi Library was stamped with DONO DI F. SEVERI and assigned with a progressive number. INdAM's headquarters in Rome preserve a manuscript inventory, listing a series of volumes, specifying a progressive number, author, title, place, and date of publication. The progressive number matches the one reported on the corresponding book in INdAM's library, and from number 1 to number 1433 (with occasional exceptions), the list specifies that the volume is a gift (*dono*). From number 1434 onwards, the books in the list are classified as purchases (*acquisto*). The first purchased book is recorded on January 16, 1961. This is consistent with the following remark from a manuscript note about Severi Library, filed in Beniamino Segre's personal archive (now at Caltech): "In 12 casse si trovano i volumi di matematica (circa 1.400)," that is, "Books on mathematics (about 1,400) are gathered in 12 boxes."[7] Therefore, it seems plausible to conjecture that the mathematical volumes belonging to Severi's personal library approximately correspond to the first 1433 entries of the manuscript inventory, which we are reproducing below.[8]

---

[5] See Leonard Roth, "Francesco Severi. Obituary note," *Journal of the London Mathematical Society*, 38, 1, 1963, pp. 282–307. https://londmathsoc.onlinelibrary.wiley.com/doi/10.1112/jlms/s1-38.1.282, p. 304.

[6] See Gino Roghi, "Materiale per una Storia dell'Istituto Nazionale di Alta Matematica dal 1939 al 2003," *Bollettino dell'Unione Matematica Italiana*, 8, 8-A (2005), pp. 3–301, http://www.bdim.eu/item?id=BUMI_2005_8_8A_3-2_3_0, pp. 63–64.

[7] Caltech Archives, Beniamino Segre Papers, 007.12, p. 5.

[8] We are grateful to Giorgio Patrizio (President, INdAM), who kindly granted us the permission to include the list of Severi's book.

3 Mathematical Library of Francesco Severi[9]

Rome, INdAM

[1243] A. C. (ed.). *Soluzionario dei problemi contenuti nel trattato elementare d'aritmetica,* Turin 1867.

[1307] Abbagnano, Nicola et al. *Fondamenti logici della scienza,* Turin 1947.

[189] Abetti, Giorgio. *Scienza d'oggi. Dal cielo alla terra,* Milan 1941.

[82] Accademia del Cimento. *Saggi di naturali esperienze fatte nell'Accademia del Cimento sotto la protezione del Ser.^{mo} principe Leopoldo di Toscana e descritte dal segretario di essa Accademia,* Florence 1841.

[786] Achieser, Naum I. *Vorlesungen über Approximationstheorie,* Berlin 1953.

[318] Adamo, Giuseppe. *Le geometrie a curvatura costante secondo l'indirizzo metrico-proiettivo,* La Spezia 1952.

[678] Agostini, Amedeo. *Analisi matematica*, vol. 1, Livorno 1939.

[453] Agostini, Amedeo. *Esercizi di geometria analitica,* vol. 2.1, Bologna 1926.

[454] Agostini, Amedeo. *Esercizi di geometria analitica*, vol. 2.2, Bologna 1926.

[701] Agostini, Amedeo. *Lezioni di analisi matematica*, vol. 2, Livorno 1937.

[726] Agostini, Amedeo. *Nomografia,* Livorno 1942.

[452] Agostini, Amedeo; Bortolotti, Enea. *Esercizi di geometria analitica*, vol. 1, Bologna 1925.

[81] Albareda Herrera, José Maria. *Consideraciones sobre la investigación científica,* Madrid 1951.

[1372] Aleksandrov, Aleksandr Danilovič; Kolmogorov, Andrej Nikolaevič; Lavrent'ev, Michail Alekseevič. [*Mathematics: Its Content, Methods and Meaning*], Moscow 1956.

[896] Aliprandi, Giuseppe. *Matematica finanziaria e attuariale,* Padua 1931.

[1256] Alliata, Giulio. *Unzulänglichkeiten und Irrtümer der Physik,* Bern 1934.

[954] Almansi, Emilio. *Esercizi di meccanica razionale,* Turin 1899.

[400] Amaldi, Edoardo. *Fisica sperimentale*, vol. 2, Rome 1938.

[1304] Amaldi, Ugo. *Elementi di geometria,* Bologna 1942.

[1323] Amaldi, Ugo. *I gruppi continui di trasformazioni puntuali dello spazio a tre dimensioni*, vol. 2, Modena 1912.

[1403] Amaldi, Ugo. *I gruppi continui infiniti di trasformazioni puntuali dello spazio a tre dimensioni*, vol. 1: *Gruppi che ammettono una schiera di ∞^1 superficie,* Modena 1910.

[1405] Amaldi, Ugo. *I gruppi continui reali di trasformazioni conformi dello spazio,* Turin 1905.

[9] Volumes are listed in alphabetical order by author; the numbers in square brackets at the beginning of each entry are the progressive inventory numbers assigned to the volumes at the time of their registration in the handwritten *Registro inventario*, available at INdAM. Collective volumes are listed under VV. AA. ("various authors"); bibliographies and encyclopedias are listed at the end.

[544] Amaldi, Ugo. *Introduzione alla teoria dei gruppi continui infiniti di trasformazione*, vol. 1, Rome 1942.

[636] Amaldi, Ugo. *Introduzione alla teoria dei gruppi continui infiniti di trasformazione*, vol. 2, Rome 1944.

[700] Amaldi, Ugo. *Lezioni di analisi infinitesimale,* Rome 1928.

[779] Amaldi, Ugo. *Lezioni di analisi infinitesimale*, vol. 1, Padua 1930.

[68] Amaldi, Ugo. *Lezioni di geometria descrittiva con applicazioni,* Padua 1920.

[1207] Amaldi, Ugo. *Nozioni di geometria,* Bologna 1940.

[1404] Amaldi, Ugo. *Sui gruppi continui infiniti di trasformazioni di contatto dello spazio,* Turin 1906.

[1322] Amaldi, Ugo. *Sulla classificazione dei gruppi continui di trasformazioni di contatto dello spazio,* Rome 1918.

[683] Amerio, Luigi. *Funzioni analitiche e trasformazione di Laplace,* Milan 1951.

[277] Amodeo, Federico. *Origen y desarrollo de la geometría proyectiva,* Rosario 1939.

[163] Amodeo, Federico. *Origine e sviluppo della geometria proiettiva,* Naples 1939.

[289] Amoroso, Luigi. *Meccanica economica. Lezioni tenute nell'Anno Accademico 1940–41,* Città di Castello 1942.

[443] Amoroso, Luigi. *Esercizi di geometria analitica e proiettiva,* Rome 1911.

[388] Amoroso, Luigi; Bompiani, Enrico. *Esercizi di geometria analitica e proiettiva,* Rome 1913.

[908] Andrissi, Giovanni L. *Il cosmografo Vincenzo Coronelli e l'astronomia nel seicento,* Rome 1951.

[699] [Anonymous]. *Faisceaux analytiques. Étude du faisceau des relations entre p fonctions holomorphes,* [s.l.] [s.d.].

[1398]. [Anonymous]. *Catálogo de la Biblioteca de la Facultad de Ciencias Exactas, Físicas y Naturales de la Universidad de Buenos Aires,* Buenos Aires 1930–1931.

[1319]. [Anonymous]. *Il regolo calcolatore logaritmico e il modo di usarlo,* Lahr [s.d.].

[403]. [Anonymous]. *Riassunto di mineralogia e geologia,* Padua 1919.

[1123]. [Anonymous]. *Trattato di trigonometria,* vol. 1, [s.l.] [s.d.].

[221] Appell, Paul. *Selecta. Cinquantenaire scientifique,* Paris 1927.

[616] Appell, Paul; Goursat, Édouard. *Théorie des fonctions algébriques et de leurs intégrales,* Paris 1895.

[524, 525, 1259, 1270] Appert, Antoine. *Propriétés des espaces abstraits les plus généraux,* Paris 1934. (4 copies)

[1428] Apraiz y Arias, Félix. *Tratado de electricidad y magnetismo,* Barcelona 1927.

[412] Argentieri, Domenico. *Ottica industriale. Il progetto analitico dei sistemi di lenti, la verifica trigonometrica dei sistemi di diottri, l'alberazione d'onda e le tolleranze ottiche, la diffrazione e il potere risolutivo, teoria e dimostrazioni, tavole numeriche e logaritmiche,* Milan 1954.

[955] Armellini, Giuseppe. *Trattato di astronomia siderale*, vol. 1: *Parte generale,* Bologna 1928.
[956] Armellini, Giuseppe. *Trattato di astronomia siderale*, vol. 2: *Le stelle,* Bologna 1931.
[957] Armellini, Giuseppe. *Trattato di astronomia siderale*, vol. 3: *Le nebulose,* Bologna 1936.
[958] Armellini, Giuseppe. *I fondamenti scientifici della astrofisica,* Milan 1952.
[959] Armellini, Giuseppe. *Corso di meccanica razionale,* Padua 1921.
[1315] Arzeliés, Henri. *La cinématique relativiste,* Paris 1955.
[21] Ascoli, Guido. *Elementi di teoria dei numeri, funzioni razionali intere, equazioni algebriche,* Turin 1954.
[11] Ascoli, Guido. *Lezioni di matematiche complementari,* Turin 1954.
[1158] Ascoli, Guido. *Lezioni elementari di analisi matematica,* Turin 1924.
[509] Baer, Reinhold. *Automorphismen von Erweiterungsgruppen,* Paris 1935.
[1169] Baffi, Contardo. *Aritmetica e geometria,* Turin-Milan 1929.
[564] Bagnera, Giuseppe. *Lezioni di analisi algebrica,* Rome [s.d.].
[572] Bagnera, Giuseppe. *Lezioni di calcolo infinitesimale,* Padua 1924.
[745] Bagnera, Giuseppe. *Lezioni sopra la teoria delle funzioni analitiche,* Rome 1927.
[1109] Bagnera, Giuseppe. *Teoria dei numeri reali. Calcolo dei radicali e dei logaritmi,* Palermo 1916.
[617] Bagnera, Giuseppe; Strazzeri, Vittorio. *Lezioni sul calcolo delle variazioni,* Palermo 1913.
[597] Baker, Henry F. *An Introduction to the Theory of Multiply Periodic Functions,* Cambridge 1907.
[31] Baker, Henry F. *Principles of Geometry*, vol. 1: *Foundations*, Cambridge 1922.
[404] Baker, Henry F. *Principles of Geometry*, vol. 1: *Foundations,* Cambridge 1929.
[32] Baker, Henry Frederick. *Principles of Geometry*, vol. 2: *Plane Geometry. Conics, Circles, Non-Euclidean Geometry*, Cambridge 1922.
[405] Baker, Henry F. *Principles of Geometry*, vol. 2: *Plane Geometry. Conics, Circles, Non-Euclidean Geometry,* Cambridge 1930.
[33] Baker, Henry Frederick. *Principles of Geometry*, vol. 3: *Solid Geometry. Quadrics, Cubic Curves in Space, Cubic Surfaces*, Cambridge 1923.
[406] Baker, Henry F. *Principles of Geometry*, vol. 3: *Solid Geometry. Quadrics, Cubic Curves in Space, Cubic Surfaces,* Cambridge 1931.
[407] Baker, Henry F. *Principles of Geometry*, vol. 4: *Higher Geometry. Being Illustrations of the Utility of the Consideration of Higher Space,* Cambridge 1932.
[408] Baker, Henry F. *Principles of Geometry*, vol. 5: *Analytical Principles of the Theory of Curves,* Cambridge 1933.
[409] Baker, Henry Frederick. *Principles of Geometry*, vol. 6: *Introduction to the Theory of Algebraic Surfaces and Higher Loci,* Cambridge 1933.
[1255] Baldassarri, Mario. *Algebraic Varieties,* Berlin 1956.

[342] Banti, Angelo. *Il primo trasporto di energia elettrica a distanza, Tivoli-Roma. Nel quarantesimo anniversario 1892–1932,* Rome 1932.

[741] Barbensi, Gustavo. *Elementi di matematica generale,* Florence 1936.

[219] Barbensi. Gustavo. *Introduzione alla bioemetria,* Florence 1952.

[87] Barbensi, Gustavo. *Paolo Ruffini,* Modena 1956.

[551] Barthel, Ernst. *Einführung in die Polargeometrie,* Leipzig 1932.

[376] Battelli, Angelo; Battelli, Federico. *Trattato pratico per le misure e ricerche elettriche,* Rome 1902.

[748] Behnke, Heinrich. *Theorie der analytischen Funktionen,* [s.l.] 1932.

[747] Behnke, Heinrich; Thullen, Peter. *Theorie der Funktionen mehrerer Komplexer Veränderlichen,* Berlin 1934.

[753] Belardinelli, Giuseppe. *Esercizi di algebra complementare,* Bologna 1923.

[698] Belardinelli, Giuseppe. *Esercizi di analisi matematica,* vol. 1, Milan 1942.

[780] Belardinelli, Giuseppe. *Esercizi di analisi matematica,* vol. 2, Milan 1947.

[493] Bell, Eric T. *I grandi matematici,* Florence 1950.

[326] Bellavitis, Giusto. *Lezioni di geometria descrittiva,* Padua 1851.

[317] Bencivenga, Ulderico. *Geometria e trigonometria iperboliche e fondamenti d'una geometria a metri variabili,* Rome 1940.

[93] Berker, Ratib. A. *Sur quelques cas d'intégration des équations du mouvement d'un fluide visqueux incompressible,* Paris-Lille 1936.

[1031] Berlese, Tommaso. *Corso di topografia*, vol. 1: *Topografia generale. Planimetria, agrimensura, altimetria, celerimensura, fotogrammetria,* Padua 1951.

[1032] Berlese, Tommaso. *Corso di topografia*, vol. 2: *Costruzioni stradali, calcoli topografici,* Padua 1951.

[583] Bernstein, Vladimir. *Leçons sur les progrès récents de la théorie des séries de Dirichlet,* Paris 1933.

[116] Bertini, Eugenio. *Introduzione alla geometria proiettiva degli iperspazi. Lezioni di geometria superiore,* Pisa 1899.

[383] Bertini, Eugenio. *Introduzione alla geometria proiettiva degli iperspazi,* Messina 1923.

[398] Bertini, Eugenio. *Complementi di geometria proiettiva,* Bologna 1927.

[1146] Bertrand, Joseph. *Trattato di algebra elementare,* Florence 1883.

[948] Bertrand, Joseph. *Trattato di algebra elementare,* Florence 1891.

[1130] Bertrand, Joseph. *Trattato di aritmetica,* Florence 1892.

[727] Berzolari, Luigi. *Algebra,* [s.l.] 1929.

[1321] Berzolari, Luigi. *Algebraische Transformationen und Korrespondenzen,* Leipzig 1932.

[65] Berzolari, Luigi. *Lezioni di geometria descrittiva,* [s.l.] 1895.

[66] Berzolari, Luigi. *Lezioni di geometria proiettiva,* [s.l.] 1897.

[864] Berzolari, Luigi; Vivanti, Giulio; Gigli, Duilio (eds.). *Enciclopedia delle matematiche elementari*, vol. 1.1, Milan 1930.

[865] Berzolari, Luigi; Vivanti, Giulio; Gigli, Duilio (eds.). *Enciclopedia delle matematiche elementari*, vol. 1.2, Milan 1932.

[866] Berzolari, Luigi; Vivanti, Giulio; Gigli, Duilio (eds.). *Enciclopedia delle matematiche elementari*, vol. 2.1, Milan 1937.
[867] Berzolari, Luigi; Vivanti, Giulio; Gigli, Duilio (eds.). *Enciclopedia delle matematiche elementari*, vol. 2.2, Milan 1938.
[868] Berzolari, Luigi; Vivanti, Giulio; Gigli, Duilio (eds.). *Enciclopedia delle matematiche elementari*, vol. 3.1, Milan 1947.
[869] Berzolari, Luigi; Vivanti, Giulio; Gigli, Duilio (eds.). *Enciclopedia delle matematiche elementari*, vol. 3.2, Milan 1950.
[1137] Bettini, Bettino; Ciamberlini, Corrado. *Elementi di geometria*, vol. 1, Livorno 1916.
[1138] Bettini, Bettino; Ciamberlini, Corrado. *Elementi di geometria*, vol. 2, Livorno 1917.
[1401] Bianchi, Luigi. *Congruenze di sfere di Ribaucour e superficie di Peterson,* Bologna 1928.
[191] Bianchi, Luigi. *Lezioni di geometria analitica. Anno 1903–1904,* Pisa 1904.
[327] Bianchi, Luigi. *Lezioni di geometria differenziale*, vol. 1, Pisa 1922.
[296] Bianchi, Luigi. *Lezioni di geometria differenziale*, vol. 2.1, Pisa 1923.
[297] Bianchi, Luigi. *Lezioni di geometria differenziale*, vol. 2.2, Pisa 1924.
[719] Bianchi, Luigi. *Lezioni sulla teoria dei gruppi continui finiti di trasformazioni,* Pisa 1903.
[548] Bianchi, Luigi. *Lezioni sulla teoria dei gruppi continui finiti di trasformazioni,* Pisa 1918.
[784] Bianchi, Luigi. *Lezioni sulla teoria dei gruppi di sostituzioni e delle equazioni algebriche secondo Galois,* Pisa 1900.
[732] Bianchi, Luigi. *Lezioni sulla teoria delle funzioni di variabile complessa e delle funzioni ellittiche*, vol. 1 *Funzioni monodrome di variabile complessa,* Pisa 1899.
[738] Bianchi, Luigi. *Lezioni sulla teoria delle funzioni di variabile complessa e delle funzioni ellittiche*, vol. 2: *Funzioni ellittiche,* Pisa 1899.
[536] Bianchi, Luigi. *Lezioni sulla teoria delle equazioni differenziali lineari (teoria di Fuchs-Riemann),* Catania 1924.
[658] Bianchi, Luigi. *Opere*, vol. 1.1, Rome 1952.
[659] Bianchi, Luigi. *Opere*, vol. 1.2, Rome 1953.
[660] Bianchi, Luigi. *Opere*, vol. 2: *Applicabilità e problemi di deformazione,* Rome 1953.
[661] Bianchi, Luigi. *Opere*, vol. 3: *Sistemi tripli ortogonali,* Rome 1955.
[662] Bianchi, Luigi. *Opere*, vol. 4.1: *Deformazioni delle quadratiche. Teoria delle trasformazioni delle superficie applicabili sulle quadratiche,* Rome 1956.
[663] Bianchi, Luigi. *Opere*, vol. 4.2: *Deformazioni delle quadratiche. Teoria delle trasformazioni delle superficie applicabili sulle quadratiche,* Rome 1956.
[664] Bianchi, Luigi. *Opere*, vol. 5: *Trasformazioni delle superficie e delle curve,* Rome 1957.
[665] Bianchi, Luigi. *Opere*, vol. 6: *Congruenze di rette e di sfere e loro deformazioni,* Rome 1957.

[45, 46] Bianchi, Luigi. *Opere*, vol. 7: *Problemi di rotolamento*, Rome 1957. (2 copies)

[47, 48] Bianchi, Luigi. *Opere*, vol. 8: *Classi di superficie*, Rome 1958. (2 copies)

[49] Bianchi, Luigi. *Opere*, voll. 9: *Geometria degli spazi di Riemann* and 10: *Ricerche varie*, Rome 1958.

[1298] Bianchi, Luigi. *Opere*, vol. 11: *Corrispondenza,* Rome 1959.

[608] Bianchi, Luigi. *Teoria dei numeri algebrici,* Pisa 1925.

[346] Bieberbach, Ludwig. *Differentialgeometrie,* Leipzig-Berlin 1932.

[399] Bieberbach, Ludwig. *Projektive Geometrie,* Leipzig 1931.

[916] Bilancioni, Guglielmo. *Veteris vestigia flammae. Pagine storiche della scienza nostra,* Rome 1922.

[368] Bini, Umberto. *Elettrologia,* Rome 1934.

[69, 70] Bini, Umberto. *Lezioni di analisi matematica*, vol. 1, Florence 1947. (2 copies)

[1225] Bini, Umberto. *Lezioni di analisi matematica*, vol. 2, Florence 1947.

[1226] Bini, Umberto. *Lezioni di analisi matematica*, vol. 3, Florence 1947.

[71] Bini, Umberto. *Trigonometria piana,* Florence 1946.

[1224] Bini, Umberto. *Trigonometria piana,* Florence 1947.

[1218] Bisconcini, Giulio. *Algebra*, vol. 1, Rome 1925.

[1142] Bisconcini, Giulio. *Aritmetica e algebra,* Rome 1924.

[920] Bisconcini, Giulio. *Elementi di matematica finanziaria e attuariale,* Rome 1921.

[1143] Bisconcini, Giulio. *Elementi di trigonometria sferica,* Rome 1924.

[375] Bisconcini, Giulio. *Esercizi di meccanica razionale,* Rome 1922.

[952] Bisconcini, Giulio. *Esercizi e complementi di meccanica razionale,* Milan 1927.

[1220] Bisconcini, Giulio. *Geometria elementare*, vol. 1, Rome 1933.

[1221] Bisconcini, Giulio. *Geometria elementare*, vol. 2, Rome 1933.

[1222] Bisconcini, Giulio. *Geometria elementare*, vol. 3, Rome 1933.

[1223] Bisconcini, Giulio. *Geometria,* Rome 1934.

[1141, 1219] Bisconcini, Giulio. *Nozioni di geometria,* Rome 1931. (2 copies)

[1144] Bisconcini, Giulio; Freda, Eleonora. *Trigonometria piana,* Rome 1938.

[1287] Blanc-Lapierre, André; Casal, Pierre; Tortrat, Albert. *Méthodes mathématiques de la mécanique statistique,* Paris 1959.

[214] Blaschke, Wilhelm. *Analytische Geometrie,* Hanover 1948.

[340] Blaschke, Wilhelm. *Conferenze di geometria,* Messina 1950.

[339] Blaschke, Wilhelm. *Ebene Kinematik,* Leipzig-Berlin 1938.

[402] Blaschke, Wilhelm. *Einführung in die Differentialgeometrie,* Berlin 1950.

[1306] Blaschke, Wilhelm. *Einführung in die Geometrie der Waben,* Basel 1955.

[162] Blaschke, Wilhelm. *Geometría de los tejidos,* Barcelona 1954.

[215] Blaschke, Wilhelm. *Kreis und Kugel,* Berlin 1956.

[330] Blaschke, Wilhelm. *Nicht-euklidische Geometrie und Mechanik I-II-III,* Leipzig-Berlin 1942.

[196] Blaschke, Wilhelm. *Projektive Geometrie,* Hanover 1947.

[341] Blaschke, Wilhelm. *Vorlesungen über Integralgeometrie,* Leipzig-Berlin 1935.

[174] Blaschke, Wilhelm. *Vorlesungen über Integralgeometrie,* Berlin 1955.

[357] Blaschke, Wilhelm; Bol, Gerrit. *Geometrie der Gewebe. Topologische Fragen der Differentialgeometrie,* Berlin 1938.

[739] Bochner, Salomon; Martin, William T. *Several Complex Variables,* Princeton 1948.

[1168] Bodmer, Johann J. *Aufgaben für den Unterricht in der Arithmetik und Algebra an Sekundarschulen*, vol. 1.3, Zürich 1893.

[750] Boggio, Tommaso. *Lezioni di analisi matematica*, vol. 1, Turin 1926.

[751] Boggio, Tommaso. *Lezioni di analisi matematica*, vol. 2, Turin 1926.

[923, 924] Bombelli, Raffaele. *L'algebra. Libri IV e V comprendenti la parte geometrica inedita tratta dal manoscritto B. 1569, Biblioteca dell'Archiginnasio di Bologna*, a cura di Ettore Bortotti, Bologna 1929. (2 copies)

[34] Bompiani, Enrico. *Applicazioni di geometria descrittiva,* Rome 1928.

[72] Bompiani, Enrico. *Applicazioni di geometria descrittiva*, Rome 1933.

[469] Bompiani, Enrico. *Geometria analitica,* Rome 1937.

[848] Bompiani, Enrico. *Geometria degli elementi differenziali*, vol. 1: *Elementi differenziali regolari piani rispetto al gruppo proiettivo,* Rome 1955.

[26] Bompiani, Enrico. *Geometria descrittiva*, Rome 1928.

[22] Bompiani, Enrico. *Geometria descrittiva,* Rome 1938.

[166] Bompiani, Enrico. *Metriche non-euclidee,* Rome 1951.

[1139] Bonfantini, Giuseppe. *Compendio d'algebra elementare*, vol. 1, Novara 1926.

[1140] Bonfantini, Giuseppe. *Compendio d'algebra elementare*, vol. 2, Novara 1924.

[1159] Bonfantini, Giuseppe. *Compendio di algebra elementare,* Novara 1930.

[1147] Bonfantini, Giuseppe. *Compendio di algebra elementare,* Novara 1931.

[308] Bonola, Roberto. *La geometria non-euclidea,* Bologna 1906.

[907] Bonomi, Lino. *Naturalisti, medici e tecnici trentini. Contributo alla storia della scienza in Italia,* Trento 1930.

[434] Bordiga, Giovanni. *Lezioni di geometria descrittiva*, vol. 1.1: *Rappresentazione mediante coppie di punti,* Padua 1907.

[435] Bordiga, Giovanni. *Lezioni di geometria descrittiva*, vol. 1.2: *Rappresentazione mediante coppie di rette,* Padua 1907.

[436] Bordiga, Giovanni. *Lezioni di geometria descrittiva*, vol. 1.3 *Rappresentazione dello spazio su di un piano mediante terne di punti,* Padua 1907.

[437] Bordiga, Giovanni. *Lezioni di geometria descrittiva*, vol. 1.4: *Fondamenti geometrici della prospettiva e della fotogrammetria,* Padua 1907.

[1334] Bordoni, Ugo. *Fondamenti di fisica tecnica*, vol. 1, Rome 1932.

[641] Borel, Émile. *Les nombres inaccessibles,* Paris 1952.

[115] Borel, Emile. *L'imaginaire et le réel en mathématiques et en physique,* Paris 1952.

[172, 222] Borel, Émile. *Selecta. Jubilé scientifique,* Paris 1940. (2 copies)

[974] Borel, Émile; Chéron, André. *Théorie mathématique du bridge a la portée de tous,* Paris 1940.
[271] Borgogelli, Guido. *Projezioni e prospettiva delle ombre. Trattato teorico-pratico,* Rome 1921.
[269] Borgogelli, Guido. *Prospettiva delle ombre,* Rome 1910.
[1418] Borgogelli, Guido. *Prospettiva lineare. Trattato teorico-pratico,* Rome 1921.
[270] Borgogelli, Guido. *Prospettiva teorico-pratica,* Rome 1921.
[345] Borsari, Raffaele. *Logica concreta,* Florence 1950.
[421] Bortolotti, Enea. *Geometria descrittiva,* Padua 1939.
[1254] Bortolotti, Enea. *Geometria analitica e proiettiva,* Padua 1938.
[344] Bortolotti, Enea. *Spazi a connessione proiettiva,* Rome 1941.
[1393] Bortolotti, Ettore. *I cartelli di matematica disfida e la personalità psichica e morale di Girolamo Cardano,* Imola 1933.
[912] Bortolotti, Ettore. *La scuola matematica di Bologna. Cenno storico,* Bologna 1928.
[369] Bortolotti, Ettore. *Studi e ricerche sulla storia della matematica in Italia nei secoli XVI e XVII,* Bologna 1928.
[315, 1278] Bouligand, Georges. *La causalité des théories mathématiques,* Paris 1934. (2 copies)
[506] Bouligand, Georges; Giraud, Georges; Delens, Paul C. *Le problème de la dérivée oblique en théorie du potentiel,* Paris 1935.
[1284] Bourbaki, Nicolas. *Espaces vectoriels topologiques,* Paris 1955.
[505] Bourbaki, Nicolas. *Fonctions d'une variable réale,* Paris 1951.
[1282] Bourbaki, Nicolas. *Théories des ensembles,* Paris 1954.
[849] Bourbaki, Nicolas. *Topologie générale,* Paris 1953.
[1265] Brauer, Richard. *Über die Darstellung von Gruppen in Galoisschen Feldern,* Paris 1935.
[1399] Brelot, Marcel. *Éléments de la théorie classique du potentiel,* Paris 1959.
[1261] Brelot, Marcel. *Étude des fonctions sous-harmoniques au voisinage d'un point singulier,* Paris 1934.
[225] Brill, Alexander. *Das Relativitätsprinzip. Eine Einführung in die Theorie,* Leipzig 1918.
[1015] Brill, Alexander. *Vorlesungen über allgemeine Mechanik,* Munich 1928.
[576] Brill, Alexander. *Vorlesungen über ebene algebraische Kurven und algebraische Funktionen,* Braunschweig 1925.
[800] Broda, Engelbert. *L'energia atomica,* Milan 1957.
[155] Broglie, Louis de. *La mécanique ondulatoire des systèmes de corpuscules,* Paris 1939.
[154] Broglie, Louis de. *L'électron magnétique,* Paris 1934.
[156] Broglie, Louis de. *Théorie générale des particules a spin,* Paris 1943.
[1316] Broglie, Louis de. *Théorie générale des particules a spin,* Paris 1954.
[1373] Bruevich, N. G.; Artobolevskiĭ, I. I. (eds.). [*The Scientific Legacy of P. L. Chebyshev*], Moscow 1945.
[489] Burali-Forti, Cesare. *Geometria analitico-proiettiva,* Turin 1926.

[1148] Burali-Forti, Cesare; Boggio, Tommaso. *Esercizi di matematica,* Turin 1924.
[301] Burali-Forti, Cesare; Boggio, Tommaso. *Espaces courbes. Critique de la relativité,* Turin 1924.
[949] Burali-Forti, Cesare; Marcolongo, Roberto. *Analisi vettoriale generale e applicazioni,* vol. 1: *Trasformazioni lineari,* Bologna 1929.
[1149] Burali-Forti, Cesare; Marcolongo, Roberto. *Corso di matematica,* vol. 2: *Geometria,* Naples 1920.
[302] Burali-Forti, Cesare; Marcolongo, Roberto. *Elementi di calcolo vettoriale,* Bologna 1921.
[1150] Burali-Forti, Cesare; Marcolongo, Roberto. *Elementi di trigonometria,* Naples 1929.
[757] Burau, Werner. *Grundmannigfaltigkeiten der projektiven Geometrie,* Barcelona 1950.
[951] Burgatti, Pietro. *Analisi vettoriale generale e applicazioni,* vol. 3: *Teoria matematica della elasticità,* Bologna 1931.
[950] Burgatti, Pietro; Boggio, Tommaso; Burali-Forti, Cesare. *Analisi vettoriale generale e applicazioni,* vol. 2: *Geometria differenziale,* Bologna 1930.
[1184] Burnengo, Giuseppe. *Algebra elementare razionale panoramica. Antologia matematica,* vol. 1, Turin 1930.
[1185] Burnengo, Giuseppe. *Algebra elementare razionale panoramica. Antologia matematica,* vol. 2, Turin 1930.
[264] Burzio, Filippo. *Lagrange,* Turin 1942.
[1011] Burzio, Filippo. *Complementi di balistica esterna,* vol. 1: Il secondo problema balistico (rotazione dei proietti), vol. *2: La resistenza dell'aria in balistica,* Rome 1934.
[1004] Butty, Enrique. *Introducción a la física matemática,* vol. 1: *Cálculo vectorial. Cálculo tensorial (para transformaciones lineales ortogonales y complementos de análisis vectorial). Estudio particular del tensor del segundo rango,* Buenos Aires 1931.
[1005] Butty, Enrique. *Introducción a la física matemática,* vol. 2: *Elementos de geometría diferencial. Cálculo tensorial,* Buenos Aires 1934.
[229] Butty, Enrique. *Introducción filosófica a las teorías de la relatividad,* Buenos Aires 1924.
[153] Butty, Enrique. *La duración de Bergson y el tiempo de Einstein,* Buenos Aires 1937.
[1003] Butty, Enrique. *La ingeniería. Enseñanza, profesión, función social,* Buenos Aires 1932.
[798] Buzano, Piero. *Lezioni di analisi matematica,* Turin 1950.
[208] C. W. W. *La relatività e il signor Robinson,* Milan 1945.
[190] Cafiero, Federico. *Funzioni additive d'insieme ed integrazione negli spazi astratti,* Naples 1953.
[1347] Caligo, Domenico; de Schwarz, Maria Josepha. *Manuali per applicazioni tecniche del calcolo,* vol. 4: *Calcolo delle,* volte cilindriche circolari sottili, Rome 1960.
[242] Calvi, Gerolamo. *Vita di Leonardo,* Brescia 1949.

[401] Candido, Giacomo. *Conferenze,* Galatina 1945.
[361] Candido, Giacomo. *Scritti matematici,* Florence 1948.
[74] Canella, Renzo. *Nozioni sugli stili dell'ornato ed architettura,* Padua 1912.
[18, 1424] Cantelli, Francesco Paolo. *Alcune memorie matematiche,* Milan 1958. (2 copies)
[775] Carafa, Mario. *Lezioni sulla teoria delle equazioni differenziali,* Rome 1954.
[963] Carathéodory, Constantin. *Elementare Theorie des Spiegelteleskops von B. Schmidt,* Leipzig-Berlin 1940.
[837] Carathéodory, Constantin. *Reelle Funktionen,* vol. 1, Berlin 1939.
[838] Carathéodory, Constantin. *Variationsrechnung,* Berlin 1935.
[650] Carruccio, Ettore. *Matematica e logica nella storia e nel pensiero contemporaneo,* Turin 1958.
[915] Carruccio, Ettore. *Matematica e logica nella storia e nel pensiero contemporaneo. Corso di storia delle matematiche,* Turin 1951.
[314, 516] Cartan, Élie. *La méthode du repère mobile, la théorie des groupes continus et les espaces généralisés,* Paris 1935. (2 copies)
[549] Cartan, Élie. *Leçons sur la géométrie projective complexe,* Paris 1950.
[515] Cartan, Élie. *Les espaces de Finsler,* Paris 1934.
[1415] Cartan, Élie. *Les espaces métriques fondés sur la notion d'aires,* Paris 1933.
[213] Cartan, Élie. *Selecta. Jubilé scientifique,* Paris: Gauthier-Villars 1939.
[518] Cartan, Henri. *Sur les classes de fonctions définies par des inégalités portant sur leurs dérivées successives,* Paris 1940.
[517, 1267] Cartan, Henri. *Sur les groupes de transformations analytiques,* Paris 1935. (2 copies)
[1289] Casari, Ettore. *Computabilità e ricorsività. Problemi di logica matematica,* Milan 1959.
[1386] Casarini, Angelo; Rossi, Nicola. *Cellule cianofile e cellule mucoide dell'adenoipofisi nelle disendocrinie umane e sperimentali,* Vatican 1953.
[1178, 1179] Caselli, Vincenzo. *Elementi di trigonometria piana,* Lanciano 1934. (2 copies)
[251] Caselli, Vincenzo. *L'aritmetica intuitiva,* Palermo 1950.
[731] Cassina, Ugo. *Approssimazioni numeriche,* Milan 1948.
[937] Cassina, Ugo. *Calcolo numerico,* Turin 1922.
[938] Cassina, Ugo. *Calcolo numerico,* Bologna 1928.
[272] Cassinis, Gino; Solaini, Luigi. *Lezioni di fotogrammetria. Testo,* Milan 1936.
[273] Cassinis, Gino; Solaini, Luigi. *Lezioni di fotogrammetria. Figure,* Milan 1936.
[231] Castelfranchi, Gaetano. *Fisica moderna. Visione sintetica, pianamente esposta, della fisica d'oggi,* Milan 1931.
[444] Castelfranchi, Gaetano. *Televisione. Le basi fisiche del "radiovedere",* Milan 1930.
[1152] Castelnuovo, Emma. *Geometria intuitiva,* Rome 1949.
[305] Castelnuovo, Guido. *Calcolo delle probabilità,* Milan 1919.
[294] Castelnuovo, Guido. *Calcolo delle probabilità,* vol. 1, Bologna 1925.
[295] Castelnuovo, Guido. *Calcolo delle probabilità,* vol. 2, Bologna 1928.

[359] Castelnuovo, Guido. *La probabilité dans les différentes branches de la science,* Paris 1937.
[901] Castelnuovo, Guido. *Le origini del calcolo infinitesimale nell'era moderna,* Bologna 1938.
[192] Castelnuovo, Guido. *Lezioni di geometria analitica,* Milan 1909.
[187] Castelnuovo, Guido. *Lezioni di geometria analitica e proiettiva,* Rome 1905.
[498] Castelnuovo, Guido. *Memorie scelte,* Bologna 1937.
[410] Castelnuovo, Guido. *Spazio e tempo,* Bologna 1923.
[749] Cattaneo, Paolo. *Corso elementare di matematiche superiori*, vol. 1: *Analisi infinitesimale,* Bologna 1923.
[541] Cattaneo, Paolo. *Elementi di teoria dei numeri,* Padua 1920.
[913] Cazalas, J. J. A. Eutrope. *Carrés magiques au degré n.,* Paris 1934.
[651] Cecioni, Francesco. *Lezioni sui fondamenti della matematica*, vol. 1: *Premesse e questioni generali,* Padua 1958.
[1019] Cesari, Lamberto; Conforto, Fabio; Minelli, Carlo. *Travi continue inflesse e sollecitate assialmente,* Rome 1941.
[332] Cesaro, Ernesto. *Lezioni di geometria intrinseca,* Naples 1896.
[181] Chatelet, Albert. *Arithmétique et algèbre modernes,* Paris 1954.
[58] Cherubino, Salvatore. *Calcolo delle matrici,* Roma 1957.
[386] Cherubino, Salvatore. *Lezioni di geometria analitica*, vol. 1: *Coordinate nel piano e nello spazio, elementi di calcolo di matrici, teoria generale delle coniche e delle quadriche, proiettività nelle forme di prima e di seconda specie, con applicazioni alle coniche,* Genoa 1940.
[708] Chini, Mineo. *Esercizi di calcolo infinitesimale,* Livorno 1914.
[415] Chisini, Oscar. *Lezioni di geometria analitica e proiettiva,* Bologna 1931.
[183] Chisini, Oscar; Masotti Biggiogero, Giuseppina. *Elementi di geometria descrittiva,* Milan 1956.
[423] Chisini, Oscar; Masotti Biggiogero, Giuseppina. *Lezioni di geometria descrittiva,* Milan 1946.
[109] Chisini, Oscar; Masotti Biggiogero, Giuseppina. *Lezioni di geometria descrittiva,* Milan 1955.
[1238] Ciamberlini, Corrado. *Nozioni di aritmetica,* Florence 1919.
[472] Ciani, Edgardo. *Il metodo delle coordinate proiettive omogenee nello studio degli enti algebrici,* Turin 1928.
[490] Ciani, Edgardo. *Il metodo delle coordinate proiettive omogenee nello studio degli enti algebrici. Sèguito alle lezioni di geometria proiettiva ed analitica,* Pisa 1915.
[847] Ciani, Edgardo. *Introduzione alla geometria algebrica. Lezioni di geometria superiore,* Padua 1931.
[471] Ciani, Edgardo. *Lezioni di geometria proiettiva ed analitica,* Pisa 1912.
[855] Ciani, Edgardo. *Scritti geometrici scelti*, vol. 1, Padua 1937.
[856] Ciani, Edgardo. *Scritti geometrici scelti*, vol. 2, Padua 1937.
[500] Cinquini, Silvio. *Funzioni quasi-periodiche,* Pisa 1949.
[718] Cipolla, Michele. *Analisi algebrica,* Palermo 1921.

[752] Cipolla, Michele. *La matematica elementare nei suoi fondamenti, nei riguardi didattici e negli sviluppi superiori. Conferenze,* Palermo 1927.

[894] Cipolla, Michele. *La matematica elementare nei suoi fondamenti, nei riguardi didattici e negli sviluppi superiori,* Palermo 1929.

[895] Cipolla, Michele. *La matematica elementare nei suoi fondamenti, nei riguardi didattici e negli sviluppi superiori,* Palermo 1936.

[1228] Cipolla, Michele; Mignosi, Gaspare. *Lezioni di analisi matematica elementare*, vol. 1, Catania 1930.

[1229] Cipolla, Michele; Mignosi, Gaspare. *Lezioni di analisi matematica elementare*, vol. 1, Turin 1931.

[1227] Cipolla, Michele; Mignosi, Gaspare. *Lezioni di analisi matematica elementare*, vol. 2, Catania 1925.

[690] Cisotti, Umberto. *Analisi matematica,* Milan 1940.

[691] Cisotti, Umberto. *Analisi matematica,* Milan 1943.

[303] Cisotti, Umberto. *Cenni sui fondamenti del calcolo tensoriale con applicazioni alla teoria dell'elasticità,* Milan 1932.

[607] Cisotti, Umberto. *Lezioni di analisi matematica,* Milan 1926.

[304] Cisotti, Umberto. *Lezioni di calcolo tensoriale,* Milan 1928.

[977] Cisotti, Umberto. *Lezioni di meccanica razionale,* Milan 1926.

[978] Cisotti, Umberto. *Meccanica razionale,* Milan 1939.

[979] Cisotti, Umberto. *Meccanica razionale,* Milan 1942.

[853, 854] CNRS. *Algèbre et théorie des nombres (Paris, 25 septembre-1er octobre 1949),* Paris 1950. (2 copies)

[582] Coble, Arthur B. *Algebraic Geometry and Theta Functions,* New York 1929.

[1303] Cockcroft, John D.; Robson, Anthony E. *Problemi dell'energia nucleare,* Varese 1959.

[252] Colangeli, Cesare. *Materia e radiazione: origine e struttura. Teoria unitaria dell'universo fisico,* Milan 1950.

[1252] Colonnetti, Gustavo. *L'équilibre des corps déformables,* Paris 1955.

[609] Comessatti, Annibale. *Lezioni di analisi algebrica,* Padua 1921.

[492] Comessatti, Annibale. *Lezioni di geometria analitica e proiettiva*, vol. 1, Padua 1926.

[29] Comessatti, Annibale. *Lezioni di geometria analitica e proiettiva*, vol. 1, Padua 1930.

[175] Comessatti, Annibale. *Lezioni di geometria analitica e proiettiva*, vol. 1, Padua 1941.

[479] Comessatti, Annibale. *Lezioni di geometria analitica e proiettiva*, vol. 2, Padua 1927.

[30] Comessatti, Annibale. *Lezioni di geometria analitica e proiettiva*, vol. 2, Padua 1931.

[176] Comessatti, Annibale. *Lezioni di geometria analitica e proiettiva*, vol. 2, Padua 1942.

[53] Comessatti, Annibale. *Lezioni di geometria descrittiva*, vol. 2, Padua 1923.

[23] Comessatti, Annibale. *Metodi di rappresentazione,* Padua 1923.

[680] Conforto, Fabio. *Abelsche Funktionen und algebraische Geometrie,* Berlin 1936.
[765] Conforto, Fabio. *Funzioni abeliane e matrici di Riemann*, vol. 1, Rome 1942.
[684] Conforto, Fabio. *Funzioni abeliane modulari*, vol. 1, Rome 1951.
[587] Conforto, Fabio. *Le superficie razionali,* Bologna 1939.
[424, 491] Conforto, Fabio. *Lezioni di geometria descrittiva,* Milan 1946. (2 copies)
[973] Conforto, Fabio. *Meccanica razionale,* Milan 1946.
[591] Coolidge, Julian. *A Treatise on Algebraic Plane Curves,* Oxford 1931.
[316] Corbino Orso, Mario. *Conferenze e discorsi,* Roma 1938.
[881] Corbino, Orso Mario. *Nozioni di fisica,* vol. 1: *Meccanica, acustica, cosmografia,* Palermo 1925.
[882] Corbino, Orso Mario. *Nozioni di fisica,* vol. 2: *Calore, ottica, elettricità e magnetismo,* Palermo 1925.
[451] Coriselli, Cesare. *Geometria descrittiva ad uso degli istituti tecnici e per lo studio privato,* Trento 1914.
[202] Costa, Manoel Amoroso. *Introducção á theoria da relatividade,* Rio de Janeiro 1922.
[458] Courant, Richard; Robbins, Herbert. *Che cos'è la matematica?,* Turin 1950.
[1020] Crocco, Arturo G. *Elementi di aviazione*, vol. 1, Rome 1930.
[1024] Crocco, Arturo G. *Problemi aeronautici,* Rome 1931.
[1023] Crocco, Arturo G. *Sulla stabilità laterale degli aeroplani,* Rome 1912.
[1022] Crocco, Arturo G. *Superartiglieria e superaviazione,* Rome 1926.
[310] Crudeli, Umberto. *Lezioni sulle probabilità,* Naples 1943.
[571] Crudeli, Umberto. *Calcolo differenziale e calcolo integrale. Fondamenti, compendio, esercizi,* Rome 1923.
[1273] Curie, Marie. *Les rayons α, β, γ des corps radioactifs en relation avec la structure nucléaire,* Paris 1933.
[799] D'Arbon, André. *Une doctrine de l'infini,* Paris 1957.
[177] D'Ovidio, Enrico. *Geometria analitica,* Turin 1903.
[561] D'Ovidio, Enrico. *Compendio di algebra complementare,* Turin 1900.
[1108] Da Rios, Luigi Sante. *Elementi d'algebra*, vol. 2, Padua 1927.
[1383] Dal Borgo, Vittorio. *Studio fisico dell'aorta normale e patologica,* Vatican 1952.
[279] Darboux, Gaston. *Éloges académiques et discours,* Paris 1912.
[946] Dassen, Claro C. *Elementos de cosmografía,* Buenos Aires 1930.
[1181] Dassen, Claro Cornelio. *Geometría del espacio,* Buenos Aires 1916.
[1180] Dassen, Claro Cornelio. *Geometría plana,* Buenos Aires 1925.
[1390] Dassen, Claro Cornelio. *Las matemáticas en la Argentina,* Buenos Airees 1924.
[298] Dassen, Claro Cornelio. *Metafísica de los conceptos matemáticos fundamentales (espacio, tiempo, cantidad, límite) y del análisis llamado infinitesimal,* Buenos Aires 1901.
[283] Dassen, Claro Cornelio. *Réflexions sur quelques antinomies et sur la logique empiriste,* Buenos Aires 1933.
[207] De Broglie, Louis. *Materia e luce,* Milan 1943.

[1271] De Broglie, Louis. *Une nouvelle conception de la lumière,* Paris 1934.
[485] De Donder, Théophile. *Théorie nouvelle de la mécanique statistique,* Paris 1938.
[13] De Finetti, Bruno. *Lezioni di matematica attuariale,* Rome 1957.
[12] De Finetti, Bruno. *Lezioni di matematica finanziaria,* Rome 1956.
[939] De Finetti, Bruno. *Matematica logico intuitiva,* Trieste 1944.
[360] De Finetti, Bruno. *Probabilismo,* Naples 1931.
[1107] De Franchis, Michele. *Geometria elementare,* Palermo 1909.
[480] De Franchis, Michele. *Lezioni di geometria analitica e proiettiva,* Palermo 1921.
[226] De Franchis, Michele. *Lezioni di geometria analitica e proiettiva,* Palermo 1939.
[961] De Galdeano, Zoel G. *Geometría elemental,* Toledo 1888.
[223] De la Vallée Poussin, Charles-Jean. *Jubilé Professoral,* Louvain 1928.
[877] De Marchi, Luigi. *Memorie scientifiche 1883–1932,* Padua 1932.
[806] De Menezes Corrêa Acciaiuoli, Luís. *Bibliografia hidrológica do império português,* vol. 1, Lisbon 1949.
[807] De Menezes Corrêa Acciaiuoli, Luís. *Bibliografia hidrológica do império português,* vol. 2, Lisbon 1950.
[79, 805] De Menezes Corrêa Acciaiuoli, Luís. *Geologia de Portugal. Ensaio bibliográfico,* Lisbon 1957.
[556] De Rham, Georges; Kodaira, Kunihiko. *Harmonic Integrals,* Princeton 1950.
[276] De Vonderweid, Édouard. *È realtà il muoversi della terra? Il sistema del mondo,* Genoa 1931.
[1369] Dekanosidze, Elena Nikolaevna. [*Tables of Cylindrical Functions in Two Variables*], Moscow 1956.
[470] Del Pezzo, Pasquale. *Principi di geometria projettiva,* Naples 1913.
[24] Del Re, Alfonso. *Lezioni di geometria descrittiva,* vol. 2: *Metodi di rappresentazione. Costruzione delle curve e delle superficie,* Naples 1906.
[1420] Del Re, Alfonso. *Lezioni di algebra della logica,* Naples 1907.
[1394] Del Re, Alfonso. *Lezioni sulle forme fondamentali dello spazio rigato, sulla dottrina degli immaginari e sui metodi di rappresentazione nella geometria descrittiva,* Naples 1905.
[313] Delens, Paul C. *La métrique angulaire des espaces de Finsler et la géométrie différentielle projective,* Paris 1934.
[888] Della Francesca, Piero. *De prospectiva pingendi,* Florence 1942.
[889] Della Francesca, Piero. *De prospectiva pingendi (disegni),* Florence 1942.
[1263] Delsarte, Jean. *Sur les ds^2 d'Einstein à symétrie axiale,* Paris 1934.
[507] Destouches, Jean-Louis. *Le rôle des espaces abstraits en physique nouvelle,* Paris 1935.
[1433] Deutsche Akademie der Wissenschaften zu Berlin (ed.). *Max Planck. Zum Gedenken,* Berlin 1959.
[1121] Di Noi, Salvatore. *Elementi di geometria,* vol. 1, Rome 1949.
[1189] Di Noi, Salvatore. *Geometria,* vol. 1, Milan 1943.
[1190] Di Noi, Salvatore. *Geometria,* vol. 2, Milan 1946.

[526] Dieudonné, Jean A. *Sur quelques propriétés des polynômes,* Paris 1934.
[527] Dieudonné, Jean A. *Sur les groupes classiques,* Paris 1948.
[874] Dini, Ulisse. *Opere*, vol. 1: *Algebra, geometria differenziale,* Rome 1953.
[104] Dini, Ulisse. *Opere*, vol. 2: *Funzioni di variabile reale e sviluppi in serie, problema di Dini-Neumann, funzioni analitiche*, Rome 1954.
[105] Dini, Ulisse. *Opere*, vol. 3: *Equazioni differenziali ordinarie e alle derivate parziali*, Rome 1955.
[1300] Dini, Ulisse. *Opere*, vol. 4: *Serie di Fourier,* Rome 1959.
[1301] Dini, Ulisse. *Opere*, vol. 5: *Sviluppi in serie,* Rome 1959.
[51] Dixmier, Jacques. *Les algèbres d'operateurs dans l'espace hilbertien,* Paris 1957.
[712] Dubreil, Paul. *Algèbre*, vol. 1, Paris 1946.
[1296] Dubreil, Paul. *Algèbre*, vol. 1, Paris 1954.
[510] Dubreil, Paul. *Quelques propriétés des variétés algébriques se rattachant aux théories de l'algèbre moderne,* Paris 1935.
[841] Dubreil-Jacotin, Marie Louise; Lesieur, Leonice; Croisot, Robert. *Leçons sur la théorie des treillis des structures algébriques ordonnées et des treillis géométriques,* Paris 1953.
[845] Dulaey, Maurice. *Construction des abaques,* Paris 1951.
[442] Eddington, Arthur S. *Espace, temps et gravitation,* Paris 1921.
[206] Eddington, Arthur S. *La filosofia della scienza fisica,* Bari 1941.
[802] Einstein, Albert. *Lettres à Maurice Solovine,* Paris 1956.
[281] Emanuelli, Pio. *Il cielo e le sue meraviglie. Atlante di 150 tavole,* Milan 1934.
[331] Enriques, Federigo. *Conferenze sulla geometria non-euclidea*, a cura di Olegario Fernandez, Bologna 1918.
[236] Enriques, Federigo. *Il significato della storia del pensiero scientifico,* Bologna 1936.
[904] Enriques, Federigo. *Le matematiche nella storia e nella cultura,* Bologna 1938.
[627] Enriques, Federigo. *Le superficie algebriche,* Bologna 1949.
[373] Enriques, Federigo. *Lezioni di geometria descrittiva,* Bologna 1902.
[167] Enriques, Federigo. *Lezioni di geometria proiettiva,* Bologna 1920.
[635] Enriques, Federigo. *Lezioni sulla teoria delle superfici algebriche*, vol. 1, Padua 1932.
[642] Enriques, Federigo. *Lezioni sulla teoria geometrica delle equazioni e delle funzioni algebriche*, a cura di Oscar Chisini, vols. 1–2, Bologna 1918.
[643] Enriques, Federigo. *Lezioni sulla teoria geometrica delle equazioni e delle funzioni algebriche*, a cura di Oscar Chisini, vol. 3, Bologna 1924.
[666] Enriques, Federigo. *Memorie scelte di geometria*, vol. 1: *1893–1898,* Bologna 1956.
[647] Enriques, Federigo. *Memorie scelte di geometria*, vol. 2: *1899–1910*, Bologna 1959.
[652] Enriques, Federigo. *Natura, ragione e storia,* Turin 1958.
[232] Enriques, Federigo. *Per la storia della logica. I principii e l'ordine della scienza nel concetto dei pensatori matematici,* Bologna 1922.

[280] Enriques, Federigo. *Problemi della scienza,* Bologna 1906.
[233] Enriques, Federigo. *Scienza e razionalismo,* Bologna 1912.
[588] Enriques, Federigo. *Sulle classificazioni delle superficie algebriche particolarmente di genere zero,* Rome 1934.
[858] Enriques, Federigo (ed.). *Questioni riguardanti le matematiche elementari,* vol. 1: *Critica dei principi,* Bologna 1912.
[860] Enriques, Federigo (ed.). *Questioni riguardanti le matematiche elementari,* vol. 1.1: *Critica dei principi,* Bologna 1924.
[861] Enriques, Federigo (ed.). *Questioni riguardanti le matematiche elementari,* vol. 1.2: *Critica dei principi,* Bologna 1925.
[859] Enriques, Federigo (ed.). *Questioni riguardanti le matematiche elementari,* vol. 2: *Problemi classici della geometria, numeri primi e analisi indeterminata, massimi e minimi,* Bologna 1914.
[862] Enriques, Federigo (ed.). *Questioni riguardanti le matematiche elementari,* vol. 2: *I problemi classici della geometria e le equazioni algebriche,* Bologna 1926.
[863] Enriques, Federigo (ed.). *Questioni riguardanti le matematiche elementari,* vol. 3: *Numeri primi e analisi indeterminata, massimi e minimi,* Bologna 1927.
[1266] Enriques, Federigo. *Signification de l'histoire de la pensée scientifique,* Paris 1934.
[1124] Enriques, Federigo; Amaldi, Ugo. *Elementi di geometria,* Bologna 1903.
[1136] Enriques, Federigo; Amaldi, Ugo. *Elementi di geometria,* Bologna 1909.
[1200] Enriques, Federigo; Amaldi, Ugo. *Elementi di geometria,* Bologna 1928.
[1205] Enriques, Federigo; Amaldi, Ugo. *Geometria elementare*, vol. 1, Bologna 1924.
[1204] Enriques, Federigo; Amaldi, Ugo. *Geometria elementare*, vol. 1, Bologna 1926.
[1202] Enriques, Federigo; Amaldi, Ugo. *Geometria elementare*, vol. 1, Bologna 1928.
[1206] Enriques, Federigo; Amaldi, Ugo. *Geometria elementare*, vol. 2, Bologna 1921.
[1165, 1203] Enriques, Federigo; Amaldi, Ugo. *Geometria elementare,* vol. 2, Bologna 1928. (2 copies)
[1119] Enriques, Federigo; Amaldi, Ugo. *Nozioni di geometria,* Bologna 1910.
[1199] Enriques, Federigo; Amaldi, Ugo. *Nozioni di geometria,* Bologna 1933.
[1201] Enriques, Federigo; Amaldi, Ugo. *Nozioni di matematica*, vol. 1, Bologna 1914.
[600] Enriques, Federigo; Chisini, Oscar. *Lezioni sulla teoria geometrica delle equazioni e delle funzioni algebriche*, vol. 4: *Funzioni ellittiche e abeliane,* Bologna 1934.
[411] Esnault-Pelterie, Robert. *Analyse dimensionnelle et métrologie. Le système Giorgi,* Lausanne 1950.
[654] Evans, Griffith C. *Infinitely Multiple Valued Harmonic Functions in Space with Two Base Branch Curves of Order One,* vol. 2, Berkeley 1959.

[1422] Evans, Griffith C. *Infinitely Multiple Valued Harmonic Functions in Space with Two Base Branch Curves of Order One*, vol. 3, Berkeley 1960.
[885] Fadini, Angelo; Vitale, Darwin. *Fisica*, vol. 1, Florence 1948.
[886] Fadini, Angelo; Vitale, Darwin. *Fisica*, vol. 2, Florence 1948.
[1151] Faifofer, Aureliano. *Elementi di algebra,* Venice 1893.
[1126] Faifofer, Aureliano. *Elementi di geometria,* Venice 1898.
[1244] Faifofer, Aureliano. *Il primo libro d'Euclide,* Venice 1885.
[235] Famà, Frank. *Cronache di scienza e di filosofia,* Catania 1915.
[198] Famà, Frank. *I campi vettoriali e la teoria elettromagnetica della luce. Saggio di filosofia scientifica,* Catania 1915.
[857] Fano, Gino. *Complementi di geometria,* Turin 1935.
[306] Fano, Gino. *Geometria non-euclidea,* Bologna 1935.
[28] Fano, Gino. *Lezioni di geometria descrittiva*, [s.l.] 1903.
[468] Fano, Gino. *Lezioni di geometria descrittiva,* Turin 1910.
[481] Fano, Gino; Terracini, Alessandro. *Lezioni di geometria analitica e proiettiva,* Turin 1930.
[478] Fano, Gino; Terracini, Alessandro. *Lezioni di geometria analitica e proiettiva,* Turin 1948.
[705] Fantappiè, Luigi. *I funzionali analitici,* Città di Castello 1930.
[737] Fantappiè, Luigi. *Teoría de los funcionales analíticos y sus aplicaciones,* Barcelona 1943.
[50] Favard, Jean. *Cours de géométrie différentielle locale,* Paris 1957.
[161] Favard, Jean. *Espaces et dimension,* Paris 1950.
[249] Fermi, Enrico. *Conferenze di fisica atomica,* Rome 1950.
[25] Ferrari, Camillo; Mutinelli, Giuseppe. *Lezioni di geometria descrittiva*, Pisa 1904.
[200] Ferretti, Bruno. *Fisica atomica,* Rome, stampa 1953.
[197] Ferretti, Bruno. *La fisica contemporanea,* Rome 1946.
[178]. Ferrobeton. *Ferrobeton: Opere eseguite*, vol. 1: *Ponti e costruzioni marittime,* Rome 1954.
[10] Fichera, Gaetano. *Lezioni sulle trasformazioni lineari,* Trieste 1954.
[655] Fichera, Gaetano. *Premesse ad una teoria generale dei problemi al contorno per le equazioni differenziali,* Rome 1958.
[390] Filiasi Carcano, Paolo. *La filosofia d'oggi,* Rome 1950.
[460] Filiasi Carcano, Paolo. *Problematica della filosofia odierna,* Rome 1953.
[1234] Finzi, Aldo. *Complementi di algebra e geometria,* Naples 1918.
[742] Finzi, Bruno; Pastori, Maria. *Istituzioni di matematica per chimici. Geometria analitica, analisi infinitesimale—meccanica razionale—termodinamica—teoria delle probabilità,* Milan 1940.
[781] Finzi, Bruno; Pastori, Maria. *Lezioni di matematica per chimici. Geometria analitica, analisi infinitesimale, meccanica razionale, termodinamica, teoria delle probabilità,* Milan 1930.
[244] Flora, Francesco (ed.). *Leonardo,* Verona 1952.
[782] Flores d'Arcais, Francesco. *Lezioni di analisi infinitesimale,* Padua 1921.

[1162] Florio, Fortunato. *Logaritmi a 10 decimali e valori naturali delle funzioni trigonometriche,* Naples 1934.

[324] Fortet, Robert. *Calcul des probabilités,* Paris 1950.

[898] Frajese, Attilio. *La matematica nel mondo antico,* Rome 1951.

[467] Frantappiè, Luigi. *Principi di una teoria unitaria del mondo fisico e biologico,* Rome 1944.

[1163] Frattini, Giovanni. *Lezioni di algebra, geometria e trigonometria piana e sferica,* vol. 1, Rome 1921.

[1164] Frattini, Giovanni. *Lezioni di algebra, geometria e trigonometria piana e sferica,* vol. 2, Rome 1921.

[522, 523] Fréchet, Maurice. *L'Arithmétique de l'infini,* Paris 1934. (2 copies)

[364] Frege, Gottlob. *Aritmetica e logica,* Turin 1948.

[852] Fubini, Guido. *Introduzione alla teoria dei gruppi discontinui e delle funzioni automorfe,* Pisa 1908.

[601] Fubini, Guido. *Lezioni di analisi matematica,* Turin 1919.

[602] Fubini, Guido. *Lezioni di analisi matematica,* Turin 1925.

[649] Fubini, Guido. *Opere scelte,* Rome 1958.

[873] Fubini, Guido. *Opere scelte,* vol. 1, Rome 1957.

[291] Fubini, Guido; Čech, Eduard. *Geometria proiettiva differenziale,* vol. 1, Bologna 1926.

[292] Fubini, Guido; Čech, Eduard. *Geometria proiettiva differenziale,* vol. 2, Bologna 1927.

[299] Fubini, Guido; Čech, Eduard. *Introduction à la géométrie projective différentielle des surfaces,* Paris 1931.

[743] Fubini, Guido; Vivanti, Giulio. *Esercizi di analisi matematica. Calcolo infinitesimale, con speciale riguardo alle applicazioni,* Turin 1920.

[568] Fubini, Guido; Vivanti, Giulio. *Esercizi di analisi matematica. Calcolo infinitesimale, con speciale riguardo alle applicazioni,* Turin 1930.

[259] Fumagalli, Giuseppina (ed.). *Leonardo omo sanza lettere,* Florence 1943.

[286]. Galilei, Galileo; Torricelli, Evangelista. *Due insigni autografi di Galileo Galilei e di Evangelista Torricelli,* Florence 1908.

[1155] Gallucci, Generoso. *Algebra elementare,* Naples 1936.

[976] Gallucci, Generoso. *Applicazioni dell'algebra alla risoluzione dei problemi geometrici,* Naples 1931.

[1215] Gallucci, Generoso. *Aritmetica pratica,* Milan 1936.

[159] Gallucci, Generoso. *Complementi di geometria proiettiva,* Naples 1928.

[463] Gallucci, Generoso. *Elementi di geometria proiettiva ed esercizi di analitica,* Naples 1938.

[1216] Gallucci, Generoso. *Elementi di geometria razionale,* vol. 1, Milan 1935.

[1217] Gallucci, Generoso. *Elementi di geometria razionale,* vol. 2, Milan 1936.

[1212] Gallucci, Generoso. *Geometria intuitiva,* Milan 1935.

[337] Gallucci, Generoso. *Nuovo saggio su l'infinito. Contributo allo studio dei problemi della logica,* Naples 1931.

[1213] Gallucci, Generoso. *Prime lezioni di algebra,* Milan 1937.

[1214] Gallucci, Generoso. *Trigonometria piana,* Milan 1936.

[240] Galvani, Luigi. *Il taccuino di Luigi Galvani. Riproduzione in fac-simile dell'autografo conservato nella Biblioteca dell'Archiginnasio,* Bologna 1937.
[902] Gambioli, Dionisio. *Breve sommario della storia delle matematiche, colle due appendici sui matematici italiani e sui tre celebri problemi geometrici dell'antichità,* Palermo 1929.
[75] Ganot, Adolphe. *Trattato elementare di fisica*, vol. 1, Milan 1864.
[76] Ganot, Adolphe. *Trattato elementare di fisica*, vol. 2, Milan 1864.
[363] Garbasso, Antonio. *Scienza e poesia,* Florence 1934.
[165] García de Galdeano y Yanguas, Zoel. *Geometría general*, vol. 1: *Teoremas, problemas y métodos geométricos,* Zaragoza 1895.
[433] García, Vicente I. *Ensayo de materiales,* Montevideo 1930.
[903] Gardiner, William. *Tavole logaritmiche,* Florence 1810.
[1170] Garneri, Augusto. *Corso elementare di disegno geometrico,* Rome 1893.
[1027] Garnier, René. *Cours de cinématique*, vol. 1, Paris 1941.
[1028] Garnier, René. *Cours de cinématique*, vol. 2, Paris 1949.
[1029] Garnier, René. *Cours de cinématique*, vol. 3, Paris 1951.
[2] Garnier, René. *Cours de cinématique*, vol. 1, Paris 1954.
[3] Garnier, René. *Cours de cinématique*, vol. 2, Paris 1955.
[550] Garnier, René. *Cours des mathématiques générales*, vol. 1, Paris 1945.
[530] Garnier, René. *Cours des mathématiques générales*, vol. 2, Paris 1948.
[531] Garnier, René. *Leçons d'algèbre et de géométrie*, vol. 1, Paris 1935.
[532] Garnier, René. *Leçons d'algèbre et de géométrie*, vol. 2, Paris 1936.
[533] Garnier, René. *Leçons d'algèbre et de géométrie*, vol. 3, Paris 1937.
[1430] Gauthier, Vincenzo. *Nozioni di idrologia moderna. Crenoterapis,* Naples 1919.
[540] Gazzaniga, Paolo. *Gli elementi della teoria dei numeri,* Verona 1903.
[1232] Gazzaniga, Paolo. *Libro di aritmetica generale e di algebra elementare,* Verona-Padua 1904.
[333] Geppert, Harald; Koller, Siegfried. *Erbmathematik. Theorie der Vererbung in Bevölkerung und Sippe,* Leipzig 1938.
[311] Gerbaldi, Francesco; Loria, Gino (eds.). *Scritti matematici offerti ad Enrico d'Ovidio in occasione del suo LXXV genetliaco, 11 agosto 1918,* Milan 1918.
[247] Geymonat, Ludovico. *Saggi di filosofia neorazionalistica,* Turin 1953.
[595] Ghizzetti, Aldo. *Calcolo simbolico,* Bologna 1943.
[1292] Giacomelli, Raffaele. *Galileo Galilei giovane e il suo "De Motu"*, Pisa 1949.
[1001] Giacomelli, Raffaele. *Gli scritti di Leonardo da Vinci sul* volo, Rome 1936.
[427] Giannelli, Aristide. *Lezioni sui telai elastici piani,* Rome 1932.
[725] Gigli, Duilio. *Aritmetica generale,* Milan 1929.
[1241] Giorgi, Angelo. *Elementi di geometria,* Trieste 1937.
[204] Giorgi, Giovanni. *Che cos'è l'elettricità,* Rome 1928.
[900] Giorgi, Giovanni. *Compendio di storia delle matematiche,* Turin 1948.
[201] Giorgi, Giovanni. *L'etere e la luce. Dall'etere cosmico alle moderne teorie della luce,* Rome 1938.
[438] Giorgi, Giovanni. *Lezioni di meccanica razionale*, vol. 1, Rome 1931.
[439] Giorgi, Giovanni. *Lezioni di meccanica razionale*, vol. 2, Rome 1934.

[1014] Giorgi, Giovanni. *Meccanica razionale,* Rome 1946.

[1332] Giorgi, Giovanni. *Metrologia elettrotecnica antica e nuova,* Milan 1937.

[1326] Giorgi, Giovanni. *Pubblicazioni scientifiche e tecniche,* Rome 1894–1929.

[1013] Giorgi, Giovanni. *Sull'uso delle batterie di accumulatori negli impianti di trazione elettrica,* Rome 1894.

[205] Giorgi, Giovanni; Rosati, Maria. *I colori e la cromatica moderna,* Rome 1930.

[440] Gnesotto, Tullio. *Fisica tecnica. Termologia ed ottica fisica,* Padua 1919.

[542] Godeaux, Lucien. *Géométrie algébrique*, vol. 1, Liège 1948.

[543] Godeaux, Lucien. *Géométrie algébrique*, vol. 2, Liège 1949.

[546] Godeaux, Lucien. *La géométrie,* Liège 1931.

[513, 1269] Godeaux, Lucien. *La théorie des surfaces et l'espace réglé,* Paris 1934. (2 copies)

[514] Godeaux, Lucien. *Les involutions cycliques appartenant à une surface algébrique,* Paris 1935.

[168] Godeaux, Lucien. *Leçons de géométrie projective,* Paris 1933.

[575] Godeaux, Lucien. *Leçons de géométrie supérieure,* Liège 1933.

[628] Godeaux, Lucien. *Leçons élémentaires sur les équations différentielles et le calcul des variations,* Brussels 1926.

[511, 512] Godeaux, Lucien. *Les surfaces algébriques non rationnelles de genres arithmétique et géométrique nuls,* Paris 1934. (2 copies)

[1281] Godeaux, Lucien. *Questions non résolues de géométrie algébrique. Les involutions de l'espace et les variétés algébriques à trois dimensions,* Paris 1933.

[534] Godeaux, Lucien; Garnier, René. *Géométrie algébrique. Théorie des courbes et des surfaces,* Paris 1952.

[195] Godeaux, Lucien; Rozet, Octave. *Leçons de géométrie projective,* Liège 1952.

[528] Goursat, Édouard. *Cours d'analyse mathématique*, vol. 1: *Dérivées et différentielles, intégrales définies, développements en série, applications géométriques*, Paris 1927.

[821] Goursat, Édouard. *Cours d'analyse mathématique*, vol. 2: *Théorie des fonctions analytiques, équations différentielles, équations aux dérivées partielles du premier ordre,* Paris 1929.

[529] Goursat, Édouard. *Cours d'analyse mathématique*, vol. 3: *Intégrales infiniment voisines, équations aux dérivées partielles du 2. ordre, équations intégrales, calcul des variations*, Paris 1927.

[420] Grillo, Vincenzo (ed.). *5 anni del museo, 1953–1958,* Milan 1958.

[645] Gröbner, Wolfgang. *Idealtheoretischer Aufbau der Algebraischen Geometrie,* vol. 1, Leipzig 1941.

[1] Gröbner, Wolfgang. *Matrizenrechnung,* Munich 1955.

[626] Gröbner, Wolfgang. *Moderne algebraische Geometrie,* Vienna 1949.

[703] Gröbner, Wolfgang; Hofreiter, Nikolaus. *Integraltafel*, vol. 1: *Unbestimmte Integrale,* Vienna 1949.

[704] Gröbner, Wolfgang; Hofreiter, Nikolaus. *Integraltafel*, vol. 2: *Bestimmte Integrale,* Vienna 1950.

[702] Grosrey, Adrien. *Éléments de calcul infinitésimal,* Paris 1945.

[577] Grüß, Gerd. *Variationsrechnung,* Leipzig 1938.

[428] Guidi, Camillo. *Memorie raccolte e ripubblicate a cura di un gruppo di antichi allievi in occasione dell'ottantesimo genetliaco del maestro, 24 luglio 1933,* Palermo 1929.

[365] Haack, Wolfgang. *Differentialgeometrie*, vol. 1, Hanover 1949.

[366] Haack, Wolfgang. *Differentialgeometrie*, vol. 2, Hanover 1948.

[212] Hadamard, Jacques. *Selecta. Jubilé scientifique,* Paris 1935.

[560] Hadamard, Jacques. *Poincaré i la teoria de les equacions diferencials. Conferències,* Barcelona 1921.

[810] Hartzler Fox, Ralph (ed.). *Algebraic Geometry and Topology: A Symposium in Honor of Solomon Lefschetz,* Princeton 1957.

[419] Heffter, Lothar. *Grundlagen und analytischer Aufbau,* Leipzig 1940.

[769] Hasse, Helmut. *Höhere Algebra*, vol. 2, Berlin 1937.

[1392] Henry, Charles. *Psychobiologie et énergétique,* [s.l.] 1911.

[755] Hasse, Helmut. *Über die Klassenzahl abelscher Zahlkörper,* Berlin 1952.

[508] Hasse, Helmut. *Über gewisse Ideale in einer einfachen Algebra,* Paris 1934.

[716] Hasse, Helmut. *Zahlentheorie,* Berlin 1949.

[1242] Hessenberg, Gerhard W. *Trigonometría plana y esférica,* Barcelona-Buenos Aires 1926.

[448] Hessenberg, Gerhard W. *Vorlesungen über Darstellende Geometrie,* Leipzig 1929.

[278] Hilbert, David; Ackermann, Wilhelm. *Grundzüge der theoretischen Logik,* Berlin 1928.

[640, 790] Hodge, William V. D. *The Theory and Applications of Harmonic Integrals,* Cambridge 1941. (2 copies)

[557] Hodge, William V. D.; Pedoe, Daniel. *Methods of Algebraic Geometry*, vol. 1: *Book 1. Algebraic Preliminaries; Book 2. Projective Space*, Cambridge 1947.

[772] Hodge, William V. D.; Pedoe, Daniel. *Methods of Algebraic Geometry*, vol. 2: *Book 3. General Theory of Algebraic Varieties in Projective Space; Book 4. Quadric and Grassmann Varieties,* Cambridge 1952.

[773] Hodge, William V. D.; Pedoe, Daniel. *Methods of Algebraic Geometry*, vol. 3: *Book 5. Birational Geometry,* Cambridge 1954.

[389] Hoenen, Peter. *Filosofia della natura inorganica,* Brescia 1949.

[382] Hopf, Heinz. *Capitoli scelti della teoria delle varietà topologiche,* Rome 1953.

[210] Humbert, Georges. *Œuvres*, vol. 1, Paris 1929.

[171] Humbert, Georges. *Œuvres*, vol. 2, Paris 1936.

[922] INPS. *Tecnica delle assicurazioni sociali,* Rome 1951.

[919] Insolera, Filadelfo. *Corso di matematica finanziaria,* Turin 1937.

[1264] Iyanaga, Shōkichi. *Sur les classes d'idéaux dans les corps quadratiques,* Paris 1935.

[965] Jadanza, Nicodemo. *Elementi di geodesia,* Turin 1894.

[964] Jadanza, Nicodemo. *Teorica dei cannocchiali esposta secondo il metodo di Gauss,* Turin 1885.

[789] Jahnke, Eugen; Emde, Fritz. *Funktionentafeln, mit Formeln und Kurven,* Leipzig-Berlin 1933.

[1006] Japanese National Commission on the Teaching of Mathematics. *Divisional Reports on Present Tendencies in the Development of Mathematical Teaching in Japan,* Tokyo 1936.

[293] Jeans, James Hopwood. *Les nouvelles bases philosophiques de la science,* Paris 1935.

[441] Joffé, Abram F. *Semi-conducteurs électroniques,* Paris 1935.

[1279] Joliot, Frédéric; Curie, Irène. *Radioactivité artificielle,* Paris 1934.

[1245] Jordan, Wilhelm. *Mathematische und geodätische Hilfstafeln,* Stuttgart 1878.

[422] Jorini, Antonio F. *Geometria analitica,* Milan 1922.

[1416] Julia, Gaston. *Cours de géométrie infinitésimale*, vol. 2.1: *Généralités*, Paris 1955.

[1313] Julia, Gaston. *Cours de géométrie infinitésimale,* vol. 3.1: *Méthodes générales, théorie des courbes,* Paris 1955.

[584] Julia, Gaston. *Exercices d'analyse*, vol. 2, Paris 1933.

[585] Julia, Gaston. *Exercices d'analyse*, vol. 3, Paris 1933.

[502] Julia, Gaston. *Introduction mathématique aux théories quantiques*, vol. 1, Paris 1936.

[157] Julia, Gaston. *Introduction mathématique aux théories quantiques*, vol. 2, Paris 1938.

[501] Julia, Gaston. *Leçons sur la représentation conforme des aires multiplement connexes,* Paris 1934.

[496] Jung, Heinrich W. E. *Algebraische Flächen,* Hanover 1925.

[754] Jung, Heinrich W. E. *Einführung in die Theorie der algebraischen Funktionen einer Veränderlichen,* Berlin 1951.

[1384, 1385] Junkes, Joseph. *Ein Vorschlag zur empirischen Reduktion von Spektralverteilungen,* Vatican 1952. (2 copies)

[1305] Kähler, Eric. *Einführung in die Theorie der Systeme von Differentialgleichungen,* Leipzig-Berlin 1934.

[1417] Kanitani, Jôyô. *Géométrie différentielle projective des hypersurfaces,* Ryojun 1931.

[1367] Karnmazina, Liudmila N.; Kurockina, L. V. [*Tables of Interpolation Coefficients*], Moscow 1956.

[1365] Karpov, Konstantin Adrianovič; Razumovskij, Spartak Nikolaevič. [*Tables of Logarithmic Integrals*], Moscow 1956.

[914] Klein, Felix. *Conferénces sur les mathématiques,* Paris 1898.

[379] Klein, Felix. *Riemann'sche Flächen,* Göttingen 1894.

[624] Klein, Felix. *Über lineare Differentialgleichungen der zweiten Ordnung,* Leipzig 1906.

[828] Knopp, Konrad; Mangoldt, H. von. *Einführung in die Höhere Mathematik,* vol. 1, Leipzig 1942.

[829] Knopp, Konrad; Mangoldt, H. von. *Einführung in die Höhere Mathematik,* vol. 2, Leipzig 1942.

[830] Knopp, Konrad; Mangoldt, H. von. *Einführung in die Höhere Mathematik,* vol. 3, Leipzig 1942.

[352] Kodaira, Kunihiko. *On Compact Analytic Surfaces,* [s.l.] 1955.
[1293] Kodaira, Kunihiko. *The Theory of Harmonic Integrals and Their Applications to Algebraic Geometry,* Princeton 1955.
[1012] Kodaira, Kunihiko; Spencer, D. C. *Existence of Complex Structure on a Differentiable Family of Deformations of Compact Complex Manifolds,* [s.l.] 1958.
[353] Kodaira, Kunihiko; Spencer, Donald C. *On Deformations of Complex Analytic Structures,* Part I, Princeton 1958.
[1294] Kodaira, Kunihiko; Spencer, Donald C. *On Deformations of Complex Analytic Structures,* Part III, Princeton 1959.
[998] Krall, Giulio. *La diga di sbarramento a,* volta-*cupola per laghi artificiali. Nuovo metodo di calcolo,* Rome 1951.
[1327] Krall, Giulio. *Meccanica tecnica delle vibrazioni,* vol. 1: *Sistemi discreti,* Bologna 1940.
[1328] Krall, Giulio. *Meccanica tecnica delle vibrazioni,* vol. 2: *Sistemi continui,* Bologna 1940.
[1349] Krall, Giulio. *Ricerche di meccanica,* vol. 1, Rome 1922–1942.
[1350] Krall, Giulio. *Ricerche di meccanica,* vol. 2, Rome 1922–1932.
[85] Kues, Nikolaus von. *Die mathematischen Schriften,* Hamburg 1952.
[1370] Kufaradze, Grigan Zosimovich. [*Tables of Protocols of numbers of 3 Digits by Numbers of 3 Digits up to 999x999*], Moscow 1952.
[756] Kurosch, Alexandr G. *Gruppentheorie,* Berlin 1953.
[1236] La Barbera, Alberto. *I numeri reali. Calcolo dei radicali e dei logaritmi,* Velletri 1923.
[241] La Stella, Mario. *Guglielmo Marconi. Mago dell'invisibile, dominatore degli spazi,* Milan 1937.
[1285] Lalesco, Trajan. *La géometrie du triangle,* Paris 1937.
[787] Lampariello, Giovanni. *Esercizi e complementi di calcolo infinitesimale,* Rome 1932.
[982] Lampariello, Giovanni. *Lezioni di fisica matematica,* Messina 1947.
[980] Lampariello, Giovanni. *Lezioni di meccanica razionale,* Messina 1946.
[981] Lampariello, Giovanni. *Lezioni di meccanica razionale,* vol. 1: *Nozioni introduttive, cinematica, principii della dinamica,* Rome 1942.
[632] Landau, Edmund. *Grundlagen der Analysis,* Leipzig 1930.
[1272] Langevin, Paul. *La notion de corpuscules et d'atomes,* Paris 1934.
[679] Laureati, Ferdinando. *Elementi di calcolo infinitesimale,* Rome 1883.
[464] Lebesgue, Henri. *Les coniques,* Paris 1942.
[504] Lebesgue, Henri. *Leçons sur l'intégration et la recherche des fonctions primitives,* Paris 1928.
[804] Lebesgue, Henri. *Sur la mesure des grandeurs,* Paris 1935.
[811] Lefschetz, Solomon. *Algebraic Geometry,* Princeton 1953.
[813] Lefschetz, Solomon. *Algebraic Topology,* New York 1942.
[814] Lefschetz, Solomon. *Introduction to Topology,* Princeton 1949.
[574, 822] Lefschetz, Solomon. *L'analysis situs et la géométrie algébrique,* Paris 1924.

[499] Lefschetz, Solomon. *Lectures on Differential Equations,* Princeton 1946.
[812] Lefschetz, Solomon. *Topology,* New York 1930.
[1000] Legendre, Adrien-Marie. *Elementi di geometria,* Naples 1843.
[953] Legendre, Adrien Marie. *Elementi di geometria,* Florence 1899.
[367] Legnazzi, Enrico Nestore. *Aggiunte illustrative alla commemorazione del professore conte Giusto Bellavitis. Brevi cenni sulle equipollenze, immaginari, risoluzione delle equazioni, quaternioni, logismografia,* Padua 1881.
[20] Lehto, Olli. *Commentationes in honorem R. Nevanlinna LXXX annos nato,* Helsinki 1975.
[243] Leonardo da Vinci. *L'occhio nell'universo*, a cura di Giuseppina Fumagalli, Florence 1943.
[5] Leonardo Pisano (Fibonacci). *Le livre des nombres carrés*, traduit pour la première fois du latin médiéval en français avec une introd. et des notes par Paul Ver Eecke, Paris-Bruges 1952.
[1276] Leprince-Ringuet, Louis. *Rayons cosmiques. Les mésons,* Paris 1934.
[615] Levi-Civita, Tullio. *Caractéristiques des systèmes différentiels et propagation des ondes,* Paris 1932.
[629] Levi-Civita, Tullio. *Caratteristiche dei sistemi differenziali e propagazione ondosa,* Bologna 1931.
[430] Levi-Civita, Tullio. *Corso di meccanica razionale*, vol. 1: *Teorie introduttive e cinematica,* Padua 1918.
[431] Levi-Civita, Tullio. *Corso di meccanica razionale*, vol. 2: *Principi. Statica elementare,* Padua 1918.
[432] Levi-Civita, Tullio. *Corso di meccanica razionale*, vol. 3: *Dinamica. Complementi di statica, Idromeccanica,* Padua 1918.
[348] Levi-Civita, Tullio. *Lezioni di calcolo differenziale assoluto,* Rome 1925.
[101] Levi-Civita, Tullio. *Opere matematiche*, vol. 1, Bologna 1954.
[102] Levi-Civita, Tullio. *Opere matematiche*, vol. 2, Bologna 1956.
[224] Levi-Civita, Tullio. *Opere matematiche*, vol. 4, Bologna 1960.
[152] Levi-Civita, Tullio. *Questions de mecanica classica i relativista,* Barcelona 1921.
[347] Levi-Civita, Tullio. *The Absolute Differential Calculus,* London 1927.
[710] Levi-Civita, Tullio; Amaldi Ugo. *Lezioni di meccanica razionale,* Padua 1922.
[967] Levi-Civita, Tullio; Amaldi, Ugo. *Lezioni di meccanica razionale*, vol. 1: *Cinematica, principi e statica,* Bologna 1923.
[970] Levi-Civita, Tullio; Amaldi, Ugo. *Lezioni di meccanica razionale*, vol. 1: *Cinematica, principi e statica,* Bologna 1950.
[968] Levi-Civita, Tullio; Amaldi, Ugo. *Lezioni di meccanica razionale*, vol. 2.1: *Dinamica dei sistemi con un numero finito di gradi di libertà. Parte prima,* Bologna 1926.
[971] Levi-Civita, Tullio; Amaldi, Ugo. *Lezioni di meccanica razionale*, vol. 2.1: *Dinamica dei sistemi con un numero finito di gradi di libertà. Parte prima,* Bologna 1951.

[969] Levi-Civita, Tullio; Amaldi, Ugo. *Lezioni di meccanica razionale*, vol. 2.2: *Dinamica dei sistemi con un numero finito di gradi di libertà. Parte seconda,* Bologna 1927.

[972] Levi-Civita, Tullio; Amaldi, Ugo. *Lezioni di meccanica razionale*, vol. 2.2: *Dinamica dei sistemi con un numero finito di gradi di libertà. Parte seconda,* Bologna 1952.

[966] Levi-Civita, Tullio; Amaldi, Ugo. *Nozioni di balistica esterna,* Bologna 1935.

[569] Levi, Beppo. *Analisi matematica e algebra infinitesimale,* Bologna 1937.

[1407] Levi, Eugenio Elia. *Opere*, vol. 1, Rome 1959.

[1408] Levi, Eugenio Elia. *Opere*, vol. 2, Rome 1960.

[445] Lévy, Paul. *Processus stochastiques et mouvement brownien,* Paris 1948.

[180] Levy, Paul. *Théorie de l'addition des variables aléatoires,* Paris 1954.

[746] Lévy, Paul P; Pellegrino, Franco M. *Problèmes concrets d'analyse fonctionnelle,* Paris 1951.

[287] Liapunov, Aleksandr M. *Sur certaines séries de figures d'équilibre d'un liquide hétérogène en rotation*, vol. 1, Leningrad 1925.

[288] Liapunov, Aleksandr M. *Sur certaines séries de figures d'équilibre d'un liquide hétérogène en rotation*, vol. 2, Leningrad 1925.

[52] Lichnerowicz, André. *Théorie globale des connexions et des groupes d'holonomie,* Paris 1955.

[839] Lindelöf, Ernst. *Einführung in die Höhere Analysis,* Berlin 1934.

[554] Lo Voi, Antonino. *Curve algebriche*, Palermo 1939.

[553] Lo Voi, Antonino. *Lezioni di geometria algebrica,* Palermo 1942.

[487] Lo Voi, Antonino. *Lezioni di geometria descrittiva con disegno,* Palermo 1951.

[319] Lobačevskij, Nikolaj Ivanovič. *Nuovi principi della geometria,* Turin 1955.

[1371] Lobačvskij, Nikolaj Ivanovič. [*Selected Works on Geometry*], Moscow 1966.

[95] Loedel, Enrique. *Fisica relativista,* Buenos Aires 1955.

[653] Lombardo Radice, Lucio. *Piani grafici finiti non desarguesiani,* Palermo 1959.

[1246] Lörcher, Otto; Löffler, Eugen. *Methodischer Leitfaden der Geometrie,* Stuttgart 1914.

[883] Loria, Gino. *Il passato ed il presente delle principali teorie geometriche. Storia e bibliografia,* Turin 1907.

[884] Loria, Gino. *Il passato ed il presente delle principali teorie geometriche. Storia e bibliografia,* Padua 1931.

[307] Loria, Gino. *Scritti, conferenze, discorsi sulla storia delle matematiche,* Padua 1936.

[646] Loria, Gino. *Vorlesungen über darstellende Geometrie*, vol. 2: *Anwendungen auf eben flächige Gebilde, Kurven und Flachen,* Leipzig-Berlin 1913.

[1274] Lotka, Alfred J. *Théorie analytique des associations biologiques*, vol. 1: *Principes,* Paris 1934.

[687] Ludovico, Domenico. *L'aeroplano cosa è. Soluzione ed evoluzione del problema del volo,* Rome 1949.

[19] Lupattelli, Astorre. *L'Università italiana per stranieri di Perugia, 1925–1943,* Perugia 1947.
[1262] Lusin, Nikolai. *Sur les suites stationnaires,* Paris 1934.
[1268] Lusternik, Lazar A.; Schnirelmann, Lev G. *Méthodes topologiques dans les problèmes variationnels*, vol. 1, Paris 1934.
[984] Luvini, Giovanni. *Tavole di logaritmi a sette decimali,* Turin 1876.
[1131] Luvini, Giovanni. *Compendio di geometria piana e solida,* Turin 1867.
[147] Maccone, Adriano; Mascalchi, Maria. *Corso di fisica sperimentale*, vol. 1, Turin 1948.
[148] Maccone, Adriano; Mascalchi, Maria. *Corso di fisica sperimentale*, vol. 2, Turin 1948.
[149] Maccone, Adriano; Mascalchi, Maria. *Corso di fisica sperimentale*, vol. 3, Turin 1949.
[1026] Maggi, Gian Antonio. *Selecta. Raccolta di scritti matematici dal 1880 al 1931,* Milan 1932.
[320] Magni, Pasquale (ed.). *Scienza e mistero*, vol. 1, Rome 1948.
[581] Mammana, Gabriele. *Sistemi differenziali,* Naples 1937.
[248] Mancini, Niccolò. *Errore della scienza,* Florence 1950.
[887] Mancini, Niccolò. *Energia universale e reazione della materia,* Florence 1948.
[55] Manfredi, Franco. *Ortoprospettiva. Teoria sulla quarta proiezione ortogonale ad uso delle scuole professionali e industriali,* Naples 1928.
[1186, 1188] Marcolongo, Roberto. *Algebra,* Naples 1924. (2 copies)
[1187] Marcolongo, Roberto. *Complementi di algebra e di analisi,* Naples 1930.
[1167] Marcolongo, Roberto. *Corso di matematica*, vol. 1: *Algebra,* Naples 1920.
[1400] Marcolongo, Roberto. *La Meccanica di Leonardo da Vinci,* Naples 1932.
[261] Marcolongo, Roberto. *Leonardo da Vinci artista-scienziato,* Milan 1939.
[258] Marcolongo, Roberto. *Memorie sulla geometria e la meccanica di Leonardo da Vinci,* Naples 1937.
[426] Marcolongo, Roberto. *Relatività,* Messina 1923.
[909] Marcolongo, Roberto; Marzella, Lena. *Temi di matematica assegnati nei concorsi per le R. scuole medie. Svolgimento e relativi richiami teorici*, vol. 1, Napoli 1937.
[910] Marcolongo, Roberto; Marzella, Lena. *Temi di matematica assegnati nei concorsi per le R. scuole medie. Svolgimento e relativi richiami teorici*, vol. 2, Napoli 1939.
[911] Marcolongo, Roberto; Marzella, Lena. *Temi di matematica assegnati nei concorsi per le R. scuole medie. Svolgimento e relativi richiami teorici*, vol. 3, Napoli 1941.
[193] Marconi, Guglielmo. *Scritti,* Rome 1941.
[336] Marconi, Guglielmo. *Per la ricerca scientifica,* Rome 1935.
[1145] Marletta, Giuseppe. *Trattato di geometria elementare,* Catania 1912.
[1193] Maroni, Arturo. *Trigonometria,* Florence 1947.
[1194] Maroni, Arturo. *Elementi di analisi matematica,* Florence 1948.
[84] Maros Dell'Oro, Angiolo. *La teoria fisica,* Padua 1955.

[1129] Marrese, Pietro. *Aritmetica e geometria,* Naples 1917.
[1154] Marseguerra, Vincenzo; Zappalà, Attilio. *Elementi di geometria,* Turin 1934.
[170] Martinelli, Enzo. *Lezioni di geometria,* Genoa 1950.
[15] Martinelli, Enzo. *Lezioni di geometria con esercizi per il secondo anno di studi di Matematica, Fisica, Ingegneria,* Genoa 1954.
[107] Martinelli, Enzo. *Lezioni di topologia,* Rome 1956.
[108] Martinelli, Enzo. *Lezioni di topologia,* Rome 1957.
[64] Marzolo, Francesco. *Idraulica generale,* Padua 1920.
[113] Masotti Biggiogero, Giuseppina. *Lezioni di geometria proiettiva,* Milan 1958.
[1348] Masotti, Arnaldo. *Note idrodinamiche,* Milan 1935.
[253] Massignon, Daniel. *Mécanique statistique des fluides. Fluctuations et propriétés locales,* Paris 1957.
[256] Mastronardi, Letizia. *Il dominio degli Sforza e l'opera di Leonardo da Vinci a Vigevano,* Vigevano 1952.
[960] Mazzoni, Enrico. *Materia e movimento, sole entità di ogni fenomeno fisico,* Vicenza 1954.
[184] Medici, Giuseppe. *Conoscere per amministrare,* Rome 1957.
[1406] Medici, Giuseppe. *Introduzione al piano di sviluppo della scuola,* Rome 1959.
[1318] Mercier, André. *Principes de mécanique analytique,* Paris 1955.
[329] Mineo, Corradino. *Conferenze sulla geometria non euclidea,* Palermo 1937.
[1009] Minetti, Silvio. *Appunti di meccanica razionale*, vol. 1: *Cinematica. Nozioni generali e complementi,* Bari 1950.
[203] Mochi, Alberto. *Scienza e scientismo,* Siena 1945.
[429] Modugno, Francesco. *Teoria e costruzione degli ingranaggi ad assi paralleli. Con applicazione ai riduttori marini,* Spoleto 1940.
[457] Modugno, Francesco. *Progetto e costruzione delle macchine*, vol. 1: *Calcolo dei solidi di rivoluzione e degli alberi,* Rome [s.d.].
[1010] Modugno, Francesco. *Ingranaggi cilindrici. Calcolo, disegno e costruzione delle ruote dentate per importanti trasmissioni di potenze, studio del profilo dei denti e dei suoi eventuali difetti, avarie nelle dentature e loro riparazione,* Milan 1951.
[265] Moneti, Guglielmo; Borgogelli, Guido. *Geometria descrittiva elementare pratica*, vol. 1, Rome 1920.
[266] Moneti, Guglielmo; Borgogelli, Guido. *Geometria descrittiva elementare pratica*, vol. 2, Rome 1920.
[267] Moneti, Guglielmo; Borgogelli, Guido. *Geometria descrittiva elementare pratica*, vol. 3, Rome 1920.
[268] Moneti, Guglielmo; Borgogelli, Guido. *Geometria descrittiva elementare pratica*, vol. 4, Rome 1920.
[598] Morin, Ugo. *Algebra astratta e geometria algebrica*, vol. 1: *Algebra astratta,* Padua 1953.
[16] Morin, Ugo. *Elementi di geometria analitica,* Padua 1953.
[91] Morin, Ugo. *Lezioni di geometria*, vol. 2: *Curve piane,* Padova 1954.

[92] Morin, Ugo. *Lezioni di geometria*, vol. 3: *Elementi di geometria proiettiva,* Padova 1955.

[1310] Mott-Smith, Geoffrey. *Mathematical Puzzles, for Beginners and Enthusiasts,* New York 1954.

[1317] Mourier, Edith. *Éléments aléatoires dans un espace de Banach,* Paris 1953.

[343] Mukhopadhyaya, Syamadas. *Collected Geometrical Papers,* Calcutta 1931.

[1122] Muñoz Oribe, Rodolfo. *Cuestiones de matemática elemental,* vol. 1: *El método en la enseñanza,* Montevideo 1930.

[372] Naccari, Andrea. *Lezioni di fisica sperimentale,* Turin 1898.

[573] Nalli, Pia. *Esposizione e confronto critico delle diverse definizioni proposte per l'integrale definito di una funzione limitata o no,* Palermo 1914.

[300] Nalli, Pia. *Lezioni di calcolo differenziale assoluto,* Catania 1952.

[791] Natanson, Isidor P. *Theorie der Funktionen einer reellen Veränderlichen,* Berlin 1954.

[892] Natucci, Alpinolo. *Il concetto di numero e le sue estensioni. Studi storico-critici intorno ai fondamenti dell'aritmetica generale,* Turin 1923.

[928] Navarro Borrás, Francisco. *Curso general de Matemáticas aplicadas a la Física, a la Química y a las Ciencias Naturales,* Madrid 1940.

[792] Navarro Borrás, Francisco. *Curso superior de análisis matemático para ingenieros,* Madrid 1942.

[274] Navarro Borrás, Francisco. *Lecciones de mecánica teórica,* Madrid 1945.

[836] Neugebauer, Otto. *The Exact Sciences in Antiquity,* Princeton 1952.

[1111] Nicolescu, Miron. *Analizâ matematicâ*, vol. 1, Bucharest 1957.

[1112] Nicolescu, Miron. *Analizâ matematicâ*, vol. 2, Bucharest 1958.

[1233] Nicoletti, Onorato; Maroni, Arturo. *Aritmetica razionale,* Genoa 1925.

[1173] Nicoletti, Onorato; Sansone, Giovanni. *Aritmetica e algebra*, vol. 1, Naples 1924.

[1197] Nicoletti, Onorato; Sansone, Giovanni. *Aritmetica e algebra*, vol. 1, Palermo 1924.

[1174, 1175] Nicoletti, Onorato; Sansone, Giovanni. *Aritmetica e algebra*, vol. 1, Naples 1932. (2 copies)

[1198] Nicoletti, Onorato; Sansone, Giovanni. *Aritmetica e algebra*, vol. 2, Palermo 1925.

[1113] Nicoletti, Onorato; Sansone, Giovanni. *Aritmetica e algebra*, vol. 2, Naples 1931.

[1176] Nicoletti, Onorato; Sansone, Giovanni. *Aritmetica e algebra*, vol. 2, Naples 1932.

[1177] Nicoletti, Onorato; Sansone, Giovanni. *Aritmetica e algebra*, vol. 3, Naples 1947.

[138] Nielsen, Niels (ed.). *Beretning om den anden Skandinaviske matematikerkongres,* Copenhagen 1911.

[1260] Noether, Emmy. *Maximalordnungen,* Paris 1934.

[1397] Nordio, Antonio. *Guida ragionata per i calcoli delle costruzioni in cemento armato*, vol. 1: *I problemi primi,* Padua 1956.

[417] Nubar, Zareh. *Le premier principe rien n'est arbitraire. La mécanique fondée sur une théorie des chocs durs,* Paris 1930.
[610] Occhipinti, Roberto. *Elementi di matematiche per gli studenti in chimica e scienze naturali,* Palermo 1929.
[797] Onicescu, Octav; Galburā, Gheorge. *Algebrā*, vol. 1, Bucharest 1948.
[358] Onicescu, Octav; Mihoc, Gheorghe. *Les chaînes de variables aléatoires,* Bucharest 1943.
[578] Ostrowski, Alexander. *Vorlesungen über differential und Integralrechnung*, vol. 2: *Differentialrechnung auf dem Gebiete mehrerer Variablen,* Basel 1951.
[962] Pacinotti, Antonio; Banti, Angelo. *Raccolta degli scritti di Antonio Pacinotti sulla priorità dell'invenzione della dinamo elettrica e comunicazione al Congresso della Società Italiana per il Progresso delle Scienze. Ottobre 1932,* Rome 1932.
[1110] Padoa, Alessandro. *Aritmetica intuitiva*, vol. 1, Palermo 1923.
[1166] Palatini, Francesco. *Aritmetica e algebra,* Turin 1915.
[1157] Palatini, Francesco. *Geometria*, vol. 1, Turin 1926.
[1235] Palatini, Francesco. *Geometria*, vol. 2, Turin 1926.
[1391] Panow, Dimitrij J. *Formelsammlung zur numerischen Behandlung partieller Differentialgleichungen nach dem Differenzenverfahren,* Berlin 1955.
[413] Pantaleo, Mario. *L'assoluto nella teoria di Einstein,* Naples 1923.
[999] Pantaleo, Mario. *Elementi di meccanica,* Florence 1934.
[623] Pascal, Ernesto. *Repertorium der höheren Geometrie*, vol. 2.2: *Raumgeometrie,* Leipzig-Berlin 1922.
[173] Pascali, Justo. *Geometria proyectiva,* Buenos Aires 1935.
[449] Pasini, Claudio. *Orologi solari. Costruzioni grafiche e calcolo degli orologi solari e delle meridiane a tempo medio loro pratica esecuzione e storia,* Padua 1900.
[1338] Pasini, Claudio. *Metodo dei minimi quadrati per la compensazione degli errori di osservazione. Appendice al trattato di topografia,* Bologna 1921.
[1017] Pasini, Claudio. *Trattato di topografia,* Bologna 1921.
[1016] Pasini, Claudio. *Tratado de topografía,* Barcelona 1924.
[605] Peano, Giuseppe. *Lezioni di analisi infinitesimale*, vol. 1, Turin 1893.
[606] Peano, Giuseppe. *Lezioni di analisi infinitesimale*, vol. 2, Turin 1893.
[6] Peano, Giuseppe. *Opere scelte,* vol. 1: *Analisi matematica, calcolo numerico,* Rome 1957.
[648] Peano, Giuseppe. *Opere scelte*, vol. 2: *Logica matematica, interlingua ed algebra della grammatica*, Rome 1958.
[1299] Peano, Giuseppe. *Opere scelte*, vol. 3: *Geometria e fondamenti, meccanica razionale, varie,* Rome 1959.
[905] Pelseneer, Jean. *Esquisse du progrès de la pensée mathématique. Des primitifs au IXe Congrès international des Mathématiciens,* Paris 1935.
[1308] Pelseneer, Jean. *L'Evolution de la notion de phénomène physique des primitifs à Bohr et Louis de Broglie,* Bruxelles 1947.
[1237] Pensa, Angelo. *Elementi di geometria,* Turin 1912.
[21, 54] Perazzo, Umberto. *Lezioni di geometria descrittiva,* Turin 1910. (2 copies)

[67] Peri, Giuseppe. *Corso elementare di geometria descrittiva, libri tre con atlante di 24 tavole. Seguiti da un'appendice sul metodo delle projezioni quotate,* Pistoia 1869.

[565] Perna, Alfredo. *Lezioni di algebra,* Rome 1921.

[1314] Pernot, Pierre. *Introduction au calcul des systèmes hyperstatiques. Exercices traités avec l'application de trois méthodes: méthode de Hardy Cross, méthode de Bresse, méthode de Castigliano,* Paris 1956.

[483] Perrin, Jean. *Grains de matière et de lumière,* Paris 1935.

[1275] Perrin, Jean. *Grains de matière et de lumière,* vol. 3: *Noyaux des atomes,* Paris 1935.

[482] Persico, Enrico. *Lezioni di meccanica ondulatoria,* Padua 1930.

[263] Pession, Giuseppe. *Guglielmo Marconi,* Turin 1941.

[199] Pession, Giuseppe. *La bomba atomica,* Milan 1945.

[1331] Pession, Giuseppe. *Manuale di astronomia pratica. Astronomia geodetica,* Rome 1943.

[1329] Pession, Giuseppe. *Misure radiotecniche e formulario,* Milan 1939.

[450] Piazzolla Beloch, Margherita. *Elementi di fotogrammetria terrestre ed area,* Padua 1934.

[1395] Picard, Émile. *La vie et l'œuvre de Pierre Duhem,* Paris 1922.

[503] Picard, Émile. *Leçons sur quelques types simples d'équations aux dérivées partielles. Avec des applications à la physique mathématique,* Paris 1927.

[150] Picard, Emile. *Mélanges de mathématiques et de physique,* Paris 1924.

[823] Picard, Émile. *Quelques applications analytique de la théorie des courbes et des surfaces algébriques,* Paris 1931.

[211, 220] Picard, Émile. *Selecta. Cinquantenaire scientifique,* Paris 1928. (2 copies)

[497] Picard, Émile. *Théorie des fonctions algébriques et de leurs intégrales,* vol. 1, Paris 1897.

[818] Picard, Émile. *Traité d'analyse,* vol. 1: *Intégrales simples et multiples, l'équation de Laplace et ses applications, développements en séries, applications géométriques du calcul infinitésimal,* Paris 1922.

[819] Picard, Émile. *Traité d'analyse,* vol. 2: *Fonctions harmoniques et fonctions analytiques, introduction a la théorie de équations différentielles, intégrales abéliennes et surfaces de Riemann,* Paris 1922.

[820] Picard, Émile. *Traité d'analyse,* vol. 3: *Des singularités des intégrales des équations différentielles. Étude du cas, ou la variable reste réelle. Des courbes définies par des équations différentielles. Équations linéaires. Analogies entre les équations algébriques et les équations linéaires,* Paris 1922.

[638] Picard, Émile; Simart, Georges. *Théorie des fonctions algébriques de deux variables indépendantes,* Paris 1897.

[589] Picone, Mauro. *Analisi superiore,* Rome 1946.

[535] Picone, Mauro. *Appunti di analisi superiore,* Naples 1940.

[596] Picone, Mauro. *Corso di analisi superiore,* Rome.

[1253] Picone, Mauro. *Criteri necessari per un estremo di alcuni funzionali,* Rome 1959.

[539] Picone, Mauro. *Fondamenti di analisi funzionale lineare,* Rome 1943.
[759] Picone, Mauro. *Lezioni di analisi infinitesimale*, vol. 1.1: *La derivazione e l'integrazione,* Catania 1923.
[760] Picone, Mauro. *Lezioni di analisi infinitesimale*, vol. 1.2: *La derivazione e l'integrazione,* Catania 1923.
[378] Picone, Mauro. *Teoria dell'integrazione lebesguiana*, a cura di Aldo Ghizzetti, Rome 1942.
[7] Picone, Mauro. *Trattato di analisi matematica*, vol. 1, Rome 1954.
[9] Picone, Mauro; Fichera, Gaetano. *Trattato di analisi matematica*, vol. 2, Rome 1954.
[696] Picone, Mauro; Tortorici, Paolo. *Trattato di matematiche generali*, vol. 1, Rome 1947.
[552] Picone, Mauro; Viola, Tullio. *Lezioni sulla teoria moderna dell'integrazione,* Turin 1952.
[1134] Pincherle, Salvatore. *Gli elementi dell'aritmetica,* Bologna 1894.
[657] Pincherle, Salvatore. *Gli elementi della teoria delle funzioni analitiche*, vol. 1, Bologna 1922.
[570] Pincherle, Salvatore. *Lezioni di calcolo infinitesimale,* Bologna 1919.
[788] Pincherle, Salvatore. *Lezioni di algebra complementare,* Bologna 1909.
[603] Pincherle, Salvatore. *Lezioni di algebra complementare,* Bologna 1920.
[604] Pincherle, Salvatore. *Lezioni di algebra complementare,* Bologna 1921.
[103] Pincherle, Salvatore. *Opere scelte*, vol. 1, Rome 1954.
[871] Pincherle, Salvatore. *Opere scelte*, vol. 2, Rome 1954.
[1114, 1115] Pinto, Olga (ed.). *A Union List of American Periodicals in Italy,* Rome 1958. (2 copies)
[27] Pittarelli, Giulio. *Lezioni di geometria descrittiva,* Rome 1926.
[146] Plans y Freyre, José Maria. *Nociones fundamentales de mecánica relativista,* Madrid 1921.
[328] Poincaré, Henri. [*Selected Works*], Introduzione di Francesco Severi, Florence 1949.
[774] Poincaré, Henri. *Œuvres*, vol. 6: *Géometrie,* Paris 1953.
[844] Polacco, Giuseppe. *Lo spazio piano. Il primo libro dei principi,* Rome 1955.
[484] Polara, Virgilio. *L'esperienza nelle teorie di Maxwell e di Lorentz e l'interpretazione meccanica dei fenomeni elettrici,* Catania 1911.
[897] Polvani, Giovanni. *Alessandro Volta,* Pisa 1942.
[250] Polvani, Giovanni. *Elementi di metrologia teoretica,* Milan 1947.
[321] Pompilj, Giuseppe. *Complementi di calcolo delle probabilità*, Rome, 1949.
[1309] Pompilj, Giuseppe; Dall'Aglio, Giorgio. *Piano degli esperimenti,* Turin 1959.
[80] Pompilj, Giuseppe; Napolitani, Diego. *Piano degli esperimenti ed elaborazione probabilistica dei risultati. Con particolare riguardo alla sperimentazione in biologia,* Rome 1954.
[986] Pope, Francis; Otis, Arthur S. *Elements of Aeronautics,* New York 1941.

[1426] Porchetti, Ricciardo. *Il problema economico-sociale nelle prospettive scientifiche e pratiche della matematica filosofica-sociologica-economica e finanziaria*, vol. 1: *Soluzione teorica del problema fondamentale,* Terni 1939.

[1336] Porro, Francesco. *Manuale di cosmografia,* Bologna 1925.

[983] Porro, Francesco. *Problemi dell'universo,* Bologna 1934.

[1333] Porro, Francesco. *Trattato di astronomia*, vol. 1, Bologna 1920.

[1183] Predella, Pilo. *Algebra e aritmetica,* Turin 1921.

[1182] Predella, Pilo. *Geometria,* Turin 1921.

[290] Prym, Friedrich; Rost, Georg. *Theorie der Prym'schen Funktionen erster Ordnung. Im Anschluss an die Schöpfungen Riemanns,* Leipzig 1911.

[384] Puig Adam, Pedro. *Curso de geometría métrica*, vol. 1 *Fundamentos,* Madrid 1947.

[385] Puig Adam, Pedro. *Curso de geometría métrica*, vol. 2 *Complementos,* Madrid 1948.

[234] Puma, Marcello. *Elementi per una teoria matematica del contagio,* Rome 1939.

[447] Puma, Marcello. *Nuova teoria quantistica della luce,* Rome 1947.

[475] Puma, Marcello. *Nuovo trattato di geometria e meccanica quantistiche,* Rome 1945.

[1231] Purgotti, Sebastiano. *Elementi di aritmetica ragionata,* Perugia 1847.

[1251] Purgotti, Sebastiano. *Elementi di geometria,* Perugia 1958.

[1230] Purgotti, Sebastiano. *Prospetto ragionato delle più interessanti proposizioni tratte in numero di cinquecento dagli elementi di aritmetica algebra e geometria,* Perugia 1843.

[1002] Raethjen, Paul. *Einführung in die Physik der Atmosphäre,* Leipzig 1942.

[1135] Raganti, Bernardo. *Aritmetica fondamentale,* Sarzana 1907.

[456] Ragno, Saverio. *Tecnologia superiore*, vol. 1 *Prove e classifiche industriali dei materiali metallici,* Milan 1920.

[1387] Revelli, Paolo (ed.). *Cristoforo Colombo e la scuola cartografica genovese*, vol. 1: *La partecipazione italiana alla mostra oceanografica internazionale di Siviglia 1929-VII E. F.,* Genoa 1937.

[1388] Revelli, Paolo (ed.). *Cristoforo Colombo e la scuola cartografica genovese*, vol. 2: *Cristoforo Colombo e la scuola cartografica genovese,* Genoa 1937.

[1389] Revelli, Paolo (ed.). *Cristoforo Colombo e la scuola cartografica genovese*, vol. 3: *Carte topografiche e corografiche manoscritte della Liguria e delle immediate adiacenze conservate nel R. Archivio di stato di Genova,* Genoa 1937.

[586] Rey Pastor, Julio. *Fundamentos de la geometría proyectiva superior,* Madrid 1916.

[927] Rey Pastor, Julio. *La ciencia y la técnica en el descubrimiento de América,* Buenos Aires 1945.

[925] Rey Pastor, Julio. *La matemática superior. Métodos y problemas de siglo XIX,* Buenos Aires 1951.

[926] Rey Pastor, Julio. *Los matemáticos españoles del siglo XVI,* [s.l.] 1926.

[1411] Rey Pastor, Julio. *Teoría de la representació conforme. Conferencies donades el Juny de 1915,* Barcelona 1915.

[850] Rey Pastor, Julio. *Teoría de los algoritmos lineales de convergencia y de sumación,* Buenos Aires 1931.
[824] Rey Pastor, Julio. *Teoría geométrica de la polaridad,* Madrid 1929.
[1118] Rey Pastor, Julio; Puig Adam, Pedro. *Elementos de geometria,* Madrid 1928.
[1250] Rey Pastor, Julio; Puig Adam, Pedro. *Elementos de aritmética,* Madrid 1928.
[349] Reye, Théodor. *Die Geometrie der Lage*, vol. 1, Hanover 1877.
[350] Reye, Théodor. *Die Geometrie der Lage*, vol. 2, Hanover 1880.
[387] Reye, Théodor. *Die Geometrie der Lage,* Stuttgart 1907.
[1280] Rham, Georges de. *Variétés différentiables. Formes, courants, formes harmoniques,* Paris 1955.
[1128] Riboni, Gaetano. *Elementi di geometria,* Bologna 1894.
[566] Ricci-Curbastro, Gregorio. *Lezioni di analisi algebrica ed infinitesimale,* Padua 1918.
[870] Ricci Curbastro, Gregorio. *Opere*, vol. 1: *Note e memorie,* Rome 1956.
[106] Ricci-Curbastro, Gregorio. *Opere*, vol. 2: *Note e memorie. Teoria dell'elasticità,* Rome 1957.
[416] Righi, Augusto. *I fenomeni elettro-atomici sotto l'azione del magnetismo. Narrazione di ricerche sperimentali sui fenomeni elettrici prodotti nel campo magnetico,* Bologna 1918.
[1346] Ringel, Gerhard. *Färbungsprobleme auf Flächen und Graphen,* Berlin 1959.
[377] Rios García, Sixto. *Conferencias sobre teoría de la integral,* Madrid 1940.
[667] Rivera, Vincenzo. *Malattie delle piante,* Rome 1929.
[1132] Rizzi, Giacomo. *Come diventare calcolatori,* Rome 1951.
[1239] Roberto, Giovanni. *Aritmetica e geometria,* Florence 1924.
[1320] Rohn, Karl; Berzolari, Luigi. *Algebraische Raumkurven und abwickelbare Flächen,* Leipzig 1926.
[227] Roiti, Antonio. *Elementi di fisica*, vol. 1, Florence 1891.
[228] Roiti, Antonio. *Elementi di fisica*, vol. 2, Florence 1888.
[891] Romão, Joao Antonio de Mattos. *Galileu e o método científico*, vol. 1, Lisbon 1944.
[275] Ronchi, Vasco. *Galileo e il cannocchiale,* Udine 1942.
[1171] Rosati, Carlo; Benedetti, Piero. *Geometria*, vol. 1, Naples 1927.
[1172] Rosati, Carlo; Benedetti, Piero. *Geometria*, vol. 2, Naples 1927.
[1257] Roth, Leonard. *Algebraic Threefolds,* Berlin 1955.
[947] Roth, Leonard. *Modern Elementary Geometry,* London 1948.
[622] Roth, Leonard. *Saggi di geometria algebrica classica,* Genoa 1955.
[621] Roth, Leonard. *Sistemi canonici ed anticanonici,* Genoa 1955.
[878] Ruffini, Paolo. *Opere matematiche*, vol. 1, Palermo 1915.
[879] Ruffini, Paolo. *Opere matematiche*, vol. 2, Palermo 1953.
[880] Ruffini, Paolo. *Opere matematiche*, vol. 3, Palermo 1954.
[474] Sabbatini, A. *Esercizi di geometria analitica e proiettiva*, Rome 1925.
[381] Salmon, George. *Traité de géométrie analytique à trois dimensions*, vol. 1: *Lignes et surfaces du premier et du second ordre,* Paris 1882.
[618] Salmon, George. *Analytische Geometrie des Raumes*, vol. 1, Leipzig 1922.
[619] Salmon, George. *Analytische Geometrie des Raumes*, vol. 2, Leipzig 1923.

[625] Samuel, Pierre. *Méthodes d'algèbre abstraite en géométrie algébrique,* Berlin 1955.

[833] Sansone, Giovanni. *Equazioni differenziali nel campo reale*, vol. 1, Bologna 1941.

[834] Sansone, Giovanni. *Equazioni differenziali nel campo reale*, vol. 2, Bologna 1952.

[633] Sansone, Giovanni. *Funzioni di variabili reali,* Florence 1945.

[778] Sansone, Giovanni. *Lezioni di analisi matematica*, vol. 1, Padua 1929.

[793] Sansone, Giovanni. *Lezioni di analisi matematica*, vol. 1, Padua 1943.

[724] Sansone, Giovanni. *Lezioni di analisi matematica*, vol. 1, Padua 1952.

[715] Sansone, Giovanni. *Lezioni di analisi matematica*, vol. 2, Florence 1929.

[794] Sansone, Giovanni. *Lezioni di analisi matematica*, vol. 2, Padua 1943.

[14] Sansone, Giovanni. *Lezioni di analisi matematica*, Padua 1934.

[44] Sansone, Giovanni. *Lezioni di analisi matematica*, Padua 1939.

[918] Sansone, Giovanni. *Lezioni di analisi matematica*, vol. 2, Padua 1954.

[694] Sansone, Giovanni. *Lezioni sulla teoria delle funzioni di una variabile complessa*, vol. 1, Padua 1950.

[695] Sansone, Giovanni. *Lezioni sulla teoria delle funzioni di una variabile complessa*, vol. 2, Padua 1949.

[611] Sansone, Giovanni. *Lezioni sulla teoria delle funzioni di variabile complessa*, vol. 1, Padua 1947.

[612] Sansone, Giovanni. *Lezioni sulla teoria delle funzioni di variabile complessa*, vol. 2, Padua 1947.

[1297] Sansone, Giovanni. *Lezioni sulla teoria delle funzioni di una variabile complessa*, vol. 2, Padua 1955.

[851] Sansone, Giovanni. *Lezioni sulla teoria delle funzioni di variabili reali,* Florence 1951.

[771, 832] Sansone, Giovanni. *Moderna teoria delle funzioni di variabile reale*, vol. 2: *Svluppi in serie di funzioni ortogonali,* Bologna 1935. (2 copies)

[555] Sansone, Giovanni. *Moderna teoria delle funzioni di variabile reale,* vol. 2: *Sviluppi in serie di funzioni ortogonali*, Bologna 1946.

[835] Sansone, Giovanni. *Moderna teoria delle funzioni di variabile reale*, vol. 2, Bologna 1952.

[57] Sansone, Giovanni; Conti, Roberto. *Equazioni differenziali non lineari,* Rome 1956.

[262] Sartoris, Alberto. *Léonard architecte,* Paris 1952.

[362] Savorgnan di Brazzà, Francesco. *Da Leonardo a Marconi,* Rome 1932.

[1120] Scala, Alberto; Ciamberlini, Corrado. *Geometria,* Turin 1923.

[1283] Schlick, Moritz; Vouillemin, Charles E. *Les énoncés scientifiques et la réalité du monde extérieur,* Paris 1934.

[354] Schwarz, Hermann A. *Mélanges relatifs au domaine des surfaces minima,* Pisa 1918.

[706] Scorza Dragoni, Giuseppe. *Elementi di analisi matematica*, vol. 1: *Elementi di algebra,* Padua 1952.

[935] Scorza Dragoni, Giuseppe. *Elementi di analisi matematica*, vol. 1: *Elementi di algebra,* Padua 1954.

[1410] Scorza Dragoni, Giuseppe. *Elementi di analisi matematica*, vol. 1: *Elementi di algebra,* Padova 1961.

[936] Scorza Dragoni, Giuseppe. *Elementi di analisi matematica*, vol. 2: *La continuità e la differenziabilità,* Padua 1953.

[707] Scorza Dragoni, Giuseppe. *Elementi di analisi matematica*, vol. 2: *La continuità e la differenziabilità,* Padua 1956.

[975] Scorza, Gaetano. *Complementi di geometria*, vol. 1, Bari 1914.

[785] Scorza, Gaetano. *Corpi numerici e algebre,* Messina 1921.

[194] Scorza, Gaetano. *Elementi di geometria analitica,* Messina 1925.

[656] Scorza, Gaetano. *Gruppi astratti,* Rome 1942.

[1413] Scorza, Gaetano. *Opere scelte*, vol. 1: *1899–1915,* Rome 1960.

[783] Segre, Beniamino. *Arithmetical Questions on Algebraic Varieties,* London 1951.

[675] Segre, Beniamino. *Esercizi e complementi di analisi algebrica. Anno Accademico 1930–31,* Rome 1931.

[735] Segre, Beniamino. *Forme differenziali e loro integrali*, vol. 1: *Calcolo algebrico esterno e proprietà differenziali locali,* Rome 1951.

[736] Segre, Beniamino. *Forme differenziali e loro integrali*, vol. 2: *Omologia, coomologia, corrispondenze ed integrali sulle varietà,* Rome 1956.

[169] Segre, Beniamino. *Geometria analitico-descrittiva, a cura di Luigi Muracchini,* Bologna 1950.

[599] Segre, Beniamino. *Lezioni di geometria moderna*, vol. 1 *Fondamenti di geometria sopra un corpo qualsiasi,* Bologna 1948.

[254] Segre, Beniamino. *Lezioni di geometria proiettiva e descrittiva,* Bologna 1933.

[97] Segre, Beniamino. *Proprietà locali e globali di varietà e di trasformazioni differenziabili con speciale riguardo ai casi analitici ed algebrici,* Rome 1956.

[637] Segre, Beniamino. *The Non-Singular Cubic Surfaces,* Oxford 1942.

[872] Segre, Corrado. *Opere*, vol. 1, Rome 1957.

[112] Seifert, Herbert; Threlfall, William. *Lecciones de topología,* Madrid 1951.

[425] Selvaggi, Filippo. *Problemi della fisica moderna,* Brescia 1953.

[462] Selvaggi, Filippo. *Valore e metodo della scienza,* Rome 1952.

[590] Semple, John G.; Roth, Leonard. *Introduction to Algebraic Geometry,* Oxford 1949.

[476] Sestini, Fausto; Funaro, Angiolo. *Corso di chimica*, Livorno 1896.

[41, 42] Severi, Francesco. *Applicazioni del metodo delle proiezioni quotate,* Padua 1917. (2 copies)

[43] Severi, Francesco. *Appunti di applicazioni della geometria descrittiva,* Rome [s.d.].

[230] Severi, Francesco. *Complementi di geometria proiettiva,* Bologna 1906.

[767] Severi, Francesco. *Elementos de geometría*, vol. 1, Barcelona 1946.

[768] Severi, Francesco. *Elementos de geometría*, vol. 2, Barcelona 1946.

[670] Severi, Francesco. *Elementi di geometria*, vols. 1–2, Florence 1928.

[562] Severi, Francesco. *Esercizi di algebra complementare*, Turin 1902.
[563] Severi, Francesco. *Esercizi di geometria analitica*, Turin 1902.
[160] Severi, Francesco. *Geometria proiettiva*, Padua 1921.
[672] Severi, Francesco. *Geometria* [various, volumes gathered together], Florence.
[720, 721] Severi, Francesco. *Lezioni di analisi*, vol. 1, Bologna 1933. (2 copies)
[825] Severi, Francesco. *Lezioni di analisi infinitesimale*, Rome 1928.
[35, 36, 39] Severi, Francesco. *Lezioni di geometria descrittiva, capp. 1–2*, Padua 1913. (3 copies)
[37] Severi, Francesco. *Lezioni di geometria descrittiva, cap. 2*, Padua 1913.
[38] Severi, Francesco. *Lezioni di geometria descrittiva, capp. 3–4*, Padua 1913.
[40] Severi, Francesco. *Lezioni di geometria descrittiva, capp. 3–4–5–6*, Padua 1913.
[614, 630, 631] Severi, Francesco (ed.). *Teorema di Riemann-Roch e questioni connesse*, Rome 1955. (3 copies)
[558] Severi, Francesco; Bini, Umberto. *Aritmetica razionale*, Florence 1947.
[56] Severi, Francesco; Comessatti, Annibale. *Lezioni di geometria analitica e proiettiva*, Padua 1919.
[669] Severi, Francesco et al. *Algebra per vari tipi di scuola*, Florence 1937.
[668] Severi, Francesco et al. *Aritmetica per vari tipi di scuola*, Florence 1947.
[671] Severi, Francesco *et al. Geometria, aritmetica e algebra* [various, volumes gathered together], Florence.
[795] Severi, Francesco; Scorza Dragoni, Giuseppe. *Lezioni di analisi*, vol. 1: *Determinanti, equazioni lineari, limiti derivate e differenziali, funzioni ed equazioni algebriche*, Bologna 1938.
[796] Severi, Francesco; Scorza Dragoni, Giuseppe. *Lezioni di analisi*, vol. 2.1: *Serie di funzioni, applicazioni geometriche, integrali rettilinei, funzioni di più variabili, derivazione e integrazione ad esse inerenti*, Bologna 1948.
[722] Severi, Francesco; Scorza Dragoni, Giuseppe. *Lezioni di analisi*, vol. 3: *Equazioni differenziali ordinarie e loro sistemi, problemi al contorno, relativi, serie trigonometriche, applicazioni geometriche*, Bologna 1951.
[418] Severini, Gino. *Du cubisme au classicisme. Esthétique du compas et du nombre*, Paris 1921.
[995] Signorini, Antonio. *Lezioni di fisica matematica*, vol. 1, Rome 1950.
[996] Signorini, Antonio. *Lezioni di fisica matematica*, vol. 2, Rome 1950.
[997] Signorini, Antonio. *Lezioni di fisica matematica*, vol. 2, Rome 1951.
[987] Signorini, Antonio. *Meccanica razionale con elementi di statica grafica*, vol. 1, Naples 1936.
[989] Signorini, Antonio. *Meccanica razionale con elementi di statica grafica*, vol. 1, Naples 1938.
[991] Signorini, Antonio. *Meccanica razionale con elementi di statica grafica*, vol. 1, Rome 1947.
[993] Signorini, Antonio. *Meccanica razionale con elementi di statica grafica*, vol. 1, Rome 1952.
[988] Signorini, Antonio. *Meccanica razionale con elementi di statica grafica*, vol. 2, Naples 1937.

[990] Signorini, Antonio. *Meccanica razionale con elementi di statica grafica*, vol. 2, Naples 1939.
[992] Signorini, Antonio. *Meccanica razionale con elementi di statica grafica*, vol. 2, Rome 1948.
[994] Signorini, Antonio. *Meccanica razionale con elementi di statica grafica*, vol. 2, Rome 1954.
[1018] Silla, Lucio; Teofilato, Pietro. *Aerodinamica,* Pavia 1939.
[1330] Silla, Lucio; Teofilato, Pietro. *Lezioni di aerodinamica*, vol. 1, Rome 1933.
[374] Sobrero, Luigi. *Lezioni di fisica matematica,* Rome 1936.
[322] Solari, Luigi. *Marconi e l'invenzione della radio,* Florence 1941.
[1025] Somigliana, Carlo. *Memorie scelte,* Turin 1936.
[940] Spallanzani, Lazzaro. *Opere*, vol. 1: *Circolazione, digestione, respirazione animale,* Milan 1932.
[941] Spallanzani, Lazzaro. *Opere*, vol. 2: *Respirazione animale (II), Microbiologia,* Milan 1933.
[942] Spallanzani, Lazzaro. *Opere*, vol. 3: *Respirazione delle piante, rigenerazioni animali, fecondazione naturale e artificiale, sistema nervoso, varia,* Milan 1934.
[943] Spallanzani, Lazzaro. *Opere*, vol. 4: *Viaggi sull'Appennino, al lago Ventasso, sulle Alpi lombarde, sui Grigioni, nella Svizzera, nel Mediterraneo,* Milan 1935.
[944] Spallanzani, Lazzaro. *Opere*, vol. 5.1: *Viaggio a Costantinopoli,* Milan 1936.
[945] Spallanzani, Lazzaro. *Opere*, vol. 5.2: *Viaggi alle due Sicilie,* Milan 1936.
[182] Spampinato, Nicolò. *Lezioni di geometria superiore*, vol. 9: *Teoria delle curve ellittiche ed elementi di geometria ideale sopra una superficie. Determinazione corpoproiettiva e birazionale di una verità algebrica. Prolungamenti di enti complessi in enti supercomplessi,* Naples 1952.
[391] Spampinato, Nicolò. *Lezioni di geometria superiore*, vol. 1: *Nozioni introduttive,* Naples 1950.
[392] Spampinato, Nicolò. *Lezioni di geometria superiore*, vol. 3: *Fondamenti di geometria algebrica delle varietà virtuali, ideali e generali nel campo complesso,* Naples 1948.
[393] Spampinato, Nicolò. *Lezioni di geometria superiore*, vol. 4: *Fondamenti di geometria moderna in un'algebra doppia,* Naples 1949.
[394] Spampinato, Nicolò. *Lezioni di geometria superiore*, vol. 5: *Geometria delle funzioni in un'algebra del II o III ordine,* Naples 1947.
[395] Spampinato, Nicolò. *Lezioni di geometria superiore*, vol. 7: *Elementi di geometria algebrica e corpogeometria sopra una verità, le superficie di Riemann e gl'integrali abeliani,* Naples 1950.
[396] Spampinato, Nicolò. *Lezioni di geometria superiore*, vol. 8: *Enti iperalgebrici e geometrie fondamentali nell'S2 complesso, rappresentazioni complesse delle corpoproiettività dell'S2 biduale, derivazione ed integrazione nel campo ipercomplesso,* Naples 1952.
[921] Speiser, Andreas. *Die mathematische Denkweise,* Zürich 1932.
[466] Spicacci, Pasquale. *Della materia. Le nuove visioni del mondo atomico e una nuova concezione dell'universo,* Naples 1931.
[681] Spiegel, Izak Willem van. *Geometry of Aggregates,* Assen 1957.

[644] Stahl, Hermann. *Abriss einer Theorie der algebraischen Funktionen einer Verbandlichen in neuer Fassung,* Leipzig 1911.

[1195] Strazzeri, Vittorio. *Geometria elementare,* Palermo 1919.

[1196] Strazzeri, Vittorio. *Geometria elementare,* Palermo 1926.

[77] Strazzeri, Vittorio. *Lezioni di geometria descrittiva*, vol. 2, Palermo 1931.

[808, 1402] Strazzeri, Vittorio. *Lezioni di geometria differenziale,* Palermo 1924. (2 copies)

[809] Strazzeri, Vittorio. *Lezioni di geometria differenziale proiettiva,* Palermo 1928.

[397] Study, Eduard. *Geometrie der Dynamen,* Leipzig 1903.

[579] Study, Eduard. *Vorlesungen über ausgewählte Gegenstande der Geometrie,* vol. 2 *Das imaginäre in der ebenen Geometrie,* Bonn 1933.

[1277] Swings, Polydore. *Travaux récents sur les molécules dans le Soleil, les planètes et les étoiles,* Paris 1934.

[473] Taibo, Angel. *Geometría descriptiva y sus aplicaciones*, vol. 2: *Curvas y superficies,* Madrid 1943.

[370] Takasu, Tsurusaburo. *Differentialgeometrien in den Kugelraumen*, vol. 1, Tokyo 1938.

[371] Takasu, Tsurusaburo. *Differentialgeometrien in den Kugelraumen*, vol. 2, Tokyo 1939.

[1302] Tartaglia, Niccolò. *Quesiti et inventioni diverse,* Brescia 1959.

[985] Tatin, Victor. *Théorie et pratique de l'aviation,* Paris 1910.

[1429] Teixeira, Francisco Gomez. *História das matemáticas em Portugal,* Lisbon 1934.

[284] Teixeira, Francisco Gomez. *Obras sobre matemática*, vol. 4, Coimbra 1908.

[285] Teixeira, Francisco Gomez. *Obras sobre matemática*, vol. 5, Coimbra 1909.

[4] Terracini, Alessandro (ed.). *In memoria di Giuseppe Peano,* Cuneo 1955.

[1030] Tessarotto, Mario. *Teoria e tecnica delle vibrazioni meccaniche. Fondamenti ed applicazioni alla tecnica moderna,* Rome 1940.

[1127] Testi, Giuseppe Maria. *Corso di aritmetica con numerosi esercizi e problemi,* Livorno 1891.

[842, 1312] Thébault, Victor. *Parmi les belles figures de la géométrie dans l'espace. Géométrie du tétraèdre,* Paris 1955. (2 copies)

[1423] Thimm, Walter. *Theorie der Potenzreihenringe,* Bonn [s.d.].

[465] Thovez, Ettore. *La meccanica dell'universo. La materia, il moto, l'energia, la gravitazione, l'inerzia, la carica elettrica, l'atomo, il vortice, l'irradiazione, l'astronomia atomica e l'astronomia celeste,* Turin 1930.

[803] Tis, Jacques. *Sur certaines classes d'espaces homogènes de groupes de Lie,* Paris 1935.

[723] Tocchi, Luigi. *Sui fondamenti della teoria delle serie doppie,* Naples 1937.

[917] Toja, Guido. *Il calcolo delle tariffe di una compagnia di assicurazioni sulla vita,* Rome 1925.

[151] Tommasina, Thomas. *La physique de la gravitation et la dynamique de l'univers,* Paris 1928.

[537] Tonelli, Leonida. *Fondamenti di calcolo delle variazioni*, vol. 1, Bologna 1921.
[538] Tonelli, Leonida. *Fondamenti di calcolo delle variazioni*, vol. 2, Bologna 1923.
[73] Tonelli, Leonida. *Opere scelte*, vol. 1, Roma 1960.
[547] Tonelli, Leonida. *Serie trigonometriche,* Bologna 1928.
[446] Tonini, Valerio. *Epistemologia della fisica moderna,* Milan 1953.
[843] Tonnelat, Marie-Antoinette. *La théorie du champ unifié d'Einstein et quelques-uns de ses développements,* Paris 1955.
[934] Tonolo, Angelo. *Corso di analisi,* Padua 1919.
[717] Tonolo, Angelo. *Lezioni di analisi algebrica e infinitesimale,* Padua 1941.
[846] Toraldo di Francia, Giuliano. *Electromagnetic Waves,* New York-London 1953.
[567] Torelli, Gabriele. *Lezioni di calcolo infinitesimale,* Naples 1921.
[88] Torricelli, Evangelista. *De infinitis spiralibus*, a cura di Ettore Carruccio, Pisa 1955.
[355] Torroja y Caballé, Eduardo. *Teoría geométrica de las líneas alabeadas y de las superficies desarrollables,* Madrid 1904.
[158] Torroja y Caballé, Eduardo. *Tratado de geometría de la posición y sus aplicaciones a la geometría de la medida,* Madrid 1899.
[1191] Tortorici, Pietro. *Elementi di analisi matematica,* Palermo 1933.
[728] Tortorici, Pietro. *Esercitazioni matematiche*, vol. 1, Palermo 1929.
[677] Tortorici, Pietro. *Esercitazioni matematiche*, vol. 2, Palermo 1931.
[758] Tortorici, Pietro. *Lezioni di analisi matematica*, vol. 1, Palermo 1927.
[1192] Tortorici, Pietro; Marseguerra, Vincenzo. *Elementi di analisi matematica,* Palermo 1930.
[1161] Trainito, Nunzio. *Elementi di algebra,* Gela 1938.
[620] Tricomi, Francesco G. *Equazioni differenziali,* Turin 1948.
[697] Tricomi, Francesco G. *Esercizi e complementi di analisi matematica,* Padua 1925.
[777] Tricomi, Francesco G. *Esercizi e complementi di analisi matematica*, vol. 1, Padua 1951.
[688] Tricomi, Francesco G. *Funzioni analitiche,* Bologna 1936.
[689] Tricomi, Francesco G. *Funzioni ellittiche,* Bologna 1937.
[682] Tricomi, Francesco G. *Funzioni ellittiche,* Bologna 1951.
[673] Tricomi, Francesco G. *Funzioni ipergeometriche confluenti,* Rome 1954.
[929] Tricomi, Francesco G. *Lezioni di analisi matematica*, vol. 1, Padua 1925.
[931] Tricomi, Francesco G. *Lezioni di analisi matematica*, vol. 1, Padua 1928.
[933] Tricomi, Francesco G. *Lezioni di analisi matematica*, vol. 1, Padua 1935.
[763] Tricomi, Francesco G. *Lezioni di analisi matematica*, vol. 1, Padua 1939.
[930] Tricomi, Francesco G. *Lezioni di analisi matematica*, vol. 2, Padua 1925.
[932] Tricomi, Francesco G. *Lezioni di analisi matematica*, vol. 2, Padua 1928.
[776] Tricomi, Francesco G. *Lezioni di analisi matematica*, vol. 2, Padua 1935.
[764] Tricomi, Francesco G. *Lezioni di analisi matematica*, vol. 2, Padua 1939.
[477] Università di Pisa. *Leonida Tonelli. In memoriam,* Pisa 1952.

[770] Usai, Giuseppe. *Complementi di matematiche generali,* Catania 1939.
[179] Vagnetti, Luigi. *La Facoltà di Architettura di Roma nel suo trentacinquesimo anno di vita,* Rome 1955.
[239] Vailati, Giovanni. *Scritti (1863–1909),* Leipzig-Florence 1911.
[1382] Valentini, Ernesto. *Tendenza aggressiva e accertamento precoce del sesso nel pavoncello,* Vatican 1951.
[188] Valiron, Georges. *Fonctions analytiques,* Paris 1954.
[94] Van der Corput, Johannes G. *Asymptotic Expansions*, vol. 1: *Fundamental Theorems of Asymptotics*, Berkeley 1954.
[89] Van der Corput, Johannes G. *Aymptotic Expansions*, vol. 2: *Elementary Methods*, Berkeley [s.d.].
[111] Van der Corput, Johannes G. *Asymptotic Expansions*, vol. 3: *The Asymptotic Behaviour of the Real Solutions of Certain Second Order Differential Equations*, Berkeley, 1955.
[486] Van der Waerden, Bartel L. *Die Gruppentheoretische Methode in der Quantenmechanik,* Berlin 1932.
[592] Van der Waerden, Bartel L. *Einführung in die algebraische Geometrie,* Berlin 1939.
[815] Veblen, Oswald. *Analysis situs,* New York 1931.
[709] Vegas, Luis; Navarro Borrás, F. *Manual de matemáticas para biólogos,* Madrid 1936.
[1116] Vergerio, Attilio. *Geometria elementare*, vol. 1, Turin 1930.
[1117] Vergerio, Attilio. *Geometria elementare*, vol. 2, Turin 1930.
[1240] Vergerio, Attilio. *Geometria intuitiva,* Turin 1931.
[61] Veronese, Gino. *Fisica dei corsi d'acqua,* Padua 1920.
[60] Veronese, Gino. *Origine dei corsi d'acqua: Introduzione,* Padua 1920.
[62] Veronese, Gino. *Sistemazione dei corsi d'acqua,* Padua 1920.
[63] Veronese, Gino. *Trasporto delle acque,* Padua 1920.
[1153] Veronese, Giuseppe. *Complementi di algebra e geometria,* Padua 1915.
[1208] Veronese, Giuseppe. *Elementi di geometria*, vol. 1, Padua 1904.
[1210] Veronese, Giuseppe. *Elementi di geometria*, vol. 1, Padua 1909.
[1209] Veronese, Giuseppe. *Elementi di geometria*, vol. 2, Padua 1905.
[380] Veronese, Giuseppe. *Fondamenti di geometria,* Padua 1891.
[1211] Veronese, Giuseppe. *Nozioni elementari di geometria intuitiva,* Padua 1902.
[1156] Veronese, Giuseppe. *Nozioni elementari di geometria intuitiva,* Padua 1912.
[8] Villa, Mario. *Geometria analitica con elementi di proiettiva,* Padua 1957.
[893] Villa, Mario (ed.). *Repertorio di matematiche,* Padua 1951.
[1366] Vinogradov, Ivan Matveevich. [*The Method of Trigonometric Sums in the Theory of Numbers*], Moscow 1947.
[1380] Vinogradov, Ivan Matveevich. [*Selected Works*], Moscow 1952.
[209] Viscardini, Mario. *L'hyper. Théorie mathématique de l'univers comme cellule vivante,* Brussels 1943.
[351] Viscardini, Mario. *La struttura dell'Universo,* Milan 1953.
[356] Vitali, Giuseppe. *Geometria nello spazio Hilbertiano,* Bologna 1929.

[831] Vitali, Giuseppe; Sansone, Giovanni. *Moderna teoria delle funzioni di variabile reale*, vol. 1, Bologna 1935.
[580] Vitali, Giuseppe. *Moderna teoria delle funzioni di variabile reale*, vol. 1: *Aggregati, analisi delle funzioni, integrazione, derivazione*, Bologna 1943.
[676] Vivanti, Giulio. *Esercizi di analisi infinitesimale,* Turin 1935.
[761] Vivanti, Giulio. *Lezioni di analisi matematica*, vol. 1: *Analisi algebrica e principii della Teoria delle funzioni. Derivazione e integrazione,* Turin 1930.
[762] Vivanti, Giulio. *Lezioni di analisi matematica*, vol. 2: *Applicazioni geometriche, equazioni differenziali, appendice,* Turin 1930.
[217] Voigt, Max. *Bildnisse Göttinger Professoren aus zwei Jahrhunderten (1737–1937)*, Göttingen 1937.
[164] Vojtěch, Jan. *Geometrie projektivní,* Prague 1932.
[1008] Volterra, Vito. *Corso di meccanica razionale*, vols. 2–3, Turin 1898.
[312] Volterra, Vito. *Leçons sur la théorie mathématique de la lutte pour la vie,* Paris 1931.
[1007] Volterra, Vito. *Lezioni di meccanica. Prime nozioni di cinematica,* Livorno 1896.
[98] Volterra, Vito. *Opere matematiche. Memorie e note*, vol. 1: *1881–1892*, Rome 1954.
[99] Volterra, Vito. *Opere matematiche. Memorie e note*, vol. 2: *1893–1899*, Rome 1956.
[100] Volterra, Vito. *Opere matematiche. Memorie e note*, vol. 3: *1900–1913*, Rome 1957.
[1409] Volterra, Vito. *Opere matematiche. Memorie e note*, vol. 4: *1914–1925,* Rome 1960.
[744] Volterra, Vito. *Teoría de los funcionales y de las ecuaciones integrales e integro-diferenciales,* Madrid 1927.
[766] Volterra, Vito. *Theory of Functionals and of Integral and Integro-Differential Equations,* London 1930.
[729] Vrănceanu, Gheorghe. *Lecţii de geometrie diferenţiala*, vol. 1, Bucharest 1951.
[730] Vrănceanu, Gheorghe. *Lecţii de geometrie diferenţiala*, vol. 2, Bucharest 1952.
[1324] VV. AA. *Algèbre et théorie des nombres 1954–1955. Séminaire P. Dubåreil,* Paris 1955.
[827] VV. AA. *Atti del Congresso di studi metodologici promosso dal Centro di Studi Metodologici, Torino 17–20 dicembre 1952,* Turin 1954.
[126] VV. AA. *Atti del Congresso Internazionale dei Matematici*, vol. 1, Bologna 1928.
[127] VV. AA. *Atti del Congresso Internazionale dei Matematici*, vol. 2, Bologna 1928.
[128] VV. AA. *Atti del Congresso Internazionale dei Matematici*, vol. 3, Bologna 1928.
[129] VV. AA. *Atti del Congresso Internazionale dei Matematici*, vol. 4, Bologna 1928.

[130] VV. AA. *Atti del Congresso Internazionale dei Matematici*, vol. 5, Bologna 1928.
[131] VV. AA. *Atti del Congresso Internazionale dei Matematici*, vol. 6, Bologna 1928.
[238] VV. AA. *Atti del Congresso Internazionale di Telegrafia e Telefonia,* Como 1927.
[143] VV. AA. *Atti del Convegno Internazionale di Ultracustica,* Rome 1951.
[141] VV. AA. *Atti del Convegno Matematico,* Rome 1942.
[132] VV. AA. *Atti del I Congresso dell'Unione Matematica Italiana,* Florence 1937.
[133] VV. AA. *Atti del II Congresso dell'Unione Matematica Italiana,* Bologna 1940.
[134] VV. AA. *Atti del III Congresso dell'Unione Matematica Italiana,* Pisa 1948.
[119] VV. AA. *Atti del IV Congresso Internazionale dei Matematici*, voll. 1–2, Rome 1908.
[120] VV. AA. *Atti del IV Congresso Internazionale dei Matematici*, vol. 3, Rome 1908.
[135] VV. AA. *Atti del IV Congresso dell'Unione Matematica Italiana,* vol. 1, Taormina 1951.
[136] VV. AA. *Atti del IV Congresso dell'Unione Matematica Italiana,* vol. 2, Taormina 1951.
[137] VV. AA. *Atti del V Congresso dell'Unione Matematica Italiana,* Pavia 1955.
[218] VV. AA. *Atti del V Congresso Internazionale di Filosofia,* Naples 1924.
[1339] VV. AA. *Atti del VIII convegno annuale dell'Associazione Italiana di Geofisica,* Rome 1959.
[1425]. VV. AA. *Bulletin of the National Research Council* no. 63, Washington 1928.
[325] VV. AA. *Celebrazione in luogo del centenario della nascita di Gregorio Ricci Curbastro, 2 maggio 1954,* Lugo 1954.
[78, 86] VV. AA. *Centenaire de la naissance de Émile Picard,* Paris 1956. (2 copies)
[1335] VV. AA. *Cinquième rapport de la Commission pour l'étude des relations entre les phénomènes solaires et terrestres,* Florence 1939.
[594] VV. AA. *Colloque de géométrie algébrique,* Liège 1949.
[1290] VV. AA. *Colloquio su questioni di analisi numerica,* Rome 1958.
[1291] VV. AA. *Colloquio sul trattamento numerico delle equazioni a derivate parziali con caratteristiche reali,* Rome 1958.
[711] VV. AA. *Colloquio sulle moderne macchine calcolatrici,* Rome 1956.
[734] VV. AA. *Colloque sur les fonctions de plusieurs variables. Bruxelles 1953,* Brussels 1953.
[110] VV. AA. *Contributi alla storia dell'Università di Pavia pubblicati nell'XI centenario dell'Ateneo,* Pavia 1925.
[139] VV. AA. *Convegno internazionale sui Reticoli e le Geometrie proiettive,* Palermo-Messina 1957.
[674] VV. AA. *Convegno internazionale sulle equazioni lineari alle derivate parziali, Trieste 1954,* Rome 1955.

[140] VV. AA. *Convegno italo-francese di algebra astratta,* Padua 1956.
[257] VV. AA. *Dalla terra alla luna,* Turin 1952.
[686] VV. AA. *Deuxième colloque de géométrie algébrique,* Liege 1952.
[323] VV. AA. *Études sur Léonard De Vinci, savant et philosophe,* Turin 1953.
[216] VV. AA. *Festschrift Rudolf Fueter,* Zurich 1940.
[1419] VV. AA. *Festschrift zur Feier des 100. Geburtstages Eduard Kummer. Mit Briefen an seine Mutter und an Leopold Kronecker,* Leipzig 1910.
[1414] VV. AA. *Festschrift zur Feier des zweihundertjährigen Bestehens des Akademie der Wissenschaft in Göttingen,* vol. 1: *Mathematisch-physikalische Klasse,* Berlin 1951.
[309] VV. AA. *Gli scienziati italiani dall'inizio del Medio Evo ai nostri giorni. Repertorio biobibliografico dei filosofi, matematici, astronomi, fisici, chimici, naturalisti, biologi, medici, geografi italiani,* Rome 1923.
[613] VV. AA. *Homenaje a Beppo Levi,* Buenos Aires 1955.
[906] VV. AA. *Il primo congresso dei dotti a Pisa, Ottobre 1839,* Pisa 1839.
[338] VV. AA. *In memoria di Temistocle Calzecchi Onesti nel X anniversario della morte 1922–1932,* Rome 1932.
[461] VV. AA. *I problemi filosofici del mondo moderno,* Rome 1949.
[1396] VV. AA. *Jubilé scientifique de M. Jacques Hadamard. Allocutions prononcées à la cérémonie du 7 janvier 1936,* Paris 1937.
[142] VV. AA. *Kolmastoista Skandinaavinen Matemaatikkokongressi / Trettonde Skandinaviska Matematikerkongeressen / Treizième Congrès des mathématiciens scandinaves,* Helsinki 1957.
[1288] VV. AA. *La didattica della matematica nella scuola primaria. Atti del convegno nazionale, Roma 1956,* Rome 1956.
[90] VV. AA. *La ricostruzione della Scuola Italiana,* Florence 1950.
[1325] VV. AA. *La Scuola di oggi e domani,* Milan 1959.
[875, 876] VV. AA. *Le livre du centenaire de la naissance de Henri Poincaré 1854–1954,* Paris 1955. (2 copies)
[245] VV. AA. *Leonardo,* Milan 1939.
[255] VV. AA. *Leonardo. Saggi e ricerche* a cura del Comitato Nazionale per le onoranze a Leonardo da Vinci nel quinto centenario della nascita, 1452–1952, Rome 1954.
[237] VV. AA. *L'Europa nel secolo XIX*, vol. 3.1 *Le scienze teoriche,* Padua 1932.
[282] VV. AA. *L'Europa nel secolo XIX*, vol. 3.2: *Le scienze applicate,* Padua 1932.
[1412] VV. AA. *Lezioni di metodologia statistica per ricercatori*, vol. 1, Rome 1960.
[114] VV. AA. *L'universo illustrato,* Milan 1867.
[1368] VV. AA. [*Mathematics of Computation*], 1, Moscow 1957.
[123] VV. AA. *Matematiker Kongressen,* Copenhagen 1925.
[144] VV. AA. *Mathematiker Kongress*, vol. 1, Zurich 1932.
[145] VV. AA. *Mathematiker Kongress*, vol. 2, Zurich 1932.
[185] VV. AA. *Monografie delle Università e degli Istituti superiori*, vol. 1, Rome 1911.

[186] VV. AA. *Monografie delle Università e degli Istituti superiori*, vol. 2, Rome 1913.

[1431] VV. AA. *Nel centenario della morte di Alessandro Volta*, Milan 1927.

[335] VV. AA. *Nel terzo centenario della morte di Galileo Galilei. Saggi e conferenze*, Milan 1942.

[1421] VV. AA. *Origines et aspects de l'industrie chimique bâloise*, Lausanne 1959.

[96] VV. AA. *Partie complémentaire: demi-groupes* (Séminaire Albert Chatelet et Paul Dubreil, 7e année: 1953–54), Paris 1956.

[124] VV. AA. *Proceedings of the Fifth International Congress of Mathematicians*, vol. 1, Cambridge 1912.

[125] VV. AA. *Proceedings of the Fifth International Congress of Mathematicians*, vol. 2, Cambridge 1912.

[121] VV. AA. *Proceedings of the International Congress of Mathematicians*, vol. 1, Cambridge 1950.

[122] VV. AA. *Proceedings of the International Congress of Mathematicians*, vol. 2, Cambridge 1950.

[117] VV. AA. *Proceedings of the International Congress of Mathematicians*, vol. 1, Amsterdam 1954.

[118] VV. AA. *Proceedings of the International Congress of Mathematicians*, vol. 2, Amsterdam 1954.

[685] VV. AA. *Proceedings of the International Congress of Mathematicians*, vol. 3, Amsterdam 1956.

[692] VV. AA. *Proceedings of the International Mathematical Congress*, vol. 1, Toronto 1924.

[693] VV. AA. *Proceedings of the International Mathematical Congress*, vol. 2, Toronto 1924.

[1340] VV. AA. *Ricerca e scienza*, Turin 1959.

[459] VV. AA. *Saggi di critica delle scienze*, Turin 1950.

[1295] VV. AA. *Scritti matematici in onore di Filippo Sibirani*, Bologna 1957.

[801] VV. AA. *Scritti matematici offerti a Luigi Berzolari*, Pavia 1936.

[83]. VV. AA. *Scritti matematici offerti a M. Picone nel suo settantesimo compleanno*, Bologna 1955.

[334] VV. AA. *Scritti varii di Francesco Porro*, Genoa 1936.

[1381] VV. AA. *Semaine d'étude sur le problème biologique du cancer*, Vatican 1949.

[59] VV. AA. *Semaine d'étude sur le problème des microséismes*, Roma 1952.

[1286] VV. AA. *Simposio internacional de topología algebraica*, Mexico 1958.

[260] VV. AA. *Studi su Leonardo da Vinci scienziato e filosofo*, Milan 1953.

[840] VV. AA. *Studies in Mathematics and Mechanics Presented to Richard von Mises*, New York 1954.

[639] VV. AA. *Topologie algébrique*, Paris 1949.

[1311] VV. AA. *Trabalhos do seminário de análise general*, Lisbon 1941.

[1432] VV. AA. *Vittorio Fossombroni. Nel primo centenario della morte*, Arezzo 1947.

[1374]. VV. AA. [*Works of the "V. A. Steklov" Institute of Mathematics*], nos. 40–49, Moscow 1953–1955.
[1375]. VV. AA. [*Works of the "V. A. Steklov" Institute of Mathematics*], no. 12, Moscow 1945.
[1376]. VV. AA. [*Works of the "V. A. Steklov" Institute of Mathematics*], no. 16, Moscow 1945.
[1377]. VV. AA. [*Works of the "V. A. Steklov" Institute of Mathematics*], no. 17, Moscow 1945.
[1378]. VV. AA. [*Works of the "V. A. Steklov" Institute of Mathematics*], no. 20, Moscow 1947.
[1379]. VV. AA. [*Works of the "V. A. Steklov" Institute of Mathematics*], no. 50, Moscow 1957.
[1427] Wachsmuth, Günther. *Die ätherischen Bildkräfte in Kosmos, Erde und Mensch. Ein Weg zur Erforschung des Lebendigen*, Dornach 1926.
[713] Waerden, Bartel Leendert van de. *Moderne Algebra*, vol. 1, Berlin 1930.
[714] Waerden, Bartel Leendert van de. *Moderne Algebra*, vol. 2, Berlin 1931.
[1258] Waismann, Friedrich. *Einführung in das mathematische Denken. Die Begriffsbildung der modernen Mathematik*, Vienna 1947.
[816] Walker, Robert J. *Algebraic Curves*, Princeton 1950.
[519] Weil, André. *Arithmétique et géométrie sur les variétés algébriques*, Paris 1935.
[545] Weil, André. *Foundations of Algebraic Geometry*, New York 1946.
[520] Weil, André. *L'intégration dans les groupes topologiques et ses applications*, Paris 1940.
[1021] Weil, André. *Sur les courbes algébriques et les variétés qui s'en déduisent*, Paris 1948.
[521] Weil, André. *Variétés abéliennes et courbes algébriques*, Paris 1948.
[494] Weinberg, Julius R. *Introduzione al positivismo logico*, Turin 1950.
[634] Weise, Karl H. *Gewöhnliche Differentialgleichungen*, Hanover 1948.
[559] Wiener, Hermann; Truetlein, Peter. *Verzeichnis mathematischer Modelle Sammlungen*, Leipzig 1912.
[246] Wiener, Norbert. *Introduzione alla cibernetica*, Turin 1953.
[1247] Xiberta Roqueta, Manuel. *Elementos de aritmética*, Madrid 1928.
[1248] Xiberta Roqueta, Manuel. *Elementos de geometría*, Madrid 1928.
[890] Yoshino, Yuji. *The Japanese Abacus Explained*, Tokyo 1937.
[1249] Young, Grace C.; Young, William H. *Geometria per i piccoli*, Turin 1911.
[1160] Young, Jacob W. A. *I concetti fondamentali dell'algebra e della geometria*, Naples 1919.
[1133] Young, Jacob W. A. *L'insegnamento delle matematiche nelle scuole elementari e secondarie*, Milan 1924.
[414] Young, John W. A. *Projective Geometry*, Chicago 1930.
[1125] Zaccaria, Angelo. *Nozioni di aritmetica, geometria e computisteria pratica*, Turin 1925.
[455] Zaccherini, Giampietro. *Letture di economia matematica*, vol. 1 *L'economia stazionaria uniforme*, Rome 1946.

[17] Zamansky, Marc. *Introduction à l'algèbre et l'analyse modernes,* Paris 1958.
[733] Zappa, Guido. *Gruppi, corpi, equazioni,* Naples 1950.
[899] Zappa, Guido. *La matematica, oggi,* Rome 1952.
[488] Zappa, Guido. *Lezioni di geometria descrittiva,* Rome 1947.
[826] Zappa, Guido. *Reticoli e geometrie finite,* Naples 1952.
[495] Zariski, Oscar. *Algebraic Surfaces,* Berlin 1935.
[817] Zariski, Oscar. *Introduction to the Problem of Minimal Models in the Theory of Algebraic Surfaces,* Tokyo 1958.
[593] Zeuthen, Hieronymus Georg. *Lehrbuch der abzählenden Methoden der Geometrie,* Leipzig 1914.
[740] Zondadari, Enrico. *Integrazione grafica e studio delle equazioni differenziali ordinarie del primo ordine coi metodi della geometria descrittiva,* Milan 1917.
[1341] *Bibliografia matematica italiana,* vol. 1: *1950,* Rome 1951.
[1342] *Bibliografia matematica italiana,* vol. 3: *1952,* Rome 1954.
[1343] *Bibliografia matematica italiana,* vol. 4: *1953,* Rome 1955.
[1344] *Bibliografia matematica italiana,* vol. 6: *1955,* Rome 1957.
[1345] *Bibliografia matematica italiana,* vol. 8: *1957,* Rome 1959.
[1351] *Bibliografia matematica italiana,* vol. 2: *1951,* Rome 1953.
[1352] *Bibliografia matematica italiana,* vol. 5: *1954,* Rome 1956.
[1353] *Bibliografia matematica italiana,* vol. 7: *1956,* Rome 1958.
[1354] *Bibliografia matematica italiana,* vol. 9: *1958,* Rome 1959.
[1033] *Encyklopädie der mathematischen Wissenschaften,* vol. II-1.4.
[1034] *Encyklopädie der mathematischen Wissenschaften,* vol. II-1.5.
[1035] *Encyklopädie der mathematischen Wissenschaften,* vol. II-1.7.
[1036] *Encyklopädie der mathematischen Wissenschaften,* vol. II-1.8.
[1037] *Encyklopädie der mathematischen Wissenschaften,* vol. II-1.9.
[1038] *Encyklopädie der mathematischen Wissenschaften,* vol. II-2.1.
[1039] *Encyklopädie der mathematischen Wissenschaften,* vol. II-2.4.
[1040] *Encyklopädie der mathematischen Wissenschaften,* vol. II-3.1.
[1041] *Encyklopädie der mathematischen Wissenschaften,* vol. II-3.2.
[1042] *Encyklopädie der mathematischen Wissenschaften,* vol. II-3.3.
[1043] *Encyklopädie der mathematischen Wissenschaften,* vol. II-3.4.
[1044] *Encyklopädie der mathematischen Wissenschaften,* vol. III-1.1.
[1045] *Encyklopädie der mathematischen Wissenschaften,* vol. III-1.2.
[1046] *Encyklopädie der mathematischen Wissenschaften,* vol. III-1.3.
[1047] *Encyklopädie der mathematischen Wissenschaften,* vol. III-1.4.
[1048] *Encyklopädie der mathematischen Wissenschaften,* vol. III-1.5.
[1049] *Encyklopädie der mathematischen Wissenschaften,* vol. III-1.6.
[1050] *Encyklopädie der mathematischen Wissenschaften,* vol. III-1.7.
[1051] *Encyklopädie der mathematischen Wissenschaften,* vol. III-2.2.
[1052] *Encyklopädie der mathematischen Wissenschaften,* vol. III-2.3.
[1053] *Encyklopädie der mathematischen Wissenschaften,* vol. III-2.4.
[1054] *Encyklopädie der mathematischen Wissenschaften,* vol. III-2.5.
[1055] *Encyklopädie der mathematischen Wissenschaften,* vol. III-2.6.
[1056] *Encyklopädie der mathematischen Wissenschaften,* vol. III-2.7.

[1057] *Encyklopädie der mathematischen Wissenschaften*, vol. III-2/3.
[1058] *Encyklopädie der mathematischen Wissenschaften*, vol. III-3.4.
[1059] *Encyklopädie der mathematischen Wissenschaften*, vol. III-3.5.
[1060] *Encyclopédie des sciences mathématiques*, vol. I-1.1.
[1061] *Encyclopédie des sciences mathématiques*, vol. I-1.2.
[1062] *Encyclopédie des sciences mathématiques*, vol. I-1.3.
[1063] *Encyclopédie des sciences mathématiques*, vol. I-1.4.
[1064] *Encyclopédie des sciences mathématiques*, vol. I-2.1.
[1065] *Encyclopédie des sciences mathématiques*, vol. I-2.2.
[1066] *Encyclopédie des sciences mathématiques*, vol. I-2.3.
[1067] *Encyclopédie des sciences mathématiques*, vol. I-2.4.
[1068] *Encyclopédie des sciences mathématiques*, vol. I-3.1.
[1069] *Encyclopédie des sciences mathématiques*, vol. I-3.2.
[1070] *Encyclopédie des sciences mathématiques*, vol. I-3.3.
[1071] *Encyclopédie des sciences mathématiques*, vol. I-3.4.
[1072] *Encyclopédie des sciences mathématiques*, vol. I-3.5.
[1073] *Encyclopédie des sciences mathématiques*, vol. I-4.1.
[1074] *Encyclopédie des sciences mathématiques*, vol. I-4.2.
[1075] *Encyclopédie des sciences mathématiques*, vol. I-4.3.
[1076] *Encyclopédie des sciences mathématiques*, vol. I-4.4.
[1077] *Encyclopédie des sciences mathématiques*, vol. II-1.1.
[1078] *Encyclopédie des sciences mathématiques*, vol. II-1.2.
[1079] *Encyclopédie des sciences mathématiques*, vol. II-2.1.
[1080] *Encyclopédie des sciences mathématiques*, vol. II-3.1.
[1081] *Encyclopédie des sciences mathématiques*, vol. II-4.1.
[1082]. *Encyclopédie des sciences mathématiques*, vol. II-4.2.
[1083] *Encyclopédie des sciences mathématiques*, vol. II-5.1.
[1084] *Encyclopédie des sciences mathématiques*, vol. II-5.2.
[1085] *Encyclopédie des sciences mathématiques*, vol. II-6.1.
[1086] *Encyclopédie des sciences mathématiques*, vol. II-6.2.
[1087] *Encyclopédie des sciences mathématiques*, vol. III-1.1.
[1088] *Encyclopédie des sciences mathématiques*, vol. III-1.2.
[1089] *Encyclopédie des sciences mathématiques*, vol. III-2.1.
[1090] *Encyclopédie des sciences mathématiques*, vol. III-3.1.
[1091] *Encyclopédie des sciences mathématiques*, vol. III-3.2.
[1092] *Encyclopédie des sciences mathématiques*, vol. III-4.1.
[1093] *Encyclopédie des sciences mathématiques*, vol. IV-1.1.
[1094] *Encyclopédie des sciences mathématiques*, vol. IV-2.1.
[1095] *Encyclopédie des sciences mathématiques*, vol. IV-2.2.
[1096] *Encyclopédie des sciences mathématiques*, vol. IV-5.1.
[1097] *Encyclopédie des sciences mathématiques*, vol. IV-5.2.
[1098] *Encyclopédie des sciences mathématiques*, vol. IV-6.1.
[1099] *Encyclopédie des sciences mathématiques*, vol. V-1.1.
[1100] *Encyclopédie des sciences mathématiques*, vol. V-2.1.
[1101] *Encyclopédie des sciences mathématiques*, vol. V-3.1.

[1102] *Encyclopédie des sciences mathématiques*, vol. V-4.1.
[1103] *Encyclopédie des sciences mathématiques*, vol. VI-1.1.
[1104] *Encyclopédie des sciences mathématiques*, vol. VI-2.1.
[1105] *Encyclopédie des sciences mathématiques*, vol. VII-1.1.
[1106] *Encyclopédie des sciences mathématiques*, vol. VII-1.2.

Francesco Severi and the Fascist Regime

Angelo Guerraggio

Abstract Francesco Severi was an eminent leader of the Italian school of algebraic geometry at the beginning of the twentieth century. A biographic profile of his is here traced but concerning his political and intellectual activity. Severi began his political career in Padua as a member of the Socialist Party. Then, in the middle of the 1920s, he converted to the Fascist regime. Severi remained close to Mussolini even in the following darkest years. After the end of the Second World War and the fall of the Fascism, he underwent the procedure of "purge" in the University and in the Accademia dei Lincei for his important activity in favor of the Fascist regime. But he got away essentially unscathed.

Keywords Biography · Severi as a member of the Socialist Party · Severi and the Fascist regime · The "purge" after WWII

1 Severi: The Mathematician and the Politician

This chapter will not discuss the importance of Severi's work as a mathematician, for whom I refer to the long essay "Geometria algebrica" by Brigaglia, A. and Ciliberto, C. [1] and the most recent "Francesco Severi: il suo pensiero matematico e politico prima e dopo la Grande Guerra" by Ciliberto, C. and Sallent Del Colombo, E. [2] I will limit myself to recalling, in an extremely succinct way, that Severi

[1] Brigaglia, A. and Ciliberto, C. "Geometria algebrica" in *La matematica italiana dopo l'Unità*, Di Sieno, S., Guerraggio, A., Nastasi, P., (Eds), Marcos y Marcos, Milan, 1998, pp. 185–320.

[2] Ciliberto, C. and Sallent Del Colombo, E., "Francesco Severi: il suo pensiero matematico e politico prima e dopo la Grande Guerra," in *Serva di due padroni*, Cogliati, A. (Ed), EGEA, Milan, 2019.

A. Guerraggio (✉)
Università Bocconi, Milano, Italy
e-mail: angelo.guerraggio@unibocconi.it

G. Bini (ed.), *Algebraic Geometry between Tradition and Future*,
Springer INdAM Series 53, https://doi.org/10.1007/978-981-19-8281-1_3

obtained his chair in 1904 and went then to teach at the universities of Parma, Padua, and Rome. His scientific production, especially in the years preceding the outbreak of the First World War, included particularly significant contributions in the field of projective varieties, enumerative geometry, and classification of algebraic surfaces, with specific attention to irregular ones. His research activity would be soon recognized and valued, beyond the university professorship. In the period between 1905 and 1915, Severi received the XL Medal (1906), the Prix Bordin of the Institut de France with Federigo Enriques (1907), the Guccia Medal on the occasion of the International Congress of Mathematicians (1908), and the Royal Prize for Mathematics by the Accademia dei Lincei (1913). In the first postwar period, Severi's scientific activity was mainly oriented towards systematizing the results obtained in the previous period and taking care of his students. The most original results concerned the complex analysis. In 1921, his *Lezioni di Geometria Algebrica* of 1908 were translated in German as *Vorlesungen uber Algebraische Geometrie*. In 1926, the first (and only) volume of the *Treatise on Algebraic Geometry* was published, followed—already during the Second World War—by the lectures held at the National Institute of Higher Mathematics.

Here, I will deal with Severi as a politician and an intellectual, framing his work and vicissitudes in the more general theme of relations between Italian intellectuals and politicians during the Fascist regime. In the period between the two World Wars, Severi achieved the greatest intellectual substance and renown among the Italian mathematicians, following the steps of Vito Volterra albeit with completely different political outcomes. The intellectual was conceived as someone that—despite enjoying a good, often notable, reputation in his or her professional field—chooses to place himself or herself above the sectoral interests of his or her profession, using his or her skills to speak on behalf of society as a whole. Intellectuals would deal with topics and problems that do not concern them directly and would try—with their judgments—to influence the public opinion. They would put aside their university careers to get actively involved in social life as columnists and activists. It is a category, that of the "committed" intellectual, which today appears less frequent. Starting from the 1980s and 1990s, social, political, and economic macrophenomena have indeed transformed the figure of the intellectual into that of an expert, an interpreter, and a communicator who facilitates the translation of concepts formulated within a given knowledge. We have gone from one universalistic to many particular perspectives, where the search for meaning within one's own discipline has replaced the critical attention to the general society framework.

When Severi was active—from the beginning of the twentieth century to the Cold War period—the characteristics of the typical intellectual which I have outlined fit perfectly on his person. Severi started to deal with problems that "do not concern him," in the sense that they were extraneous to his life as a mathematician, even well before the Fascist period. In 1909, Severi became president of Mathesis, the association that brought together mathematics teachers at national level, founded about 10 years earlier by Rodolfo Bettazzi. While teaching and working on the improvement of the school system could be seen as a commitment to society shared by all researchers, the young Severi—who had been at that time a professor for a few years only—was the first university professor to take on the role of president of Mathesis. His election highlighted that not only he was not avoidant of extra-university positions, but he was almost seeking for them too. His interest would later bring him to be a member of both the Board and the Superior Council of the Ministry of Education and President of the National Association of University Professors.

More striking was the 1913 episode of the poisonous exchange of views with Benedetto Croce on philosophical issues, in the wake of the more famous controversy that Federigo Enriques had had with Croce in 1911 after the International Congress of Philosophy in Bologna. At that time, the relations between the two mathematicians could not have been better. Immediately after graduation, Severi had been an assistant to Enriques, who had oriented his interests towards the study of algebraic surfaces, starting a fruitful season of scientific collaboration. The controversy between Severi and Croce yet began when Severi—who had moved to Padua after a short stay at the University of Parma—chose "Rationalism and spiritualism" as the title and theme of his inaugural lecture for the academic year 1913–1914, going far beyond the necessary interdisciplinarity required on occasions like this. In his lecture, he wrote: "I think that the reaction against rationalism and science, based on the belief that they distance us from human principles, risks bringing us back towards harmful mental habits. The recent idealism of Croce and Gentile does not lead us to correcting errors, but to the persecution of those who err." [3] Such a harsh comment could not leave Croce and Gentile silent. The sharp reply was entrusted to Croce in an already very sarcastic intervention titled "And if they talked about mathematics?": "To Professor Severi, who is an intellectual, I would like to address a prayer to avoid discussing concepts that belong to fields that are foreign to him, and for which I do not know if he has an aptitude, but certainly does not have the preparation." [4]

After Croce's reply, Severi would no longer venture into philosophical debates. To express his interests outside the realm of mathematics, he chose politics, first as a militant of the Socialist Movement and then as a powerful management figure. He was not only an intellectual then, but also a man of power. Severi would be president of the "Banca Popolare" di Arezzo, his hometown, and of the Vallecchi publishing house. He liked to study and debate, but also to take action and translate his ideas into facts. He was a mathematician, a politician, and a man of power with broad horizons, who came to achieve ambitious goals and managed to overcome with ease the difficulties he encountered.

[3] *Annuario della R. Università di Padova*, 1913–4, pp. xxvi–lvii
[4] *La Critica*, 1914, pp. 79–80

2 Severi in Padua

Severi's political debut occurred in 1910 in the City Council of Padua, to which he participated as an elected member of the Socialist Party. While Severi was only the ninth elected councilor in terms of number of preferences, he had only received 100 votes less than the first of those elected. On August 6, Severi was also appointed as councilor for public education. In fact, Severi had already occupied an administrative position in Padua before, when in 1906 the progressive junta chaired at the time by Giacomo Levi-Civita (Tullio's father) had appointed him—newcomer to town—as President of the Administrative Commission supervising the water and gas distribution by municipal companies for water and gas. After 2 years, however, Severi resigned due to a workload that had proved excessive and unequally distributed among the members of the board of directors.

Severi's debut in the city council was immediately marked by very clear and controversial positions. In greeting the other councilors, Severi continued to call the members of the minority party "clerical, even though they call themselves Catholics," [5] inducing angry reactions expressed through various interruptions. His experience as a city councilor was very productive. The acts of the council recorded his intense efforts to develop and rationalize the school system in Padua. His interventions often featured polemical and anticlerical tones, including at the time when the municipality decided to name an elementary school after the philosopher and pedagogist Roberto Ardigò, an illustrious representative the University of Padua, to celebrate his 80th birthday. Severi's speech explicitly referred to secular and positivist principles. He stated emphatically that it was necessary to guide the child to the observation of facts and to shape his or her education solely on what is scientifically proven, and not on precepts which are outside the objective reality and which, for that very reason, cannot be the object of affirmation or negation.

Severi's experience in the City Council was intense but short. Severi remained in office until the end of 1911, resigning after the "political incident" of September 27. That day, the socialists had proclaimed a national strike against the war in Libya and Giolitti's politics. In Padua too, there had been several demonstrations with strikes affecting the functioning of the tramway and other public services. It had been a day of strong political tensions: in the late morning, a delegation of nationalists came to the town hall demanding that the municipality display the tricolor flag to distance itself from socialist demonstrations and express its closeness to the Italian troops. The mayor, absent at that moment, had made it known that he would have not considered the display of the tricolor to be appropriate but had left the final decision to Severi. While declaring that the socialists were against the war and that for him "the flag remained the symbol of the nation and not of the nationalists," Severi agreed to have the tricolor displayed on the facade of the town

[5] This and the following quotes of this section are taken from documents kept in the General Archives of the Municipality of Padua. I wish to particularly thank Dr. Elena Ferraro for her collaboration and kindness.

hall. The decision triggered many tensions within the junta, on which the minority speculated, speaking of an evident difference of opinion between the mayor and one of his councilors. Moreover, Severi had not shared the mayor's decision not to pay the day of strike to the municipal employees, as he did not support the practice of striking in public sector provision. After this accident, Severi resigned as Member of the Council. It is yet apparent that his decision was a sign of the different positions among the progressive forces in Padua, and also within the Socialist Party, on the war in Libya. These differences were explicitly recalled in the other resignation letter, this time from the city councilor, which Severi reviewed a few weeks later, at the beginning of 1923: "Following a difference of appreciation on an important political issue which arose between me and the majority of the members of my party, and convinced as I am that whoever is invested with political or administrative representation must be able to faithfully interpret the views of the majority of their Party, I consider it my duty to resign from Councilor of the Municipality." The Socialist Party remained opposed to the intervention in Libya, but lost an important member when Leonida Bissolati was expelled because of his support to the military action in Libya. Benito Mussolini—then director of the party's newspaper *L'Avanti*—accused the dissidents of "Libyan ministerialism and warfare." Severi became instead close to Bissolati's reformist positions and increasingly cold towards his own party, now led by the revolutionary current.

The contrast exploded sharply in 1914 with the outbreak of the First World War. As the socialists were searching for a difficult balance between the nationalistic sentiments of large sectors of the youth population as well as of many intellectuals and the solidarity with the European proletarians that were on the other side, the slogan "Neither adhere, nor sabotage"—which Severi did not agree with—was forged. For Severi, the war against the Prussian militarism and for the defense and consolidation of democracy was just. The unfolding of the war events—with the invasion of Belgium and the brutal violation by Germany of people's rights—led to the war with a further characterization in terms of conflict between two civilizations, and Severi had no doubts in taking sides with the attacked countries and the nationalities to which the central empires denied recognition. Severi left the party, and the article he published in March 1915 in the newspaper *L'Adriatico*, supporting democratic interventionism, represents his substantial farewell to the Socialist Party. There he wrote: "While it is true that the Socialist Party, as a political organism, could never be the promoter of a war-like intervention, it seems much better to me if the socialist opposition to the war was always limited to being a theoretical assumption (...). A less assertive attitude by our party's leadership would also be politically very important, leaving each member the freedom to evaluate the situation (...). Just as there is no Socialist who—living in this bourgeois society—does not adapt to what the environment imposes on him while still working for a better tomorrow, so I do not find any substantial contradiction between the faith of our ideals and what today's historical reasons—superior to our will—can prescribe us to do (...). Moreover, has the Socialist Party perhaps not recognized that in daily life it is better to adapt to a minimum program and not to avoid contacts with the most enlightened fractions of the bourgeoisie when it is, for example, necessary to oppose the parties which perniciously threaten political freedoms, which constitute the prerequisite for the economic conquests of the proletariat? And why should we lock ourselves in a form of intransigent denial, when the importance of something transcends so much the everyday politics, and it is basically a vital matter of freedom?" For Severi, this is a personal choice too. He enlisted as a volunteer, earning a promotion and two war crosses for the contribution provided in responding to the technical needs of the army. As a mathematician, Severi's specific contribution concerned the artillery with the revision of the firing tables, depending on the geography of the Dolomite area where the conflict was taking and the Italian mountain phonotelemetry system to identify the enemy batteries.

3 Moving to Rome

In 1921, Severi moved to the University of Rome, winning over the competition of other important mathematicians, such as Leonida Tonelli and Federigo Enriques, who aspired to the same position. Severi arrived in Rome with the label of "Socialist," even if he was no longer a member of the Party. His appointment in 1923 as Rector of the University of Rome by Giovanni Gentile, who would become the Minister of Education in the first Mussolini Fascist Government, was thus met with some surprise. To be expected was his decision to sign the so-called *Manifesto Croce* as a democratic intellectual in 1925.

The Fascist movement still faced, at the time, the issue of convincing the high intellectuals, still very suspicious—if not decidedly contrary—to the political course begun with the March on Rome in October 1922. The "Battle of Posters" began with the first national conference of "Fascist Cultural Institutions" promoted by Gentile in Bologna in March 1925. That was a meeting which featured no discussion but only written communications, previously presented to the organizing committee, due to fear that "any theoretical overrun would not be inconclusive but would also divert the attention away from the initiative's immediate positive purposes, for which it was believed necessary to gather in Bologna the most influential Italian thinkers." From the conference came the appeal by Gentile to solicit the intellectuals' adherence to the Fascist movement—what became immediately known as the *Manifesto Gentile* was released to the press on April 21. The response of the anti-fascist intellectuals was not long in coming, entrusted to Benedetto Croce. His counter-manifesto, published on 1 May, was widely supported, including by Severi, together with other illustrious mathematicians such as Leonida Tonelli, Vito Volterra, Guido Castelnuovo, and Tullio Levi-Civita.

For the leaders of the Fascist movement, that was too much. They had already experienced with some disappointment Gentile's choice to entrust the position of rector of the University of Rome, a highly regarded position, to a socialist. Now, the rector even gets to sign the anti-fascist manifesto! The "jar was full" for them, and it was time for Severi to pay the right price for his political past (and present), also taking advantage of the fact that he could no longer count on the presence of

Gentile (and his Undersecretary Balbino Giuliano, replaced at the beginning of 1925 by Pietro Fedele) at the Ministry of Education. A few months after Severi's public adhesion to the Croce Manifesto, Severi was investigated for alleged administrative irregularities. As illustrated in his defensive memoir, Severi immediately understood the gravity of the situation, without excluding that "orders may have come from above to make the rectorial life impossible for me, because I am not politically close to the government. But even on this point, my conscience is calm, because I have always contained the manifestations of my political ideas within the limits that the importance of my role imposed on me." [6] He replied to all the allegations, with the belief that "my work has been assiduous, prudent, and thrifty to the point of unbelievable," [7] deciding not to resign until the conclusion of the investigation. It was only when the minister grudgingly recognized the correctness of his actions that Severi resigned. In the memorial written during the purge procedure (which we will discuss later), speaking of himself in the third person, he noted: "With this act, Severi's active participation in political life was closed forever. He remained on the sidelines, frowned upon and systematically kept out of positions, competition commissions, etc."

4 The Oath

The reconstruction I have just mentioned, made by Severi 20 years later, is functional to his defense in the purge process. Things had actually turned out differently. Severi came back to politics, joining in 1932 the National Fascist Party (NFP) with an act that, for a former socialist, cannot be considered merely formal, particularly as he went straight back to the "high places" of power, where the academic and political world grant each other mutual favors.

After his resignation as rector, Severi actually experienced a short period of isolation, which was very heavy on him: he felt useless, away from all discussions and people that matter, unable to use the power by which he was so attracted and with which he identified. He spent a few months abroad, hosted by some Spanish universities, but yet looking forward to returning and regaining his place in the national elite to which he felt he belonged. He began to think that Fascism was a reality to live with and that any important change would only happen from within. Political ideals gave way to more pragmatic views. He focused on the pleasure of personal affirmation and the ability to manage, in any possible way, what he was able to achieve.

---

[6] This quotation is taken from a letter from Severi to Gentile, dated July 31, 1925. The whole letter is published in Guerraggio, A., and Nastasi, P., *Gentile e i matematici italiani*, Bollati Boringhieri, Turin, 1993.

[7] *Ibid.*

It is from these considerations that his march towards Fascism originated. In reality, Severi was fast in changing affiliation. He continued to have a cordial relationship with Gentile, who was then politically his friend and who agreed to write the presentation for a school text such as the *Elementi di geometria*. He even started communicating personally with Mussolini. In January 1929, he addressed a memorandum directly to the Duce, which confirmed an already established personal relationship and the beginning of Severi's political parabola. The memo was about university professors. According to Severi, their initial opposition to Fascism has been greatly attenuated, both in terms of number and intensity, and which must be acknowledged with satisfaction. Continuing to remember some unorthodox political demonstrations of previous years with the consequent recurrent threat of dismissal for teachers who had not immediately joined the new regime "would be fatal to Italian culture and science, and would result in moral and material damage for the nation, with serious repercussions near and far." Severi was thinking about himself, in the hope of being able to "silently and faithfully serve the Fascist state, in my current function. To serve with pure disinterest, but to serve without technical limitations, so that our work can be efficiently carried out; to serve, not expecting prizes or distinctions or leadership positions." [8]

How could the Fascist regime acknowledge that the many critical positions of the first years of power had changed and trust that they could be used without danger to the benefit of the nation and of political stability? Severi had an idea, which he expressed in a letter to Gentile shortly after the memorandum to the Duce, sent on February 15 from Barcelona where he was for a series of seminars. Severi knew that the question of intellectuals was being discussed by the Grand Council of Fascism and that there, or in the Council of Ministers, a new oath of allegiance would soon be put on the table, for university professors to sign. He wrote: "It would be necessary that the oath be represented as an act of direct intransigence to obtain the much requested fascistization of the Universities; as an appeal to the loyalty of professors, who could not avoid taking the oath without incurring into far more serious measures than the dismissal of authority. But at the same time as an amnesty of political acts that had occurred long time ago, so that the State could benefit without limitation of the technical skills of professors who had signed the oath, thus eliminating the absurd current situation of so many professors not even being able to be part of judging commissions!" [9]

The oath, precisely with the characteristics invoked by Severi and owned by Gentile, would become a reality in 1931. Compared to the previous formulations, that of 1931 conceptualized the Fascist party away from a simple party that had won the elections but as the very backbone of the State, equated with the monarchy. People had to swear to be faithful to the regime: "I swear to be faithful to the King,

[8] The entire memorandum is published in Guerraggio, A. and Nastasi, P., *Matematica in camicia nera*, Bruno Mondadori, Milan, 2005.

[9] The entire letter was published in Guerraggio, A. and Nastasi, P., *Matematica in camicia nera*, Bruno Mondadori, Milan, 2005.

to his Royal successors and to the Fascist regime (. . .) with the intention of forming industrious, upright and devoted citizens to the country and the Fascist regime." The stick and the carrot: the carrot for professors who, by swearing their loyalty to the regime, showed that they had repented and could therefore fully reenter the great family of the (Fascist) nation; the stick, in the form of immediate dismissal, for diehards who chose not to bend over. Severi, Gentile, and the Fascist regime bet that few would refuse, and they were right. The oath passed and became a success for the regime with a plebiscite adhesion. All but 12 intellectuals (1% of university professors!) did swear. Among them, the only mathematician was Vito Volterra.

The reward for the Severi's "conversion" and for his help with the political characteristics of the oath arrived immediately. What he gained was entering the *Academy of Italy*, the new cultural institution wanted by the Fascist regime to replace the Lincei, which had been too reluctant to align. When the moment came to choose new academics, and only one place was reserved for mathematicians, at last the name of Severi replaced that of Enriques. This meant that, for the regime, Severi was the leader of Italian mathematicians, adding an important layer to his image and prestige. All this occurred in March 1929, only a few weeks after the memo and the letter to Gentile previously mentioned.

5 1938

With his appointment as the "Italian academic" and his subsequent registration with the NFP, Francesco Severi's long journey into Fascism began. In the 1930s, our mathematician represented a perfect expression of the intertwining of politics and the world of universities: always present in the front row in the ceremonies with which Fascism showed its attention to science, and always ready to embark on trips abroad to magnify the awakening and the bright future towards which the new Italy has set out.

This "partnership" is perhaps best exampled by the behavior held when the racial laws of 1938 were promulgated. Italians were discovered to be of Aryan race with the publication, on July 15, of the so-called "Manifesto of racist scientists," which sought to provide a scientific justification for an eminently political operation. Immediately after, in September, the persecution of Jews began with the expulsion of hundreds of professors and thousands of students from universities and schools, the "purification" of school manuals from Jewish contamination, the implementation of measures hindering the presence of Jews in public office as well as commercial/industrial activities and in the free professions, and a whole series of hateful acts of harassment to make life in Italy as difficult as possible for the Jews.

The anti-Semitic legislation also affected the world of mathematicians with devastating results. Among others, Federigo Enriques, Gino Fano, Guido Fubini, Beppo Levi, Tullio Levi-Civita, and Alessandro Terracini were removed from teaching. Somewhat surprisingly, the Italian Mathematical Union (UMI) did not protest, nor it asked any easing of the measures taken, which would have been an understandable ask given the delicate situation and the repressive climate established in the country. In the document that summarized the discussion held on December 10, the Scientific Commission of UMI took note of the decisions taken at a political level, almost supporting them by observing that "the Italian mathematical school, which has acquired a wide reputation throughout the scientific world, is almost entirely the creation of Italic (Aryan) scientists." The note only asked that "none of the chairs for mathematicians made vacant by the measures established for the integrity of the race be removed." One of the blackest and most shameful pages in the history of the association was thus written, as UMI itself has recognized in recent years with great intellectual honesty.

At the meeting of December 10, Severi—present as a member of the Scientific Committee—did not raise any objections to the final text. Given his personality and undisputed scientific value, his silence was particularly heavy. In this regard, it is important to remember that Severi did not express any negative comment even in 1935, when the "Fascist reclamation" by Minister De Vecchi had also demanded that the appointment of the UMI President, the Vice President, and members of the Scientific Commission took place only after the approval of the Ministry of National Education, thus leading to the replacement of Giulio Vivanti and Vito Volterra who had been duly elected to the Scientific Commission. Similarly, Severi would never dissociate himself from the discriminatory measures introduced by the racial laws in universities. Actually, as soon as these were promulgated, he intervened to ensure that colleagues such as Castelnuovo, Enriques, and Levi-Civita were prohibited from entering the library of the Mathematical Institute in Rome. He was also not sorry that the geometry textbooks for secondary schools authored by Enriques became banned, widening widely the school market for his texts. Severi also took advantage of the racial laws to "Aryanize" the editorial board of the most prestigious Italian mathematical journal, the *Annali di Matematica Pura ed Applicata*, removing its Jewish components and being named the sole director. In short, Severi capitalized on his adhesion to Fascism, with significant benefits coming to him in 1939 with the foundation of the Istituto Nazionale di Alta Matematica, of which he became President and through which he prepared to counterbalance the work by the Istituto Nazionale per le Applicazioni del Calcolo by Mauro Picone, which he had always considered as too biased in favor of applications.

6 The End of 20 Years of Fascism

At the basis of Severi's behavior was the same motivation that had brought him, in the second half of the 1920s, to abandon his democratic ideals to enlist directly under the Duce and Gentile: a considerable sense of self and of his abilities—which, if gone used, would have been a waste for the nation—combined with the pleasure and ambition of participating in the construction of the new Fascist state and the power games that revolved around it, as well as the opportunity to develop his career and gain ever-greater power. His path was common to that of many Italians and most intellectuals. The historical judgment on their attitude towards Fascism is, however, unanimous by now: their initial suspicion towards the innovations introduced by the "black shirts" and their squad methods were soon replaced by a resignation for what seemed inevitable, an acceptance of the "less worst," the sympathetic acquiescence, and the realistic wish to protect their own families. Eugenio Garin, in an interview in 1988, stated: "Our clerics have been very acquiescent. Nobody was saved from Fascism. I say that only those who died, those who ended up in jail, those who went out of Italy were saved; (... The others) genuflected or remained silent." [10]

Like most Italians, intellectuals understood very late the gravity of the situation and the consequences of the attitude held over 20 years. War was the detonator of a new awareness: for some (very few), it was the war in Ethiopia with a conflict still very distant and the seduction of the Empire; for many, it was the outbreak of the Second World War, and all the dramatic events that occurred in the weeks from 25 July to 8 September 1943. Only then the intellectuals started moving away from the regime, with many joining the Resistance. Gustavo Colonnetti, Francesco Tricomi, Giuseppe Zwirner, Ugo Morin, Enrico Magenes, Lucio Lombardo Radice, and the young Mario Fiorentini (who would participate in the organization of the attack in via Rasella) are some of the mathematicians whose life choices anticipated the maturation of a general anti-fascist awareness that then spread through the academic world.

Severi, however, remained close to the Duce, to Gentile, and to Fascism even in the darkest months. In March 1944, in the middle of the civil war, with Italy split in two also geographically, as the allied troops were forcing the German army to retreat and a few months later they would liberate Rome, Severi participated in Florence at the commemoration by the Accademia d'Italia of Gian Battista Vico on the second centenary of his death. The initial speech of Gentile, who had forcefully reentered the political scene with the so-called *Discorso agli Italiani* ("Speech to the Italians") held in the Campidoglio on 24 June 1943 and who had become President of the Academy, opened with an all-political preamble in which he highlighted "how, as a result of the betrayal, Italy was reduced to a flock without a leader, a displaced multitude without a soul, humiliated and scarred by the stranger, as if the dishonor of a gesture wanted by a few unconscious had canceled 25 centuries of history sparkling with genius, virtue, work and daring. But the voice of Mussolini, who people wanted to believe had disappeared forever with that great Italy that he had created, still echoed through all the districts of Italy and the world, bringing together the dispersed multitude. With the resurrection of Mussolini, a young, loyal, generous, daring Italy was reborn, confident in its own strength, anxious for justice for all." Not many academics participated in this extreme tribute to the regime in Florence but, among them, there was Severi, who did not miss to acknowledge those who continued to raise the banner of Fascism, to praise Hitler as the leader of great Germany, and to invite young people to enlist and to follow Mussolini in his last attempt at resistance. Carlo Alberto Biggini, former rector of the University of Pisa

[10] Ajello, N., Interview to Eugenio Garin, *La Repubblica*, June 16, 1992.

and Minister of National Education in the Republic of Salò, in a moving portrait of Gentile (killed 1 month after the speech, on April 15) recalled how many of those present in Florence could not hold back their tears when the philosopher once again reaffirmed his love for the Fascist homeland for which "if necessary, we want to die, because without it, we would not know what to do with the wreckage of the miserable shipwreck." [11]

7 The Purge

The showdown was near. For Severi, it came with "the purge," the usual procedure with which—in a sudden and often violent transition from one political system to another—the holders of the new power condemn or remove from top positions the elements that have colluded with the previous regime to make sure that they can no longer harm, that they do not have a chance to regroup, and that they somehow pay for their choices. In Italy, the purge following the fall of the Fascist regime was a complex phenomenon for a whole series of legal, administrative, and political reasons. It actually began immediately with the arrival of the allied troops and continued on 25 July 1943 with the defenestration of Mussolini through the vote of the Grand Council of Fascism to end with Togliatti's amnesty in June 1946. It was not an easy operation to reconstruct because, in previous months, Italy had actually witnessed three different purges: one by the Anglo-American army that went up the peninsula from Sicily; one by the Italian Government which progressively extended its power over the peninsula and the liberated territories; and the third one operated in Northern Italy by the partisan forces. The purge was also a complex phenomenon from a legal point of view because the new state wanted to adopt rigorous measures that would definitively leave behind seasons of revenge and arbitrariness. It was thus difficult to define what conduct was liable to punishment. At the end, three levels of judgment were defined: the first instance procedure, the appeal before a central commission, and the last appeal to the Council of State. No wonder then if this time was featured by a flood of decrees and laws that tried to clarify the terms of the purge, also in function of a political situation that was very rapidly changing. The so-called Magna Carta of the purge of 27 July 1944 represented the most organic answer to the legal problems, but it too had to deal with the country still struggling with a civil war and a battle against the German army that affected numerous regions.

For Severi, the purge procedure began very early, in the summer of 1943 with the reorganization commission of the university in Rome preparing the initial report that examined the behavior of the mathematician. The report included a two-faced judgement: while Severi's colleagues had no doubts about his participation in the

[11] The quotations from Gentile's speech in Firenze are taken from Guerraggio, A., *Matematici da epurare*, EGEA, Milan, 2018.

regime's politics, they felt almost embarrassed to accuse a scientist of his level and declared that "the task of judging his political-moral activity was particularly serious and painful." After that, however, the evaluation was clear: Severi had "also expressed, in contrast with his inner attitudes, an open action of propaganda for the Fascist regime; and, after the advent of the Italian Social Republic, he personally attended the first meeting in Florence (March 20, 1944) of the Academy of Italy." [12] The report closed with the proposed removal of Severi from the university but also with the underlining of the "damage that this provision could cause to Italian science, from a purely doctrinal point of view." The report was dated 25 July. On July 31, the Minister of Education removed Severi from teaching as a precaution.

On November 3, the First Instance Commission decided to challenge Severi with two specific charges: "having made repeated manifestations of apology for Fascist politics in numerous conferences" and "with his participation in the solemn political demonstrations called by the Fascist republican pseudo-government for the start of the work of the Accademia d'Italia in Florence, having substantially collaborated the government itself." Severi had to defend himself and he did so with a memorial of about 20 pages that he sent to the first-degree purge commission on November 18, and in which he reconstructed the stages of his scientific and political career.

Apologia of Fascism in conferences held outside the more strictly scientific sphere? But when? It is a false accusation, according to Severi, who explains it to his judges with didactic patience. First of all, "those conferences were all cultural and scientific, never political." Although they contained expressions of support for the regime and appreciation of the Duce's talents, they were "the usual expressions of praxis, which anyone who spoke in solemn public events had to adopt, especially in relation to their own functions, and which must therefore be evaluated in that context." "We all behaved in the same way," Severi insinuated, including many of those who perhaps were then judging him and had only recently changed faction. These were seen, in any case, as venial sins, in which nothing was considered to have a judicial interest, and for which what was happening to him was deemed completely out of place and beyond any reasonable measure. This rhetoric highlighted "characteristics and results of Fascism that, even now, could not be dismissed" and never, even remotely, included "the apologetic extremes of Fascism and, more specifically, the aspects of Fascism that are to be condemned and constitute the worst and most sectarian aspects, i.e. its arrogance and denial of elementary freedoms." "Was it a crime to talk about the ongoing war and hope that it would end with the victory of Italy? This was certainly not an apologia of Fascism: it was only respect for the sacrifice of the people, the fighters, and the dead, a basic duty towards Italy."

More problematic was his presence in Florence at the commemoration of Vico, organized by the Academy of Italy chaired by Gentile. As his appointment had occurred after 25 July 1943, when Mussolini was no longer the Head of

[12] All the quotations of this section are taken from Guerraggio, A., *Matematici da epurare*, EGEA, Milan, 2018.

Government, and it was then attributable to the Republic of Salò, Severi could thus be accused of collaborating with the Germans, like all the officials who followed Mussolini in his last attempts. Also in this case, as for the accusation of apologia, Severi's defense focused entirely on the cultural value of the event in Florence, declaring that he could not refuse the invite he had received given the personal relationship with Gentile who lived in his same building in Rome and who had sent a car to pick him up in Arezzo: "the political connotation that Gentile would have given to the commemoration was not foreseeable and was completely unknown to him."

In general, Severi judged his behavior as irreproachable. His support to the regime was justified on the basis that, at the beginning of the 1930s, the opposition to Fascism was only for those few who saw "the dangers that the regime, in an overwhelming nationalistic exaltation, could reserve for Italy in the future" and that "the many who did not have, in full good faith, the same vision cannot be blamed." The message was clear: if there were so few anti-fascists, in all probability it meant that those who now stood as his judges had been part of the Fascist majority and shared its orientations. If there was a crime, they were accomplices. In any case, Severi (who in the Memorial continues to talk about himself in the third person) broke off all relations with Fascism after 25 July, when the Italian Social Republic was created. Neither in Rome nor in Arezzo, he maintained any contact with the republicans and the German occupation forces.

The verdict closing the First Instance proceedings was reached just before Christmas, on 23 December 1944. For the first accusation—that of apologia of Fascism—the Commission believed that the evidence collected was sufficient to configure an execrable behavior and that the laudable relief action to help civilians in Arezzo could not be considered a sufficiently mitigating factor. The conclusion was clear: removal from the university, dismissal, or dispensation from service. As for the accusation of collaboration with the enemy through the Republic of Salò, some doubts remained, as the justifications put forward by Severi appeared plausible, although it was still disconcerting to think that a personality of his experience was really surprised by the meeting's evident political character. In conclusion, "the Commission declares Professor Francesco Severi guilty of the charge of repeated manifestations of apologia of Fascism and therefore proposes that he be dispensed from the service. It acquits Professor Severi from the other accusation of collaboration with the self-styled republican government."

It is not necessary to be a sophisticated lawyer to appreciate the sentence as weak, leaving—one way or another—many openings for recourse or a change of judgement. By acquitting Severi of the most serious accusation of collaboration with the enemy, the First Instance Commission also reduced the weight of the other accusation (that for apologia), whose gravity was clear in the path that had led to intelligence collaboration with both last fascists and Germans. Severi presented his appeal immediately, on December 30, a week after the First Instance sentence, without many new details added beyond stressing the consequences to which his removal and that of other teachers in the same conditions would lead: "But do you, men of the new Italy, really wish to exclude your best talents from managing a

State that must be rebuilt from the rubble, and rather keep those mediocre who have stayed in shadows? Do you think that it would perhaps be advantageous for the future State, which needs to be rebuilt with so much energy, to expel its morally and technically best elements, and instead keep the timid, the mediocre, the immoral, who in the intellectual milieu of today boast of having betrayed fascism, without having had no need to do it, (...) and who always rushed overhead without ever making a gesture of personal independence?" The second-degree sentence, by the Central Commission, arrived on 9 May 1945. Only 4 months had passed since the appeal—a very short period considering the general slowness of the Italian judiciary. In 1945, however, time ran quickly and the passage from winter to spring also signaled a different political climate. The new orientation of the judges was a clear signal that the wind was changing, at least in Rome and at least in the institutions, and that the more radical positions were giving way to a more understanding attitude towards the behavior of many Italians: "A careful examination, then, of the various apologetic ideas, in said speeches contained, leads to the conclusion that the ideas themselves were dictated not by a specific animus to spread the Fascist propaganda, but by a generic patriotic aim to highlight Italy's contribution in the field of science, culture and work (. . .); he (Severi), yet had allowed his distinguished name, his moral integrity, his high scientific merits and his past as an anti-fascist intellectual, to serve—beyond the good of Italy—the political purposes of the regime. Responsibility all the more serious, since said adhesion, for a personality like that of Severi, actually constituted, for the regime, a notable strengthening; and this cannot remain without sanction (. . .). But only a minor penalty (. . .). This sanction, considering all the circumstances, can only be the minimum; and that is the censorship."

8 The Purge at the Lincei

The acquittal obtained at the end of the purge process allowed the accused greater legitimacy: they had been investigated, their behavior during the 20 years of Fascism was subject to a detailed examination, and in the end they had been acquitted. Therefore, they were right to defend themselves and reject the accusations! At least this was the attitude taken by Severi in his other purge process, which started after the (almost) acquittal sentence of the Central Commission, i.e., by the Accademia dei Lincei which, after it risked being closed and replaced by Academia d'Italia during the Fascist period, wanted to settle all the disputes with the academics that had given up to political power, behaving in an abject and shameful way.

The Academy's purge commission, which saw the participation of Guido Castelnuovo and Benedetto Croce among others, first established that belonging to the Academy of Italy did not in itself constitute a sufficient reason to be expelled from the Lincei. However, it also decided that "having participated in the session held in Florence by the former Academy of Italy, under the presidency of Giovanni

Gentile, constitutes in itself a reason for expulsion, having to be interpreted as a form of adhesion and collaboration with the enemy." [13]

This particularly concerned Severi who had gone to Florence and who, on the immediate eve of the meeting of the Commission that would deal with his case, on 1 June 1945, addressed directly Castelnuovo: "I learned a little while ago that my case should be decided today at the Lincei (...). I have no reason to correct any page of the book of my life and I will certainly not be afraid that my frankness may have aroused any aversion against me that in this place, with people like you, I do not think can find any resonance." Shameful—how to define it otherwise?—is the postscript that closes the letter: "I thought I did something nice for you by not mentioning among the significant examples of my behavior in front of the racial law the interest I showed in 1940 for your book on the origins del Calcolo not to be taken out of circulation, something you nudged me to with a friendly note." Severi sent all the members of the Commission a memorandum with an almost triumphant incipit: "Francesco Severi, acquitted by the Central Purge Commission, is about to resume his office as a university professor!" The memo did not spare jibes and insinuations against the more radical elements of the Commission and those accusing it of anti-Semitism: "Severi's favorite pupil was a Jew, Beniamino Segre, currently a professor at the University of Manchester. Severi helped Segre even while he was in England making sure that Segre was paid his pension wherever he was and that he was allowed to transfer a certain sum to England."

Castelnuovo, however, remained firm in supporting his expulsion, in line with the general position adopted by the Academy, and 2 days later he wrote to his colleague: "Dear Severi, I received your Memorial, but it did not matter. From the very first session, the Commission to which the Government entrusted the unfortunate task of purifying the Academy of Lincei, paying particular attention to the members who had belonged to the Academy of Italy, unanimously decided to exclude all academics who had taken part in the session in Florence in March 1944 (...). However, I hope that in the future, when the partisan issues—painful but inevitable after the catastrophe to which the fascist regime has brought the country—will be over, the Academy will be able to embrace again, with new elections, those among the now excluded members who will be considered more deserving thanks to their scientific work." Castelnuovo's wish will soon come true. The Lincei yearbook reports next to Severi's name "Expelled through D.L. on January 4, 1946. Re-elected as National Member from July 15, 1948." Severi had thus to wait a little more than 2 years to finish paying the price for his relationship with Fascism.

[13] The quotes of this section too are taken from Guerraggio, A., *Matematici da epurare*, EGEA, Milan, 2018.

Fabio Conforto (1909–1954): His Scientific and Academic Career at the University of Rome

Maria Giulia Lugaresi

Abstract The mathematician Fabio Conforto (1909–1954) played a significant role in the framework of the historical research about the Italian school of algebraic geometry in the decades before, during, and after WWII. The 1930s of the twentieth century represent the year of Conforto's scientific maturation. During the 1940s, Conforto, besides the difficult war experience, achieves his national and international success. He was professor of analytical and descriptive geometry at the University of Rome, he kept lectures at INdAM, and he was one of the most brilliant collaborators of Mauro Picone inside the INAC.

Unfortunately, Conforto's untimely death suddenly interrupted his professional career. In this work, we want to outline a scientific biography of Conforto in the light of his varied mathematical production.

Keywords Fabio Conforto · Italian school of algebraic geometry · Collaboration with INAC and Mauro Picone · Lectures at INdAM

1 Introduction

> Fabio Conforto gave a lot to his students, university, and science, with his deep and many-sided work in the most different fields of scientific research and teaching.[1]

[1] *Molto ha dato Fabio Conforto ai suoi allievi, alla scuola, alla scienza, con la sua intensa e multiforme opera nei campi più diversi della ricerca scientifica e dell'insegnamento.* This is the testimony of Mario Rosati (1928–2018), who was a student of Conforto. Rosati commemorated his teacher on the occasion of the donation of the personal library of Conforto to the Library of the Department of Mathematics and Computer Science of the University of Ferrara (November 2008). See [72]. Rosati graduated with Conforto in Rome in July 1950 defending a thesis about "projective geometry of the abelian varieties." Rosati was assistant to the chair of geometry. He

M. G. Lugaresi (✉)
University of Ferrara, Ferrara, Italy
e-mail: lgrmgl@unife.it

G. Bini (ed.), *Algebraic Geometry between Tradition and Future*,
Springer INdAM Series 53, https://doi.org/10.1007/978-981-19-8281-1_4

Fabio Conforto has been one of the most relevant figures of the Italian mathematical landscape in the difficult historical period that goes from the 1930s to the beginning of the 1950s of the twentieth century. Besides the educational activity inside the University of Rome, Conforto carried out a significant research activity in the field of algebraic geometry but also of the study of applied mathematics. He was in fact one of the collaborators of Mauro Picone (1885–1977) at the National Institute for the Applications of Calculus (Istituto Nazionale per le Applicazioni del Calcolo, from now on INAC) and one of the professors of the National Institute of Higher Mathematics (Istituto Nazionale di Alta Matematica, from now on INdAM), managed by Francesco Severi (1879–1961). During his life, Conforto collected a wide selection of books and scientific papers that ranged from pure to applied mathematics, without neglecting works about logics, philosophy, and foundations. Conforto's collection of books and papers reflects his multiple interests and represents a useful instrument for reconstructing his scientific biography.[2] In this work, we are going to point out the most relevant events of Conforto' short but intense academic career in Rome.

2 Early Career of Conforto

Fabio Conforto was born on 13 August 1909 in Trieste, which at that time was part of the Austro-Hungary Empire. Soon after his birth, his family moved to Vienna. Conforto's parents were Italian speaking, so at home, the young Fabio spoke Italian. When he began attending elementary school, he learnt to speak, read, and write in German. His bilingualism would be very useful for Conforto after some years.[3]

Conforto spent the first 10 years of his life in Vienna. After the end of WWI, Conforto's family returned to Trieste, where Fabio completed his secondary education. He attended the Gymnasium, and then he completed his schooling at scientific high school (*Liceo Scientifico*), undertaking the work of the final 2 years in a single year and taking the State Examination 1 year early.

In 1926, Conforto entered the Faculty of Engineering at the Polytechnic University of Milan. Oscar Chisini, who was one of his teachers at the Polytechnic, immediately recognized Conforto's mathematical inclination and suggested him to

also attended the lessons given by Severi and Conforto to INdAM. In 1966, Rosati won the chair of geometry in Padua.

[2] In 2008, Conforto's heirs gave the collection of books and papers, belonged to Fabio Conforto, to the Library of the Department of Mathematics and Computer Science of the University of Ferrara. This is a particularly rich collection: it contains more than 700 books and about 2600 scientific papers. The volumes of the *Fondo Conforto* have been catalogued and can be found in the national OPAC catalogue. The cataloguing work of the scientific papers is still in progress.

[3] Conforto's biographical references can be found in [3, 6], and [73].

devote himself to pure mathematics, instead of practical applications.[4] So Conforto was discharged from the Milan Polytechnic in order to go to the University of Rome, and in the autumn of 1928, he and his family moved to Rome, which was going to become his second "native land."[5]

3 Fabio Conforto in Rome

When Conforto arrived in Rome, the city was one of the most important centers as regards mathematics. Thanks to the presence of mathematicians like Guido Castelnuovo (1865–1952), Federigo Enriques (1871–1946), Vito Volterra (1860–1940), and Tullio Levi-Civita (1873–1941), the program of the so-called *Rome of scientists*—conceived since 1870 by Luigi Cremona (1830–1903) and Quintino Sella (1827–1884)—was going to be pursued, despite the difficulties. Castelnuovo was called to Rome in 1891 to become full professor. At the beginning of his Roman period, he met Enriques, who had just graduated at the *Scuola Normale Superiore* in Pisa and had moved to Rome to attend Cremona's courses. Castelnuovo oriented Enriques' study toward algebraic surfaces.[6] After the end of WWI, the Italian school of algebraic geometry was growing, and Rome represented the main site.[7] Besides Castelnuovo, the leading representatives were Enriques and Severi, who were called to Rome in 1921.[8]

Conforto's scientific education was carried out in Rome since the end of the 1920s. In the autumn of 1928, Conforto entered the University of Rome where he

---

[4] Oscar Chisini (1889–1967) graduated in Mathematics at the University of Bologna in 1912 under the supervision of Federigo Enriques. Chisini, who has obtained the lecturing post in 1918, was professor of algebra and analytical and projective geometry at the University of Milan since 1925. During his stay in Milan, Chisini also held the chair of analytical and projective geometry at the Polytechnic University. See [66].

[5] In Conforto's dossier that is preserved at the Archive of Milan Polytechnic, we can read: *Ingegneria Congedato in data 23 novembre 1928 per Regia Università di Roma*. See http://www.archimistaweb.polimi.it/groups/Polimi-ArchiviStorici.

[6] See [7, p. 25] and [69].

[7] In the first decades of the twentieth century, many students (both from Italy and from abroad) came to Rome in order to study mathematics. Thanks to the Rockefeller Foundation, many foreign students came to Italy as scholarship. In September 1924, funded by a Rockefeller fellowship, Dirk Struik (1894–2000) spent 9 months in Rome, working with Levi-Civita. Oscar Zariski (1899–1986) took a Rockefeller fellowship in Rome during the years 1925–1927. Alexander Weinstein (1897–1979) was Rockefeller bursar in Rome in 1926–1927, Harald Geppert (1902–1945) in 1928–1929, and Paul Dubreil (1904–1994) in 1930–1931 (with Enriques). See [77, p. 200] and also [68, p. 102].

[8] Enriques spent more than 20 years (1894–1921) in Bologna, where he was called to teach projective and descriptive geometry. Then he moved to Rome to teach complementary mathematics and, since 1923, also higher geometry. Enriques' period in Bologna is described in [61]. In 1921, Severi arrived in Rome to teach in the chair of algebraic analysis.

attended lectures by Tullio Levi-Civita, Guido Castelnuovo, and Federigo Enriques. Conforto was particularly influenced by the lectures of higher mechanics of Vito Volterra.

Following Volterra's theories, Conforto's first works, appeared in 1930 and 1931, dealt with absolute differential calculus in a continuous function space. The three works, prepared under Volterra's guidance, were presented by Volterra in the reports of the Accademia dei Lincei.[9] In the period July–September 1930 and 1931, Conforto attended the cadet school in the city of Lucca. Here he met the physicist Gilberto Bernardini (1906–1995), who was assistant of Giuseppe Occhialini (1907–1993) in the chair of physics at the University of Florence. Conforto graduated in pure mathematics at the University of Rome *summa cum laude* on 3 July 1931.[10]

Soon after the degree, Conforto obtained a scholarship of £8000, supported by the National Research Council, in order to spend a period of study at the University of Göttingen, in Germany. Conforto stayed in Göttingen from January to June 1932. Thanks to his bilingualism (Italian and German), he could benefit and be influenced by two different mathematical ways of study.[11] In Göttingen, Conforto could deepen his research in the field of geometry in a continuous function space. In this subject, he prepared a complete exposition of the absolute differential calculus in continuous function space.[12] Conforto could also devote himself to items related with physical mathematics. He dealt with the problem of impulses in elastic isotropic bodies.[13]

When he came back to Italy, Conforto fulfilled military obligations, as second lieutenant of artillery, and he was attached to the 13th Artillery Regiment and stationed in Orvinio, near the city of Rieti, during the months of July and August 1932.

The 1930s of the nineteenth century marked the beginning of Conforto's scientific growth. After a further period of study and training at the *Fondazione Beltrami*, since the academic year 1933–1934, Conforto began his teaching activity at the University of Rome. He initially was temporary assistant (*assistente incaricato*) of Guido Castelnuovo in the chair of analytical and descriptive geometry. Then, in October 1934, Conforto competed for an assistant position for the chair of analytical

---

[9] See [9, 10], and [11].

[10] *Il sottoscritto ha conseguito nella sessione di luglio dell'anno 1930–1931 presso la Regia Università di Roma la laurea in Matematica pura, ottenendo il massimo dei punti e la lode*. This testimony was given by Conforto in a letter (Rome, 10 November 1931) to the Mathematical Commission of the National Research Council (CNR). See http://media.accademiaxl.it/pubblicazioni/Matematica/link/conforto.pdf.

[11] In some letters written to Enrico Bompiani, Conforto described the rather different approach to mathematics in Germany as compared to Italy. These letters are kept in the *Fondo Bompiani* of the *Accademia Nazionale delle Scienze* (*Accademia dei XL*) in Rome and can be freely consulted online. See http://media.accademiaxl.it/pubblicazioni/Matematica/link/conforto.pdf.

[12] The paper was published in the Annals of the Normale Superiore di Pisa, Science Class. See [14].

[13] Thanks to Tullio Levi-Civita, Conforto published the results of his discoveries in the reports of the Accademia dei Lincei: see [12] and [13].

and projective geometry. Conforto won the competition and gained the assistant position. He was tenured assistant (*assistente di ruolo*) in the chair of analytical geometry from the academic year 1934–1935 to the academic year 1938–1939. Conforto's educational activity was always very appreciated, as it is confirmed in the report of the Exam Commission: "Brilliant and very precise speaker, completely familiar with the subject, he was able to confidently guide students' knowledge."[14]

After Castelnuovo's retirement (1935), Conforto became Enrico Bompiani (1889–1975)'s assistant in the same chair. In Rome, Conforto was also in touch and strongly influenced by the geometers Federigo Enriques and Francesco Severi. Thanks to Enriques and Severi, Conforto approached algebra and algebraic geometry, which represented the new mathematical research fields.[15]

In January 1936, Enriques proposed Conforto as volunteer assistant for the chair of higher geometry. After having attended the lessons given by Enriques, Conforto collected them in the book *Le superficie razionali*, which was published some years later in 1939.[16] The handbook was intended for university students of mathematics, and it can be considered a sort of "natural continuation" of the handbook *Teoria geometrica delle equazioni e delle funzioni algebriche* [63] by F. Enriques and O. Chisini, which is often quoted.[17] Conforto took inspiration from Enriques' lessons of higher geometry but also from the classical theorems by Luigi Cremona, Alfred Clebsch (1833–1872), and Eugenio Bertini (1846–1933), revised according to a more modern point of view.[18] As regards Conforto's original contribution, it dealt with some researches that he had published some years before and that he used for the writing of the handbook. He referred to two works about Halphen pencils (*fasci di Halphen*): one note about surfaces common to two families of rational quartics with double points, discovered by Noether (*superficie comuni alle due famiglie di quartiche razionali con punto doppio, scoperte da Noether*), a critical addition to the reduction of rational double planes (*complemento critico portato alla riduzione dei piani doppi razionali*), and one note about the bisection of the canonical series over entities of fourth genre (*la nota sulla bisezione della serie canonica sopra gli enti*

[14] *Durante il periodo predetto il Dott. Conforto ha svolto il suo lavoro con intelligenza, capacità e zelo. Espositore brillante e chiarissimo, perfettamente padrone della materia, di modi signorili che riflettono l'elevatezza dei suoi sentimenti, ha saputo ogni anno coltivarsi la simpatia delle numerose classi di studenti e guidare con sicurezza la preparazione.* Members of the Exam Commission were Guido Castelnuovo, Enrico Bompiani, and Mauro Picone. See Archivio Storico Università La Sapienza, *Fascicolo personale di Fabio Conforto.*

[15] Some problems related to the foundations of algebraic geometry began to be taken into account, especially by Severi. See [7, pp. 41-55].

[16] The handbook was republished in 1945 with the title *Le superficie razionali nelle lezioni del Prof. F. Enriques*. See [21] and [25].

[17] The work, whose first draft dated back to 1915, consisted of four volumes. It collected the main results in the theory of algebraic curves, presented from various points of view. See [7, p. 103].

[18] Another volume completed Enriques' trilogy of treatises devoted to the exposition and systematization of algebro-geometric knowledge, i.e., *Le superficie algebriche* [62], that was published posthumously in 1949. Such treatise reworked and extended an earlier edition compiled by Enriques in collaboration with Luigi Campedelli (1903–1978). See [7, p. 103].

di genere quattro).[19] Inside the handbook, at the end of some chapters, the reader can find wide and useful historical notes that prove Conforto's interest for historical aspects in the teaching of mathematics.[20] As regards the classification of the fourth-degree surfaces, he quoted international but also Italian works, starting from the second half of the nineteenth century: A. Cayley (1864), L. Cremona (1868), A. Clebsch (1868, 1870), M. Noether (1871), G. Darboux (1873), C. Segre (1884), G. Castelnuovo (1894), H. Mohrmann (1923), and G. Gherardelli (1936).

3.1 Conforto at INAC

During the 1930s, Conforto began to collaborate with INAC. The Institute, originally founded by Mauro Picone in Naples (1927), was later moved to Rome, where it began its scientific activity in October 1933.[21]

> The Institute is a scientific institution able to subsidize the experimental sciences and the technique, in the quantitative mathematical analysis of their problems. It carries out its research aimed at perfecting or creating mathematical analysis methods that respond to the fulfillment of the aforementioned task. Upon request, it provides study, collaboration and consultancy services for mathematical investigations in various applications, including industry.[22]

At the Institute, mathematical applications were studied in order to try to solve concrete problems, i.e., problems about aerodynamics, elasticity, and building science. As a consultant of the Institute, Conforto was involved in such kind of studies, also in collaboration with other researchers. Together with Tullio Viola (1904–1985), Conforto dealt with the numerical solution of a seismology problem, which was proposed by the geodynamic observatory of Padua (1936).[23] Other topics were the vibration of aircrafts (1937)[24] and the elastic deformation of a

[19] Conforto [21, p. VIII]. The papers quoted by Conforto are [15–17, 19], and [20].

[20] In the handbook, it's clear Enriques' influence on Conforto, not only for the writing of the handbook but also for the importance given by Conforto to the history in the teaching of mathematics. According to Enriques, historical notes have an important didactic role because they represent an instrument that helps students to understand better the origin of mathematical disciplines but also the main concepts and their genesis. As a consequence, history of mathematics is not to be seen only as a rigid succession of events but also as a key to understand achievements. See [65].

[21] In 1932, Mauro Picone was called at the University of Rome in the chair of higher analysis, in the place of Vito Volterra, who refused the oath of allegiance to fascism.

[22] *L'Istituto per le Applicazioni del Calcolo è un organo scientifico atto a sussidiare le scienze sperimentali e la tecnica, nell'analisi matematica quantitativa dei loro problemi. Compie ricerche proprie rivolte al perfezionamento od alla creazione di metodi di analisi matematica rispondenti all'adempimento del sopradetto compito. Fornisce, su richiesta, opera di studio, di collaborazione e di consulenza, per le indagini matematiche nelle varie applicazioni anche industriali.* See [67, p. 89].

[23] Conforto and Viola [59].

homogeneous and isotropic dihedron (1941).[25] This last research solved a problem related to the fixing of great lenses, which was proposed by the National Optical Institute of Florence.

One of the Institute's activities was the preparation of a handbook for the calculus of a continuous flexed beam subject to an axial thrust. The handbook, written by Conforto jointly with Lamberto Cesari (1910–1990) and Carlo Minelli (1898–1954), was prepared for aeronautical constructions.[26] All these works, despite the evident theoretical part, gave also important practical contribution and proved the close connection between technology and pure science.

3.2 *The Scientific Relationship with Francesco Severi*

When, in the autumn of 1936, Conforto obtained a lecturing post (*libera docenza*) in analytical geometry with elements of projective and descriptive geometry with drawing at the University of Rome, he was going to almost definitively orient his scientific interest toward algebraic geometry. This scientific tendency was stimulated by the presence at the University of Rome of Guido Castelnuovo, Federigo Enriques, and, most of all, Francesco Severi, to whom Conforto got closer.

In those years, Severi became the main reference point for Italian school of algebraic geometry. Conforto attended the lessons about algebraic varieties given by Severi from the a.a. 1936–1937 until 1940–1941, first at the University of Rome and later at INdAM.[27]

During the 1930s of the twentieth century, Conforto's academic career developed along different directions, not only inside the University of Rome but also at INdAM. Since 1939, Conforto was regular professor of analytical and descriptive geometry at the Faculty of Science of the University of Rome, together with Enrico Bompiani.[28]

From 1939, the year of its foundation, to 1953, Conforto taught many courses at INdAM on different topics about algebraic geometry: abelian functions, geometry of

[24] On this subject, Conforto published a paper, together with Carlo Minelli: [52]. Minelli was the link between the INAC and the Ministry of Aeronautics. Another work about the aircraft stress is [18]. These works proved Conforto's collaboration with the Ministry of Aeronautics.

[25] Conforto [23].

[26] Cesari et al. [8]. See [67, p. 120].

[27] All these lessons were collected by Conforto and Enzo Martinelli (1911–1999) and published in a volume: [74]. The treatise dealt with a systematic exposition of the theory of series of equivalence. The first volume appeared in 1942, and the other two volumes were published in 1958 and 1959.

[28] In his academic career, Conforto kept also the chair of history of mathematics (1938–1939), number theory (1940–1941, 1944–1945, 1945–1946, 1946–1947, 1947–1948), and topology (1950–1951, 1951–1952, 1952–1953, 1953–1954).

algebraic surfaces, and abelian modular functions.[29] Severi positively remembered Conforto's collaboration to the courses of INdAM.

> Fabio Conforto was one of our best colleagues at INdAM. He in fact began his collaboration at the moment the Institute began its life and its activities in the year 1939–40. Conforto's lessons represented a model of the teaching in an Institute aimed to information and scientific routing. They always contain a personal and re-elaborated version of theories and results of researches as well as recommendation of new problems for the researchers. I [Francesco Severi] remembered the international appeal that were gained by Conforto's courses, not only those that were published by the Institute, but also those, like a course about the theory of the automorphic functions, that have never been published.[30]

Conforto published a lithographed edition of his lessons about abelian functions and Riemann matrices (1942). The book *Funzioni abeliane e matrici di Riemann*, which reproduced Conforto's lessons held at INdAM in 1941, is divided into two parts: in the first one, the author developed the theory of abelian functions, generalizing a theorem stated by Appell in 1891. In the second chapter, Conforto established the links with algebraic geometry.[31] Another later volume of lessons on abelian modular functions (1951) was edited by his pupil Mario Rosati. In his exposition, Conforto followed on from previous development made by Carl Ludwig Siegel (1896–1981) and provided a reconstruction of all the subjects known at the time.[32]

[29] Francesco Severi founded the National Institute of Higher Mathematics (INdAM) in 1939 (Legge 13 luglio 1939, n. 1129). Many students came to Rome in order to attend the lessons given at the Institute. See [71]. Conforto kept his lectures from 1939 to 1953, with some breaks due to WWII or to other academic tasks. The topic of his lectures were algebraic parametrically representable surfaces (a.a. 1939–1940), abelian functions and Riemann matrices (a.a. 1940–1941), particular classes of abelian functions (a.a. 1941–1942), issues related to the theory of abelian functions and to the rationality of varieties (a.a. 1942–1943), introduction to the theory of abelian functions (a.a. 1944–1945), geometry of algebraic surfaces (a.a. 1946–1947), abelian modular functions (a.a. 1950–1951, 1951–1952, and 1952–1953).

[30] *Fabio Conforto fu il primo fra i migliori dei nostri collaboratori all'Istituto di Alta Matematica. Egli infatti iniziò avanti di ogni altro la sua collaborazione con noi nel momento stesso in cui l'Istituto, nell'a. 1939–40, cominciò la sua vita e la sua attività. I corsi che egli professò presso l'Istituto di Alta Matematica costituiscono un modello di quello che può essere un insegnamento in un Istituto come il nostro, che ha un fine di informazione e di pretto istradamento scientifico. Essi contengono sempre una rielaborazione del tutto personale delle teorie e dei risultati di ricerche, che si può dire sono state fatte gradualmente nella immediata precedenza della esposizione, nonché indicazione dei problemi nuovi che si affacciano per i ricercatori. Ricordai [...] la risonanza internazionale che hanno avuto questi corsi. Ma non sono soltanto quelli che tutti possono conoscere attraverso le pubblicazioni dell'Istituto, giacché alcuni di pur notevole interesse, come un corso sulla "Teoria delle funzioni automorfe," Conforto non ha avuto mai la opportunità di pubblicarli* [6, p. 217].

[31] Conforto [24]. A review of the handbook can be found in [5]. The topic was broadly expanded in the work [49].

[32] Conforto [43]. The handbook *Funzioni abeliane modulari* collected the lessons given by Conforto at INdAM in the a.a. 1950–1951 and 1951–1952. The book was reviewed by Aldo Andreotti in [1]. Conforto would have liked to continue his editorial project with a second volume, but his premature death stopped it.

4 Conforto and WWII

The Italian political-military events soon intersected with the personal and professional ones of Conforto. After the outbreak of WWII (1 September 1939), Conforto was very soon called to arms with the rank of Lieutenant of Artillery and assigned to the First Regiment of Artillery of Foligno (16 November 1939). On 1 December 1939, he received the news that he won the academic competition for the chair of extraordinary professor of analytical geometry with elements of projective and descriptive geometry with drawing at the University of Rome. The chair was previously held by Gaetano Scorza (1876–1939).[33]

> Conforto's scientific work shows an original researcher, gifted with open-mindedness and suitable abilities to put them in place. The matters he poses himself have always a real interest and are always brilliantly carried out. [...] We can also add Conforto's deep knowledge with the most recent and highest fields of algebraic geometry, created by Severi, and the sharpness with which he knows how to pass from geometric matters to others of mathematical physics and analysis, bringing a precious set of knowledge that let him achieve the proposed purpose. Conforto's excellent teaching skills are known.[34]

Soon after the winning of the academic competition, Conforto was sent on leave on request of the University of Rome. There were two biennial courses of analytical and projective geometry at the University of Rome, whose regular teachers were Fabio Conforto and Enrico Bompiani. During the month of November 1939, Bompiani held both the chairs, because of Conforto's military task. Since 7 December 1939, Conforto obtained an extraordinary license of 30 days in order to carry on his university course. Considering the importance of this teaching, which was aimed at about 500 students, the Rector Pietro De Francisci asked later for an unlimited license for the rest of the academic year.[35] Conforto's teaching activity both at the

---

[33] The judging board of the academic competition for the chair of analytical geometry in the University of Turin was composed of the professors Francesco Severi, Enrico Bompiani, Giuseppe Marletta, Nicolò Spampinato, and Renato Calapso. The works of the commission took place at the University of Rome between the end of October and the beginning of November 1939. There were 12 candidates: Maria Ales, Giorgio Aprile, Achille Bassi, Pietro Buzano, Ugo Cassina, Fabio Conforto, Antonino Lo Voi, Ugo Morin, Margherita Piazzolla Beloch, Luigi Tocchi, Tullio Turri, and Mario Villa. The grade triplet was composed of Fabio Conforto, Mario Villa, and Margherita Piazzolla Beloch.

[34] *La produzione del Conforto rivela in lui un ricercatore originale, dotato di larghe vedute e di mezzi adeguati a tradurle in atto. Le questioni che si pone hanno sempre un effettivo interesse e sono sempre nettamente condotte a termine. [...] A ciò si aggiunga la profonda dimestichezza del Conforto con i rami più recenti e più elevati della geometria algebrica creati dal Severi; e l'agilità con cui il Conforto sa passare da questioni di geometria ad altre, di fisica matematica e di analisi, portandovi un prezioso corredo di conoscenze che in ogni caso gli consentono di raggiungere lo scopo propostosi. Le ottime attitudini didattiche del Conforto sono note.* See [70, p. 1684].

[35] Some documents, preserved at the Archive of the University of Rome, provide useful information about Conforto's calls to arms and academic licenses. The letters sent by the Rector of the University Pietro De Francisci to the War Minister (5 December 1939) and the Director of Mathematical School Enrico Bompiani to the Personnel Office of the University of Rome (7 December 1939) highlighted the importance of Conforto's teaching activity. *All'inizio delle lezioni,*

University and at INdAM was very relevant and appreciated. That's why he was often asked for military leave.

Despite military commitments, in November 1942, Conforto attended the Mathematical Congress of Rome, which took place between 8 and 12 November, when the fighting of WWII was at its most intense. The Congress was promoted by INdAM, organized and chaired by Severi, who held the opening lecture about contemporary mathematics and mathematicians (*Matematica e matematici d'oggi*). Although described as an international congress, only a few specially invited foreigners attended the Congress. Among the 137 participating mathematicians, there were 17 foreigners, who came from "not enemy" nations.[36]

One month later (on 1 December 1942), Conforto was appointed full professor in the chair of analytical geometry (with elements of projective and descriptive geometry) at the University of Rome.[37]

Military tasks slowed down Conforto's scientific activity. During the years 1942–1943, he wrote a joint paper together with Annibale Comessatti (1886–1945).[38] Because of WWII, Conforto was unable to fully practice his university

il 6 novembre 1939-XVIII, il Prof. Bompiani si assunse il gravoso incarico di iniziare anche il corso del Prof. Conforto; a partire dal 7 dicembre [1939] il corso fu continuato dal Prof. Conforto stesso, il quale ottenne una licenza straordinaria di giorni 30, con scadenza 6 gennaio 1940-XVIII. Quando le lezioni ricominceranno, dopo le vacanze natalizie, questa R, Università sarà costretta a sospendere il corso di Geometria Analitica del Prof. Conforto, se non potrà disporre del detto insegnante. Invero non può pensarsi di affidare continuativamente il corso del Prof. Conforto al Prof. Bompiani, il quale trovasi già gravato da molti altri insegnamenti. [. . .] Bisogna infatti tener presente che l'insegnamento è rivolto contemporaneamente a circa 500 allievi ingegneri e che si tratta di uno dei corsi di importanza capitale per la preparazione scientifica dei futuri ingegneri. Letter of the Rector De Francisci to the War Minister (2 December 1940). See Archivio Storico Università La Sapienza, *Fascicolo personale di Fabio Conforto.*

36 The invited foreign mathematicians came from Germany (Wilhelm Blaschke, Constantin Caratheodory, Helmut Hasse), Bulgaria (Nicola Obreshkoff, Kiril Popoff, Ljubomir Chakaloff), Romania (Gheorghe Galbura, Miron Nicolescu, Gheorghe Vranceanu, Grigore Constantin Moisil), Norway (Poul Heegaard), Sweden (Torsten Carleman), Croatia (Rudolf Cesarec), Switzerland (Rudolf Fueter, Andreas Speiser), and Hungary (Bela Kerekjarto). The list of all the participants together with the lectures of the speakers were published in the proceedings of the Congress: *Atti del Convegno matematico tenuto in Roma dall'8 al 12 novembre 1942* [2]. Conforto gave a talk about the theory of the systems of equivalence and of the correspondences between algebraic varieties. See [26].

37 *Con Decreto in corso di registrazione, il professore Fabio Conforto è nominato ordinario della Cattedra di Geometria analitica con elementi di proiettiva e geometria descrittiva con disegno di codesta Regia Università a decorrere dal 1 dicembre 1942 XXI.* See Archivio Storico Università La Sapienza, *Fascicolo personale di Fabio Conforto.*

38 Conforto and Comessatti [51]. Comessatti graduated at Padua in 1908 under the guidance of Severi. In 1920, he was appointed professor of algebraic analysis and analytical geometry at the University of Cagliari. In 1922, he moved to Padua, where he stayed until his death in 1945.

tasks because he had to alternate between military service and the studies aimed at military applications at INAC. During the first part of 1943, Conforto again worked at the University of Rome, at INdAM, and at INAC. In this period, Picone's Institute was undertaking military applications as part of the war effort. In July 1943, Conforto went to Germany together with Picone. The journey was aimed to develop scientific relationship with German mathematicians.[39] Conforto and Picone visited some important German universities: Jena, Berlin, Hamburg, Heidelberg, Darmstadt, and Braunschweig. They were in touch with the main mathematicians of those universities.[40] Picone hoped for collaboration between INAC and the German institute of Braunschweig.[41]

4.1 *Difficult Years: Rome, Reggio Calabria, Lecce*

Meanwhile in Italy the political situation was quickly changing. On 9 July 1943, Allied forces invaded Sicily. The Allied landing on mainland Italy took place on 3 September 1943 in Reggio Calabria and on 9 September in Salerno on the western coast. As regards the Roman situation, on 19 July 1943, Rome had been bombed by Allied planes, causing many damages also to University structures. On 25 July, Mussolini was deposed. On 8 August, Conforto decided to volunteer for military service. He was firstly assigned to the Artillery Regiment in Foligno, and then, on

[39] Since some years, Picone hoped for a collaboration with the German armed forces. His main interlocutors were his pupil, Wolfgang Gröbner (1899–1980), and the German mathematician Gustav Doetsch (1892–1977). But still in 1942, Picone's project wasn't realized. Soon after the Mathematical Congress of Rome, Picone was invited for some conferences in Germany. Picone, who didn't speak German, asked Conforto for coming with him. See [67, pp. 132-133].

[40] From 1934 to 1943, Friedrich Karl Schmidt (1901–1977) and Robert König (1885–1979) represented pure mathematics in Jena. Helmut Hasse (1898–1979) was professor in Göttingen. From 1939 until 1945, Hasse was on war leave from Göttingen. He was an officer in the German Navy and he worked in Berlin on problems in ballistics. At the beginning of WWII, Hermann Schmid (1908–1986), pupil of Hasse, undertook a short spell of military service before moving to the University of Berlin in 1940 as an assistant to Harald Geppert. Wilhelm Blaschke (1885–1962) was professor at the University of Hamburg since 1919. He had a leading role in German mathematics until the end of WWII. In 1942, Blaschke made a lecture tour of Italy, and in November, he attended the Mathematical Congress in Rome. Hans Zassenhaus (1912–1991) was professor at Hamburg. After 1940, because of his refuse to join the Nazi Party, he resigned and joined the navy, working on weather forecasting during the remainder of WWII. Alwin Walther (1898–1967) was full professor at Technical University of Darmstadt since 1927 and director of the Institute for Applied Mathematics, which he built. Walther paid great attention to the practical application of mathematics, especially for the engineers. Wilhelm Schlink (1875–1968) was appointed assistant of mechanics at Darmstadt Technical University in 1900 and habilitated there for mechanics in 1903. In 1907, he was appointed associate professor and in 1908 full professor at Braunschweig Technical University. In 1921, he accepted a call to Darmstadt Technical University and worked there until 1949 when he retired.

[41] During WWII, Braunschweig was home to the Aeronautical Research Institute (Luftfahrtforschungsanstalt, LFA), a secret facility for airframe, aeroengine, and aircraft weapon testing.

16 August, he was moved to Reggio Calabria, where he joined the Italian Army. After the Allied landing in Reggio Calabria (3 September 1943), Conforto was taken prisoner. Because of war events, for 10 months, he had no contact with his family in Rome.

After the Armistice (signed on 8 September 1943), Conforto was released and sent to Lecce, where he worked for the Minister of War and at the Military Academy.[42] Conforto taught descriptive geometry and rational mechanics. During this period, he also collected and published the lectures he gave at the Military Academy.[43] On 4 June 1944, Rome fell to the Allies and Conforto could contact his family. In August, he was able to return to Rome and be reunited with his family. On 20 December 1944, the period of military leave ended. Conforto came back to his civil life, and he could resume his research and teaching activity on an ongoing basis.[44] During the years 1946 and 1947, he was also particularly active on the editorial side, publishing many textbooks for university and secondary school, often in collaboration with academic colleagues.[45]

In 1948, Conforto oriented his research interest toward quasi-abelian functions, soon after the publication of the work by Severi (*Funzioni quasi abeliane* [75]). Conforto's idea was to create for the quasi-abelian case a theory that was similar to that of Riemann matrices for the abelian functions. The paper *Sopra le trasformazioni in sé della varietà di Jacobi ...* contained the first construction of an arithmetical theory of the quasi-abelian functions.[46]

5 Conforto's Scientific Journeys

The years following the end of WWII were particularly important for Conforto's international achievement. He took part in many congresses, and he was invited to lecture both in Italy and abroad. In the summer of 1947, Conforto was sent by CNR to the International Congress on Engineering Education in Darmstadt as Italian delegate. In the section "Mathematics and Physics," presented by Prof.

[42] The Military Academy of Modena resumed its function in May 1944 at the barracks of Pico in Lecce as a Special Commando Royal Military Academy. After the end of the war and the fall of the monarchy, the Military Academy came back in Modena (1947).

[43] The handbook *Nozioni di geometria analitica, proiettiva e descrittiva ad uso degli allievi dell'Accademia militare di Lecce* was written jointly with Emilio Tomassi. See [55].

[44] During the period of WWII, thanks to some leave periods, besides the chair of analytical and descriptive geometry, Conforto kept the chair of number theory in the a.a. 1940–1941. He resumed the teaching of number theory in the a.a. 1944–1945.

[45] The handbooks [27–30, 64], and [31] were prepared for university students. Together with Giuseppe Vaccaro (1917–2004), Conforto wrote some textbooks for secondary school: [56, 57], and [58].

[46] See [32]. A second work dealt with univocal correspondences between points of a Picard quasi-abelian variety: [33]. Other works about the quasi-abelian functions were [37, 38], and [41]. Conforto also gave a talk about the theory of quasi-abelian functions and varieties at the Third Congress of UMI (Pisa, 23–26 September 1948).

Alwin Walther, Conforto gave a talk about the latest contributions gained at INAC, thanks to the researches made by Picone and his pupils, in the field of integration of partial differential equations. Conforto underlined that these new methods required powerful calculating machines for their numerical development. These kinds of machines, recently built in America, were still lacking in Italy.

1950 was probably the year of Conforto's consecration on an international level. Between January and February, he was invited by many European universities to give a series of conferences about algebraic geometry.[47]

At the end of summer 1950, Conforto took part to the International Congress of Mathematicians at Cambridge (Massachusetts, USA). Conforto travelled to the USA together with Beniamino Segre (1903–1977). After the Congress, the two mathematicians spent almost a semester as invited professors at the Institute for Advanced Studies of Princeton (October–November 1950). In a letter of introduction, prepared for Conforto in order to go to the USA, Guido Castelnuovo hoped for collaboration between Conforto and American mathematicians.[48]

> Dr. Fabio Conforto was for a period of many years my Assistant at the University of Rome. He is now full Professor at that University where he teaches the same Course I used to teach before my retirement from the University. He is an excellent teacher, a mathematician with a very large cultural background, an author of very good works in analytical geometry, in analysis (theory of Abel's functions) and in applied mathematics. He has taught very well many courses in the different branches of high mathematics. He is a person which I highly recommend, and I firmly believe that if he could sojourn for some time in an American Institution of high learning, it would be of a very great advantage for establishing more strict

[47] At the University of Basel, Conforto gave a conference about "New researches on the theory of abelian functions." Then he moved to Holland where he spent 2 weeks. At Gröningen University, he held two conferences. The first one dealt with historical item ("Historical evolution of mathematics in Italy"), and the second one was about algebraic geometry of surfaces and still open problems. As regards algebraic geometry, Conforto also presented three short talks on "Spirit and typical methods of Italian algebraic geometry." By invitation of the Mathematisch Centrum in Amsterdam, Conforto gave two conferences: "Some new methods for integration of linear partial differential equation used at INAC" and "Recent researches of algebraic geometry in the field of abelian and quasi-abelian functions." At Leiden University, he presented a historical conference about "Historical evolution and significance of projective geometry," while at Utrecht University, he talked about "The role of continuity in projective and algebraic geometry." In West Germany, Conforto was invited by the universities of Hamburg and Münster to hold conferences on "Irregular surfaces and abelian functions" and on "New researches on the theory of abelian functions." See Boll. Un. Mat. Ital. **(3)**, 5 (1950), n. 1, p. 98.

[48] At Princeton, Conforto worked closely with Carl Ludwig Siegel, who had arrived at the Institute for Advanced Study in 1940. Siegel was appointed to a permanent professorship at Princeton in 1946. In his lectures on functions of several complex variables at Princeton, Siegel developed the theory of abelian functions along the lines laid down by Conforto. Siegel's lectures were published in 1949: [76]. In 1966, Siegel prepared a revised version of his lectures. Through the works of Conforto and Siegel, Frobenius' theory of Jacobian functions had become a basic element in the classical treatment of abelian functions and varieties.

> relations between the work of the Italian and the American mathematicians, and moreover it would be of a great advantage for the progress of our Science.[49]

During the years 1951 and 1952, Conforto travelled extensively in Europe. He took part to international congresses and gave lectures by invitation of some European scientific societies in Austria, Switzerland, Germany, and Belgium.[50] This was a successful period. Conforto was internationally appreciated.[51]

However, back in Italy, Conforto became seriously ill in February 1953. He gave some lectures at INdAM on abelian modular functions. In autumn 1953, his health conditions were getting worse. Even when he was taken to a clinic, he still had his books with him as he struggled to work until the end. Conforto died in Rome on 24 February 1954.[52]

Following Conforto's death, some works based on his lecture notes were published by his colleagues and students. In 1956, the Austrian mathematician Wolfgang Gröbner published a revised version of Conforto's lectures about abelian functions. The volume "Abelian functions and algebraic geometry" [49] was based on lecture notes for courses given between 1940 and 1951. The lecture notes were collected by Conforto's pupils Mario Rosati and Aldo Andreotti.[53]

---

[49] Letter of Guido Castelnuovo, President of the National Lincean Academy. Rome, 20 January 1950.

[50] In March 1951, Conforto and Picone took part to the Congress of the Society for Applied Mathematics and Mechanics (Freiburg im Breisgau). Conforto gave a talk in German about some results of experiments for periodical analysis that were carried out in the INAC. Then Conforto moved to Vienna where he was invited by the Austrian Mathematical Society for two conferences about the relationship between algebraic geometry and abelian functions. At the Swiss Mathematical Society (Neuchâtel, May 1952), Conforto gave a talk about algebraic manifolds that allow birational transformations in themselves (*Uber algebraische Mannigfaltigkeiten, die birationale Transformationen in sich gestatten*). In June 1952, Conforto, together with Aldo Andreotti, Oscar Chisini, and Mario Villa, took part to a conference of algebraic geometry, Belgium Center of Mathematical Research (Liège). Conforto's talk (*Problèmes résolus et non résolus de la théorie des fonctions abéliennes dans ses rapports avec la géométrie algébrique*) was published in the proceedings of the conference [46]. Then Conforto moved to Germany. He was invited by the universities of Hamburg and Berlin for some conferences. In September 1952, Conforto took part to the Third Mathematical Congress of the Austrian Mathematical Society in Salzburg giving a talk on abelian varieties. See Boll. Un. Mat. Ital. **(3)**, 7 (1952), n. 2, pp. 230–232; Boll. Un. Mat. Ital. **(3)**, 7 (1952), n. 3, pp. 357–361.

[51] The University of Hamburg would have liked to appoint him as the successor of Wilhelm Blaschke. See [6, pp. 212, 216].

[52] See [6, pp. 206-207].

[53] Aldo Andreotti (1924–1980) graduated in Mathematics in 1947 at the University of Pisa. He spent the next 3 years in Rome, first as "research student" at INdAM and then as assistant in the chair of geometry, where he could collaborate with Severi. In 1951, he was a visiting scholar at the Institute for Advanced Study at Princeton. In the same year, he won a chair of geometry at the University of Turin, and then in 1956, he moved to the University of Pisa. During the next 20 years, Andreotti often travelled abroad, and he worked and taught both in Pisa and in other foreign universities.

Introduzione alla topologia [50] consisted of the lectures Conforto had given at the University of Rome in the 5 years before his death. The lectures were collected by Mario Benedicty, another student of Conforto.[54]

6 Conclusions

In just over 20 years of scientific activity and despite the prolonged stops due to military service, WWII, and imprisonment, Fabio Conforto published a hundred works about algebraic geometry, algebra, analysis, theoretical and applied mathematics, and history of mathematics. These works were evidence of his extraordinary versatility.

Finally, we cannot fail to mention Conforto's commitment as an editor. Since 1939, he was member of the editorial board of the journal *Rendiconti del Seminario Matematico della R. Università di Roma* (now *Rendiconti di matematica e delle sue applicazioni*).[55] Soon after the end of WWII, Conforto took part in the editorial project of the *Enciclopedia Italiana*. He collaborated with Mario Niccoli (1904–1964), preparing both mathematical and biographical entries. Besides it, Conforto wrote an encyclopedic article for the *Enciclopedia delle Matematiche Elementari*, together with Severi, and an article about the postulates of Euclidean and non-Euclidean geometry in the *Repertorio di matematiche*.[56]

The vastness of Conforto's scientific interests is evidenced not only by many reviews, published in national and international journals, like *Bollettino dell'Unione Matematica Italiana*, *Zentralblatt für Mathematik und ihre Grenzgebiete*, and *Mathematical Reviews*, but also by various historical and educational papers. As regards Conforto's contributions about the history of mathematics, he wrote three collective articles on Italian scientific research and the development of algebraic geometry in Italy.[57] He devoted two papers to Bonaventura Cavalieri and Evangelista Torricelli, their life and scientific works.[58] The centenary of the publication of Riemann's thesis occasioned the publication of a short note.[59]

---

[54] Mario Benedicty (1923–2011), born in Trieste, received his PhD in mathematics from the University of Rome in 1946 and taught mathematics at Pontifical Gregorian University in the Vatican, La Sapienza University of Rome, and the University of British Columbia before going to Pittsburgh in 1958. Benedicty also collected some unpublished notes by Conforto: [4].

[55] The journal was founded in 1913 and started its publication in 1914 under the direction of Vito Volterra. Since 1939, the name of Conforto appeared in the editorial board of the *Rendiconti*. Editors in chief were Enrico Bompiani and Francesco Severi.

[56] See [53] and [45].

[57] Conforto [22], Conforto and Sobrero [54] and Conforto and Zappa [60].

[58] Conforto [34]. The Torricelliana Society of Science and Literature of Faenza was founded in 1948, and Conforto was elected corresponding fellow of the society. During the first congress of the society, on 25 October 1948, Conforto gave a commemorative lecture about Torricelli's life. See [35].

[59] Conforto [44].

Conforto was also interested in issues related to the teaching of mathematics, as confirmed by some articles published in the journal *Archimede*, addressed to teachers and connaisseurs of pure and applied mathematics.[60]

After having recalled Conforto's scientific work, it's evident that

> he never left the idea of a unified vision of mathematical sciences. As far as possible he aimed at connecting the most varied theories, as witnessed by his works.[61]

References

1. Andreotti, A.: Review of: Fabio Conforto, Funzioni Abeliane Modulari, Vol. I. Boll. Un. Mat. Ital. **3**, 8, 460–462 (1953)
2. Atti del Convegno matematico tenuto in Roma dall'8 al 12 novembre 1942, Tipografia Bardi, Roma (1945)
3. Benedicty, M.: Fabio Conforto. Boll. Un. Mat. Ital. **(3)**, 9, 227–228 (1954)
4. Benedicty, M.: Sopra alcuni appunti inediti di Fabio Conforto. Cremonese, Roma (1955)
5. Berzolari, L.: Review of: Fabio Conforto, Funzioni abeliane e matrici di Riemann (Parte prima). Corsi del Reale Istituto di Alta Matematica. Boll. Un. Mat. Ital. **3**, 3, 172–175 (1948)
6. Bompiani, E., Severi, F. et al.: In memoria di Fabio Conforto. Rend. Mat. Appl. **4**, 13, 199–218 (1954)
7. Brigaglia, A., Ciliberto, C.: Italian Algebraic Geometry between the Two World Wars. Queen's University, Kingston (1995)
8. Cesari, L., Conforto, F., Minelli, C.: Travi continue inflesse e sollecitate assialmente. Pubblicazioni dell'INAC, Roma (1941)
9. Conforto, F.: Metrica e fondamenti di Calcolo differenziale assoluto in uno spazio funzionale continuo. Atti Accad. Naz. Lincei Rend. Cl. Sci. Fis. Mat. Nat. **(6)**, 12, 547–552 (1930)
10. Conforto, F.: Parallelismo negli spazi funzionali. Atti Accad. Naz. Lincei Rend. Cl. Sci. Fis. Mat. Nat. **(6)**, 13, 173–178 (1931)
11. Conforto, F.: Formalismo matematico in uno spazio funzionale continuo retto da un elemento lineare di seconda specie. Atti Accad. Naz. Lincei Rend. Cl. Sci. Fis. Mat. Nat. **(6)**, 13, 562–568 (1931)
12. Conforto, F.: Considerazioni sugli impulsi nei corpi elastici isotropi. Atti Accad. Naz. Lincei Rend. Cl. Sci. Fis. Mat. Nat. **(6)**, 15, 130–135 (1932)
13. Conforto, F.: Sugli impulsi nei corpi elastici isotropi. Atti Accad. Naz. Lincei Rend. Cl. Sci. Fis. Mat. Nat. **(6)**, 15, 148–156 (1932)
14. Conforto, F.. Sopra il calcolo differenziale assoluto negli spazi funzionali continui. Ann. Sc. Norm. Super. Pisa Cl. Sci. **(2)**, 2, 309–324 (1933)
15. Conforto, F.: Sui fasci d'Halphen i cui punti base appartengono ad una cubica ellittica degenere. Mem. Accad. naz. Lincei, Cl. Sci. fis. mat. nat. **(6)**, 6, 355–383 (1936)

[60] In [39], Conforto gave a historical report about projective geometry, while the paper [40] was devoted to algebra and the theory of equations. Another work dealt with some notes about real numbers: [47]. Other educational works written by Conforto are[36, 42], and [48].

[61] *Dopo taluni lavori giovanili nel campo della teoria delle funzioni, la sua attività si è volta principalmente alla geometria algebrica, senza però abbandonare mai una visione unitaria delle scienze matematiche, come provano i suoi lavori, numerosi anche in campi più o meno lontani da quello originario, ma sempre intesi a collegare, per quanto possibile le più svariate teorie.* See [3, p. 227].

16. Conforto, F.: Sui tipi cremoniamente distinti a cui si riducono i fasci d'Halphen, che hanno i loro punti base su di una curva ellittica degenere. Rend. Sem. Mat. Roma **(4)**, 1, 120–138 (1936)
17. Conforto, F.: Sulle rigate razionali del quinto ordine. Atti I Congr. Un. Mat. Ital. Firenze, 278–281 (1937)
18. Conforto, F.: Sollecitazioni nei velivoli, provocate da determinati tipi di raffiche. Atti I Congr. Un. Mat. Ital. Firenze, 571–575 (1937)
19. Conforto, F.: Sopra la bisezione della serie canonica su di una curva di genere quattro. Rend. Sem. Mat. Roma **(4)**, 2, 216–223 (1938)
20. Conforto, F.: Sui piani doppi razionali. Rend. del Sem. Mat. dell'Univ. di Roma **(4)**, 2, 156–172 (1938)
21. Conforto, F.: Le superficie razionali. Zanichelli, Bologna (1939)
22. Conforto, F.: Il contributo italiano al progresso della geometria algebrica negli ultimi cento anni. In: Un secolo di progresso scientifico italiano: 1839–1939, vol. 1, pp. 125–153. Società Italiana per il Progresso delle Scienze, Roma (1939)
23. Conforto, F.: Sulle deformazioni elastiche di un diedro omogeneo ed isotropo. Mem. della R. Acc. delle Scienze di Torino **(2)**, 70, 163–232 (1941)
24. Conforto, F.: Funzioni abeliane e matrici di Riemann. Corsi dell'Istituto Nazionale di Alta Matematica (lit.). Libreria dell'Università di Roma, Roma (1942)
25. Conforto, F.: Le superficie razionali nelle lezioni del Prof. F. Enriques. Zanichelli, Bologna (1945)
26. Conforto, F.: Lo stato attuale della teoria dei sistemi d'equivalenza e delle corrispondenze tra varietà algebriche. Atti del Convegno matematico tenuto in Roma dall'8 al 12 novembre 1942, 49–83 (1945)
27. Conforto, F.: Meccanica razionale. Principato, Milano-Messina (1946)
28. Conforto, F.: Lezioni di geometria descrittiva. Principato, Milano-Messina (1946)
29. Conforto, F.: Complementi ed esercizi di geometria descrittiva per il I biennio universitario (lit.). Veschi, Roma (1946)
30. Conforto, F.: Lezioni di geometria analitica per il I biennio universitario (lit.). Veschi, Roma (1947)
31. Conforto, F.: Complementi ed esercizi di geometria analitica per il I biennio universitario (lit.). Veschi, Roma (1947)
32. Conforto, F.: Sopra le trasformazioni in sé della varietà di Jacobi relativa ad una curva di genere effettivo diverso dal genere virtuale, in ispecie nel caso di genere effettivo nullo. Ann. Mat. Pura Appl. **4**, 27, 273–291 (1948)
33. Conforto, F.: Sopra le corrispondenze univoche tra i punti di una varietà quasi abeliana di Picard. Atti Accad. Naz. Lincei Rend. Cl. Sci. Fis. Mat. Nat. **(8)**, 5, 369–375 (1948)
34. Conforto, F.: L'opera scientifica di Bonaventura Cavalieri e di Evangelista Torricelli. In: Atti del Convegno di Pedagogia e Didattica Matematica tra gli insegnanti degli Istituti Tecnici, Pisa, 23–27 settembre 1948, pp. 35–55 (1948)
35. Conforto, F.: Nel terzo centenario della morte di Evangelista Torricelli. Società Torricelliana di Scienze e Lettere di Faenza **1**, 7–23 (1948)
36. Conforto, F.: Vedute unitarie nel campo delle scienze matematiche. Atti del Centro di Studi Metodologici di Torino **1**, 185–187 (1948)
37. Conforto, F.: Sulla totalità delle relazioni generalizzate di Hurwitz di una matrice quasi abeliana. Ann. Mat. Pura Appl. **4**, 28, 299–315 (1949)
38. Conforto, F.: Sulla nozione di corpi equivalenti e di corpi coincidenti nella teoria delle funzioni quasi abeliane. Rend. Sem. Mat. Univ. Padova **18**, 292–310 (1949)
39. Conforto, F.: La geometria proiettiva: suo sviluppo storico e suo significato. Archimede **1**, 7–17 (1949)
40. Conforto, F.: La teoria delle equazioni nell'algebra moderna. Archimede **1**, 217–228 (1949)
41. Conforto, F.: Una proposizione sulle matrici quasi abeliane. Rend. Mat. Appl. **5**, 9, 335–345 (1950)

42. Conforto, F.: L'ordinamento degli studi di geometria per gli allievi ingegneri. Boll. Un. Mat. Ital. **(2)**, 2, 356–358 (1950)
43. Conforto, F.: Funzioni abeliane modulari. Vol. I (Lezioni raccolte dal Dott. Mario Rosati). Corsi dell'Istituto Nazionale di Alta Matematica, Docet, Edizioni Universitarie, Roma (1951)
44. Conforto, F.: Per il centenario della dissertazione di Bernhard Riemann. Experientia, **7**, 5 (1951)
45. Conforto, F.: Postulati della geometria euclidea e geometria non euclidea. In: Villa, M. (ed.) Repertorio di matematiche, vol. 2, pp. 194–224. Cedam, Padova (1951)
46. Conforto, F.: Problèmes résolus et non résolus de la théorie des fonctions abéliennes dans ses rapports avec la géométrie algébrique. In: Deuxième Colloque de Géométrie Algébrique. Centre Belge de Recherches Mathématiques, pp. 90–110. Thon, Liège (1952)
47. Conforto, F.: Alcune considerazioni sui numeri reali. Archimede **4**, 133–142 (1952)
48. Conforto, F.: Razionalità ed intuizione nella matematica. In: Didattica della matematica. Saggi e conferenze a cura del Movimento Circoli della Didattica, pp. 169–174. Angelo Signorelli Editore, Roma (1956)
49. Conforto, F.: Abelsche Funktionen und algebraische Geometrie, Gröbner, W. (ed.). Springer, Berlin (1956)
50. Conforto, F., Benedicty, M.: Introduzione alla topologia. Cremonese, Roma (1960)
51. Conforto, F., Comessatti, A.: Sulla deduzione delle relazioni bilineari tra i periodi di un corpo di funzioni abeliane. Atti del R. Ist. Veneto di Scienze, Lettere e Arti **102**, 541–549 (1943)
52. Conforto, F., Minelli, C.: Indagini sulle vibrazioni dei velivoli. L'Aereotecnica, **24**, 16–17 (1937)
53. Conforto, F., Severi, F.: Caratteri ed indirizzi nella matematica moderna. In: Berzolari, L., Gigli, D., Vivanti, G. (eds.) Enciclopedia delle Matematiche elementari, vol. 3, part II, pp. 753–813. Hoepli, Milano (1950)
54. Conforto, F., Sobrero, L.: I più importanti risultati conseguiti nell'anno XIII E. F. nel campo delle matematiche pure. In: XXIV Riunione della Società Italiana per il Progresso delle Scienze. Palermo, 12–18 ottobre 1935, pp. 18. Societa italiana per il progresso delle scienze, Roma (1936)
55. Conforto, F., Tomassi, E.: Nozioni di geometria analitica, proiettiva e descrittiva ad uso degli allievi dell'Accademia militare di Lecce. Roma (1947)
56. Conforto, F., Vaccaro, G.: Algebra ad uso degli Istituti Magistrali Superiori. Principato, Milano-Messina (1946)
57. Conforto, F., Vaccaro, G.: Algebra ad uso dei Ginnasi Superiori, Principato, Milano-Messina (1946)
58. Conforto, F., Vaccaro, G.: Aritmetica razionale per gli Istituti Magistrali Superiori. Principato, Milano-Messina (1947)
59. Conforto, F., Viola, T.: Sul calcolo di un integrale doppio che interviene nella determinazione della profondità degl'ipocentri sismici. Consiglio Nazionale delle ricerche, Istituto per le applicazioni del calcolo, Roma (1936)
60. Conforto, F., Zappa, G.: La geometria algebrica in Italia (dal 1939 a tutto il 1950). Pontificia Academia Scientiarum, Relationes de auctis scientiis tempore belli, pp. 1–43 (1946)
61. D'Alterio, C.: Federigo Enriques a Bologna (1894–1921) tra ricerca e didattica: le *Lezioni di geometria proiettiva*. Master's Thesis. Università di Bologna (2020)
62. Enriques, F., Campedelli, L.: Le superficie algebriche. Zanichelli, Bologna (1949)
63. Enriques, F., Chisini, O.: Lezioni sulla teoria geometrica delle equazioni e delle funzioni algebriche. Zanichelli, Bologna (1924)
64. Lezioni di geometria descrittiva per il I biennio universitario (lit.). Veschi, Roma (1946)
65. Lugaresi, M. G., D'Alterio, C.: Federigo Enriques (1871–1946): a critical study of *Lezioni di geometria proiettiva*. Hist. Math. **60 C**, 15–52 (2022). https://doi.org/10.1016/j.hm.2022.03.001
66. Manara, C. F.: Ricordo Oscar Chisini. Periodico di matematiche **4**, 46, 1–20 (1968)
67. Nastasi, P.: I primi quarant'anni di vita dell'Istituto per le Applicazioni del Calcolo "Mauro Picone". Boll. Un. Mat. Ital. (8) **9-A**, 3–244 (2006)

68. Nastasi, P., Guerraggio, A.: Italian Mathematics Between the Two World Wars. Birkhäuser Verlag, Basel-Boston-Berlin (2006)
69. Nastasi, P., Rogora, E.: From internazionalization to autarky: Mathematics in Rome between the two world wars. Rend. Mat. Appl. **7**, 41, 1–50 (2020)
70. Relazione della Commissione giudicatrice del concorso a professore straordinario alla cattedra di geometria analitica con elementi di proiettiva e geometria descrittiva con disegno nella R. Università di Torino. In: Bollettino Ufficiale, Ministero dell'Educazione nazionale, pp. 1673–1694 (1940)
71. Roghi, G.: Materiale per una Storia dell'Istituto Nazionale di Alta Matematica dal 1939 al 2003. Boll. Un. Mat. Ital. (8) **VIII-A**, 3–301 (2005)
72. Rosati, M.: Commemorazione di Fabio Conforto. University of Ferrara, 21 November 2008
73. Segre, B.: L'opera scientifica di Fabio Conforto. Rend. Mat. Appl. **4**, 14, 48–74 (1954)
74. Severi, F.: Serie, sistemi d'equivalenza e corrispondenze algebriche sulle varietà algebriche, Conforto, F., Martinelli E. (eds.). Edizioni Cremonese, Roma (1942)
75. Severi, F.: Funzioni quasi abeliane. Pontificia Academia Scientiarum, Scripta Varia. Roma (1947)
76. Siegel, C. L.: Analytic Functions of Several Complex Variables: Lectures Delivered at the Institute for Advanced Study. Princeton (1949)
77. Siegmund-Schultze, R.: Rockefeller and the Internationalization of Mathematics Between the Two World Wars. Birkhäuser, Basel (2001)

Alessandro Terracini (1889–1968): Teaching and Research from the University Years to the Racial Laws

Livia Giacardi

Abstract Alessandro Terracini's life is divided into three main phases; each of them has different characteristics and is framed in different historical and social scenarios. He was trained in the School of Corrado Segre in Turin, and from the time of his degree dissertation, he was directed by him towards differential projective geometry, the field in which he would make the most significant contributions. In this essay, the scientific biography of Terracini during the first phase of his life is reconstructed, based on extensive unpublished documentation: scientific results and contacts, his participation in World War I, his teaching experiences in Modena and Catania, his return to Turin, and, finally, the dramatic experience of racial laws and the consequent decision to emigrate to Argentina with his family.

Keywords Alessandro Terracini · Corrado Segre's School · Differential projective geometry · Terracini's teaching at the University of Turin · World War I · Racial laws

Abbreviations

ABTT	Archivio privato famiglia Benedetto Terracini, Torino
ACS-Roma	Archivio Centrale dello Stato Roma
ASUCT	Archivio Storico dell'Università degli Studi di Catania
ASUT	Archivio Storico dell'Università di Torino
BSMT	Biblioteca Speciale di Matematica 'G. Peano', Dipartimento di Matematica, Università di Torino

Livia Giacardi is a member of GNSAGA, Group of INdAM. The essay was originally published in Italian in 2020 in Quaderno 36 of the Academy of Science of Turin, cf. [13]. The entire volume is available in open access on the Academy website.

L. Giacardi (✉)
Department of Mathematics, University of Turin, Turin, Italy
e-mail: livia.giacardi@unito.it

G. Bini (ed.), *Algebraic Geometry between Tradition and Future*,
Springer INdAM Series 53, https://doi.org/10.1007/978-981-19-8281-1_5

SPSL, *AT* Archive of the Society for the Protection of Science and Learning, Bodleian Library, University of Oxford, I. Correspondence relating to individual scholars, I.12 Mathematics, *Terracini, Professor Alessandro (1888–)*, File 1938–46, 285/5, fols. 340–87

1 Introduction

> A serious world, almost ready for grandeur, of silent streets and palaces of the last century, very aristocratic ... High civilization of cafes, ice cream, Turin chocolate. Trilingual libraries. University, good library ... City with splendid boulevards; incomparable landscapes on the banks of the Po. ([32], III5, p. 313)

With these words, Friedrich Nietzsche, who settled in Turin in 1888, described with almost pictorial tones the austere beauty of the city that hosted him and the strong cultural dimension that characterized not only the academic sites, but also the libraries and discussions in cafes.

Indeed, the scenario that presented itself to anyone traveling through the history of the subalpine capital in the last 20 years of the nineteenth century was that of a city in full turmoil in both the scientific sector and in the humanities: schools of thought developed, new journals were created, growing need to integrate scientific research and popularization fostered debates, and collaboration between universities and industry began.

Among the avant-garde sectors at the end of the century, mathematics certainly occupied a prominent place: in the Piedmont capital, three scientific schools of international importance flourished simultaneously, the Italian school of geometry of Corrado Segre (1863–1924), that of mathematical logic of Giuseppe Peano (1858–1932), and that of mathematical physics of Vito Volterra (1860–1940).

These stimulating years are the setting for Alessandro Terracini's training at the school of Corrado Segre at Turin University (Fig. 1).

2 Training, First Works, and First Scientific Contacts

Terracini was born in Turin on October 19, 1889, in a wealthy Jewish family. The Turin Jewish community at the end of the nineteenth century was a large, cultured, and productive group that included representatives of the rich bourgeoisie, of the world of culture—university professors, teachers, and managers—and of the publishing industry and was an active part in the impetuous scientific development and cultural heritage of the city [24]. Terracini's father, Benedetto, was a wealthy merchant, but died when Alessandro was just 10, leaving the family management in the hands of his wife Eugenia Levi, a woman of great energy.

Alessandro began his secondary studies in 1899 at the Ginnasio Liceo D'Azeglio and continued them at the Ginnasio Liceo Cavour, where he had Rodolfo Bettazzi,

Fig. 1 ABTT, Alessandro Terracini in 1924

the founder of the National Mathesis Association of mathematics teachers, as his math teacher. He was a brilliant student and collected prizes throughout his school career, which ended in 1907 with a school leaving certificate with honors.

As a student, he already showed a strong propensity for mathematics and engaged in the problems proposed by three journals, *Supplemento al Periodico di matematica*, *Il Pitagora*, and *Mathesis*, a Belgian periodical, earning the praise of the editors of the columns for some of the solutions sent by him.[1] Among these, the extensive treatment of a particular problem on triangles aroused the admiration of Cristoforo Alasia, known precisely for his research on the geometry of the triangle:

[1] Questions solved in the *Supplemento al Periodico di matematica*: from January 1905 to December 1906, the young Terracini sent the solutions to the following: games 200, 198, 202, 206, 203, 210, 212, 219, 220, 221, 227, 197, 205, 208. 225. 226, 222, 228, 232, 229, 233, 242, 243, 246, 244, 245, and 252; questions sent for competitions 56, 57, 59, 60, 64, 62, 65, 66, and 67; and questions 622, 624, 627, 628, 643, 644, 633, 634, 666, 672, 674, 676, 682, 680, 702, 704, 705, 808, 706, 712, 714, 707, 717, 720, 724, 727, 729, 722, 723, 733, 735, 725, 736, 739, 709, 715, and 716. Issues resolved in *Il Pitagora*: in 1904–1905 (year XI), he sent the solutions to questions no. 750, 753, 754, 755, 766, 772 (in issues 3–4–5), 776, 777, 778, 782, 783, 784, and 785 (in issues 6–7). Questions solved in *Mathesis*: from December 1906 to March 1908, he solved questions 1587, 1573, 1591, 1604, 1682, and 1641 and two of his notes, "Note sur les coniques" (VII, 1907, pp. 42–44) and "Théorèmes sur les transversales" (VII, 1907, p. 127), were published.

> The work of Mr. Terracini is the most extensive, as he wrote about 80 pages and in various points is not without merits that demonstrate that he has a broad spirit of observation and great ease of deduction, showing he must profitably have read more than one work on the recent geometry of the triangle.[2]

Many years later, his friend Eugenio Togliatti, referring to this youthful contribution by Terracini, wrote: "I then felt a sense of admiration for this then unknown companion who had been able to find so many properties of triangles that were so particular that they almost filled a volume."[3]

Right from the time of high school, Bettazzi, recognizing Alessandro's qualities, had put him in contact with Corrado Segre, the professor of Higher Geometry at the University of Turin, who in the early twentieth century, together with his students, had already acquired international fame. It is therefore natural that, having finished his secondary studies on October 31, 1907, Terracini enrolled in the degree course in mathematics at Turin University. Here, he attended the courses of E. D'Ovidio, G. Fano, G. Sannia, G. Peano, C. Somigliana, T. Boggio, and E. Laura, and for his personal interests, he attended Segre's higher geometry courses for three consecutive years, devoted, respectively, to *Rassegna di concetti e metodi della Geometria moderna* [Review of concepts and methods of modern geometry] (1908–1909); *Superficie del 3° ordine e curve piane del 4° ordine* [third-order surfaces and fourth-order plane curves] (1909–1910); and *Le curve e le superficie algebriche, dal punto di vista della Geometria delle trasformazioni birazionali* [Algebraic curves and surfaces, from the point of view of the geometry of birational transformations] (1910–1911).[4] In his autobiography, Terracini recalls:

> CORRADO SEGRE's lectures were usually only attended from the third year of university, but I began to attend them in the second year, that is, in 1908–1909. They were held on Tuesday, Thursday and Saturday morning from 10 to 11, formerly on the first floor in the hall that occupied the place then taken by the room that currently precedes the *aula magna*, and later, I believe, in room XVII on the second floor of the University Building in Via Po, around the walls of which there ran the glass cabinets with the geometrical models of BRILL which later, I think, were destroyed in a bombing.[5]

---

[2] Cf. [1]: *Il lavoro del sig. Terracini è il più voluminoso avendo egli scritto circa 80 facciate e in diversi punti non è privo di pregi che dimostrano come in lui è un largo spirito di osservazione e molta facilità di deduzione, e come debba aver letto con profitto più d'un lavoro sulla recente geometria del triangolo.*

[3] Cf. [61] on p. 398: *Provai allora un senso di ammirazione per questo compagno allora sconosciuto che aveva saputo trovare tante proprietà di triangoli così particolari da riempire quasi un volume.*

[4] ASUT: Scienze MFN, Registro Carriera, matr. 1201–1402, *AT* p. 178; see also [58], p. 11, and http://www.corradosegre.unito.it/I21_30.php, in [17].

[5] Cf. [58], p. 10: *Le lezioni di CORRADO SEGRE—ricorda Terracini—si seguivano di regola soltanto a partire dal terzo anno di università, ma io incominciai a frequentarle nel secondo anno, cioè nel 1908–1909. Esse avevano luogo il martedì, giovedì e sabato mattina dalle 10 alle 11, anticamente al primo piano nell'aula che occupava il posto preso poi dall'attuale antiaula magna, e più tardi, credo, in quell'aula XVII del secondo piano del Palazzo Universitario di via Po, alle cui pareti correvano gli armadi a vetri coi modelli geometrici di BRILL che poi, penso, andarono distrutti in un bombardamento.*

His passion for geometry also led him to choose from the two subsidiary courses held by Gustavo Sannia, devoted, respectively, to analytical geometry and non-Euclidean geometries. Since the first months of university, the desire to try his hand at research led Terracini to send Alfredo Capelli, the editor of the *Giornale di Matematiche*, his first scientific note dedicated to an "extension (spontaneous, but certainly not peculiar) of the well-known hemi-symmetric determinants, to which the determinants in question can be traced back when the 'modulus' λ becomes equal to zero," [6] a note published in 1909 [41]. However, he was soon attracted to a new field of research for which Segre was laying the foundations in those years, hyperspatial differential projective geometry.[7] Segre, in addition to being a "great geometrician," was also an attentive and conscientious teacher and "assigned degree theses in writing with a long and detailed explanation of the state-of-the-art of the issue that the graduate student had to deal with ... he examined them quite often, always formulating in writing his criticisms and any advice for continuation."[8]

So, Terracini asked Segre for a thesis and was assigned a topic of differential projective geometry as the subject of study; in particular, it was a question of establishing a connection between two classes of hyperspatial varieties, thus connecting together two problems that, in particular cases, had already been studied by Francesco Severi and Gaetano Scorza. He solved the problem brilliantly and arrived at an important result, known as the *Terracini Lemma*.[9]

In the spring of 1910, Guido Castelnuovo, who was teaching at the University of Rome, informed Segre that his pupil Enrico Bompiani, same age as Alessandro, was preparing a thesis on hyperspatial differential projective geometry, and so the two young people came into contact through their thesis supervisors, although they only met in 1912 on the occasion of the International Congress of Mathematicians in Cambridge. On 5 July 1911, Terracini got his degree with 100/100 *cum laude*[10] presenting a thesis *Sulla teoria delle varietà luoghi di spazi* [On the theory of the varieties of linear spaces], part of which was published in the same year in *Rendiconti del Circolo matematico di Palermo* [42]. This was soon followed by other works on differential projective geometry (Fig. 2).

Barely a year after graduating, Terracini, then an assistant to Gino Fano, took part in the International Congress of Mathematicians held in Cambridge (UK) from 22

---

6 *Ibidem*, p. 8: *estensione (spontanea, ma non certo peregrina) dei notissimi determinanti emisimmetrici, ai quali i determinanti in esame si riconducono quando il 'modulo' λ diventa eguale a zero.*

7 Cf. in particular the paper by Segre [38].

8 Cf. [58], pp. 12, 13, and 14: *assegnava le tesi di laurea per scritto con una lunga e particolareggiata esposizione dello stato in cui si trovava la questione che il laureando doveva trattare [...] le esaminava abbastanza spesso, formulando sempre per scritto le sue critiche ed eventuali consigli per la continuazione.*

9 See the three essays on Terracini's mathematical work [13], in particular §2 of the essay by Ciro Ciliberto [12]; §3.2 of the essay by Alessandro Verra [65]; and §4 of the one by Anna Fino [15].

10 ASUT: *Verbale dell'esame di laurea di Alessandro Terracini. Torino, 5 luglio* 1911, Faculty of Mathematical, Physical and Natural Sciences, Graduation exam reports, 1902–1921, p. 141.

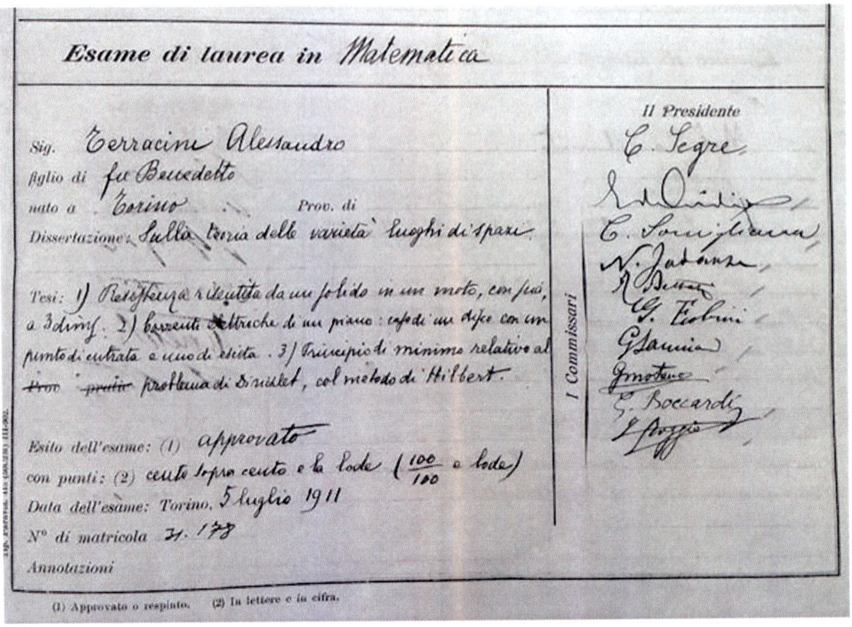

Esame di laurea in Matematica

Sig. Terracini Alessandro

figlio di fu Benedetto

nato a Torino Prov. di

Dissertazione: Sulla teoria delle varietà luoghi di spazi

Tesi: 1) [illegible] 2) [illegible] 3) Principio di minimo relativo al problema di Dirichlet, col metodo di Hilbert.

Esito dell'esame: (1) approvato

con punti: (2) cento sopra cento e la lode (100/100 e lode)

Data dell'esame: Torino, 5 luglio 1911

N° di matricola 31. 178

Annotazioni

Il Presidente C. Segre

I Commissari [illegible signatures] G. Fubini ... G. Boccardi ...

(1) Approvato o respinto. (2) In lettere e in cifra.

Fig. 2 ASUT, Minutes of the graduation exam by A. Terracini

to 28 August 1912. The fact that Segre, Castelnuovo, and Federigo Enriques were members of the International Committee was an incentive to participate. In addition, Bompiani gave a talk on *Recenti progressi nella geometria proiettiva differenziale degli iperspazi* [Recent advances in the differential projective geometry of hyperspaces] ([36], vol. II, 22–27). In it, Castelnuovo's young disciple extensively cites Segre's works and also Terracini's research deriving from his degree thesis. There thus began a friendship between the two mathematicians that was to last a lifetime, despite the break due to the racial laws, and one which was to extend to their families. During the congress, Alessandro also had the opportunity to personally meet Enriques, Castelnuovo, and Francesco Severi, the formidable trio of the Italian School of Algebraic Geometry, and other illustrious mathematicians. He wrote:

> Among the mathematicians I met in Cambridge I was particularly struck by the venerable MITTAG-LEFFLER (1846–1927) and Prof. EDMUND LANDAU (1877–1938) and, as for the places, the solemn environment rich in historical memories, and the austere rooms of the Colleges left an indelible memory.[11]

[11] Cf. [58], p. 69: *Tra i matematici visti a Cambridge mi fecero particolarmente impressione la veneranda figura di MITTAG-LEFFLER (1846–1927) e il prof. EDMUND LANDAU (1877–1938) e, quanto ai luoghi, mi lasciarono un ricordo incancellabile l'ambiente solenne e ricco di ricordi storici, e le sale austere dei Colleges.*

The common interests with Bompiani stimulated the beginning of a scientific correspondence in which the two young people shared their research ([33], pp. 98–106). They thus realized that these presented some points of contact. Bompiani wrote in this regard:

> I note from your letter that the results we have achieved are identical; but the simultaneity, the difference in method and the probable differences in development give us the same merit and we shouldn't be too sorry about having met.[12]

They met in Bologna to discuss their research in person,[13] and they decided to publish their works together. These were presented by Segre for publication in the *Atti della Accademia delle scienze di Torino* in November 1913.[14] A new field of research was being developed, hyperspatial differential projective geometry, in which Segre, Bompiani, and Terracini gave internationally known results, which Terracini was to summarize in 1927 in Appendix III to the second volume of the treatise by G. Fubini and E. Čech, *Geometria proiettiva differenziale* [Differential projective geometry] [49]. Many years later, Bompiani in a lecture given at a meeting of the Mathematical Association of America was proud to evoke these first steps: "In this new field, in which almost nothing was known, began to work Terracini and myself."[15]

In the curriculum that Terracini presented to the Society for the Protection of Science and Learning in London to find a place to go after the enactment of the racial laws, he writes about the group of his works in this area:

> The moment I began to undertake my researches on projective differential Geometry happened to coincide with the years in which this branch had just left its initial period. Some of the methods were already formed and had been put to the test through the easier problems which always present themselves at the dawn of a new theory; the opportunity of contriving other methods was still kept for the future. Among the first the most important was doubtless the method based on the use of linear partial differential equations. This method was already classical for the curves, and Wilczynski had successfully employed it in the theory of surfaces. But it is with the consideration of loci in hyperspaces that the most interesting problems arise. I have endeavoured to use such a method to confront the manifold new problems which presented themselves. If the number of linearly independent equations (of second order) exceeds a certain limit, Corrado Segre had shown that the V_k belongs to certain well determined classes, and precisely that the dimension of the locus W of its tangent spaces S_k is $<2k$. But this condition remained, so to say, merely nominal, until the V_k for which the dimension of the locus W is $2k$ were effectively known. I succeeded in

[12] E. Bompiani, to A. Terracini, [Roma] 25 February 1913 [33], p. 101: *Rilevo dalla tua lettera l'identità dei risultati da noi raggiunti; ma la contemporaneità, la differenza di metodo e le probabili diversità di sviluppo ce ne lasciano ugual merito e non c'è da dolersi troppo d'esserci incontrati.*

[13] See in particular E. Bompiani to A. Terracini, Roma, 5 March 1913; A. Terracini to E. Bompiani, Torino, 21 June 1913 in [33], pp. 102, 103–104.

[14] Cf. the paper [5] by Bompiani and the paper [43] by Terracini, work completed by *Nota II e III* in vols. 51 and 55 of the *Atti della Accademia delle scienze di Torino*.

[15] Cf. [7], p. 94. On Bompiani's first research in the field of differential projective geometry, cf. [11].

> specifying the whole class of such V_k. Moreover, the same condition even being necessary is not always sufficient that the number of linearly independent partial equations exceed the mentioned limit. What are the cases when it is not sufficient? This is the main question which I have studied in 8,15. But to this purpose it has been necessary for me to reach many fundamental results about the manner of interfering of the structure of the system of partial equations with the geometric nature of the V_k. I have also observed that the number of linearly independent equations represented both by a V_k and by its generical prime sections may present some irregularities: for what loci does it happen so? (8,19).[16]

It was Segre who had pushed Terracini towards this sector of research and from Segre he took the mentality, the method, and the lines of research, but he also considered himself a pupil of Fubini even though he had never attended his courses ([58], p. 53). Although Fubini was predominantly an analyst in geometry too, "he did extremely important work," as Terracini writes, "discovering new geometrical facts with analytical methods ... he was such a skilled tailor that, at the same time as the garment [analytic treatment], he created the person for whom it was intended [new geometrical fact]."[17]

While making use of the analytical tool (differential forms, systems of partial differential equations), Terracini, like Segre, always gives a lot of weight to the geometrical vision of problems, "to which there is linked," as Togliatti writes, "in harmonious collaboration, the analytical tool which for him has not only a control function, but also has a constructive part in research,"[18] creating a fruitful fusion between the analytical method and the synthetic method. Terracini honored his two masters throughout his life with articles and courses and by editing part of their works.

3 Participation in the Great War and First Appointments

In the years between 1912 and 1914, Terracini traveled, met various mathematicians, and established new relationships. As he tells us in his scientific autobiography ([58], chapter IX), in February 1912 in Erlangen, he met Max and Emmy Noether, and in the summer of the same year, he began his friendship with Eugenio Togliatti, who had just graduated with Segre as his supervisor and with whom he shared research interests in the field of differential projective geometry. In that period, he met Castelnuovo in Rome several times to discuss his current research and, during a stay in Parma, he met Beppo Levi and Gaetano Scorza. In July 1914, he went

[16] SPSL, *AT*: Terracini, A., Some Indications on my scientific papers, folios 369–377. Notes 8, 15, and 19, to which Terracini is referring, are those cited in note 16.

[17] Cf. [56], p. 101: *ha fatto opera estremamente importante scoprendo fatti geometrici nuovi con metodi analitici [...] era un sarto tanto abile che, allo stesso tempo che la veste [trattazione analitica], creava la persona a cui era destinata [fatto geometrico nuovo]*; see also [58], p. 59.

[18] Cf. [62], p. 147: *alla quale si collega in armonica collaborazione lo strumento analitico con un ufficio che per lui non è solo di controllo, ma ha anche parte costruttiva nelle ricerche.*

to Athens, where he visited Cyparissos Stephanos, who was known to the Turin School[19] and whom Terracini had already met on the occasion of the International Congress of Mathematicians in Cambridge in 1912.

In 1914, the First World War broke out. In May 2015, after the London Pact, which marked the adhesion of Italy to the Entente (England, France, and Russia), Italy went to war against Austria. Terracini, like other important Italian mathematicians, such as Vito Volterra and E. Elia Levi, was "clearly in favor of Italy's intervention in the war."[20] In his autobiography, he tells us in great detail about his and his brother Benvenuto's participation in the war events. After spending the first months in Rome as a private at the Battalion of the Railway Engineers, he attended one of the courses for reserve officer cadets (*allievi ufficiali di complemento*) in the engineering and artillery corps, held at the Military Academy of Turin. Assigned to the 22nd Miners Company and destined for Gorizia, he collaborated in the construction of a line of fortifications and developed a variant of the periscope that "obviated the drawback of being able to observe a field that was too low by arranging three mirrors in it instead of two, having an inclination with respect to the horizontal planes that was appropriately greater than 45 degrees." [21] In the spring of 1916, he was transferred to Gemona, where he attended a course in geodesy, but the most significant period dates back to autumn 1917 when, destined for the artillery command of the fifth Army Corps in Schio, he met Mauro Picone. Terracini already knew Picone, albeit only superficially, because he had come to Turin in 1913 as Fubini's assistant at the Polytechnic. In February 1918, both were assigned, together with General Roberto Segre, to the Artillery Command of the Sixth Army in Breganze, and here Picone involved him in the renewal of the firing charts for the use of heavy artillery in the mountains so that the shooting data were assigned according to range and height difference.

The work undertaken led to the publication, edited by Picone, *Tavole di tiro da montagna. Teoria e metodi di compilazione* [Mountain firing charts. Compilation theory and methods]. In the preface, General Segre highlights a specific mathematical contribution by Terracini to the compilation of the charts ([34], p. 5 and p. 43).

Picone was to recall this collaboration on various occasions, also due to the influence it had on his future activity. In the report presented for the mathematical analysis competition at the University of Cagliari, as well as in his speech in 1951 at the IV Congress of the UMI, indeed he recalls that at the time, together with Terracini, he had created a real calculation laboratory, the germ of the future Istituto Nazionale per le Applicazioni del Calcolo [National Institute for Computing Applications]:

---

[19] Cf. http://www.corradosegre.unito.it/Quaderni/Quad27/1_27.php, p. 6.

[20] Cf. [58], p. 78: *nettamente favorevole all'intervento dell'Italia in guerra*; on the role of Italian mathematicians in the First World War cf. [31] and [27].

[21] Cf. [58], p. 82: *ovviava all'inconveniente di poter osservare un campo troppo poco alto disponendovi tre specchi anziché due, aventi un'inclinazione rispetto ai piani orizzontali convenientemente maggiore di 45 gradi.*

Fig. 3 M. Picone and A. Terracini in 1917

> In our little office, located in the attic of a country farm, I and the collaborators I had been able to obtain worked: Prof. Terracini (Lieutenant of Engineers) ... and later Prof. Signorini (Lieutenant of Artillery) ... I had five calculating machines, ... We worked day and night.[22]
>
> Then and for a long time, one of my collaborators—a friend and a genius—was ALESSANDRO TERRACINI, ... with whom, in the 6th Army, supported by the commander of the artillery general ROBERTO SEGRE, we founded a real Calculation Institute.[23]

It is probably due to frequenting Picone that Terracini wrote an article [52] many years later, aimed at obviating the drawbacks that can occur in practice in the numerical solution of systems of linear equations with multiple unknowns, when their number increases. The procedure presented by Terracini is based on the idea of simultaneously eliminating more unknowns from the system and has the advantage of being particularly rapid with the use of the calculating machine (Fig. 3).

---

[22] Quoted in [30], p. 22: *Nel nostro ufficietto, sito in una soffitta di una fattoria di campagna, lavoravamo io e i collaboratori che avevo potuto ottenere: il Prof. Terracini (Tenente del Genio) [...] e più tardi il Prof. Signorini (Tenente d'Artiglieria) [...]. Ebbi cinque macchine calcolatrici, [...] Il lavoro era diurno e notturno.*

[23] Cf. [35], p. 27: *Fu, allora, a lungo, mio collaboratore—amico e geniale—ALESSANDRO TERRACINI, [...] col quale, alla 6ª Armata, sostenuti dal comandante l'artiglieria generale ROBERTO SEGRE, fondammo un vero e proprio Istituto di Calcolo.*

4 At the Universities of Modena and Catania

After the war, Terracini returned to Turin, where in 1919 he gave lectures[24] integrating Segre's Higher Geometry course on *Complessi di rette di 1° e 2° grado* [first- and second-degree line complexes], but in the autumn of the same year, Ermenegildo Daniele, Professor of Rational Mechanics in Modena, proposed that he moved to that university where, in addition to continuing to be an assistant, he could be in charge of the course of algebraic analysis.[25] Terracini accepted and immediately entered the university milieu, maintaining relationships with professors of all faculties and, among mathematicians, he frequented Oscar Chisini and Enea Bortolotti, with whom he shared research interests in differential geometry. A favorite meeting point was the Caffè San Carlo or the back shop of the grocery store of Telesforo and Giuditta Fini ([58], pp. 91–92).

The Modena period was productive not only for the contacts he maintained with his colleagues, but also from the point of view of research: indeed, a dozen articles date back to these years, among which Terracini particularly stressed *Sulle superficie le cui asintotiche dei due sistemi sono cubiche sghembe* [On surfaces whose asymptotics of the two systems are twisted cubic curves] ([44], also in [59] I, pp. 130–155):

> Particular examples of surfaces, whose asymptotic lines are twisted cubic (that is the most simple twisted lines, from the algebraical point of the view) were still known. I succeeded in a general method for finding the whole class of these surfaces. More generally I found (in finite terms) all the surfaces whose asymptotic lines of both systems belong to linear complexes (a particular case is that of the Tzitzeica-Wilczynski surfaces).[26]

The Modena years were also those that saw the rise of fascism, with all the repression and violence that characterized them. Terracini recalls this period in his autobiography:

> Those were terrible years in which fascism triumphed, also accompanied by violent aspects among which I cannot fail to mention the fire—in central Piazza Mazzini—which destroyed the house of the communist deputy Pio Donati, the brother of Prof. Mario.[27]

---

[24] Cfr. QUADERNI. 1 in BMP—*Fondo Terracini*. The lecture note can be accessed at the site http://www.corradosegre.unito.it/fondo_terracini_q.php. In [17].

[25] From 1919–1920 to 1921–1922, he was assistant to the Chair of Descriptive Geometry and in 1922–1923 to the Chair of Algebraic Analysis. From 1919–1920 to 1921–1922, he was in charge of the course of algebraic analysis, and in 1922–1923 of the descriptive geometry course (cf. ACS-Roma, Fondo del Ministero della Pubblica Istruzione, Fascicoli personali dei professori universitari. 3° Versamento Busta 452, AT).

[26] SPSL, *AT*: Terracini, A. Some indications ....

[27] Cf. [58], p. 92: *Erano quelli gli anni roventi in cui si affermò il fascismo, accompagnato anche da aspetti violenti tra i quali non posso non rammentare l'incendio—nella centrale piazza Mazzini—con cui fu distrutta la casa del deputato comunista Pio Donati, fratello del prof. Mario.*

Terracini spent a brief period in Turin, during which he married Giulia Sacerdote,[28] who was to give him three children, Lore, Cesare, and Benedetto. Then, having won chairs in both Cagliari and Catania, he chose Catania.

The milieu was stimulating, with a good scientific tradition. In the first decade of the twentieth century, excellent mathematicians had taught in Catania who laid the foundations for a prestigious mathematical institute, Mario Pieri, Giuseppe Lauricella, Guido Fubini, and Michele de Franchis, and from 1919 to 1921, Mauro Picone was the professor in charge of the course of mathematical analysis. Furthermore, in 1921, the Circolo Matematico di Catania was created, directed by Nicolò Spampinato. It published two journals, one, *Note e memorie*, edited by Gaetano Scorza, containing specialized research, and the other, aimed at a wider audience, *Esercitazioni matematiche*, edited by Michele Cipolla.[29] Terracini fit very well into this university too, as shown by the numerous offprints he received in the following years from his Catania friends.[30] Here in 1924–1925, he held the course of analytical geometry and, on a temporary basis, that of higher geometry *Sui fondamenti della geometria differenziale* [On the foundations of differential geometry], and in 1925–1926, he held only two lectures of the higher mathematics course dedicated to *geometria differenziale* [differential geometry].[31] Indeed, at the end of 1925, he was called to Turin to take up the Chair of Analytical Geometry, where he would also teach the higher geometry course until 1937–1938, to resume his post in 1947–1948 after the caesura of the exile.[32] The result of Terracini's teaching in Catania is the *Lezioni di Geometria analitica, Anno Accademico 1924–25* [Lectures in Analytical Geometry, Academic Year 1924–25], lithographed by the *Circolo Matematico* of Catania.

5 Return to Turin: From Tenure to the Racial Laws

Returning to Turin was clearly the greatest of Terracini's aspirations: there his old friends and colleagues had just been joined by Francesco Tricomi, who was already known for his studies on partial differential equations of the second order of mixed type and soon became "a frequent visitor" to his home. Thus, a friendship grew between the two mathematicians, of which Tricomi gave ample proof during the

---

[28] The wedding took place in Rome on April 16, 1924 with Eugenio Artom and Enrico Bompiani as witnesses.

[29] On the history of the University of Catania, cf. [40].

[30] The offprints are kept at the Biblioteca speciale di matematica "Giuseppe Peano" of the Department of Mathematics of the University of Turin: see [26].

[31] Cf. [2], p. 97. See also ASUCT: Registri delle lezioni, Facoltà di Scienze fisiche, matematiche e naturali, n. 30: Alessandro Terracini, Registro di Geometria superiore (1924–1925) e Registro di Matematiche superiori (1925–1926).

[32] In ASUT, there are 32 registers of Terracini's lectures from 1936–1937 to 1961–1962.

period of racial persecutions. He recalls the arrival of our geometrician in Turin with these words:

> After Terracini and I came to Turin, the mathematics section of the Faculty of Sciences reached—at least from a numerical point of view—a level never exceeded. Indeed, we mathematicians, including the geodesist and the astronomer, were then 7 out of a total of 15, while today they are not many more out of a total of 35![33]

The Turin mathematicians at the time faced off in two groups, observes Tricomi: "on the one hand, the 'Jewish' or 'rich men' one, which was headed by the illustrious Corrado Segre (1863–1924) who died prematurely the year before, and it was then reduced to Gino Fano (1871–1952) and Guido Fubini (1879–1943) [. . .]; and on the other hand the 'vectorialist' group which, in addition to Giuseppe Peano (1858–1932) [...], included Tommaso Boggio (1877–1963) and the intemperate Cesare Burali-Forti (1861–1931)."[34] The former was more conservative and the latter more progressive. As soon as he arrived in Turin, Tricomi, as he himself recounts, *toto corde* [wholeheartedly] joined the Jewish group, which Terracini also joined.

5.1 Research and Teaching

Before obtaining tenure in 1928, Terracini held two higher geometry courses dedicated to differential geometry and differential geometry of hyperspaces, the research area he favored,[35] and in the same period, he published about ten works. Among the other results, we can mention the unitary definition of the three normals, metric, affine, and projective, of a plane curve, a geometrical interpretation of the projective linear element of a surface in ordinary space introduced by Fubini, simpler than those provided by Čech and by Bompiani. A third result is related to the geometrical meaning of the invariants of a Laplace equation. Terracini's paper on this topic was followed by two notes: the first by Fubini, who from Terracini's interpretation derives a simple geometrical proof of a theorem on *W*-congruences,

[33] Cf. [63], p. 34: *Con la mia venuta a Torino e del Terracini la sezione matematica della Facoltà di Scienze raggiunse—almeno da punto di vista numerico—un'altezza mai più superata. Invero noi matematici, comprendendo anche il geodeta e l'astronomo, eravamo allora in 7 su un totale di 15, mentre oggi essi non sono molto di più su di un totale di 35!*. For the relationships between Terracini and Tricomi, see the letters in BSMT *Fondo corrispondenza* and *Carte Terracini*. A selection of the letters is present in Milanese, D. Alessandro Terracini (1889–1968) grande organizzatore culturale. Alcune corrispondenze inedite, Dissertation in Mathematics, supervisor L. Giacardi, University of Turin, academic year 1916–1917.

[34] Cf. [63], pp. 32–33: *da un lato quello 'ebraico' o 'dei ricchi', che era stato capeggiato dall'illustre Corrado Segre (1863–1924) prematuramente scomparso l'anno prima, ed era allora ridotto a Gino Fano (1871–1952) e a Guido Fubini (1879–1943) [. . .]; e dall'altro lato il gruppo 'dei vettorialisti' che oltre a Giuseppe Peano (1858–1932) [. . .], comprendeva Tommaso Boggio (1877–1963) e l'intemperante Cesare Burali-Forti (1861–1931).*

[35] Cf. QUADERNI 1, 2, and 3, in BSMT—*Fondo Terracini*. The cahiers can be accessed at the site http://www.corradosegre.unito.it/fondo_terracini_q.php. In [17].

and the second by Bompiani, who provides another geometrical interpretation.[36] In this connection, one of Terracini's favorite lines of research is that which aims to highlight the geometrical aspect of various issues in the simplest way, as Bompiani noted:

> Terracini trained in the geometrical thought of Corrado Segre, feels the value of Fubini's analytical results, but also the need to clarify the geometrical reason.[37]

For example, in the two works published in 1927 on W congruencies ([47, 50], also in [59] I, pp. 188–204), Terracini claims to have tried to construct a new theory of congruencies by combining the advantages of Fubini's theory, which is powerful but lacks geometrical significance, and that of Tzitzeica-Ribaucour, which is geometrically perspicuous but limited:

> Fubini's theory is powerful, but for a great part it lacked any geometrical signification. On the other hand, Tzitzeica-Ribaucour's theory is geometrically perspicuous, but limited in its capacity. I have tried constructing a new theory combining the advantages of both. Besides, some relations between a Laplace equation and its adjoint through integro-differential transformations.[38]

This result is among those reported by Carlo Somigliana in his report for the promotion of Terracini to full professor, a promotion that was approved in the faculty session of January 12, 1928.[39]

Like his master Segre, our mathematician also took great care over teaching and, together with Gino Fano, in 1929 he published the *Lezioni di geometria analitica e proiettiva* [Lectures in analytical and projective geometry] (Turin, Paravia, 1929) republished in a new edition in 1940 after the racial laws were passed. In a letter to Terracini, Fano explains why, despite the vetoes placed on books by Jewish mathematicians, this second edition was published:

> In 1938 the 2nd edition was started, with slight adjustments on stereotype prints In September, as soon as the storm broke out, the company wrote to me, asking if it could definitely destroy all the material! I, of course, replied that there was no hurry—[I asked them] to wait in the meantime; and in one of my trips from Lausanne to Turin, I don't remember if in November 39 or 40, I went to talk about it with Comm. Tancredi, and I pointed out that the prohibition on reprinting only concerned <u>textbooks</u>, which at the University do not exist; and that our volume, <u>a scientific treatise</u>, could therefore be reprinted, provided it suited them. He said he agreed.[40]

---

[36] Cf. [45], [46] (also in [59] vol. I, pp. 168–175, and 176–181) and [48]. See also the articles by Fubini [16] and by Bompiani [6].

[37] Cf. [8], p. 12: *Il Terracini, educato al pensiero geometrico di Corrado Segre, sente il valore dei risultati analitici del Fubini, ma anche il bisogno di chiarirne la ragione geometrica.*

[38] SPSL, *AT*: Terracini, A. Some indications ....

[39] ASUT: Scienze MFN, Adunanze 1924–32: *Relazione motivata della Facoltà di Scienze per la promozione a stabile del prof. Alessandro Terracini*, attached to the minutes of the meeting of 12 January 1928.

[40] G. Fano to A. Terracini, New York, 16 December 1947, BSMT *Carte Terracini*: *Nel 1938 fu avviata la 2a edizione, con lievi ritocchi sulle stereotipie. Nel settembre, appena scoppiata la bufera, la Ditta scrisse a me, chiedendo se poteva distruggere senz'altro tutto il materiale! Io,*

Both authors, Fano and Terracini, had already individually published the *Lezioni di Geometria analitica e proiettiva* [Analytical and projective geometry lectures], with the A. Viretto lithographer in Turin, respectively, in 1926–1927 and in 1927–1928, and just in 1927–1928, they had both taught at the Polytechnic of Turin, the former responsible for descriptive geometry with applications and the latter for analytic and projective geometry ([3], pp. 46–47); therefore, their collaboration was greatly facilitated.

The volume had a reprint in 1948 and a third edition in 1957, updated by Terracini, which enjoyed appreciation both in Italy (B. de Finetti, F. Conforto, etc.) and abroad (J. Favard, L. Godeaux, F. Marcus, D. Struik, etc.) for the completeness, clarity, and balance of the various parts.[41] In his review of the 1948 edition, Godeaux also emphasizes that on occasion the authors open up "une fenêtre sur des questions variées" ([20], p. 5). Among other things, indeed, there are hints on vectors and vector calculus, a discussion of the line geometry and, in particular, of the linear line complexes, which lends itself to important applications.

In the summer of 1928, Terracini participated in the International Congress of Mathematicians (Bologna, 3–10 September 1928), presenting a communication in the projective differential geometry section where he introduced the notion of "projective quasi-applicability" [51]. Thanks to Salvatore Pincherle, who at the time held the dual role of president of the Italian Mathematical Union (UMI) and the International Mathematical Union, the Bologna congress represented an important historical moment because it marked the resumption of scientific internationalism compromised by the First World War; at the same time, however, it also showed the first yielding of the Italian mathematical community to fascism, which was to lead to its complete subjection after the racial laws were passed [18, 19].

In the University, as in other areas of society, there was pressure for professors to join the Fascist party and also in Turin the rector Silvio Pivano urged his colleagues to do so. After some time, Terracini confessed that "our behavior, and mine in particular, was not too brilliant, in the sense that we soon followed the pressing invitation; I am ashamed to say it."[42] In August 1931, university professors were required to swear an oath of allegiance to fascism, according to a policy suggested to Mussolini by the geometrician Francesco Severi, inspired by intransigence for unrepentant people like Vito Volterra and by the amnesty to eliminate the "faults" of the ex-anti-fascists such as Severi himself. In all Italy, out of over a thousand university professors, only a dozen refused to take the oath, thus losing their professorships. Among these, the only mathematician was Volterra. Many anti-

naturalmente, risposi che non vi era fretta—aspettassero, intanto; e in una delle mie gite da Losanna a Torino, non ricordo se nel novembre 39 o nel 40, andai a parlarne con il Comm. Tancredi, e gli feci presente che la proibizione di ristampa concerneva solo i libri di testo, che all'Univ.[à] *non esistono; e che il nostro volume, trattato scientifico, poteva quindi ristamparsi, sempreché a loro convenisse. Egli si dichiarò d'accordo.*

[41] See in particular the letters addressed to Terracini in BSMT: *Carte Terracini, 1947–1948.*

[42] Cf. [58], p. 109: *il nostro, e in particolare il mio, contegno non fu troppo brillante, nel senso che presto seguimmo il pressante invito; Mi vergogno a dirlo.*

fascists also took the oath giving prevalence to concerns over the professional consequences of a refusal, i.e., dismissal, the impossibility of ensuring a future for their alumni, and the fear of leaving the way open to unscrupulous colleagues.[43]

In 1934 and then again in 1941, the statutes of scientific societies were changed, limiting more and more severely the freedom of the world of culture, almost without encountering opposition.

In his scientific autobiography, Terracini makes no particular comments on this political period of progressive "fascistization" of all institutions and all sectors of national activity, from the press to schools, the army, and professional organizations; rather, he dwells on the important research he carried out in those years—about 30 works—and on holidays or trips to the mountains with his friends, Tricomi, Togliatti, Fubini, Beniamino Segre, and Gleb Wataghin, joined in 1930 by Enrico Persico ([58], ch. XIV). Persico had come to Turin to take up the Chair of Theoretical Physics and established a strong friendship with the Terracini family, as evidenced by the beautiful correspondence dating back to the period he spent in Canada at Laval University (Québec) which we will mention below.

However, there are some hints of bitter irony, such as the comment—on the occasion of the promotion to full professor—on the term "*stabile*" (stable) then used to indicate "full professor":

> Actually, the facts did not correspond to the term, because about ten years later there came the so-called racial laws, according to which all Jewish professors, stable or not, were removed from their posts.[44]

Among the works of those years, which partly continue the previous research lines, there is a group of papers that originates from a note of 1921, in which Corrado Segre introduces the concept of approximation order in the incidence of two planes, or of any two spaces that are infinitely close.[45] Terracini was to present a summary exposition of these works several years later in two lectures held, respectively, in Louvain in 1951 and in Marseille in 1956. In this connection, Terracini wrote:

> Many years ago systems of ∞^1 (or more) planes whose two "consecutive" planes always intersect had been considered (s. 6). Having now at my disposal the notion of the order of approximation in which two "consecutive" planes may intersect, the problem has arisen to put in relation the old theory with the new notion. I could give a complete enumeration of the systems of ∞^1 or more planes whose consecutive planes always intersect in an order of approximation greater than usual. Projective applicabilities of a higher order play also a part: isothermal-asymptotic surfaces of S_3 appear from a new point of view.[46]

Terracini faces the problem of connecting the old theory with the new notion in the belief that "in science when, after long research work, a result is finally reached,

[43] Cf. e.g [29].

[44] Cf. [58], p. 110: *In realtà i fatti corrisposero poco bene al termine, perché una decina di anni dopo vennero le cosiddette leggi razziali, in base alle quali tutti i professori ebrei, stabili o no che fossero, furono rimossi dai loro posti.*

[45] See [53]; [54], also in [59] I, pp. 326–331); and [55].

[46] SPSL, *AT*: Terracini, A. Some indications ....

one must not be satisfied with the result in its brute form: it must be checked, criticized, elaborated, considered in a more general framework."[47]

Testifying to his teaching activity in this period, there remain 13 notebooks relating to the higher geometry courses[48] and 5 registers,[49] which show that Terracini, like his mentor Segre, in addition to noting his lessons in small booklets, changed the subject of the course almost every year and attached importance to the historical aspects of his discipline. In particular, he devoted the course of 1934–1935 to Segre's geometrical work and its developments.

5.2 *"Amputation" of the Italian Scientific Community and the Decision to Emigrate*

At the time, the Fascist regime had easily taken over most of the vital sectors of the nation, including that of the community of mathematicians. In 1936, the government had not granted authorization to participate in the International Congress of Mathematicians in Oslo because Norway, following the directives of the League of Nations, had sanctioned Italy for the attack on Ethiopia. Severi himself, although the regime was favorable to him, was unable to participate although he had been invited to hold a plenary lecture and was president of the international commission charged with studying international cooperation of mathematicians. Terracini was also denied authorization.[50]

The following year, he did not participate in the first Congress of the UMI (Florence, 1–3 April 1937), although he had been a member of the Union since its foundation in 1922; instead, his friend Bompiani, at the time a member of the Scientific Commission of the UMI, held a general lecture on modern addresses in differential projective geometry. The congress was characterized by opportunistic behavior towards the regime, both in the exaggeratedly celebratory tones of the inaugural speeches and in the decision to give ample space to applied mathematics, the only ones that could interest the government [4].

However, it was only in 1938, when the anti-Jewish campaign began, that Terracini seem to have realized the seriousness of the situation. On July 14, the manifesto of racist scientists, *Il Fascismo e i problemi della razza* [Fascism and the problems of race], appeared in *Il Giornale d'Italia* and was republished on August 5 by the magazine *La difesa della razza*. After that, there was a rapid crescendo that

[47] Cf. [57], p. 75: *nella scienza quando, dopo un lungo lavoro di ricerca, si giunge infine ad un risultato, non bisogna accontentarsi del risultato nella sua forma bruta: bisogna controllarlo, criticarlo, elaborarlo, considerarlo in un quadro più generale.*

[48] Cf. http://www.corradosegre.unito.it/fondo_terracini_q.php. In [17].

[49] ASUT: *Scienze MFN*, Alessandro Terracini, *Registri delle lezioni* from 1936–1937 to 1961–1962.

[50] ASUT: Alessandro Terracini, Fascicolo personale, A. Terracini to S. Pivano (rector of the University of Turin), 17 April 1936, and C. De Vecchi di Val Cismon (Minister of National Education) to S. Pivano, 30 May 1936.

led on 22nd of the same month to the census of the Jews present in the Kingdom of Italy, a survey of an "eminently political character,"[51] which provided the regime with the most effective tool for identifying those to be targeted. Terracini recalls that period with bitter words:

> The daily campaign which tended to isolate the Jews from the core of the remaining Italian population and to hold them up to public contempt and hatred, was extremely upsetting. In the second half of August the news arrived from friends, who were vacationing in Cogne at that time, that the order to take a census of Jews had already arrived there. Then came the shame of so-called discrimination—which brought with it the shameful example of some Jews who, wishing in some way to further their aspiration to change their too obviously Jewish surnames to take on "Aryan" ones, did not hesitate to declare publicly that their mothers, who in reality had never done their husbands the slightest wrong, had deceived them with "Aryans" shortly before their birth.[52]

In the autumn of the same year, the so-called racial laws were enacted, with very serious consequences for the Jewish community, which was eliminated from all the vital ganglia of the nation.

In the newspaper *Critica fascista* on September 15, 1938, we read these chilling words:

> In school, man's personality is formed, so the purge had to begin in school. If we want one hundred percent Italians, we must make them such; therefore, we must have a school that is one hundred percent Italian, and therefore such in the teachings, in the books and in the pupils. [...] Italian science was in danger of being greatly compromised by this tenacious parasitic vegetation; from today our universities are suddenly freed.[53]

Italian universities were dramatically affected. The "amputation"[54] of the national scientific community was made possible by the previous mass census operation, and it should be noted that there were no real forms of resistance in

[51] Cf. [39], p. 62: "carattere eminentemente politico."

[52] Cf. [58], p. 119–120. *La quotidiana campagna, che tendeva a isolare gli ebrei dal nucleo della rimanente popolazione italiana e ad additarli al pubblico disprezzo e odio, riusciva estremamente molesta. Nella seconda metà di agosto giunse da amici, che in quel periodo villeggiavano a Cogne, la notizia che là era già arrivato l'ordine di censire gli ebrei. Poi giunse l'onta delle cosiddette discriminazioni—che portarono con sé il vergognoso esempio di qualche ebreo, che, desiderando suffragare in qualche modo la sua aspirazione a cambiare il proprio cognome troppo manifestamente ebraico per assumerne uno «ariano», non esitò a dichiarare pubblicamente che la propria madre, la quale in realtà non aveva mai fatto il minimo torto al marito, poco prima della sua nascita lo aveva invece ingannato con un 'ariano'.*

[53] Cf. [14], p. 339: *Nella scuola si forma la personalità dell'uomo, perciò nella scuola si doveva cominciare l'epurazione. Se vogliamo italiani al cento per cento, dobbiamo formarli tali; dunque dobbiamo avere una scuola che sia italiana al cento per cento; quindi tale negli insegnamenti, nei libri e negli scolari. [...] La scienza italiana rischiava di essere molto compromessa da questa tenace vegetazione parassitaria, da oggi le nostre Università vengono di colpo liberate.*

[54] *Ivi.*

university and academic institutions, except for isolated cases such as that of Benedetto Croce.[55]

The Italian Mathematical Union itself proved to be completely subservient to the regime. All solidarity was denied to the teachers and colleagues affected by the shameful measures, and indeed people took advantage of their exclusion from the scientific and academic community. It is worth recalling once again the unspeakable statements made during the session of the scientific commission on December 10, 1938:

> The Italian school of mathematics, which has acquired vast renown throughout the scientific world, is almost entirely the creation of scientists of the Italic (Aryan) race [...]. Even after the elimination of some scholars of the Jewish race, it has retained scientists who, in number and quality, are enough to maintain the tone of Italian mathematical science at a very high level, and teachers who with their intense work of scientific proselytism ensure the Nation people worthy of holding all the necessary professorships.[56]

Many years later, making a comparison with what happened in Argentina in the summer of 1943, Terracini wrote:

> However, I believe it is worth noting that the opposition to the dictatorship in Argentina, and especially in its universities, was greater when compared with Italian resistance to fascism.[57]

In Turin, the expulsion of Jewish mathematicians from the Faculty of Sciences—in addition to Terracini, also Gino Fano, Guido Fubini, and Bonaparte Colombo[58]—created a real emergency because the chairs of higher geometry, descriptive geometry, and higher analysis and the post of complementary mathematics were left uncovered, as Tricomi pointed out:

> In any case, then things went somehow forward until 1938, when, with the mad expulsion of the Jews from teaching, our universities received a terrible blow ... In particular, here in Turin, the responsibility of the entire mathematics section of our Faculty fell practically on the shoulders of BOGGIO and the undersigned, who, while multiplying their efforts, could only slow down the now unstoppable decline.[59]

---

55 The subject has been studied extensively and has recently given rise to further research on the occasion of the 80th anniversary of the promulgation of the infamous laws. We will only mention [9, 10, 21–23, 29].

56 Bollettino della Unione Matematica Italiana, (2) 1, 89 (1939): *La scuola matematica italiana, che ha acquistato vasta rinomanza in tutto il mondo scientifico, è quasi totalmente creazione di scienziati di razza italica (ariana) [...]. Essa, anche dopo le eliminazioni di alcuni cultori di razza ebraica, ha conservato scienziati che, per numero e per qualità, bastano a mantenere elevatissimo, di fronte all'estero, il tono della scienza matematica italiana, e maestri che con la loro intensa opera di proselitismo scientifico assicurano alla Nazione elementi degni di ricoprire tutte le cattedre necessarie.* On this cf. e.g. [18], § 2.

57 Cf. [58], p. 133. *Credo tuttavia degno di essere rilevato il fatto che le opposizioni alla dittatura manifestatesi in Argentina, e soprattutto nelle sue università, furono maggiori se confrontate con le resistenze italiane al fascismo.*

58 Cf. [38] and in particular the documents in the Appendix; see also [24].

59 Cf. [63], p. 36: A*d ogni modo allora le cose andarono in qualche modo avanti sino al 1938 anno in cui, con la folle cacciata degli ebrei dall'insegnamento, le nostre università ricevettero un*

The historical archive of the University of Turin preserves extensive documentation on the census of "Personnel of Jewish race," [60] including the forms that all personnel were required to fill out, and on the effects of racial laws. The efficiency with which the offices proceeded with the analysis of the data included in the cards in order to apply these laws is sadly impressive (Figs. 4 and 5).

On 3 September 1938, Alessandro wrote to his family:

> So here it happened ... and much more than expected! You just have to absorb the blow by trying not to get too upset, and thinking about what will be necessary.[61]

His friends Tricomi, Persico, and Buzano did not abandon him, and neither did Picone, who was also openly fascist. Tricomi, in his obituary, which is anything but benevolent, acknowledges that:

> [Picone] was almost childishly vain, which even led him to strut around in the lugubrious Fascist uniform, without realizing what and how many iniquities it was a symbol of! However, when it came to more serious things than a fez or *orbace* jacket with glowing badges, he showed more dignity than some vociferous ex-anti-fascists and after the tragedy of the Jews in 1938, he courageously offered a place in his Institute to Alessandro Terracini.[62]

However, Terracini, after a few months of "voluntary isolation,"[63] made the decision to emigrate in order to regain the freedom and civil, academic, and human dignity of which he had been deprived.

Thus began a new phase in his life that was to prove fruitful from various points of view.[64]

---

colpo terribile [. . .]. In ispecie, qui a Torino, la responsabilità dell'intera sezione matematica della nostra Facoltà ricadde praticamente sulle spalle di BOGGIO e di chi vi parla che, pur moltiplicando gli sforzi, non poterono far altro che rallentare l'ormai inarrestabile decadenza.

[60] Cf. the documents in ASUT: CORRISPONDENZA—Carteggio 1938 2.1—Professori ordinari; Carteggio 1938 2.1—Professori incaricati—Pratiche generali; Carteggio 1938 2.1—Professori incaricati. Schede personali.

[61] Quoted in [60], p. 444: *Ecco dunque avvenuto [. . .] e assai più di quello che si aspettava! Non c'è che da incassare il colpo prendendosela il meno che si può, e pensare a quanto sarà necessario.*

[62] Cf. [64], p. 575: *[Picone] era di una quasi puerile vanità, che lo portò perfino a pavoneggiarsi nella lugubre uniforme fascista, senza rendersi codnto di quali e quante iniquità essa era simbolo! Però, quando si trattò di cose più serie di un fez o di una giacca d'orbace con rutilanti placche, mostrò più dignità di qualche vociferoso ex-antifascista e dopo la tragedia degli Ebrei nel 1938, offrì coraggiosamente un posto nel suo Istituto ad Alessandro Terracini.*

[63] Cf. [58], p. 121: *volontario isolamento.*

[64] On Terracini's exile in Argentina, see [25].

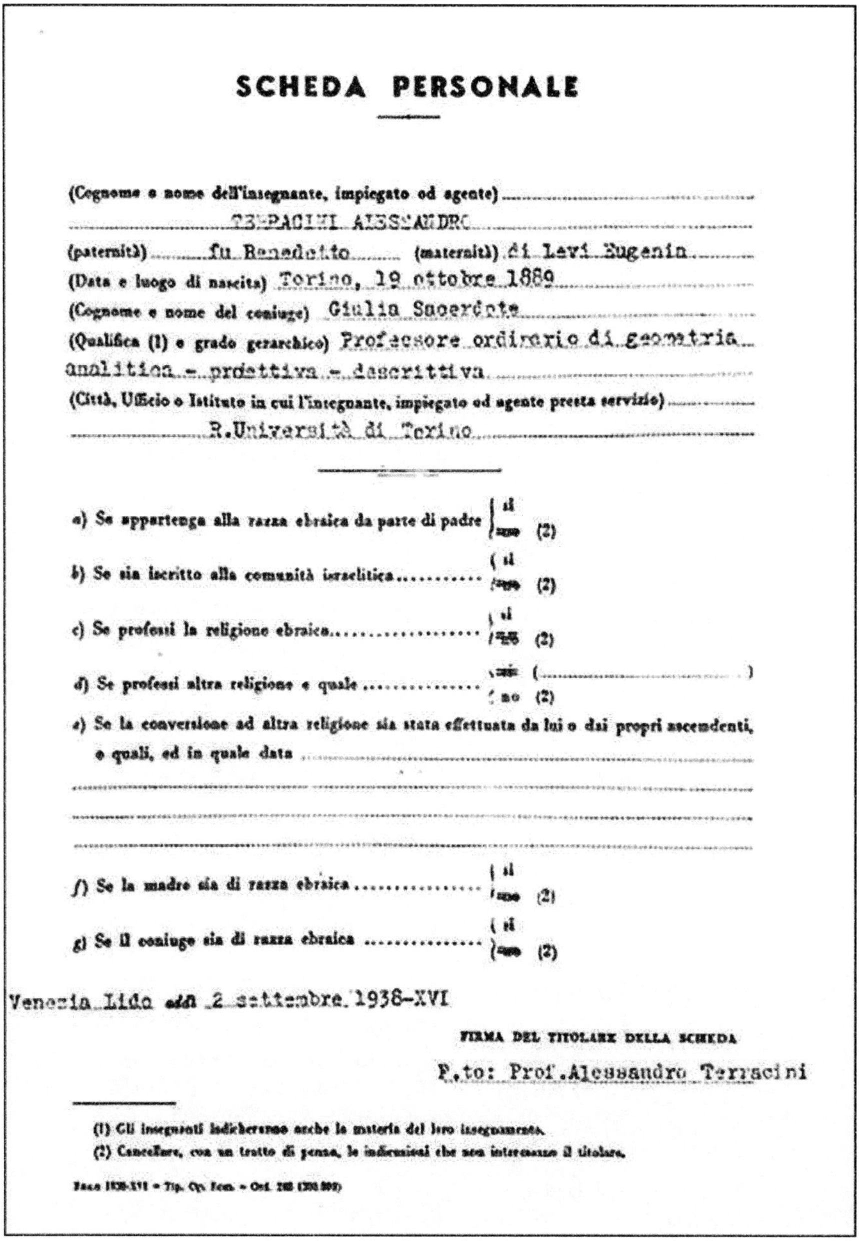

SCHEDA PERSONALE

(Cognome e nome dell'insegnante, impiegato od agente) TERRACINI ALESSANDRO

(paternità) fu Benedetto (maternità) di Levi Eugenia

(Data e luogo di nascita) Torino, 19 ottobre 1889

(Cognome e nome del coniuge) Giulia Sacerdote

(Qualifica (1) e grado gerarchico) Professore ordinario di geometria analitica - proiettiva - descrittiva

(Città, Ufficio o Istituto in cui l'insegnante, impiegato od agente presta servizio) R.Università di Torino

a) Se appartenga alla razza ebraica da parte di padre { sì / no (2)

b) Se sia iscritto alla comunità israelitica .... { sì / no (2)

c) Se professi la religione ebraica .... { sì / no (2)

d) Se professi altra religione e quale .... { sì ( ) / no (2)

e) Se la conversione ad altra religione sia stata effettuata da lui o dai propri ascendenti, e quali, ed in quale data

f) Se la madre sia di razza ebraica .... { sì / no (2)

g) Se il coniuge sia di razza ebraica .... { sì / no (2)

Venezia Lido addì 2 settembre 1938-XVI

FIRMA DEL TITOLARE DELLA SCHEDA

F.to: Prof.Alessandro Terracini

(1) Gli insegnanti indicheranno anche la materia del loro insegnamento.

(2) Cancellare, con un tratto di penna, le indicazioni che non interessano il titolare.

[illegible]

Fig. 4 ASUT, personal form of Alessandro Terracini

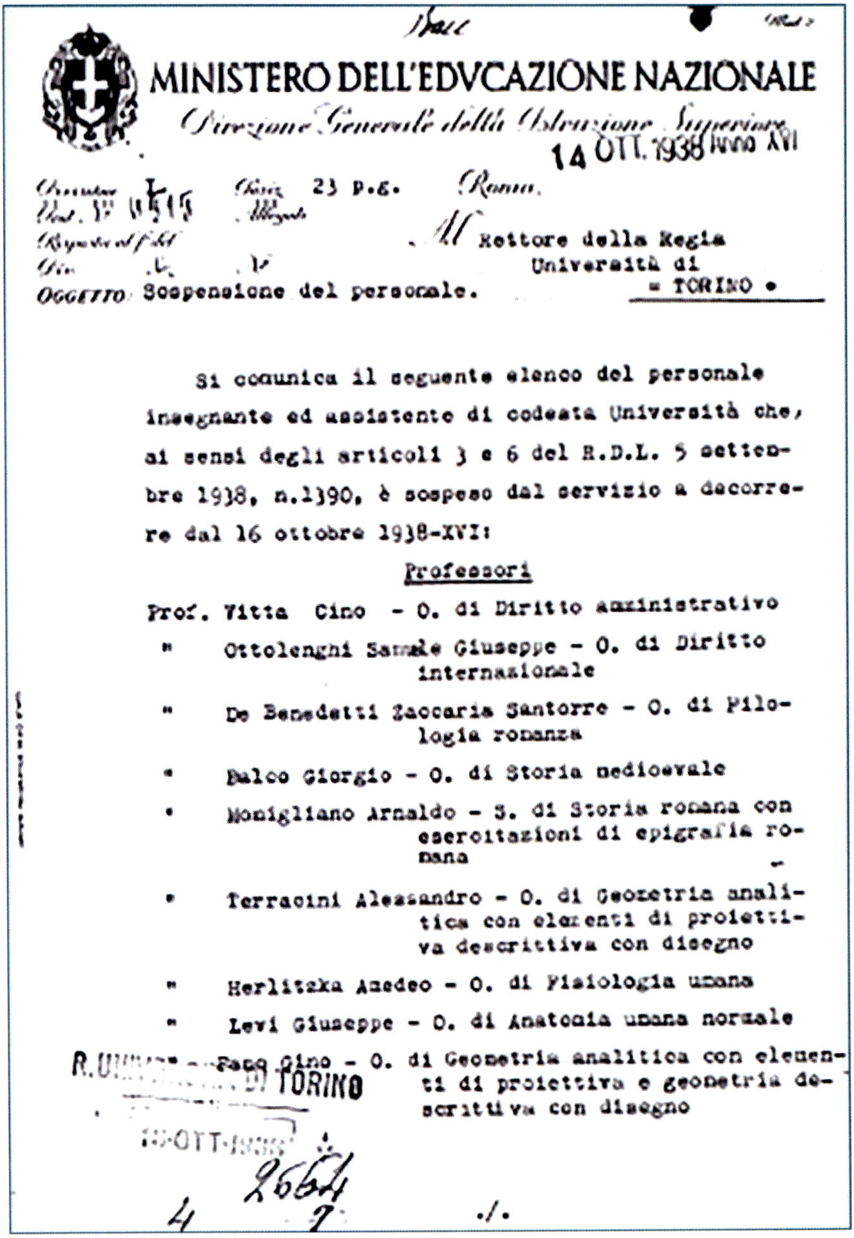

MINISTERO DELL'EDVCAZIONE NAZIONALE

Direzione Generale della Istruzione Superiore

14 OTT. 1938 ANNO XVI

Posizione I Div. 23 P.G. Roma,

Prot. N° 11515 Allegati

Risposta al foglio

Div. Sez. N°

All Rettore della Regia Università di = TORINO =

OGGETTO: Sospensione del personale.

Si comunica il seguente elenco del personale insegnante ed assistente di codesta Università che, ai sensi degli articoli 3 e 6 del R.D.L. 5 settembre 1938, n.1390, è sospeso dal servizio a decorrere dal 16 ottobre 1938-XVI:

Professori

Prof. Vitta Cino - O. di Diritto amministrativo

" Ottolenghi Samuele Giuseppe - O. di Diritto internazionale

" De Benedetti Zaccaria Santorre - O. di Filologia romanza

" Falco Giorgio - O. di Storia medioevale

" Momigliano Arnaldo - S. di Storia romana con esercitazioni di epigrafia romana

" Terracini Alessandro - O. di Geometria analitica con elementi di proiettiva descrittiva con disegno

" Herlitzka Amedeo - O. di Fisiologia umana

" Levi Giuseppe - O. di Anatomia umana normale

" Fano Gino - O. di Geometria analitica con elementi di proiettiva e geometria descrittiva con disegno

R. UNIVERSITÀ DI TORINO

OTT. 1938

2664

4 ./.

Fig. 5 Notice of suspension from teaching

References

1. Alasia, C.: Comments on the solution to Question 62. Supplemento al Periodico di matematica. **55** (1906)
2. Annuario della, R.: Università degli Studi di Catania, 1924-25, anno 481° dalla fondazione. Arti grafiche S. Monachini, Catania (1925)
3. Annuario della, R.: Scuola di Ingegneria (R. Politecnico) di Torino, Academic Year 1927-1928, Torino (1929)
4. Atti del primo Congresso dell'Unione Matematica Italiana, tenuto in Firenze nei giorni 1-2-3 aprile 1937-XV. Zanichelli, Bologna (1938)
5. Bompiani, E.: Sistemi di equazioni simultanee alle derivate parziali a caratteristica. Atti della Accademia delle scienze di Torino. **40**, 83–131 (1913–14)
6. Bompiani, E.: Postilla sull'equazione di Laplace. Bollettino della Unione Matematica Italiana. **6.2**, 61–63 (1927)
7. Bompiani, E.: Italian contribution to modern mathematics. Am. Math. Month. **88**, 83–95 (1931)
8. Bompiani, E.: Alessandro Terracini. Accademia Nazionale dei Lincei. Celebrazioni Lincee. **36**, 3–22 (1970)
9. Capristo, A.: Le accademie italiane di fronte all'espulsione dei soci ebrei. In: Le leggi antiebraiche del 1938, le società scientifiche e la scuola in Italia. Scritti e documenti XLII, pp. 157–172. Accademia Nazionale delle Scienze detta dei XL, Roma (2009)
10. Capristo, A.: Italian intellectuals and the exclusion of their Jewish colleagues from universities and academies. Telos. **164**, 63–95 (2013)
11. Ciliberto, C., Sallent, E.: Enrico Bompiani: the years in Bologna. In: Coen, S. (ed.) Mathematicians in Bologna 1861-1960, pp. 143–177. Springer, Basel (2012)
12. Ciliberto, C.: Attualità dei contributi di Alessandro Terracini su alcuni aspetti proiettivo-differenziali della geometria algebrica. In: Conte, A., Giacardi, L. (eds.) Alessandro Terracini (1889-1968). Da Torino a Torino. A 50 anni dalla morte. Quaderni. 36, pp. 111–120. Accademia delle Scienze, Torino (2020)
13. Conte, A., Giacardi (eds.): L. Alessandro Terracini (1889-1968). Da Torino a Torino. A 50 anni dalla morte. Quaderni. 36. Accademia delle Scienze, Torino (2020)
14. [Editorial Staff]: Primo: la scuola. Critica fascista. 15 settembre 1938, 338–339 (1938)
15. Fino, A.: Alessandro Terracini e la geometria differenziale proiettiva. In: Conte, A., Giacardi, L. (eds.) Alessandro Terracini (1889-1968). Da Torino a Torino. A 50 anni dalla morte. Quaderni. 36, pp. 121–132. Accademia delle Scienze, Torino (2020)
16. Fubini, G.: Un'osservazione a proposito della nota precedente. Bollettino della Unione Matematica Italiana. **6**(2), 60–61 (1927)
17. Giacardi, L. (ed.): Corrado Segre e la Scuola Italiana di Geometria Algebrica (2013–2021). Accessed 5 January 2022
18. Giacardi, L., Tazzioli, R.: Dibattiti nella comunità dei matematici italiani. L'apporto dell'Archivio dell'Unione Matematica Italiana. Atti dell'Accademia delle Scienze di Torino. **152**, 73–98 (2018)
19. Giacardi, L., Tazzioli, R.: The UMI archives – debates in the Italian mathematical community, 1922-1938. EMS Newsl. 37–44 (2019)
20. Godeaux, L.: [Compte rendu de] Fano (Gino) Et Alessandro Terracini - Lezioni Di Geometria Analitica E Proiettiva, Turin, Paravia, 1948. Bulletin des Sciences Mathématiques. **2**(73), 3–5 (1949)
21. Guerraggio, A., Nastasi, P.: Matematica in camicia nera. Il regime e gli scienziati. Bruno Mondadori, Milano (2005)
22. Israel, G.: Il fascismo e la razza. La scienza italiana e le politiche razziali del regime. Il Mulino, Bologna (2010)
23. Israel, G., Nastasi, P.: Scienza e razza nell'Italia fascista. Il Mulino, Bologna (1998)

24. Luciano, E.: From Emancipation to Persecution: Aspects and Moments of the Jewish Mathematical Milieu in Turin (1848-1938). Bollettino di Storia delle Scienze Matematiche. **38**, 127–166 (2018)
25. Luciano, E.: Alla ricerca di uno spazio di sopravvivenza intellettuale': A. Terracini, le leggi razziali e il soggiorno a Tucumán (1938-1948). In: Conte, A., Giacardi, L. (eds.) Alessandro Terracini a 50 anni dalla morte (Torino, 19.4.2018). Atti del convegno, Quaderni. 36, pp. 41–64. Accademia delle Scienze, Torino (2020)
26. Luciano, E., Scalambro, E.: Miscellanea Terracini. Schedatura (2020). http://www.corradosegre.unito.it/fondo_terracini_l.php. Accessed 5 Jan 2022
27. Mazliak, L., Tazzioli, R. (eds.): Ciascuno secondo il proprio mestiere: i matematici italiani in Guerra. Lettera Matematica PRISTEM. **92**, 5–16 (2015)
28. Milanese, D.: Alessandro Terracini (1889-1968) grande organizzatore culturale. Alcune corrispondenze inedite. Dissertation in Mathematics, supervisor L. Giacardi. University of Turin, Turin (2016–2017)
29. Nastasi, P.: La matematica italiana dal manifesto degli intellettuali fascisti alle leggi razziali. Bollettino della Unione Matematica Italiana A. **8**(3), 317–345 (1998)
30. Nastasi, P.: Un matematico alla grande guerra: Mauro Picone. Lettera Matematica PRISTEM. **92**, 17–25 (2015)
31. Nastasi, P., Tazzioli, R.: I matematici italiani e l'internazionalismo scientifico. La Matematica nella Società e nella Cultura, Rivista dell'Unione Matematica Italiana, s. **1**, VI, 355–405 (2013)
32. Nietzsche, F.: Friedrich Briefe, Januar 1887-Januar 1889. Walter de Gruyter, Berlin (1984)
33. Paoloni, G.: Il fondo "Enrico Bompiani". Quaderni PRISTEM, 2, pp. 98–106. Università Bocconi, Milano (1991)
34. Picone, M.: Tavole di tiro da montagna. Teoria e metodi di compilazione. Comando 6ª Armata, 5 and 43 (1918)
35. Picone, M.: Sull'opera matematica dell'Istituto Nazionale per le Applicazioni del calcolo nel decorso quarto di secolo della sua esistenza. In: Atti del quarto congresso dell'Unione Matematica Italiana, tenuto in Taormina nei giorni 25–31 ottobre 1951, vol. I, pp. 27–44. Cremonese, Roma (1953)
36. Proceedings of the Fifth International Congress of Mathematicians (Cambridge, 22–28 August 1912). The University Press, Cambridge (1913)
37. Rinaldelli, L.: In nome della razza. L'effetto delle leggi del 1938 sull'ambiente matematico torinese. Quaderni di Storia dell'Università di Torino. **2**, 149–208 (1997–1998)
38. Segre, C.: Preliminari di una teoria delle varietà luoghi di spazi. Rendiconti del Circolo Matematico di Palermo. **30**, 87–121 (1910)
39. Sonnino, E.: La conta degli ebrei, dalle anagrafi comunitarie al problematico censimento del 1938. In: Le leggi antiebraiche del 1938, le società scientifiche e la scuola in Italia. Scritti e documenti XLII, pp. 49–74. Accademia Nazionale delle Scienze detta dei XL, Roma (2009)
40. Tazzioli, R.: La matematica all'Università di Catania dall'Unità alla riforma Gentile. Annali di storia delle università italiane. **3**, 207–224 (1999)
41. Terracini, A.: Nota su una classe di determinanti. Giornale di matematiche. **47**, 29–32 (1909)
42. Terracini, A.: Sulle V_k per cui la varietà degli S_h (h+1) - seganti ha dimensione minore dell'ordinario. Rendiconti del Circolo Matematico di Palermo. **31**, 392–396 (1911)
43. Terracini, A.: Alcune questioni sugli spazi tangenti ed osculatori ad una varietà, Nota I. Atti della Accademia delle scienze di Torino **40**, 214–247 (1913–14)
44. Terracini, A.: Sulle superficie le cui asintotiche dei due sistemi sono cubiche sghembe. Atti della Società dei Naturalisti e Matematici di Modena. **V**, 82–107 (1919–20)
45. Terracini, A.: Sul significato geometrico della normale proiettiva. Rendiconti. Accademia dei Lincei. **3**, 584–591 (1926)
46. Terracini, A.: Sull'elemento lineare proiettivo di una superficie, Rendiconti. Accademia dei Lincei. **4**, 267–271 (1926)
47. Terracini, A.: Sulla teoria delle congruenze W. Rendiconti. R Istituto Lombardo di Scienze e Lettere. **60**, 657–674 (1927)

48. Terracini, A.: Un'osservazione sugli invarianti di un'equazione di Laplace. Bollettino della Unione Matematica Italiana 57–60 (1927)
49. Terracini, A.: Alcuni risultati di geometria proiettiva differenziale negli iperspazi. In: Fubini, G., Čech, E. (eds.) Geometria proiettiva differenziale, vol. II, Appendice III, pp. 729–769. Zanichelli, Bologna (1927)
50. Terracini, A.: Nuove ricerche sulle congruenze W. Atti. Istituto Veneto di Scienze, Lettere ed Arti. **87.2**, 179–196 (1927–28)
51. Terracini, A.: Un nuovo problema di geometria proiettiva differenziale. In: Atti del Congresso Internazionale dei matematici Bologna 3-10 Settembre 1928, vol. 4, pp. 301–304. Zanichelli, Bologna (1931)
52. Terracini, A.: Un procedimento per la risoluzione numerica dei sistemi di equazioni lineari. Ricerche di ingegneria. **III.1**, 40–48 (1935)
53. Terracini, A.: Sull'incidenza di spazi infinitamente vicini. In: Scritti Matematici offerti a L. Berzolari, Istituto matematico della R, pp. 449–478. Università, Pavia (1936)
54. Terracini, A.: Sulle varietà luoghi di $¥^1$ spazi. Rendiconti. Accademia dei Lincei. **23**, 186–191 (1936)
55. Terracini, A.: Nuove ricerche sull'incidenza di piani infinitamente vicini. Atti dell'Accademia delle Scienze di Torino. **73**, 443–459 (1937–38)
56. Terracini, A.: Guido Fubini e la geometria proiettiva differenziale. Rendiconti del Seminario Matematico. Università e Politecnico di Torino. **9**, 97–123 (1949–50)
57. Terracini, A.: I sistemi infiniti di piani nello spazio a cinque dimensioni. Rendiconti del Seminario Matematico. Università e Politecnico di Torino. **15**, 75–104 (1955–56)
58. Terracini, A.: Ricordi di un matematico. Un sessantennio di vita universitaria. Cremonese, Roma (1968)
59. Terracini, A.: Selecta. 2 vols. Cremonese, Roma (1968)
60. Terracini, L.: Cacciati dalla scuola. Carteggio ebraico '38. Belfagor. **XLV.4**, 444–450 (1931)
61. Togliatti, E.: Alessandro Terracini. Commemorazione. Atti della Accademia delle Scienze di Torino. **103**, 397–407 (1969)
62. Togliatti, E.: Alessandro Terracini. Necrologio. Bollettino della Unione Matematica Italiana **2.1** (4), 145–152 (1969)
63. Tricomi, F.: Ricordi di mezzo secolo di vita matematica torinese. Rendiconti del Seminario Matematico. Università e Politecnico di Torino. **31**, 31–43 (1972–73)
64. Tricomi, F.: Mauro Picone (1885-1977). Cenni commemorativi del Socio nazionale residente Francesco Giacomo Tricomi letti nell'adunanza dell'11 Maggio 1977. Atti dell'Accademia delle Scienze di Torino, Classe Sci. MFN. **111**, 573–576 (1977)
65. Verra, A.: Alessandro Terracini nella storia e nella matematica del suo tempo: spunti di riflessione. In: Conte, A., Giacardi, L. (eds.) Alessandro Terracini (1889-1968). Da Torino a Torino. A 50 anni dalla morte. Quaderni. 36, pp. 133–151. Accademia delle Scienze, Torino (2020)

Higher-Dimensional Geometry from Fano to Mori and Beyond

Marco Andreatta

Abstract Gino Fano's work has had a great impact on the development of modern projective geometry, in particular the studies of the varieties named after him.

Starting from Fano's results, a large number of mathematicians, often part of opposing schools, have constructed a bunch of theories in the last 50 years, which are among the most spectacular achievements of contemporary mathematics.

Keywords Fano varieties · Birational maps · Minimal model programme · Extremal rays · Rationally connected varieties

1 Introduction

The study of higher-dimensional varieties (higher than curves and surfaces) was started by B. Riemann in a remarkable lecture in 1854. Since then, the new concepts of *Mannigfaltigkeit* (variety or manifold) and of *Massverhältnisse* (metric relation) developed in various directions giving rise to different research areas in contemporary mathematics. All these theories are based on a very abstract way of thinking, similar to what happened in all arts in the same period, and they require a very strong mathematical capability and a great rigor.

The case of Algebraic Geometry was taken over soon by the Italian school at the end of 1800, for instance, by L. Cremona, G. Veronese, and C. Segre. They considered higher-dimensional projective space and properties of its linear subspaces and of its subvarieties. They studied the linear systems of divisors on these varieties, in particular the canonical system which contains information about the curvature. They understood that a classification of projective varieties should depend on the canonical divisor.

M. Andreatta (✉)
Dipartimento di Matematica, Università di Trento, Trento, Italia
e-mail: marco.andreatta@unitn.it

G. Bini (ed.), *Algebraic Geometry between Tradition and Future*,
Springer INdAM Series 53, https://doi.org/10.1007/978-981-19-8281-1_6

G. Fano, a student of C. Segre, started a systematic study of projective varieties of dimension 3 in the early 1900. His pioneering work was remarkably original and deep, although at the time the necessary mathematical tools, especially in the field of Algebra, were not well developed. It is generally accepted that his proofs are not enough rigorous for the modern standard; on the other hand, they contain many intuitions on the geometry of projective threefolds, which turned out to be correct and fundamental.

Starting from Fano's results, a large number of mathematicians, often members of opposing schools, have constructed clever theories in the last 50 years, which are among the most spectacular achievements of contemporary mathematics. A starting point for the contemporary study of Fano's legacy is the work of V. Iskovskikh and V. Shokurov. The theory of minimal models developed by the Fields medalist S. Mori gave an enormous impulse; on the one hand, it changed the approach to classification of projective varieties and on the other hand gave to the objects studied by Fano a central place in the classification. In the last 15 years, many crucial conjectures were proved, among them the feasibility of the minimal model program in any dimension, under some assumptions, in the celebrated paper by C. Birkar et al. [10].

2 Fano Varieties and Fano-Mori Contractions

We consider normal projective varieties X defined over $\mathbb{C}$; if n is the dimension of X, we sometime call X and *n-fold*. We denote by K_X the *canonical sheaf*; we assume to have good singularities such that K_X, or a multiple of it, is a line bundle (a Cartier divisor).

Let $X \subset \mathbb{P}^N$ be a projective threefold such that for general hyperplanes $H_1 and H_2$, the curve $\Gamma := X \cap H_1 \cap H_2$ is canonically embedded into $H_1 \cap H_2$ (i.e., K_Γ embeds Γ). Fano called them *Varietá algebriche a tre dimensioni a curve sezioni canoniche* [20–23].

It is not difficult to prove that a smooth threefold X (one can allow mild singularities) whose general curve section Γ is canonically embedded has the anticanonical bundle, $-K_X$, very ample. Actually the anticanonical linear system, $|-K_X|$, embeds X as a threefold of degree $2g-2$ into a projective space of dimension $g+1$, $X := X^3_{2g-2} \subset \mathbb{P}^{g+1}$, where $g = g(\Gamma)$ is the genus of Γ.

An obvious example is given by the quartic threefold in $\mathbb{P}^4$, $X_4 \subset \mathbb{P}^4$.

Fano noticed that for such varieties, the following invariants are zero:

- $h^0(X, mK_X) = 0$ for all $m \geq 1$;
  $P_m(X) := h^0(X, mK_X)$ are called *m-th plurigenera*, and if they are all zero, we say that X has *Kodaira dimension* minus infinity, $k(X) = -\infty$.
- $h^i(O_X) = 0$ for all positive i;
  in particular, the *irregularity* $q(X) = h^1(X, O_X)$ is zero.

Varieties satisfying these two conditions were called by him *Varietá algebriche a tre dimensioni aventi tutti i generi nulli*.

Fano had the insight that this class of varieties contains varieties which are non-rational, in spite of the fact that they have all plurigenera and irregularity equal to zero; they would provide a counterexample to a Castelnuovo-type rationality criteria for threefolds. None of Fano's attempts to prove non-rationality has been considered acceptable.

The first proof of the non-rationality of (all) $X_4 \subset \mathbb{P}^4$ is the celebrated Iskovskih and Manin's [32]. B. Segre constructed some unirational $X_4 \subset \mathbb{P}^4$ [55]; therefore, these unirational but not rational $X_4 \subset \mathbb{P}^4$ represent counterexamples to Lüroth problem in dimension 3, as well as to a Castelnuovo-type rationality criteria.

In the same period, Clemens and Griffiths proved the non-rationality of the cubic threefold in $\mathbb{P}^4$ [18]. Both papers gave rise to subsequent deep results and theories aimed to determine the rationality or not of Fano varieties.

Nowadays, we define a Fano manifold as follows.

Definition 1 A smooth projective variety X is called a *Fano manifold* if $-K_X$ is ample.

If $Pic(X) = \mathbb{Z}$, then X is called a *Fano manifold of the first species* or a *prime Fano manifold*. In this case, if L is the positive generator of $Pic(X)$, we have $K_X = -rL$; the integer r is called the *index of* X. □

The following is a more general "relative" definition.

Let $f : X \to Y$ be a proper surjective map between normal varieties with connected fibers; we call such an f a *contraction*. If Y is affine, we say that f is a *local contraction*. The contraction can be birational with exceptional locus a divisor; in this case, it is called a *divisorial contraction*; it can be birational with exceptional locus of codimension ≥ 2; it is called a *small contraction*; if $dimX > dimY$, f is called of *fiber type*.

Definition 2 Let $f : X \to Y$ be a contraction and assume that X is smooth or with very mild singularities; f is called a *Fano-Mori contraction* (F-M for short) if $-K_X$ is f-ample.

If $Pic(X/Y) = \mathbb{Z}$, then X is called an *elementary Fano-Mori contraction*. In this case, if L is the positive generator of $Pic(X/Y)$, we have $K_X \sim_f -rL$; the rational number r is called the *nef value of* f.

A Fano manifold can be considered as a Fano-Mori contraction with $dimY = 0$. A general fiber of a Fano-Mori contraction is a Fano manifold. The property of being a Fano variety is not a birational property. Fano varieties and Fano-Mori contractions have been playing a crucial role for 50 years in the birational and biregular study and classification of projective varieties.

The definitions of Fano manifolds and of F-M contraction could be extended to the singular case. The definitions and the studies of the appropriate setting of singularities gave rise in the last 40 years to a fundamental theory intimately related to the properties of the canonical (and anticanonical) bundle. These singularities are ordered in a hierarchy which goes from the so-called terminal and canonical singularities up to log terminal and log canonical; we omit any further details, apart from the fact that on these singular varieties one can define the canonical sheaf K_X as well as concepts of positivity and ampleness. A detailed introduction can be found in the book of J. Kollár with S. Kovacs [38].

This is a beautiful example of a typical fact of mathematical theories in which a definition contains special properties, which are not explicitly mentioned at the beginning and remain obscure for a while. Subsequent researches bring out them, displaying the intrinsic power of the original definition. It is pretty clear, however, that Fano himself was conscious that his definition should include also the case with singularities.

3 Classifications of Fano Varieties and Fano-Mori Contractions

The minimal model program (MMP) aims to classify projective varieties. The Program was initiated by S. Mori (Fields medalist in 1990 for "the proof of Hartshorne's conjecture and his work on the classification of three-dimensional algebraic varieties"), thereafter it was developed by many mathematicians including C. Hacon and J. McKernan (Breakthrough Prize in Mathematics 2018 for "transformational contributions to birational algebraic geometry, especially to the minimal model program in all dimensions") and C. Birkar (Fields medalist in 2018 for "the proof of the boundedness of Fano varieties and for contributions to the minimal model program").

According to MMP, a projective variety, smooth or with at most Kawamata log terminal singularities, is birational equivalent either to a projective variety with positive (nef) canonical bundle or to a F-M contraction, $f : X \to Y$, of fiber typer $(dimX > dimY)$.

What is even more suggestive is the fact that the birational equivalence can be obtained via a finite number of either divisorial F-M contractions or flips of small F-M contractions. The existence of the MMP was proved in dimension 3 by S. Mori [46], while for higher dimension, it has been proved in many cases by C. Birkar et al. [10].

Because of the MMP, F-M contractions became the building blocks, or the atoms, of the classification of projective varieties; as a consequence, it is worth classifying them.

Fano started a biregular classification of Fano manifolds of dimension 3 [19–23]. His work contains serious gaps and many unsatisfactory technical tools.

V.A. Iskovskih, in a series of papers, [30] and [31], has taken up the classification, and using modern tools, he has been able to justify and amplify the work of Fano, obtaining a complete classification of prime Fano threefolds. If $g := \frac{1}{2}K_X^3 + 1$ (this is equal to the genus of the curve section), he proved that $3 \leq g \leq 12$ and $g \neq 11$. For every such g, he gave a satisfactory description of the associated Fano variety.

He used Fano's method of double projection from a line; in particular, he needs the existence of a line, a delicate result proved only later by Shokurov [57].

Among his results, a nice one is the construction of the Fano manifold $X_{22} \subset \mathbb{P}^{13}$; Fano in [23] discussed the existence of X_{22}, but this was omitted by Roth in [54]. Iskovskih proved that in this case, the double projection from a line, $\pi_{2Z} : X \cdots > W \subset \mathbb{P}^6$, goes into W, a Fano threefold of index 2, degree 5, $Pic(W) = \mathbb{Z}$, and at most one singular point. The inverse is given by the linear system $3H - 2C$, where H is the hyperplane and C is a normal rational curve of degree 5. X_{22} is rational.

Some years later, S. Mukai gave a new method to classify Fano-Iskovskikh threefolds based on vector bundle constructions [50]. He provided a third description of $X_{22} \subset \mathbb{P}^{13}$ (see also [52]).

In the same period, S. Mori and S. Mukai [49] gave a classification of all Fano threefold with Picard number greater or equal than 2, and this would have concluded the classification of Fano threefold. However, in 2002, at the Fano Conference in Torino, they announced that they have omitted one of them, namely, the blow-up of $\mathbb{P}^1 \times \mathbb{P}^1 \times \mathbb{P}^1$ along a curve of tridegree $(1, 1, 3)$ (erratum in [49]). It seems now clear that there are 88 types of non-prime Fano threefolds up to deformation. Their classification is based on Iskovskihk's and on the Mori theory of extremal rays, via the so-called two-ray game.

A classification of Fano manifolds of higher dimension is an Herculean task which, however, could be done in a *finite time*. Nadel and Kollár et al., [53] and [41], proved that Fano manifolds of a given dimension form a bounded family, meaning that they are classified by the points of finitely many algebraic varieties. The same results have been proved recently by C. Birkar in the singular case [9].

Fano manifolds of index $r \geq n = dimX$ are simply the projective spaces and the quadrics, and this was proved by Kobayashi and Ochiai [35]. Fano manifolds of index $(n-1)$ are called del Pezzo manifolds; they were intensively studied by T. Fujita, who proved the existence of a smooth divisor in the linear system H generating $Pic(X)$ [26]. Mukai classified all Fano manifolds of index $= (n-2)$ under the assumption that H has an effective smooth member [50]. M. Mella proved later that this assumption is always satisfied for Fano manifolds of index $= (n-2)$, [42].

There are several projects aiming to classify singular Fano varieties in dimensions 3, 4, and 5. A very important one is carried out at Imperial College London under the guidance of A. Corti, and it is named *the periodic table of mathematical shapes*. It is estimated that 500 million shapes can be defined algebraically in 4 dimensions and a few thousand more in the fifth.

The following is a nice conjecture of Mukai [51], very useful for the classification.

Conjecture 1 Let X be a Fano manifold and ρ_X the Picard number of X, i.e., $\rho_X = dimN^1(X)$. Then

$$\rho_X(r_X - 1) \leq n.$$

More generally if $i_X = \min\{m \in \mathbb{N} \mid -K_X \cdot C = m, C \subset X \text{ rational curve}\}$ is the *pseudoindex* of X (note that $i_X = mr_X$), then

$$\rho_X(i_X - 1) \leq n \text{ with} = \text{iff } X \simeq (\mathbb{P}^{i_X-1})^{\rho_X}.$$

The conjecture holds for toric varieties [11] and in other special cases, for instance, for $n \leq 5$ [6].

In a fundamental paper, S. Mori [45], after developing his theory of extremal rays, classified all birational F-M contractions on a smooth threefold. This beautiful classification can be seen as the equivalent in dimension 3 of the Castelnuovo contraction criterion on smooth algebraic surfaces.

Later Kawamata described small local F-M contractions on a smooth fourfold [34].

Subsequently, Wisniewski and myself classified all the birational F-M contractions on a smooth fourfold [4]. All these classifications are based on a careful analysis of the deformations of rational curves contained in the fibers of the F-M contractions. The most difficult part is to construct explicit examples for all possible cases; some of them are quite peculiar and bizarre.

One can find several results on the classification of F-M contraction of fiber type on smooth threefolds and fourfolds. They range from the "classical" ones on conic bundles up to more recents which compared different birational models of F-M contractions via the so-called Sarkisov program. According to this program, every birational morphism between two fiber-type F-M contractions with the same target Y can be factorized via a finite number of few basic transformations.

In the 1980s, immediately after the introduction of the Mori theory, it appears with full evidence that the study of F-M contractions should be carried out in the singular setup. P. Francia constructed in 1981 [24] a brilliant example of commutative diagram of F-M contractions on threefolds which convinced everybody that a MMP can be performed only passing through singular cases. In particular, he showed that even on threefold with mild singularities, one can find small F-M contractions which need to be "flipped."

A careful classification of small F-M contractions on threefolds with terminal singularities, together with their flips, was given in a very deep paper by S. Mori [47] and then by S. Mori and J. Kollár [39].

Many authors, including Mori himself, are trying to obtain a complete classification of F-M contractions on threefolds with at most terminal singularities.

Based on the work of S. Mori, Y. Kawamata, Kawakita, and others on threefolds, I recently gave a characterization of birational divisorial contractions on n-fold with terminal singularities with nef value greater than $n - 3$: they are weighted blow-up of hyperquotient singularities [1].

4 Rational Curves on Fano Varieties: Rationally Connected

The name Fano variety is also used for some fundamental type of subvariety of the Grassmannian $\mathbb{G}(k, n)$ associated with a variety $X \subset \mathbb{P}^N$ (see, for instance, [28]). This is the variety of k-planes contained in X, that is,

$$F_k(X) := \{\Lambda : \Lambda \subset X\} \subset \mathbb{G}(k, n).$$

Fano studied $F_1(X)$ for some Fano manifolds X, for instance, for the cubic hypersurfaces $X_3 \subset \mathbb{P}^4$; in this case, $F_1(X_3) \subset \mathbb{G}(1, 4)$ is a surface of general type, called the Fano surface of X_3. It plays a crucial role in the proof of the irrationality of X_3 via the method of the intermediate Jacobian.

The idea of studying families of curves and not linear systems of divisors on a higher dimension variety (they coincide on surfaces), more precisely on Fano manifolds, was carried on in a spectacular way by S. Mori and developed by many other authors.

In [45], S. Mori proved the following results:

Theorem 3 *Let X be a Fano manifold. Then X contains a rational curve* $f : \mathbb{P}^1 \to D \subset X$. *In fact, through every point of X, there is a rational curve D such that*

$$0 < -(D \cdot K_X) \leq dim X + 1.$$

The proof is very nice, may be one of the nicest in the last years in algebraic geometry, and it can be quickly described, omitting some (difficult and deep) details.

Proof Take any curve C passing through the chosen point and consider its deformation space. By deformation theory and Riemann-Roch theorem, it has dimension greater or equal than

$$h^0(C, TX) - h^1(C, TX) - dim X = -C \cdot K_X - g(C) \cdot dim X.$$

Although by assumption $-C \cdot K_X$ is positive, the quantitative $-C \cdot K_X - g(C) \cdot dim X$ could not be positive, that is, the curve C may not deform. The idea of Mori at this point is to pass to a field of positive characteristic p and consider all the geometric objects over this new field, calling them X_p and C_p. There you have a new endomorphism, namely, the Frobenius endomorphism. One can change the curve C with another, which is the image of C_p via a number m of Frobenius endomorphism. Note that the genus of the curve remains $g(C)$. On the other hand, the above estimate changes by multiplying $-C_p \cdot K_{X_p}$ with p^m; in this way, one can make the quantity $-p^m \cdot C_p \cdot K_{X_p} - g(C_p) \cdot dim X_p$ positive.

Mori showed then that if a curve through a point on an algebraic variety moves, passing anyways from the point, it will "bend and break." More precisely, it will be algebraically equivalent to a reducible curve which has at least one rational component through the point. With a further step of "bend and break," he proves also that one can find a rational curve D_p with $-(D_p \cdot K_{X_p}) \leq dim X + 1$.

Having found in any characteristic a rational curve through the point, with bounded degree with respect to $-K_{X_p}$, one applies a *general principle*, based on number theory: if you have a rational curve (of bounded degree) through the point for almost all $p > 0$, then you have it also for $p = 0$.

An immediate consequence of the theorem is that a Fano variety is *uniruled*, i.e., it is covered by rational curve.

Campana [13] and Kollár et al. [40], [41] proved later that a Fano manifold is actually *rationally chain connected*, i.e., any two points can be connected by a chain of rational curves.

To be uniruled and rationally connected are birational properties.

It is straightforward to prove that if X is uniruled then $P_m(X) =$ for all $m > 0$, i.e., $k(X) = -\infty$. The converse is a long-lasting conjecture, stated by Mori in [47]:

Conjecture 2 Let X be a projective variety with canonical singularities; if $k(X) = -\infty$, then X is uniruled. □

The conjecture is false for more general singularities, for instance, for $\mathbb{Q}$-Gorenstein rational, as some examples of J. Kollár show [37]: they are rational varieties with ample canonical divisor.

Regarding rationally connectedness, we have the following conjecture of D. Mumford:

Conjecture 3 Let X be a smooth projective variety; if $H^0(X, (\Omega^1_X)^{\otimes m}) = 0$ for all $m > 0$, then X is rationally connected. □

Let me recall a curious remark of J. Harris during a school in Trento: "Mori's conjecture is well founded in birational geometry. Mumford's seems to be some strange guess, how did he come up with that?"

I think that J. Kollár was the first to notice that Mori's implies Mumford's; see [36], Chapter 4, Prop 5.7. His proof is based on the existence of the *MRC fibration* (see Theorem 9) and the *fibration theorem*, proved later by Graber-Harris-Mazur-Starr [27].

In [47], S. Mori introduced the definition of pseudo-effective divisor, i.e., a divisor contained in the closure of the cone of effective divisors in the vector space of divisors modulo numerical equivalence: $\overline{Eff(X)} \subset N^1(X)$.

He noticed that if K_X is not pseudo-effective, then $k(X) = -\infty$ and also that if X is uniruled, then K_X is not pseudo-effective. The non-pseudo-effectivity of K_X is therefore a condition in between uniruledness and negative Kodaira dimension.

The following result has been proved in [12] and in [10] using the bend and breaking theory of Mori.

Theorem 4 *Let X be a projective variety with canonical singularities; K_X is not pseudo-effective if and only if X is uniruled.* □

In a recent paper, together with C. Fontanari [2], we discuss other definitions in between uniruledness and negative Kodaira dimension which go under the title

"Termination of Adjuction." They have different levels of generality, and up to certain point, we prove the equivalence of these definitions with uniruledness. A more general definition, which has a classical flavor, was introduced by G. Castelnuovo and F. Enriques in the surface case.

Definition 5 (Termination of Adjunction in the Classical Sense) Let X be a normal projective variety and let H be an *effective Cartier divisor* on X. *Adjunction terminates in the classical sense* for H if there exists an integer $m_0 \geq 1$ such that

$$H^0(X, mK_X + H) = 0$$

for every integer $m \geq m_0$. □

It is easy to prove that uniruledness implies adjunction terminates for H and that this last condition implies that $k(X) = -\infty$.

We conjecture that if X has at most canonical singularities, then adjunction terminates for H is equivalent to uniruledness. This is true in dimension 2 by a theorem of Castelnuovo-Enriques. They proved it for *superficie adeguatamente preparate*; today, we would say for surfaces which are final objects of a MMP.

The following criteria for uniruledness were proved by Miyaoka [43]; their proof is based on a very general "bend and break technique."

Definition 6 T_X is *generically seminegative* if for every torsion-free subsheaf $E \subset T_X$, we have $c_1(E) \cdot C \leq 0$, where C is a curve obtained as intersection of high multiple of $(n-1)$ ample divisors. □

Theorem 7 *A normal complex projective variety X is uniruled if and only if T_X is not generically seminegative.* □

This criterion is a starting point to prove many nice result, including the following one of J. Wisniewski and myself [5], which is the generalization of the celebrated Frenkel-Hartshorne conjecture proved by S. Mori [44].

Theorem 8 *Let X be a projective manifold with an ample locally free subsheaf of $E \subset TX$.*
Then $X = \mathbb{P}^n$ and $E = \mathcal{O}(1)^{\oplus r}$ or $E = T_{\mathbb{P}^n}$. □

A nice conjecture in this setup has been formulated by F. Campana and T. Peternell [14].

Conjecture 4 A Fano manifold with nef tangent bundle is a rational homogeneous variety. □

Let's conclude this section with briefly mentioning two technical instruments developed in the last 30 years to study uniruled varieties. They are crucial in the proof of many deep theorems, including Theorem 8.

On a uniruled variety X, we can find a dominating family of rational curves (more precisely an irreducible component $V \subset Hom(\mathbb{P}^1, X)$ such that $Locus V = X$) having *minimal degree* with respect to some fixed ample line bundle. These families are extensively studied in the book of J. Kollár [36], and they are called *generically unsplit families*. This is a beautiful and useful extension of the concept of family of lines used by G. Fano in the study of his varieties.

For each $x \in X$, denote by V_x the family of curves from V passing through x. Let C_x be the subvariety of the projectivized tangent space at x consisting of tangent directions to curves of V_x, that is, C_x is the closure of the image of the *tangent map* $\Phi_x : V_x \to \mathbb{P}(T_x X)$. It has been considered first by S. Mori in [44] and then by many others. Hwang and Mok studied this variety in a series of papers (see, for instance, [29]) and called it *variety of minimal rational tangents* (in short, VMRT) of V.

The tangent map and the VMRT determine the structure of many Fano manifolds, for instance, of the projective space and of the rational homogeneous varieties.

Given a family of rational curves, $V \subset Hom(\mathbb{P}^1, X)$, one can define a *relation of rational connectedness with respect to* V, rcV relation for short, in the following way: $x_1, x_2 \in X$ are in the rcV relation if there exists a chain of rational curves parameterized by V which joins x_1 and x_2. The rcV relation is an equivalence relation, and its equivalence classes can be parameterized generically by an algebraic set. More precisely, we have the following result due to Campana [13] and to Kollár et al. [41].

Theorem 9 *There exist an open subset $X_0 \subset X$ and a proper surjective morphism with connected fibers $\phi_0 : X_0 \to Z_0$ onto a normal variety, such that the fibers of ϕ_0 are equivalence classes of the rcV relation.* □

We shall call the morphism ϕ_0 an *rcV fibration*. If Z_0 is just a point, then we will call X a rationally connected manifold with the respect to the family V.

More generally one can consider on a uniruled variety a rationally connectedness relation with respect to all rational curves $Hom(\mathbb{P}^1, X)$, denoted rc relation. Theorem 9 holds also in this case, and we obtain the so-called maximal rationally connected fibration (for short MRC), which we have quoted above.

The rcV and the MRC fibrations are very much connected to F-M contractions, and they are crucial tools for the study of uniruled varieties.

5 Elephants and Base Point Freeness

Let X be a Fano manifold, or more generally, let $f : X \to Y$ be a local F-M contraction. M. Reid created the neologism *general elephant* to indicate a *general element of the anticanonical system*, i.e., of the linear system $|-K_X|$.

The classification of Fano manifolds or of F-M contractions very often use and *inductive procedure* on the dimension of X, sometime called "Apollonius method", which (very) roughly speaking consists in the following steps:

1. Take a general elephant $D \in |-K_X|$, which is a variety of smaller dimension; by *adjunction formula*, it is in the special class of varieties with trivial canonical bundle.

2. *Lift up sections* of $(-K_X)_{|D}$ (or of other appropriate positive bundles) to sections of $-K_X$. This can be done via the long exact sequence associated with

$$0 \to O_X \to -K_X \to (-K_X)_{|D} \to 0.$$

This is possible thanks to the Kodaira *vanishing theorem*, which on a Fano manifolds gives $h^1(O_X) = 0$.
3. Use the sections obtained in this way to study the variety X.

More generally, one can consider a line bundle L such that either $-K_X = rL$, where r is the index of X, or $-K_X \sim_f rL$, where r is the nef value of the F-M contraction $f : X \to Y$.

Take $D \in |L|$ and do an inductive procedure on D. By adjunction formula $-K_D = (r-1)L_D$, respectively, $-K_D \sim_f (r-1)L_D$, and by Kodaira vanishing theorem sections of L_D lifts to section of L.

The procedure has classical roots and can be traced back to the Italian school of projective geometry or, as the name used above, even to classical Greek geometry. Of course, it is not as smooth as in the above rough picture, and one runs soon in many delicate problems which were handled and solved by many distinguished mathematicians in the last 50 years. Besides S. Mori and others mentioned above, we must recall V. Shokurov, Y. Kawamata, and J. Kollár.

The first crucial problem is the *existence of a general elephant*, a question unexpectedly avoided by some authors. Moreover, it is needed that the singularities of the elephant are not worse than those of X; if X is smooth, we like that also the elephant is smooth.

For the second step, it is necessary to ensure the existence of enough sections of $(-K_X)_{|D}$, more generally of L_D. This is a very delicate problem, and it goes under the name *non-vanishing theorem*. In order to get non-vanishing sections in the linear systems $|L_D|$, sometime one changes slightly the line bundle L, introducing the so-called boundary or fractional divisors. If this is the choice, then the Kodaira vanishing theorem is not sufficient, and more powerful and suitable *vanishing theorems* are needed.

The contemporary theory of MMP and of the study of F-M contractions develops as a "game" between vanishing and non-vanishing. Two "teams" were competing and/or cooperating on this. On one side, there is the group of algebraic geometers, which uses boundary and fractional divisors and the so-called Kawamata-Viehweg vanishing theorem. They refer to Shokurov as the main master of the game, and his technique has been called "spaghetti-type proofs," an attribute to the Italian origins. On the other side, there is the group of analytic geometers or complex analysts, which uses the so-called Nadel ideals and Nadel vanishing theorem; besides Nadel, the two other main active figures are Y.T. Siu and J.P. Demailly.

Maybe the most important result proved with these methods is the existence of the MMP, in dimension 3 by S. Mori [48] and later in all dimension, under some assumptions, by Birkar et al.[10].

Regarding the existence and the regularity of the elephants among the many crucial technical steps in the last 50 years, I like to recall the following ones:

- The existence of a smooth general elephant on a smooth Fano threefold (more generally of an elephant with du Val singularities on a Fano threefold with Gorenstein canonical singularities), by V.V. Shokurov [56]. This assures completeness to the proof of the classification of smooth Fano threefolds started by Fano and concluded by Iskovskih.
- The existence of a general elephant with du Val singularities on a small F-M contraction on threefold with terminal singularities, by S. Mori [48] and by S. Mori and J. Kollár [39]. This is a fundamental step to prove the existence of the flip for every small contraction on a threefold with terminal singularities and, in turn, the existence of the MMP in dimension 3.
- The existence of a general elephant with du Val singularities on a divisorial F-M contraction on threefold with terminal singularities, by M. Kawakita in a series of paper from 2001 to 2005; see, for instance, [33].
- The existence of a smooth element in the linear system $|L|$ on a Fano manifold of index $r \geq (n-2)$, where $-K_X = rL$. This is "classical" for $r \geq n$; see, for instance, [35]. It has been proved for $r = (n-1)$ by T. Fujita in 1984 (see [26]) and for $r = (n-2)$ by M. Mella in 1999 (see [42]).
- The existence of an element in $|-mK_X|$ for a positive integer m depending only on d for any d-dimensional $\mathbb{Q}$-Fano variety X, by C. Birkar in 2019 [8]. This result is the starting step to prove the boundness of the number of families of $\mathbb{Q}$-Fano variety in any fixed dimension d (BAB conjecture) [9].
- On a local F-M contraction $f : X \to Y$ such that $-K_X \sim_f rL$, the line bundle L is base point-free at every point of a fiber F with $dimF < (r+1)$; if f is birational, then the same is true also for fibers F such that $dim \leq (r+1)$. This in turn, by Bertini's theorem, will give the existence of elements in $|L|$ with singularities not worse than those of X. This was proved for varieties X with klt singularities by Wisniewski and myself in 1993 [3] and extended to log canonical singularities by O. Fujino in 2021 [25].

6 Kähler-Einstein Metrics

On a Riemannian manifold (X, g), one can consider the Einstein field equations, a set of partial differential equations on the metric tensor g which describe how the manifold X should curve due to the existence of mass or energy. In a vacuum, where there is no mass or energy, the Einstein field equations simplify. In this case, the Ricci curvature of g, Ric_g, is a symmetric $(2, 0)$ tensor, as is the metric g itself, and the equations reduce to

$$Ric_g = \lambda g$$

for a smooth function λ. A Riemannian manifold (X, g) solving the above equation is called an *Einstein manifold.* It can be proven that λ, if it exists, is a constant function.

If the Riemannian manifold has a complex structure J compatible with the metric structure (i.e., g preserves J and J is preserved by the parallel transport of the Levi-Civita connection), the triple (X, g, J) is called a *Kähler manifold.*

A *Kähler-Einstein manifold* combines the above properties of being Kähler and admitting an Einstein metric. A famous problem is to prove the existence of a Kähler-Einstein (K-E for short) metric on a compact Kähler manifold. It has been split up into three cases, depending on the sign of the first Chern class of the Kähler manifold.

If the first Chern class is negative, T. Aubin and S.T. Yau proved that there is always a K-E metric. If the first Chern class is zero, then S.T. Yau proved the Calabi conjecture, that there is always a K-E metric, which leads to the name Calabi-Yau manifolds. For this, he was awarded with the Fields medal.

The third case, which is the positive or Fano case, is the hardest. In this case, the manifold not always has a K-E metric; Y. Matsushima (1957) and A. Futaki (1983) gave necessary conditions for the existence of such metric. For instance, the blow-ups of $\mathbb{P}^2$ in one or two points do not have a K-E metric. G. Tian in [58] proposed a stability condition for a complex manifold M, called *K-stability*, connected with the existence of a K-E metric; in the same paper, he proved that there are Fano threefolds of type X_{22} which do not admit a K-E metric.

In 2012, Chen, Donaldson, and Sun proved that on a Fano manifold, the existence of a K-E metric is equivalent to K-stability. Their proof appeared in a series of articles in the *Journal of the American Mathematical Society* in 2014 [15–17].

Recently, many authors studied the existence of a K-E metric on the 105 irreducible families of smooth Fano threefolds, which have been classified by Fano, Iskovskikh, Mori, and Mukai. A very nice summary is contained in the forthcoming book by Carolina Araujo, Ana-Maria Castravet, Ivan Cheltsov, Kento Fujita, Anne-Sophie Kaloghiros, Jesus Martinez Garcia, Constantin Shramov, Hendrik Süß, and Nivedita Viswanathan; see [7]. For each family, they determine whether its general member admits a K-E metric or not; in many cases, this has been done also for the special members.

References

1. Andreatta, M.: Lifting weighted blow-ups. Rev. Mat. Iberoam., **34**(4), 1809–1820 (2018)
2. Andreatta, M., Fontanari, C.: Effective adjunction theory. Ann. Univ. Ferrara Sez. VII Sci. Mat. **64**(2), 243–257 (2018)
3. Andreatta, M., Wiśniewski, J.A.: A note on nonvanishing and applications. Duke Math. J. **72**(3), 739–755 (1993)
4. Andreatta, M., Wiśniewski, J.A.: On contractions of smooth varieties, J. Algebraic Geom. **7**(2), 253–312 (1998)

5. Andreatta, M., Wiśniewski, J.A.: On manifolds whose tangent bundle contains an ample subbundle. Invent. Math. **146**(1), 209–217 (2001)
6. Andreatta, M., Chierici, E., Occhetta, G.: Generalized Mukai conjecture for special Fano varieties. Cent. Eur. J. Math. **2**, 272–293 (2004)
7. Araujo, C., Castravet, A.M., Cheltsov, I., Fujita, K., Kaloghiros, A.S., Garcia, J.M., Shramov, C., Süß, H., Viswanathan, N.: The Calabi Problem for Fano Threefolds. Cambridge University Press, Cambridge (MPIM Preprint 2021–31)
8. Birkar, C.: Anti-pluricanonical systems on Fano varieties. Ann. Math. **190**(2), 345–463 (2019)
9. Birkar, C.: Singularities of linear systems and boundedness of Fano varieties. Ann. Math. **193**(2), 347–405 (2021)
10. Birkar, C., Cascini, P., Hacon, C.D., McKernan, J.: Existence of minimal models for varieties of log general type. J. Amer. Math. Soc. **23**(2), 405–468 (2010)
11. Bonavero, L., Casagrande, C., Debarre, O., Druel, S.: Sur une conjecture de Mukai, Comment. Math. Helv. **78**(3), 601–626 (2003)
12. Boucksom, S., Demailly, J.P., Păun, M., Peternell, T.: The pseudo-effective cone of a compact Kähler manifold and varieties of negative Kodaira dimension. J. Algebraic Geom. **22**(2), 201–248 (2013)
13. Campana, F.: Connexité rationnelle des variétés de Fano. Ann. Sci. École Norm. Sup. (4) **25**(5), 539–545 (1992)
14. Campana, F., Peternell, T.: Projective manifolds whose tangent bundles are numerically effective. Math. Ann. **289**(1), 169–187 (1991)
15. Chen, X., Donaldson, S., Sun, S.: Kähler-Einstein metrics on Fano manifolds. I: Approximation of metrics with cone singularities. J. Amer. Math. Soc. **28**, 183–197 (2015)
16. Chen, X., Donaldson, S., Sun, S: Kähler-Einstein metrics on Fano manifolds. II: Limits with cone angle less than 2π. J. Amer. Math. Soc. **28**, 199–234 (2015)
17. Chen, X., Donaldson, S., Sun, S: Kähler-Einstein metrics on Fano manifolds. III: Limits as cone angle approaches 2π and completion of the main proof. J. Amer. Math. Soc. **28**, 235–278 (2015)
18. Clemens, H., Griffiths, Ph.: The intermediate Jacobian of the cubic threefold. Ann. Math. **95**(2), 281–356 (1972)
19. Fano, G.: Sopra alcune varietà algebriche a tre dimensioni aventi tutti i generi nulli. Atti Acc. Torino **43**, 973–984 (1908)
20. Fano, G.: Su alcune varietà algebriche a tre dimensioni aventi curve sezioni canoniche. In: Scritti Matematici offerti a L. Berzolari, Pavia (1936)
21. Fano, G.: Sulle varietà algebriche a tre dimensioni aventi curve sezioni canoniche. Mem. Acc. d'Italia, **VIII**, 23–64 (1937)
22. Fano, G.: Nuove ricerche sulle varietà algebriche a tre dimensioni aventi curve sezioni canoniche. Comm Pont. Acc. Sc. **11**, 635–720 (1947)
23. Fano, G.: Su una particolare varietà algebrica a tre dimensioni aventi curve sezioni canoniche. Rendiconti Acc. Naz. Lincei, **8**(6), 151–156 (1949)
24. Francia, P.: Some remarks on minimal models. I. Compositio Math. **40**, 301–313 (1980)
25. Fujino, O.: A relative spannedness for log canonical pairs and quasi-log canonical pairs. Ann. Sc. Norm. Super. Pisa Cl. Sci. **23**(5), 265–292 (2022)
26. Fujita, T.: Classification Theories of Polarized Varieties. London Mathematical Society Lecture Note Series, vol. 155. Cambridge University Press, Cambridge (1990)
27. Graber, T., Harris, J., Mazur, B., Starr, J.: Rational connectivity and sections of families over curves. Ann. Sci. École Norm. Sup. (4) **38**(5), 671–692 (2005)
28. Harris, J.: Algebraic Geometry. A First Course. Graduate Texts in Mathematics, vol. 133. Springer, New York (1995)
29. Hwang, J.M., Mok, N.: Birationality of the tangent map for minimal rational curves. Asian J. Math. **8**(1), 51–63 (2004)
30. Iskovskih, V.A.: Fano 3-folds I. Math. U.S.S.R. Izv **11**, 485 (1977)
31. Iskovskih, V.A.: Fano 3-folds II, Math. U.S.S.R. Izv **12**, 469 (1978)

32. Iskovskih, V.A., Manin, J.I.: Three-dimensional quartics and counterexamples to the Lüroth problem. Math. USSR-Sb. **15**, 41–166 (1971)
33. Kawakita, M.: Divisorial contractions in dimension three which contract divisors to smooth points. Invent. Math. **145**(1), 105–119 (2001)
34. Kawamata, Y.: On the length of an extremal rational curve. Invent. Math. **105**(3), 609–611 (1991)
35. Kobayashi, S., Ochiai, T.: Characterizations of complex projective spaces and hyperquadrics. J. Math. Kyoto Univ. **13**, 31–47 (1973)
36. Kollár, J.: Rational curves on algebraic varieties. Ergebnisse der Mathematik und ihrer Grenzgebiete. 3. Folge. A Series of Modern Surveys in Mathematics [Results in Mathematics and Related Areas. 3rd Series. A Series of Modern Surveys in Mathematics], vol. 32. Springer, Berlin (1996)
37. Kollár, J.: Is there a topological Bogomolov-Miyaoka-Yau inequality? Pure Appl. Math. Q. **4**(2) (2008). Special Issue: In honor of Fedor Bogomolov. Part 1, 203–236
38. Kollár, J. in Collaboration with Kovács, S.: Singularities of the Minimal Model Program. Cambridge University Press, Cambridge (2013)
39. Kollár, J., Mori, S.: Classification of three-dimensional flips. J. Amer. Math. Soc. **5**(3), 533–703 (1992)
40. Kollár, J. Miyaoka, Y., Mori, S.: Rational curves on Fano varieties. In: Classification of Irregular Varieties (Trento, 1990), Lecture Notes in Mathematics, vol. 1515, pp. 100–105. Springer, Berlin (1992)
41. Kollár, J., Miyaoka, Y., Mori, S.: Rational connectedness and boundedness of Fano manifolds. J. Differ. Geom. **36**(3), 765–779 (1992)
42. Mella, M.: Existence of good divisors on Mukai varieties. J. Algebraic Geom. **8**(2), 197–206 (1999)
43. Miyaoka, Y.: The Chern classes and Kodaira dimension of a minimal variety. In: Algebraic geometry, Sendai, 1985, Advanced Studies in Pure Mathematics, vol. 10, pp. 449–476. North-Holland, Amsterdam (1987)
44. Mori, S.: Projective manifolds with ample tangent bundles. Ann. Math. **110**(3), 593–606 (1979)
45. Mori, S.: Threefolds whose canonical bundles are not numerically effective. Ann. Math. **116**(1), 133–176 (1982)
46. Mori, S.: On 3-dimensional terminal singularities. Nagoya Math. J. **98**, 43–66 (1985)
47. Mori, S.: Classification of Higher-Dimensional Varieties. Algebraic Geometry, Bowdoin, 1985. Proceedings of Symposia in Pure Mathematics, vol. 46, no. 1, pp. 269–331. American Mathematical Society, Providence (1987)
48. Mori, S.: Flip theorem and the existence of minimal models for threefolds. J. Amer. Math. Soc. **1**, 117–253 (1988)
49. Mori, S., Mukai, S.: Classification of Fano 3-folds with $B_2 \geq 2$. Manuscripta Math. **36**(2), 147–162 (1981/82). Erratum, Manuscripta Math. **110**(3), 407 (2003)
50. Mukai, S.: Biregular classification of Fano 3-folds and Fano manifolds of coindex 3. Prc. Natl. Acad. Sci. USA **86**, 3000–3002 (1989)
51. Mukai, S.: Open problems. In: Birational Geometry of Algebraic Varieties. Taniguchi Foundation, Katata (1988)
52. Mukai, S., Umemura, H.: Minimal Rational Threefolds. Algebraic Geometry (Tokyo/Kyoto, 1982). Lecture Notes in Mathematics, vol. 1016, pp. 490–518. Springer, Berlin (1983)
53. Nadel, A.M.: The boundedness of degree of Fano varieties with Picard number one. J. Amer. Math. Soc. **4**, 681–692 (1991)
54. Roth, L.: Algebraic Threefolds, Ergebnisse der Mathematik und ihrer Grenzgebiete, Heft 6. Springer, Berlin (1955)
55. Segre, B.: Variazione continua ed omotopia in geometria algebrica. Ann. Mat. Pura Appl. **50**(4), 149–186 (1960)

56. Shokurov, V.V.: Smoothness of a general anticanonical divisor on a Fano variety. Izv. Akad. Nauk SSSR Ser. Mat. **43**(2), 430–441 (1979)
57. Shokurov, V.V..: The existence of a straight line on a Fano 3-folds. Math. U.S.S.R. Izv **15**, 173 (1980)
58. Tian, G.: Kähler Einstein metrics with positive scalar curvature. Invent. Math. **130**(137), 1–37 (2021)

Gino Fano (1871–1952)

The Scientific Trajectory of an Italian Geometer Between Internationalism and Persecution

Livia Giacardi, Erika Luciano, and Elena Scalambro

Abstract Gino Fano is the first of the important group of Corrado Segre's disciples and, when he began his university studies in Turin in 1888, various circumstances favored his scientific maturation. The purpose of this essay is to highlight some less known aspects of his life and work taking into consideration the manuscripts and other unpublished documents kept in various archives in Italy and abroad. Three aspects are specially dealt with, namely:

- Fano's research, teaching, and dissemination activities in the wake of Segre's mastership and legacy, both on the national and on the international scene;
- His exile experience in Switzerland after racial discrimination;
- Fano's material and immaterial heritage in his works on threefolds.

Keywords Fano's scientific apprenticeship with C. Segre · Fano in Göttingen with Klein · Epistemological vision and teaching of mathematics · Racial laws · Swiss exile · Fano's material and immaterial heritage · Fano threefolds

Acronyms and Abbreviations

ACT Archivio Colonnetti di Torino
AMS American Mathematical Society
ASUT Archivio Storico dell'Università di Torino

The authors are members of GNSAGA, Group of INdAM.

L. Giacardi (✉) · E. Scalambro
Department of Mathematics, University of Turin, Turin, Italy
e-mail: livia.giacardi@unito.it; elena.scalambro@unito.it

E. Luciano
Department of Philosophy and Education Sciences, University of Turin, Turin, Italy
e-mail: erika.luciano@unito.it

G. Bini (ed.), *Algebraic Geometry between Tradition and Future*,
Springer INdAM Series 53, https://doi.org/10.1007/978-981-19-8281-1_7

AUL *Papers of Professor William Henry Young and Grace Chisholm Young*. Special Collections and Archives, University of Liverpool
BSMT Biblioteca Speciale di Matematica, Università di Torino
CUI Campo Universitario Italiano di Losanna
FESE Fond Européen de Secours aux Étudiants
GF Gino Fano
ICM International Congress of Mathematicians
MIT Massachusetts Institute of Technology
r recto
SPSL Society for the Protection of Science and Learning
UMI Unione Matematica Italiana
v verso

1 Introduction

Gino Fano (Mantua, January 5, 1871—Verona, November 8, 1952) is the first of the important group of Corrado Segre's disciples and, when he began his university studies in Turin in 1888, various circumstances favored his scientific maturation. Segre had just obtained the professorship of higher geometry and was broadening his research horizons under the influence of the German School, in particular of Alexander Brill, Max Noether, and Felix Klein.[1] As Fano wrote many years later, Segre was able to rapidly assimilate and make his students appreciate the research of the most important foreign mathematicians, allowing him to set going that scientific project that was to make him the new leader in the field of Italian geometry:

> He [Segre] became so, just in the moment in which Cremona's scientific activity had completely ceased, the new leader of Ital[ian] geometry, the founder of a new school. He was also able to learn, to make his own, and to let estimate by his pupils all that, for the development of his programme, was to be got from the most important foreign mathematicians (Klein, Noether, Lie, Cayley, Zeuthen, Darboux, ...); and by means of his 35 years of teaching, about all most various branches of geometry, diff[erential]. and enumerative geom[etry] (abzähl[ende] Geom[etrie]) included, he had a very great influence on the development of all geometry in Italy. ([3], c. 63*r*)

At that time in Turin, there was also Guido Castelnuovo, called by Segre in 1887 as assistant to Enrico D'Ovidio. Both soon realized Fano's mathematical skills, took his training to heart, and immediately oriented him towards the most topical research themes.

Fano came from a wealthy Jewish Mantua family,[2] and after starting his university studies as a student engineer at the University of Turin, he soon moved on to mathematical studies (Fig. 1).

[1] On Corrado Segre's work, see the recent volume by Casnati et al. [1]; on the birth of the Italian School of algebraic geometry, see [2].

[2] For Fano's biography, see inter alia [4–12].

Fig. 1 Gino Fano in 1887

While still a student, under Segre's prompting, Fano translated Klein's Erlangen program and published his first works. In 1892, he graduated[3] with a dissertation in hyperspace geometry supervised by his teacher, but also stimulated by Castelnuovo's research on curves of the highest genus in a projective space.[4]

[3] See the degree certificate in ASUT XD 193, 36.

[4] The memoir by Fano [13] is taken from the thesis. See Segre's report in *Atti dell'Accademia delle Scienze di Torino* 28 (1892–93), 865–866. Fano's mathematical work can be accessed at http://www.bdim.eu/item?id=GM_Fano.

The following year, Segre sent him to Göttingen for a period of study with Klein to complete his training. Back in Italy, in 1894, he went to Rome as an assistant to Castelnuovo, who in 1891 had obtained the professorship in that city. In 1899, he won the competition for the professorship in Messina, but in 1901, again following a competition, he obtained the chair of projective and descriptive geometry with drawing at the University of Turin, where he taught continuously until 1938. For Fano, this was a period of intense work in the field of research, teaching, and scientific dissemination. In particular, in 1904, he published his first article on three-dimensional algebraic varieties, a theme that was to occupy him throughout his life. On account of the racial laws enacted in 1938, he was forced to emigrate to Switzerland and, after returning to Italy, was to reside alternately in Italy and in the United States because his sons Ugo and Robert lived and worked overseas. The three values that inspired his whole life were, as his son Robert recalls, "his family, his country, his profession" [6].

The purpose of our essay is to highlight some lesser known aspects of Fano's life and work taking into consideration the manuscripts and other unpublished documents kept in various archives in Italy and abroad. The three points we intend to develop are the following:

- From Segre's School to achievements on the international scene: research, teaching, and dissemination of ideas;
- Racial laws, Swiss exile, and the return to Italy;
- Fano's material and immaterial heritage: the case study of his works on three-folds.

In order better to highlight the most significant features of these aspects of Fano's life and work, three different historical methodologies have been adopted. The category of "School" has been considered the most appropriate to address the issues developed in Sect. 2 and specifically to highlight the influence of Segre on Fano's work.[5] To suitably investigate the late period of Fano's trajectory, we decided to place it within a general phenomenon, i.e., the emigration of Jewish mathematicians from racist Italy. Consequently, Sect. 3 is to be considered as a case study in social history of mathematics, which narrative adopts the categories developed for the study of political emigration on racial ground from central and eastern Europe.[6] The last part (Sect. 4), in applying the heritage investigation approach, both in material and cultural dimensions, leads to a reassessment of some of the best mathematical contributions by Fano: the three-dimensional algebraic varieties today called Fano threefolds.[7]

[5] See Rowe, D.: Mathematical Schools, Communities, and Networks. In Jo Nye, M. (ed.) The Cambridge History of Science, 5, Modern Physical and Mathematical Sciences, pp. 113–132. Cambridge University Press, Cambridge (2003).

[6] See Ash, M. and Söllner, M.: Forced Migration and Scientific Change. Emigré German-Speaking Scientists and Scholars After 1933. Cambridge University Press, Cambridge (1996).

[7] See [14].

2 From Segre's School to Achievements on the International Scene

During the training phase, three mentors, Segre, Castelnuovo,[8] and Klein, oriented Fano's research in three main directions: projective geometry in a higher dimensional space, birational geometry, and Lie's theory of groups.

The sectors in which he made significant contributions are manifold: foundations of hyperspace projective geometry; algebraic curve theory; K3 surfaces and Enriques surfaces; continuous groups of transformations in projective and birational geometry; line geometry; algebraic varieties defined by linear differential equations; research on three-dimensional algebraic varieties; and birational geometry in dimension three.

Alongside this research work, Fano made a valuable contribution as a writer of treatises and with the dissemination activity that characterized the various moments in his life.

It is not one of our objectives to investigate all his varied scientific activity, on which significant literature is already available.[9] We will take into consideration only those aspects that are useful for illustrating the points we intend to develop.

2.1 *Early Research as a Student*

As already mentioned, Segre devoted himself to Fano's training starting from the second year of his university studies. In 1899, he entrusted his pupil with the translation of Klein's Erlangen program into Italian, having long recognized its relevance for the development of geometric research.[10] On that occasion, he wrote to Klein:

> I would like, for the benefit of the Italian geometers who hardly know it, to publish an Italian version [of Klein's Erlangen Program] which I would have done by one of my students (who has even already sketched it out) and which I would correct myself with the utmost care.[11]

Fano's translation, the first of the Erlangen program, was published in 1890 in the *Annali di Matematica Pura ed Applicata* [21] and was the first contact between the young researcher and Klein, a contact that was to prove important from various points of view.

---

[8] See the letters from Fano to Castelnuovo (60 documents from 1889 to 1903) in [15].

[9] On Fano's scientific work, see, inter alia, [4, 16–18] pp. 251–260.

[10] According to [16] (p. 187), the memoir by Segre [19] is "*the earliest study of geometry in the spirit of the* Erlanger Programm."

[11] C. Segre to F. Klein, Turin 19 November 1889 ([20], p. 151): *Je voudrais, pour l'avantage des géomètres italiens qui ne le connaissent presque pas, en publier une version italienne que je ferais faire par un de mes élèves (qui l'a même déjà ébauché) et que je corrigerais moi-même avec les plus grands soins.*

In that same year, during the famous course [22] in which he dealt with geometry on an algebraic curve from the triple point of view, hyperspatial, algebraic, and functional, Segre proposed the problem of assigning a system of postulates for projective hyperspace geometry:

> Define the space S_r not by means of coordinates, but rather by a series of properties from which the representation with coordinates can be deduced as a consequence.[12]

Both Fano and Federico Amodeo, who attended the course as an auditor, addressed the problem and both published separate works [23, 24], despite Segre's invitation to collaborate. Federigo Enriques, who in November 1892 had come to Turin to meet Segre, also dealt with the same problem [25]. While Fano's and Amodeo's approaches are completely abstract, i.e., make no reference to the experimental, psychological, or physiological origin of the postulates, the approach adopted by Enriques, as he himself points out, is to establish the postulates deriving from the experimental intuition of space, which are presented as the simplest ones for defining the object of projective geometry. Subsequently, his contacts with Fano, who at the time was in Göttingen, led to the publication of their correspondence on the subject [26, 27]. There thus began that interaction and comparison of methods within the School to which Segre aspired.[13]

This group of works is part of a fertile field of research, the foundations of geometry, cultivated in Italy at the time both in Segre's and Peano's Schools [29]. Fano illustrates them in a letter to Klein, who had asked him for information on the matter, and also presents a comparison with the studies of Mario Pieri, which came shortly after, highlighting differences and innovations.[14]

Fano's 1892 article, in addition to testifying to Segre's role as a mentor, is particularly significant: Fano, indeed, in order to demonstrate the independence of the postulates he established (of the n-th from the preceding $n-1$) uses a number of geometric models and so he comes to discover examples of finite projective spaces. Although these examples never became the starting points to develop a new field of research, they represent the early sources for those finite geometries that were to be developed various years later by the American School of Oswald Veblen, and in the 1930s by the German mathematicians R. Baldus, H. Liebmann, and M. Steck, authors to whom Fano was to refer in two subsequent notes on this subject published in *Rendiconti dell'Accademia dei Lincei* [30, 31].

---

[12] [23], p. 107: *Definire lo spazio* S_r *non già mediante coordinate, ma con una serie di proprietà dalle quali la rappresentazione con coordinate si possa dedurre come conseguenza.*

[13] Segre cites these papers in his 1893–1894 course, and again in his 1898–1899 course, where he presents them at greater length, also referring to an article by Mario Pieri devoted to the same subject. See BSMT, Fondo Segre, Quaderni. 5, p. 13 and Quaderni. 12, p. 3 in [28].

[14] G. Fano to F. Klein, Rome, 9 April 1897 in [20].

2.2 *Fano in Göttingen*

At the end of the nineteenth century, the favorite destinations of Italian mathematicians to perfect their studies were the German universities, which constituted an important reference point for young researchers. From 1883, Segre had constant correspondence with Klein, who a few years later was to move from the University of Leipzig to that of Göttingen. Segre himself visited the main German institutes in 1891, so it is not surprising that he sent Fano to Göttingen to strengthen his mathematical training. In his letter recommending Fano to Klein, Segre wrote:

> He is gifted with a great memory, and has a lively mind. But his tendencies are essentially directed toward geometry, pure geometry. And even though I have repeatedly encouraged him to cultivate analysis too, and in my courses, I have shown not only the synthetic methods but also analytical methods, he has remained up to now too exclusively a geometer [. . .] I believe that it is possible to strengthen him a great deal as a geometer if you can make him fully acquire the analytical tools.[15]

Fano arrived in Göttingen in mid-October 1893. Teaching at that university and German universities in general was characterized by *Lern-und Lehrfreiheit*, that is, by freedom of teaching and study and by courses on specific research topics [32].

In the winter semester and the following summer semester, Fano attended three courses held by Klein, on hypergeometric function, on second-order linear differential equations, and on elementary geometry. The latter was to converge in the volume *Vorträge über ausgewählte Fragen der Elementargeometrie* (1895), which tackles issues regarding the possibility or impossibility of performing certain geometric constructions with ruler and compass and the relative character of the concept of solving a problem, adopting a historical approach.

The courses alternated with seminars which, as Fano writes, constituted a "very important complement,"[16] because, in addition to the professors, the students—including many foreigners, especially English and American ones—were required to present a topic related or otherwise with the course. Klein devoted his summer-semester seminars to spherical functions and their applications in mathematical physics, and Fano developed two topics, one on Fourier series and another on spherical functions: *Allgemeine Bemerkungen über Fourier'sche Reihen* (June 13, 1894) and *Kugelfunktionen* (June 20, 1894)[17] (Fig. 2).

---

[15] C. Segre to F. Klein, Turin 4 October 1893: *È dotato di molta memoria ed ha un ingegno vivace. Ma le sue tendenze sono essenzialmente geometriche, per la pura geometria. E quantunque io l'abbia eccitato ripetutamente a coltivare anche l'analisi, e nei miei corsi gli abbia fatto vedere non solo i metodi sintetici ma anche quelli analitici, egli finora è rimasto troppo esclusivamente geometra* [. . .] *credo che si possa rinforzarlo di molto come geometra se si riesce a fargli acquistare pienamente gli strumenti analitici* (in [20], pp. 164–165).

[16] [32], p. 183: *complemento importantissimo.*

[17] All the presentations made in seminars from 1872 until 1913, when Klein retired, are carefully noted in his *Protocollbuch*, usually by the speaker himself See [101]. All of these reports are available online and are a valuable document regarding Klein's ambitious research and teaching

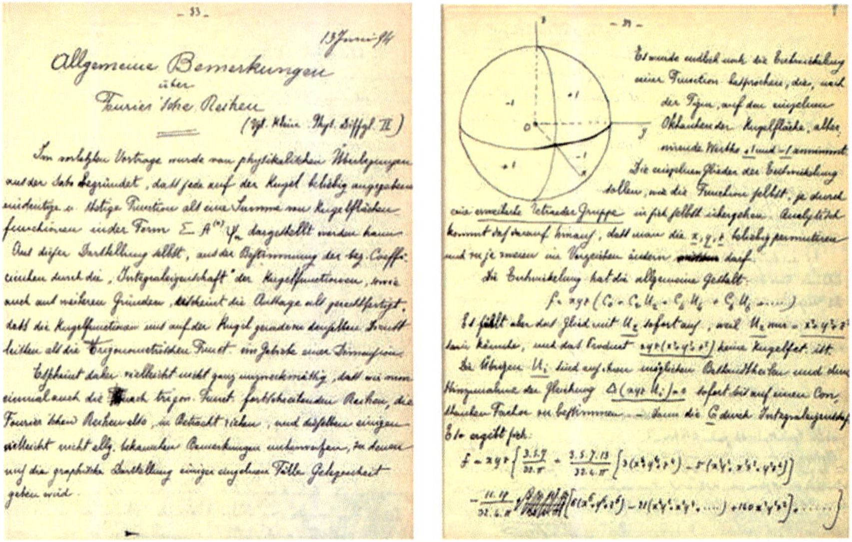

Fig. 2 The seminar by Fano in Klein's *Protocollbuch*, pp. 33 and 39

Among the students attending the seminars were Virgil Snyder, Emanuel Beke, Wilhelm Lorey, and Grace Chisholm—among other things, the latter was to come to Turin with her husband William Young in 1898–1899 to attend Segre's courses. All of them were to remain in contact with Fano, as evidenced by the various offprints they sent him over the years [97]. Relations between students and professors were favored not only by seminars, but also by extra-university meetings at the *Mathematische Gesellschaft*, where original research could be presented. Fano held two lectures here, one on recent research by Italian geometers, and the other on line geometry: *Ueber die neuesten Untersuchungen der italienischen Geometer* and *Ueber eigene Untersuchungen im Gebiete der Liniengeometrie*. Probably in the latter lecture, there was the starting point of the research on the subject that he was to expound in the 1895–1896 course in Rome and on which he was to publish important articles, where line geometry is seen, in line with Klein, as the geometry of a quadratic manifold in five-dimensional projective space (that is, the Grassmannian in a five-dimensional projective space). As David Rowe notes:

> Line geometry served here as a bridge between the older world of geometrical research and the newer style that soon led to the many more familiar achievements of Italian algebraic geometry. ([34], p. 261)

In Göttingen, Fano thus began the work of spreading Italian geometry abroad, which was to continue over the years.

---

program. Fano's seminars can be accessed at https://www.uni-math.gwdg.de/aufzeichnungen/klein-scans/klein/V12-1894-1896/V12-1894-1896.html.

Klein had had the opportunity to appreciate the work methods of the young Italian mathematician, which aimed at exploiting geometric intuition, in the style of Segre's School, and mentioned him for the 1894 prize of the Berlin Academy for research on linear differential equations. While warmly thanking him, Fano replied that at the moment his studies had different objectives:

> I am now pursuing a different but not insignificant goal, that is, the case in which, among the fundamental solutions, there are algebraic relations with constant coefficients, that is, where the periodic projective curve lies on a well determined algebraic variety (in particular, it is itself algebraic). By contrast, the Berlin Academy wants the function Z of certain variables $\frac{u_2}{u_1}, \ldots, \frac{u_n}{u_1}$ to be the object of in-depth studies.[18]

In the article that he was to publish the following year [35], Fano thanks Klein for having oriented him towards that kind of research. The German mathematician's appreciation of him was manifested again in 1899, when he offered him the chair of geometry in Göttingen previously occupied by Arthur M. Schönflies with these words:

> I conceive the chair essentially as a *geometric* chair, that is, I wish the one who holds it to extol geometric intuition and develop geometric studies in all directions. But now you know the decline of geometry in the younger German generation. I have reached the conclusion that you are precisely the man for us![19]

Fano replied very diplomatically that he was honored by such an offer but preferred a chair at an Italian university;[20] moreover, as his son Ugo recalls, he did not want to "be Germanized" ([7], p. 178). Indeed, in that same year, following a competition, he obtained a professorship at the University of Messina,[21] but his aspiration was to return to Turin. In 1901, again following a competition, he obtained the chair of Projective and Descriptive Geometry with drawing in the Piedmontese capital and here he put the multifaceted experiences of Göttingen to good use.

---

[18] G. Fano to F. Klein, Rome, 20 April 1895: *Ich jetzt ein nicht unbedeutend verschiedenes Ziel, den Fall nämlich wo zwischen den Fundamentallösungen algebraische Relationen mit const. Coeff. bestehen, d. h. wo die projectiv-periodische Curve auf einer bestimmter algebraischen Mannifaltigkeit liegt (bezw. selbst algebraisch ist. Dagegen wünscht die Berliner Akademie dass die Function Z gewisser Variablen* $\frac{u_2}{u_1}, \ldots, \frac{u_n}{u_1}$ ... *eigehend untersucht* (in [20], pp. 178–179).

[19] F. Klein to G. Fano, Göttingen 5 February 1899: *Ich fasse die Professur wesentlich als eine geometrische Professur, d. h. ich wünsche, dass der Neuzuberufende die geometrische Anschauung hervorkehrt und nach allen Richtungen die geometrischen Studien belebt. Nun kennen Sie aber den Niedergang der Geometrie in der jüngeren deutschen Generation. Ich bin also auf den Gedanken gekommen, ob nicht Sie der geeignete Mann für uns wären*! (in [20], pp. 195–196). This letter can be accessed at https://www.corradosegre.unito.it/fondofano/lettera9.pdf.

[20] G. Fano to F. Klein, Turin, 10 February 1899 (in [20], pp. 197–198).

[21] On Fano's courses and on his stay in Messina, see for example his letters to Castelnuovo: 20 November 1899, 3 December 1899, and 7 February 1900, in [15].

2.3 Research, Epistemological Vision, and Teaching

The constant reference to Segre, whom he met periodically,[22] and interactions with the other members of the School, in particular Castelnuovo and Enriques, contributed to Fano's full scientific maturation, but equally important was the period of postgraduate studies spent in Göttingen with Klein, which left a significant imprint on his research, epistemological vision, and teaching.

As is well known, the Erlangen program, which set the concept of group at the basis of the study of geometry, favored the flourishing of research on the connections between Lie's group theory and geometry and, as Hawkins ([16], p. 186) observes, Italian geometers played an especially important role. In 1893, Enriques, who among other things had spent a study period in Turin with Segre in the winter of 1893–1894, had faced the problem of determining surfaces in three-dimensional space left invariant by a continuous group of projective transformations [36], unaware that Lie had already published a work on this subject and had presented the results again in 1893 in the third volume of *Theorie der Transformationsgruppen*. On his return from Göttingen, Fano published two works [37] and [38] on the problem of determining algebraic varieties in four-dimensional space left invariant by a group of projective transformations. This research was not a mere extension of that of Enriques and Lie, because, as Hawkins writes:

> The earlier work of Lie and Enriques involved methods peculiar to 3-dimensional space and huge amounts of computation. Fano therefore looked to what would now be called the theory of the structure and representation of Lie algebras for more general methods. ([16], p. 190)

Fano therefore made use of the so-called theory of representations of Lie's algebras about 14 years before the work of Élie Cartan. Subsequently, together with Enriques, he published a work [39] on the determination of all birationally distinct types of Cremonian finite continuous groups of space and in 1897 presented the results obtained in Zurich during the International Congress of Mathematicians [103]. This work was integrated by another one [40] published in 1898.

It is not surprising that Klein invited Fano to write two articles for the *Encyklopädie der mathematischen Wissenschaften* [41, 104], one of them [41] precisely on continuous groups, which had a significant influence on Cartan, who edited a revised and expanded version in French.

We have focused on this particular line of Fano's research, which originates from Klein's work, for its scientific importance and because it is indicative of both master-disciple interactions within the School and the opening up of the School towards the outside world. Indeed, it was Segre who oriented Fano towards Klein's research and presented his 1896 and 1898 memoirs for publication in the *Memorie dell'Accademia delle Scienze di Torino*, with extensive reports in which he emphasizes the "geometric acumen" and "patient care" and highlights his

[22] See G. Fano to G. Castelnuovo, Mantova, 24 December 1901, in [15].

"considerable scientific importance."[23] It was also Segre that devoted three courses in higher geometry to continuous groups of transformations, in 1897–1898, 1906–1907, and 1911–1912,[24] where he expounded the fundamental parts of the theory, also dwelling on their classification, with an approach that privileged geometric aspects, and did not fail to refer to the contributions of his disciples.

2.3.1 The Epistemological Vision

Another aspect that clearly shows Klein's influence is Fano's epistemological vision. As is well known, for Klein, construction of mathematical knowledge takes place in three phases: gathering information derived from experience, putting the data obtained into mathematical form and proceeding to a purely mathematical treatment of the problem, and, finally, translating the mathematical results into the form most suitable for applications. Fano, like the other members of the Italian geometric School [42], takes this point of view. For the choice of the postulates at the basis of a theory, he accepts the criterion of expediency, in line with Klein, who considers postulates *vernünftige Sätze* (reasonable propositions induced by spatial intuition), and rejects the nominalist approach of Henri Poincaré, who considers them as simple conventions that are convenient but completely arbitrary as long as they are compatible with each other.

Fano also shares with Klein the unitary conception of science, the enhancement of discovery processes through intuition, "the antivenom to logic,"[25] the role of experimental procedures in mathematics too, and the distinction between *Präzisionsmathematik*, precision mathematics, and *Approximationsmathematik*, seen by Klein as the exact mathematics of approximate relations [44]. This consonance of thought is confirmed by the numerous references to the German mathematician that appear in Fano's articles on epistemology, teaching, and popularization of mathematics, in which he interprets Klein's teachings and makes them his own.

As is well known, Klein distinguished between naïve intuition and refined intuition, and highlighted the fact that the former is important in the discovery phase of a theory, while the latter intervenes in the elaboration of data furnished by naïve intuition, and in the rigorous logical development of the theory itself ([45], pp. 41–42). In harmony with this point of view, Fano attributes to intuition the role of guiding the reasoning, predicting its conclusion, and checking and experimentally confirming the result obtained ([46], p. 122). Furthermore, he distinguishes various types of intuition, all valuable: "the vision by spirit" of the Greeks; "a form of memory, developed as a consequence of past scientific work"; "the intuition of the

[23] See *Atti della R. Accademia delle Scienze di Torino*, 31, 623–624 (1895–1896) and 33, 796–797 (1897–1898): *acume geometrico, la paziente cura, la notevole importanza.*

[24] The handwritten notebooks of Segre's lessons can be found in [28], Quaderni. 11, Quaderni. 20, and Quaderni. 25.

[25] [43], p. 25: *contravveleno alla logica.*

formalist"; and "the intuitional skillfulness of a refined analyst" ([47], p. 15). As regards the relationship between intuition and logical rigor, he writes:

> With intuition [the mathematician] discovers; with logic the new discovery is broken down into the single elements, links in a chain to be reviewed in the light of criticism; with intuition [the mathematician] makes the synthesis again, precisely as the living organism is the synthesis of its cells and the former and the latter are both rich in interest; but even minute knowledge of cells is not enough to make the whole individual known. Mathematics is also a living organism, of which it could be said that logical connections constitute the skeleton; but the organism is not just a skeleton. The skeleton alone does not live![26]

For Fano, mathematics, as well as being the logical science par excellence, can be counted among the experimental sciences:

> Without the recognition of the possibility, for ourselves, of repeating the same operation several times, no mathematical knowledge would be possible; not even the greatest genius would perhaps have arrived at the notion of the natural numbers two, three, etc.[27]

By experimental procedures in mathematics, Fano means thorough and complete examination of special examples that the mathematician produces in his "intellectual laboratory" that can lead to new ideas and are a powerful method of error checking. Among experiments, he also includes the use of physics and drawing:

> Certain theories, certain calculations, can, I will say, be mentally accompanied by a geometric image, or a physical image: thanks to these, we sometimes manage to embrace in a single glance what logical deduction or calculation would show us only later.[28]

To exemplify his views, Fano mentions Klein, for whom physics was a tool of discovery: for example, to demonstrate the existence of the so-called Abelian integrals, he thought of resorting to electrical experiences; likewise, the origin of most of Riemann's ideas lies at least partially in considerations of a physical nature. The physical sciences, says Fano, have oriented mathematical research that has thus escaped the danger of becoming pure symbolism closed in on itself ([48], p. 28).

If it is true that for Fano mathematics can and must be an end in itself, almost a work of art due to its characteristics of simplicity and harmony, he is also well aware that recent developments in applied mathematics make it essential to compare

---

[26] [43], p. 27: *Coll'intuizione si scopre; colla logica si decompone il nuovo trovato nei singoli elementi, anelli di una catena per ripassarli al lume della critica; coll'intuizione di nuovo si sintetizza: appunto come l'organismo vivente è la sintesi delle sue cellule e quello e queste sono entrambi ricchi di interesse; ma la conoscenza anche minuta delle cellule non basta a far conoscere l'individuo. La matematica è anch'essa un organismo vivente, di cui si potrebbe dire che le connessioni logiche costituiscono lo scheletro; ma l'organismo non è solo scheletro. Lo scheletro da solo non vive!*

[27] [43], p. 22: *Senza la constatazione della possibilità, per noi stessi, di ripetere una medesima operazione più volte, nessuna conoscenza matematica sarebbe possibile; nemmeno il più gran genio sarebbe forse arrivato alla nozione dei numeri naturali due, tre, ecc.*

[28] [43], pp. 13–14: *Certe teorie, certi calcoli, si possono, dirò così, accompagnare mentalmente con un'immagine geometrica, o con un'immagine fisica: grazie a queste, riusciamo talvolta ad abbracciare in un sol colpo d'occhio ciò che la deduzione logica, o il calcolo ci mostrerebbero soltanto successivamente.*

"exact mathematics", which derives from a work of abstraction and idealization, and "approximate mathematics", which actually is useful in applications. The fact is that every measurement, performed with our senses or with any instruments, has a degree of approximation or precision beyond which we cannot go. In a 1911 article, Fano explicitly refers to Klein, who in his 1901 lectures in *Anwendung der Differential-und Integralrechnung auf Geometrie, eine Revision der Prinzipien*[29] stresses the importance of considering "mathematische Wissenschaft als ein zusammengehöriges Ganzes" (mathematical science as a coherent whole) and addresses the problem of the relationships between the two types of mathematics in order to bring pure mathematicians closer to technicians dealing with applications, without neglecting the importance of this approach in teaching. In his article, Fano takes up many of the examples presented by Klein, drawn from analysis, geometry, and geodesy, and concludes by emphasizing the educational value of bringing teaching closer to reality:

> it would be desirable ... to give, as soon as possible, in schools, a clear idea of the relationships between the physical world and the mathematical world, of the degree of accuracy with which mathematical concepts and formulas can represent entities and physical dependencies, and of the different ways in which the same entity or physical relationship can sometimes be idealized...; and on the other hand, if we carefully highlight... the practical value of a concept or a mathematical result, we will bring teaching close to reality, eliminating a divergence and filling a gap that are among the most fatal defects at any school level.[30]

Similar themes are also addressed by Segre in his lectures at the *Scuola di Magistero* (teacher training college) (Quaderni. 40 in [28]), which Fano had attended during his university studies; by Castelnuovo, in particular in his 1913–1914 higher geometry course [15]; by Enriques, in his epistemological writings; and by other members of the School of Segre. Among those that Fano mentions most frequently is Enriques, and in particular his dynamic vision of science seen as a process of successive approximations ([48], p. 23). Being a man broadly read in mathematics, history, and philosophy, Fano also brings in precise references to various other scholars (mathematicians, physicists, chemists, philosophers) outside the School, in addition to Klein. Among these, there are Ernst Mach, whose vision of science he shares as a minimum problem which consists of presenting the facts following the criteria of simplicity and economy of thought ([48], p. 21), as well as Borel,

---

[29] Fano carefully read this work by Klein, as we can see from the annotations on his copy of the lithograph preserved in the Biblioteca Speciale di Matematica "G. Peano" of the University of Turin. See also the letter by G. Fano to G. Castelnuovo, Colognola ai Colli, 15 July 1902 in [15].

[30] [44], p. 126: *sarebbe desiderabile che* [...] *fosse data il più presto possibile, nelle scuole, un'idea chiara delle relazioni che intercedono fra il mondo fisico e il mondo matematico, del grado di esattezza con cui i concetti e le formule matematiche possono rappresentare enti e dipendenze fisiche, dei diversi modi in cui uno stesso ente o rapporto fisico può talvolta essere idealizzato* [...]*; e d'altra parte, precisando bene* [...] *il valore pratico di un concetto o di un risultato matematico, accosteremo l'insegnamento alla realtà, eliminando una divergenza e colmando una lacuna che sono fra i difetti più fatali a ogni ordine di scuola.*

Einstein, Helmholtz, Maxwell, Peirce, Poincaré, Rignano, Riemann, and many others.

2.3.2 "Fighting Prejudices Against the Supposed Mysteries of Mathematics."[31] Fano's Commitment to Education

Fano's vision of mathematics teaching, as is natural, is closely linked to his epistemological vision. He expounds his reflections on the subject in various contexts: in the congresses of Mathesis, the national association of mathematics teachers (Florence 1908, Padua 1909, Genoa 1912, Trieste 1919, Naples 1921, Milan 1925); in the *Conferenze Matematiche torinesi* (Turin mathematical lectures) promoted by Giuseppe Peano; in various dissemination lectures, including those at the *Gabinetto di Cultura della Scuola di Guerra* (Culture Cabinet of the War School) and at the *Società di Cultura* (Culture Society) in Turin; in his institutional roles as president of the *Scuola Operaia femminile* (School of Female Workers) (1909–1937) and president of the Piedmont section (1913), later the Turin section, of Mathesis; and finally through his collaboration with the "Enciclopedia delle matematiche elementari." He was attentive to the legislative measures proposed by the government: in 1913, he commented on the syllabuses of the *Liceo moderno* (Modern High School);[32] in 1919, he expressed himself in favor of maintaining the positive aspects of the curricula of the provinces of Trento and Trieste, which had just been annexed to the Kingdom of Italy;[33] in 1921, he took sides against the suppression of the *Scuola di Magistero*; and in 1923, he highlighted positive and negative aspects of the Gentile Reform ([50], pp. 23–25).

His interest in problems connected with mathematics teaching arose from his vision of mathematics as a discipline that educates the character and accustoms people to a sense of economy of work, and to precision and clarity, but it also originated, as for Castelnuovo, from social concerns, as is evident from his commitment to tackling inequalities and illiteracy, which emerges from his work in the *Scuola Operaia femminile*. Fano writes:

> for the working classes too, education must aim not only at allowing people to acquire certain types of knowledge, but also at getting them used to a moderate intellectual discipline.[34]

---

[31] [49], p. 367: *Combattere le prevenzioni contro* [...] *i presunti misteri della matematica.*

[32] *Bollettino della Mathesis* 5.1, 46–48 (1913).

[33] *Bollettino della Mathesis* 12, 62 (1920).

[34] [46], p. 12: *anche per le classi operaie l'istruzione deve tendere, non soltanto a far acquistare determinate cognizioni, ma da abituare a una pur moderata disciplina intellettuale.*

For the multifaceted work he carried out as president and for the donations made to this school, in 1928 he was awarded the Gold Medal of Merit of Public Education.[35]

The methodological approach that guided his work concerning mathematics teaching was openly influenced by Klein, who, as is well known, was the promoter of an important teaching reform movement in Germany [53].

Like Klein, Fano repeatedly stresses the importance of establishing a bridge between secondary and university education through early introduction of the concepts of function and transformation in mathematics teaching in secondary schools ([48], p. 27). For this teaching to be profitable, it is also necessary to establish links between mathematics and reality and between mathematics and applications and to valorize experimental procedures while trying to find the right balance between rigor and intuition. In this regard, Fano takes up the simile of the tree already used by Klein:

> To ask whether ... intuition or reasoning matters most, would be like asking whether, for a tree, roots or branches and leaves, are more important and needful: the question would be badly put, because the life of the tree rests on the reciprocal action of the different organs. ([47], p. 16)

The problem of teacher training, dear to Klein, is also central to Fano. Convinced that "it is worthless to know *more* than one teaches, if this does not make the things to be taught better known,"[36] at the time of the suppression of teacher training colleges, Fano strongly supported the importance of instituting courses of "Higher views on elementary mathematics" with emphasis on the historical, critical, methodological, and teaching aspects, citing by way of example the lectures of Segre and Enriques. He also invited the faculties to accept dissertations in complementary mathematics as graduate theses and urged his colleagues to start practical training in secondary schools for prospective teachers, without waiting for ministerial decrees ([54], pp. 103 and 109).[37]

A particularly significant aspect of Fano's commitment to mathematics education is given by the numerous treatises he wrote for his university courses (in Rome and Turin, at the University and at the Polytechnic). Many of these were written in the late nineteenth and early twentieth centuries and then revised and perfected in subsequent editions, on various sectors of geometry: line geometry (1896); non-Euclidean geometry (1898, 1935); descriptive geometry (1903, 1910, 1914, 1926, 1932, 1935, 1944); projective geometry (1902, 1903, 1907); analytical geometry

[35] See G. Fano to the president of the Reale Accademia Virgiliana, Turin, 24 November 1935, in [8], pp. 149–150. For an in-depth study on this aspect of Fano's work, see [51]. For the historical background, see [52].

[36] [54], p. 102: *a nulla vale saper più di ciò che si insegna, se questo di più non fa conoscer meglio le cose da insegnare.*

[37] On the problem of teacher training in Italy and on the contribution of Italian geometers, see [55].

(1944); analytical and projective geometry (1926, 1930, 1940, and 1957 [with Terracini]); and complements of geometry (1935).[38]

Each of these treatises deserves to be studied in depth, but here we will limit ourselves to underlining the common traits that characterize them: clarity of presentation; alternation of analytical and synthetic approaches for educational purposes; presence of historical hints or a real historical approach as in those on non-Euclidean geometries; attention to applications to other scientific fields such as shadow theory, perspective, photogrammetry, and theory of relativity; and the tendency, learned from his teacher Segre, to highlight links with research in order to "allow people to presage future developments."[39]

It is sufficient to mention the example of the lesser known treatise on line geometry that originates from the free course of projective geometry that Fano held at the University of Rome in the year 1895–1896. The treatise opens with a documented historical introduction in which he illustrates the origins of line geometry from research in three different fields—geometry, mechanics, and physics—to arrive at Klein's approach, which considers the *lines* of ordinary space as *points* of Klein's quadric, which is just a smooth quadric in the five-dimensional projective space. Hence, he introduces the study of line congruences, considered as the surfaces of this quadric, and mentions the research, only sketched out, on third-order congruences, also referring to his own recent studies. Among others, he cites the contributions of Segre, who had dealt with line congruencies since his graduation dissertation and who generously made his notes on this subject available to him[40] ([57], p. 142).

Fano's lectures, as Terracini recalls, were "solemn lectures, which were prepared in every detail, but found the most effective spontaneity in the power of the arguments, in their concatenation and in the emphasis given to the fundamental ideas!"[41] Like Segre, he was a strict and demanding professor, so it is not surprising that in 1902, a few months after his return to Turin from Messina, some students attacked him in a highly controversial article that appeared in the satirical newspaper *La Campana degli studenti*.[42]

---

[38] See *Publications of Gino Fano 1890–1953*, in [4], pp. 127–137.

[39] [56], p. VI: *far presagire sviluppi futuri.*

[40] See Quaderni. 4 1891–1892: C. Segre, *Geometria della retta*, pp. 18–42, in [28].

[41] [12], p. 708: *lezioni togate, preparate in ogni particolare, ma che nella potenza delle argomentazioni, nel loro concatenamento e nel rilievo dato alle idee fondamentali ritrovavano la più efficace spontaneità!*

[42] See "La questione Fano." *La Campana degli studenti*, 27 November 1902: Fano was accused of *strage degli innocenti* (slaughter of the innocents) during the exams with rather irreverent tones. For details on the affair, see the letters: G. Fano to G. Castelnuovo, Turin, 11 and 29 November 1902 and 14 December in [15].

2.3.3 Scientific Dissemination

From the outset, Fano combined significant scientific dissemination activity in two main directions, with scientific and teaching activity. First of all, there was dissemination of the research of the Italian geometric School through various channels: stays abroad, international congresses of mathematicians, correspondence, and exchange of offprints. Fano always accompanied this activity with work of dissemination in a broad sense through various channels: the more traditional ones, such as *Mathesis*, but also the *Scuola Operaia femminile* and the *Scuola di Guerra*, articles for the "Enciclopedia Italiana," as well as journals with different readerships—teachers, philosophers, and intellectuals in general.[43] Among the themes he most often addresses are interactions between intuitive and logical aspects in the history of science, the formative value of mathematics, its role in other sciences, and the foundations of geometry. What is striking is the style he generally adopts: very clear and simple language that does not prevent him from referring to modern research in mathematics and connections with physics; a historical approach; many examples; and extensive use of metaphors to illustrate the relationship between logic and intuition and their respective roles. For Fano, metaphors serve to make a wide audience understand what he is saying by referring to aspects of real life familiar to everyone, such as chess, parts of the human body, parts of the tree, maneuvers of an army, and work of the surgeon.[44]

According to Fano, you can talk to anyone about mathematics: you just have to find the right language, talk about its applications, and make use of the history of science [102]. Only in this way can prejudices against this discipline be combated. For example, he discusses experimental demonstrations or applications of mathematics to astronomy in the *Scuola Operaia femminile*, or the educational value of mathematics, relations with physics, and the scientist's process of discovery in the *Gabinetto di Cultura della Scuola di Guerra*.

Besides, due to his sense of belonging to the Italian geometric School, Fano did not miss opportunities to disseminate its results. In 1894, still very young, he had illustrated the recent research work of Italian geometers in Göttingen and then again in 1897 during the International Congress of Mathematicians in Zurich. In 1923, the contacts with Grace Chisholm and the fame acquired earned Fano an invitation to hold a course on Italian geometry at the University College of Wales in Aberystwyth. Other cycles of lectures and conferences were later to be held in Louvain (1925), in Kazan (1926), and then in Lausanne (1942–1944).[45]

---

[43] Fano wrote articles in the following journals: *Rivista di matematica* (2), *Bollettino di bibliografia e storia delle scienze matematiche* (1), *Rivista d'Italia* (1), *Scientia* (4), *Bollettino della 'Mathesis'* (3), *Periodico di Matematiche* (1), *Nuova Antologia* (1), *Rivista di filosofia* (1), *Alere Flammam* (1), *Nuovo Cimento* (1), *Conferenze di fisica e di matematica*, then *Rendiconti del seminario matematico. Università e Politecnico di Torino* (2).

[44] See, for example [47], pp. 12 and 16, and [43], pp. 24–25, 26, 27.

[45] See Sect. 3 in this paper, and [58].

In Aberystwyth, Fano held about 20 lectures on eight themes starting from the contributions of Cremona and his successors, down to the most recent research of the Italian School of algebraic geometry, highlighting the connections with international research (Fig. 3). The themes were the following:

1. A short account of the mathematical work of Cremona [...]
2. Some notions concerning Clebsch and Noether
3. Cremona's first successors in Italy [...]
4. Summary of concepts and notions on more dimensional projective geometry (Veronese, Segre, etc.)
5. Groups, especially continuous groups of in the plane and space. [...] Birational contact-transformations in the plane
6. Geometry on an algebraic manifold, especially on an algebraic curve. Different methods [...]
7. Geometry on algebraic surfaces [...] A short account of the methods used and results obtained by Italian Geometers: their connection with Picard's theory of integrals of total differentials on surfaces
8. A short account of the new essential differences we meet with in the theory of algebraic M_3s. Irrational involutions in S_3 ([47], *Preface*)

Given the audience's poor basic knowledge of the subjects, Fano was forced to take longer than expected to explain the methods of the Italian geometers, so he was unable to deal with points 5 and 8. Perhaps some difficulties also arose from his English, defined as "eccentric" by a student who attended the courses, whose testimony we have,[46] although he was helped in the translation by his sister Maria Ettlinger, Grace Chisholm, and George A. Schott—the latter taught applied mathematics at that university.

In the Archives of the University of Liverpool (AUL), among the *Papers* of William Young and Grace Chisholm, only the typewritten lectures relating to Cremona held in February 1923 are preserved, but in the *Fondo Fano* of the Biblioteca Speciale di Matematica in Turin, there is a substantial manuscript that collects notes from different periods.[47] These notes are precious because they contain various English versions of the lectures and sometimes also the Italian version, with afterthoughts and additions, and thus show us what Fano defines as "the intellectual laboratory" (*laboratorio intellettuale*) which will be discussed later.[48] They also offer a vivid testimony to the enthusiasm that permeated the group of Segre's students in the 20 so-called golden years, to their awareness of belonging to a School, and to the collective work which Fano often refers to here and elsewhere. He writes:

---

[46] The person in question was the student Graham Sutton, who was to become the general director of the Meteorological Observatory [59].

[47] See *Appunti vari*, BSMT, Fondo Fano, Scritti. 4. The manuscript can be accessed at the website [28]. See also [106].

[48] See Sect. 4 in this paper.

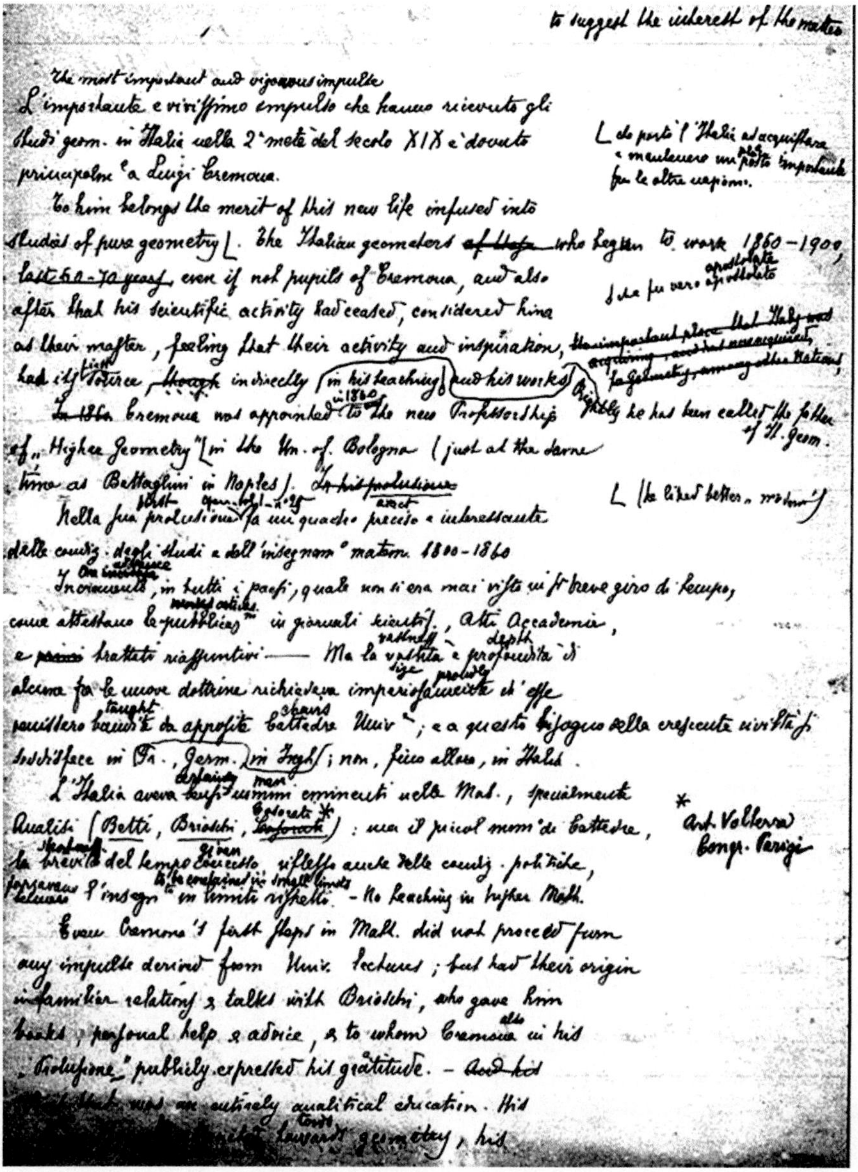

to suggest the interest of the matter

the most important and vigorous impulse

L'importante e vivissimo impulso che hanno ricevuto gli studi geom. in Italia nella 2ª metà del secolo XIX è dovuto principalmente a Luigi Cremona.

L che portò l'Italia ad acquistare e mantenere un posto importante fra le altre nazioni.

To him belongs the merit of this new life infused into studies of pure geometry L. The Italian geometers who began to work 1860–1900, even if not pupils of Cremona, and also after that his scientific activity had ceased, considered him as their master, feeling that their activity and inspiration had its first source, indirectly (in his teaching) and his works

Idea fu vero apostolato

Rightly he has been called the father of It. Geom.

Cremona was appointed in 1860 to the new Professorship of „Higher Geometry" (in the Un. of Bologna (just at the same time as Battaglini in Naples).

L (he liked better „modern")

Nella sua first prolusione (Opere vol. 1, n. 25) fa un quadro preciso e interessante delle condiz. degli studi e dell'insegnam° matem. 1800–1860

Incremento, in tutti i paesi, quale non si era mai visto in sì breve giro di tempo, come attestano le pubblicaz. in giornali scientif., Atti Accademie, e trattati riassuntivi —— Ma la vastità e profondità di alcune fra le nuove dottrine richiedeva imperiosamente ch'esse venissero bandite da apposite cattedre Univ.; e a questo bisogno della crescente civiltà si soddisface in Fr., Germ., in Inghil.; non, però allora, in Italia.

L'Italia aveva uomini eminenti nella Mat., specialmente Analisi (Betti, Brioschi, Casorati *): ma il piccol num° di cattedre, la brevità del tempo concesso, riflesso anche delle condiz. politiche, forzavano l'insegn° in limiti ristretti – No teaching in higher Math.

* Art. Volterra Congr. Parigi

Even Cremona's first steps in Math. did not proceed from any impulse derived from Univ. lectures; but had their origin in familiar relations & talks with Brioschi, who gave him books, personal help & advice, & to whom Cremona also in his „Prolusione" publicly expressed his gratitude. – … was an entirely analytical education. His … towards geometry, his

Fig. 3 Fano's draft of the lecture in Aberystwyth on Cremona's work ([3], c. 18*r*)

> It was indispensable that everything be treated and digested, that it became the blood of our blood, that we had it at our fingertips in order to be able to use it in the most advanced research ... Fecundity![49]
>
> Collective research—Segre-Castelnuovo: 1890–91 in Turin—Castelnuovo Enr[iques] (1896–900) afterwards Severi for surfaces (irreg[ular] 1904–05). Energies of investigators are summed. Their discoveries follow each other rapidly. ([3], c. 84*v*)

In addition to university lectures, Fano also held two popularizing lectures, where he sums up the various themes that characterize his vision of science.

In the first one, *Intuition in mathematics* ([47], pp. 5–17), he proposes to examine the question of whether the work of the mathematician is merely logical work. He adopts a historical approach in order to show that a rigorous treatment can only be applied to well-defined concepts and therefore to materials that our mind has already processed. He cites various examples starting from Euclid's "Elements" which with their deductive logical arrangement appear as the final result of a long period of discovery and elaboration, down to the critical studies of the nineteenth century, which led to the creation of mathematical logic and to its symbolism. He then focuses on another trend that emerged at the same time in which a "philosophic and intuitive spirit" prevails, mentioning the most representative figures, Riemann and Klein. It is clear, however, that among mathematical inventors and mathematical demonstrators, whom he defines as skilled technicians, his preferences go to the former because in his opinion it is intuition, "a pioneer of progress," that opens the way to logical developments. As Terracini observes, Fano reserved a checking function for mechanical calculation procedures, to which he attributed considerable importance. This fact "confirms that he must, consciously or not, have attributed to conceptual demonstrations such a profound intuitive origin, as to make it advisable to resort to checks of another nature."[50]

The second lecture, *All geometry is theory of relativity*, also starts from two questions: What is a geometric figure? What is a geometric property of this figure? To answer the first one, Fano observes that the geometric objects (a point, a straight line, etc.) that make up a figure are abstractions obtained from reality. To answer the second one, he introduces the concept of transformation group and cites Klein's Erlangen program, which classifies geometries according to the invariant properties for a particular group of transformations.

Hence, the meaning of the title is clear: a geometry is something related to a group of transformations. To illustrate this statement, he presents examples for the various branches of geometry—elementary, projective, and topology—and also

---

[49] [3], c. 69*r*: *era indispensabile che fossero trattate e digerite, che diventassero sangue del ns* [nostro] *sangue, averle sulla p*[unta] *delle dita, p*[er] *valersene in ricerche + elevate ... Fecondità!*

[50] [10], p. 486: *conferma che alle dimostrazioni concettuali egli doveva, consciamente o no, attribuire una così profonda origine intuitiva, da rendere consigliabile il ricorso a controlli di altra natura.*

cites non-Euclidean geometries, and to make himself clear he resorts to commonly used objects like rubber bands, ribbons, and pancakes.

The trip to Aberystwyth dates from 1923, a year after the march on Rome which marked the rise of Mussolini.

3 The Late Fano

The racial laws, which for Italian Jews determined the loss of civil and political rights and banishment from the scientific and academic arenas, triggered a series of institutional, epistemic, and social upheavals in high culture. Emigration was one of these. Unable to tolerate what G. Mortara defined "the reduction to a caste of pariahs," by the end of 1941, about 6000 Jews "packed up and left"; another 4000 would repair to Switzerland after the armistice. Among these, there were many intellectuals and scholars, about 30 men of science and 16 mathematicians, including Gino Fano, the eldest among the mathematicians uprooted from racist Italy, who left Turin for Switzerland in the winter of 1938, at the age of 70, settling in Lausanne. He would remain here until 1945, engaging in three areas: solidarity, helping Jewish aid and rescue associations to "track down people who have unfortunately disappeared forever";[51] geometric teaching, in the courses organized for Italian university students interned in the Lausanne and Huttwil camps; and dissemination, through a series of conferences on Italian algebraic geometry that he held at the Cercle Mathématique. The 7 years that he spent in Switzerland are generally dismissed as the unpleasant and unjust epilogue of a highly successful scientific life, and instead are definitely not, in the measure that they yield a broader narrative: that of Jewish scholars, trained in the Belle Époque of scientific internationalism, who saw the principles of the rule of law denied by race theories and who witnessed the perversion of collective consciences under totalitarian regimes [58, 60].

3.1 Discrimination and the Collapse of the Three Pillars of Life: Family, Country, and Profession

Born into a family where the patriotic tradition was alive and which had instilled in him "high feelings of Italianness,"[52] Fano covered the whole pathway that links Risorgimento patriotism, interventionism, and nationalism. His own, however, was not only that Garibaldian sentiment common to many Jews of the first post-Risorgimento generations. In the Great War, as an interventionist of the first hour,

[51] [10], p. 487: *rintracciare persone purtroppo scomparse per sempre.*

[52] [9], p. 262: *alti sentimenti di italianità.*

he took off his civilian clothes for the uniform and committed himself personally to directing the Industrial Mobilization Regional Committee in Piedmont [61] and making his contribution to the "spiritual assistance of the nation."[53] Disgusted by the Versailles camarilla led by "those who administrated European politics, and in particular Italy, as it was hardly conceivable that the Austria of the Holy Alliance and the prince of Metternich could do,"[54] Fano did not tolerate the barbarization of the political debate during the red 2-year period either. In the face of rising fascism:

> he was displeased but much calmer. He was involved in his scientific work, which the fascists did not disturb. He was certainly strongly nationalistic, Italy was "my country right or wrong", and my impression is that he considered Mussolini and his cohorts like a childhood disease of a very young nation, a terrible nuisance but a stage that would pass. ([7], pp. 183–184)

Perhaps convinced that there was nothing to fear as long as one "kept one's nose clean of politics", he maintained this feeling until he failed to be elected as an Academician of Italy. In fact, Mussolini himself opposed Fano's candidacy, proposed by his former pupil F. Severi, alleging Fano's belonging to the Judeo-Pluto-Masonic conspiracy. For him, it was a shame, which to some extent prepared him for the storm to come.

Of Jewish descent from either the father or the mother's lineages, but unobservant and alien to living Judaism in the communities of Mantua and Turin (Fig. 4), Fano was dismissed from his posts at the University and the Polytechnic, from November 29 and October 7, 1938, respectively; he was removed from the direction of the Special Mathematics Library and expelled from all the academic and scientific societies to which he belonged (Lincei, UMI, Virgiliana, Academy of the Sciences of Turin, etc.).

The shock faced with professional demotion, social exclusion, and complete marginalization from academia was traumatic, even if Fano did not show it outside the context of family and friends, even responding to the letter of forced resignation from the Polytechnic:

> I warmly thank the Board of this Faculty of Engineering for the mindful greeting that it was pleased to address to me through you. I always felt and I feel particularly attached to this Institute, for having lived through all its phases, from the very initial practices for its construction, to the most recent years of your enlightened and energetic Direction. The organization and gradual improvement of the two geometry courses for the first two-year level were certainly one of the performances of my long career as a teacher.[55]

[53] [62], p. 1: *assistenza spirituale della nazione.*

[54] [62], p. 10: *coloro che hanno trattato la politica europea, e in particolare l'Italia, come appena appena si può comprendere che la trattasse l'Austria della santa alleanza e del principe di Metternich.*

[55] Historical Archive of the Polytechnic of Turin, file Gino Fano: G. Fano to G. Vallauri, Colognola ai Colli, 29 October 1938: *Ringrazio vivamente il Consiglio di codesta Facoltà di Ingegneria del memore saluto che per mezzo Vostro si è compiaciuto rivolgermi. A codesto Istituto mi sono sempre sentito e mi sento particolarmente legato, per averne vissute tutte le fasi, dalle pratiche iniziali per la sua prima costruzione agli anni più recenti della illuminata ed energica Vostra Direzione.*

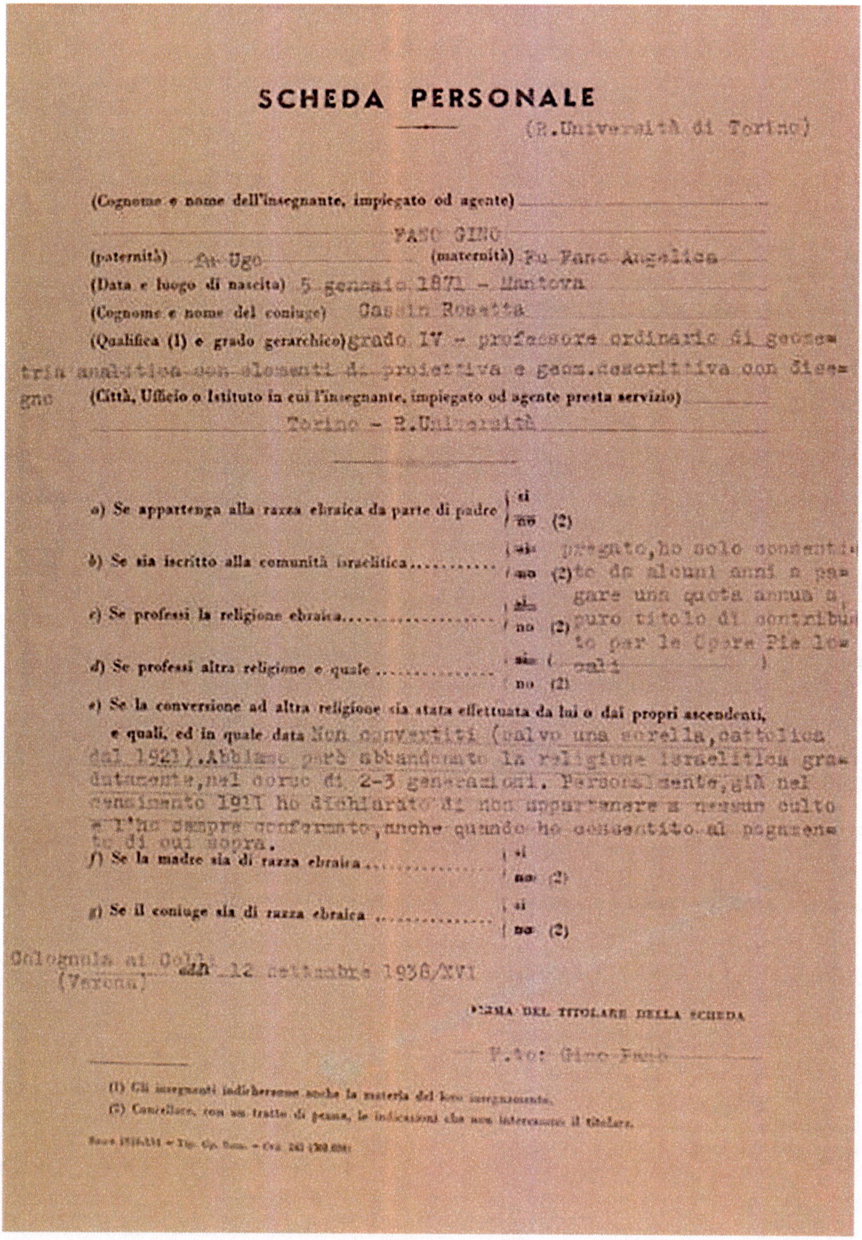

SCHEDA PERSONALE

(R.Università di Torino)

(Cognome e nome dell'insegnante, impiegato od agente) FANO GINO

(paternità) fu Ugo (maternità) fu Fano Angelica

(Data e luogo di nascita) 5 gennaio 1871 – Mantova

(Cognome e nome del coniuge) Cassin Rosetta

(Qualifica (1) e grado gerarchico) grado IV – professore ordinario di geometria analitica con elementi di proiettiva e geom.descrittiva con disegno

(Città, Ufficio o Istituto in cui l'insegnante, impiegato od agente presta servizio) Torino – R.Università

a) Se appartenga alla razza ebraica da parte di padre: si / no (2)

b) Se sia iscritto alla comunità israelitica: si / no (2) pregato, ho solo consentito da alcuni anni a pagare una quota annua a puro titolo di contributo per le Opere Pie locali

c) Se professi la religione ebraica: si / no (2)

d) Se professi altra religione e quale: si ( ) / no (2)

e) Se la conversione ad altra religione sia stata effettuata da lui o dai propri ascendenti, e quali, ed in quale data Non convertiti (salvo una sorella, cattolica dal 1921). Abbiamo però abbandonato la religione israelitica gradatamente, nel corso di 2-3 generazioni. Personalmente, già nel censimento 1911 ho dichiarato di non appartenere a nessun culto e l'ho sempre confermato, anche quando ho consentito al pagamento di cui sopra.

f) Se la madre sia di razza ebraica: si / no (2)

g) Se il coniuge sia di razza ebraica: si / no (2)

Cologncla ai Colli (Verona) addì 12 settembre 1938/XVI

FIRMA DEL TITOLARE DELLA SCHEDA

F.to: Gino Fano

(1) Gli insegnanti indicheranno anche la materia del loro insegnamento.

(2) Cancellare, con un tratto di penna, le indicazioni che non interessano il titolare.

Fig. 4 Fano's racial census form, August 1938

As a matter of fact, his situation was simpler than that of other colleagues. Firstly, in 1938, before him Fano had only 3 years left before retirement. This means that he was paid with an indemnity near to the maximum amount. His family, moreover, held a large fortune and extensive assets and estates, which allowed them to live with dignity even without income from dependent work,[56] and enjoyed such a good intellectual and social standings as to trust in the success of the reverse discrimination procedure. Fano's sister Alina Regina (1874–?) was the widow of Leo Wollenborg, senator of the Historical Left and ex-minister of Economy and Finance of the Zanardelli government; another sister, Maria Fano Ettlinger (1871–1966), was an accredited translator of novels and other texts from English; his wife Rosa was a relative of the economist Robert Michels. The Fanos' applications for reverse discrimination would indeed all be accepted, that of Gino in 1940, but the news of the positive outcome of the procedure was to arrive when the family had already dispersed.

On the other hand, it was above all young people, i.e., the sons, to be affected by racial legislation, which excluded them from any prospect of professional fulfillment. Fano had two sons: Ugo (1912–2001) and Roberto (1917–2016). The latter, on paper, had better prospects, as he was to start his fourth year at the Turin Polytechnic to graduate in 1939, which was guaranteed by law. The elder one, Ugo, graduated in Mathematics and Physics in 1934, a pupil of E. Persico and E. Fermi, and was a promising atomic physicist but was at the beginning of his career. There will be Ugo—the "Colossal Fanaccio" as he was affectionately nicknamed by the so-called Via Panisperna boys—and Roberto the first to make up their minds about emigration, encouraged by their cousin Giulio Racah and by Ugo's girlfriend, Camilla Lattes.

3.2 "So the Time Came to Flee Turin": Emigration

In one of his last interviews, Gino Fano's son, Robert, recalled as follows the family debate around the choice—staying or leaving?—which racial laws forced them to make:

FANO: And at that point my family, with the exception of my father, decided it was time to go.
INTERVIEWER: And your father?

L'organizzazione e il graduale perfezionamento dei due corsi di geometria del biennio sono stati certo una delle esplicazioni della mia non breve carriera di insegnante.

[56] Filling up the questionnaire to be submitted to the SPSL (SPSL, *GF*, c. 297), at the item "Sources of Income before dismissal," Fano stated: "About L. it. 3000 monthly from Univ. position, further, a good family position." Their properties were confiscated by a special public agency, the *Ente di Gestione e Liquidazione Immobiliare* (Agency for Real Estate Management and Liquidation), in 1943.

FANO: Well, we persuaded my father to move too.
INTERVIEWER: So you all got out?
FANO: We all got out. Yes.
INTERVIEWER: You were lucky.
FANO: [. . .] Basically there was a family emergency reunion, on my birthday as a matter of fact—November 17—in our country home near Verona. And basically we decided that we had to scramble, because that war was coming and God only knows what's going to happen.[57]

To flee from Italy, three prerequisites were needed: adequate financial assets, a network of international relationships, and a certain mentality too. The Fano family had the resources because, with highly dangerous smuggling activity, Roberto had succeeded in transferring the family's patrimony to Switzerland, and in particular the money necessary for his parents to settle in Lausanne, and for him and his brother to obtain visas for America, without asking for the help of international Jewish rescue committees (Society for the Protection of Science and Learning and Emergency Committee in Aid of Displaced Foreign Scholars).

The courage, the inner strength to rebuild one's existence and career as a stranger in a foreign country, is not for everyone. Fano, who was a nineteenth-century old gentleman, did not have it, nor, unlike some colleagues, such as Fubini, was he willing to accept any solution to keep the family together. Disagreement between him, his wife, and children mainly pivoted around "the American road." Fano could have recourse to many personal contacts in the United States, constructed along his long-term professional trajectory (J. Coolidge, V. Snyder, E.B. Stouffer, O. Veblen, S. Lefschetz), and had sojourned in America a few times: the first in St. Louis in 1904 and the last in Los Angeles in 1932. However, he belonged to the generations which identified Paris, Berlin, and Göttingen as traditional destinations of academic mobility and did not understand the young researchers, who grew up considering English as a lingua franca and looking to the United States as a new mecca for scientific studies.[58] The pragmatism, the tacit but pervasive anti-Semitism that he had perceived in some American environments, disgusted him. The desire to appropriate the so-called American way of life was inconceivable for him. Finally, there was also an ideological reason: Fano refused to take into consideration the idea of emigrating to any country likely to be at war with Italy ([6], p. 3).

Faced with his determination, his wife and children could do nothing but give up and to be contented with persuading him to take refuge in Switzerland. The choice of destination was largely against the trend. Indeed, Switzerland was only to become the "frontier of hope" for victims of persecution after the armistice of September 8, 1943 [63, 64]. In 1938, when the Fanos relocated there, very few considered it

[57] Transcript of the Interview MIT 150 | Robert M. Fano '41, ScD '47, p. 1, in https://infinite.mit.edu/video/robert-m-fano-%E2%80%9941-scd-%E2%80%9947.

[58] In the questionnaire form submitted to the SPSL (SPSL, *GF*, c. 298), as far as language knowledge is concerned, Fano assesses that he is fluent in French and German while he can read, write, and speak English "rather well."

a good country to expatriate to, and generally it was emigrants for political, not racial, reasons. The Swiss Confederation represented at most a free transit for those who intended to route to other lands. For Fano, by contrast, it was not a temporary solution, but the only one that he could accept.

Once the decision had been made, all Fano's family left—further evidence of the fact that Jewish intellectual emigration from racist Italy was largely "a family matter"—and dispersed. The first to flee Italy were Gino and Rosa, who entered Switzerland in December 1938, settling in Lausanne, at the Hotel Élite, a quite familiar environment, where in the past they had been used to sojourn for business trips or on vacation. The first son, Ugo, fled to Paris in February 1939, and at first managed to obtain a visa for Argentina; embarking in Bordeaux, he reached Buenos Aires in July 1939, thence went to New York, and finally landed in Washington, where he was offered a position at the Carnegie Institution's Department of Terrestrial Magnetism. Fano's grandson, Giulio Racah, a former professor of theoretical physics at the University of Pisa up to the time of racial measures, settled in Palestine in September 1939, after an intermediate stage in London. Roberto postponed his departure for a few months, to finish his exams, and the delay was almost fatal. Due to the outbreak of the war, he was no longer able to reach Bordeaux, one of the main French ports of embarkation to the United States. In Lausanne, where he went to say goodbye to his parents before fleeing Europe, he met his cousin Leo Wollenborg, a writer and a journalist, who, thanks to the intermediation of a high Prelate at the French embassy in Zurich, succeeded in obtaining two passes for France. He was thus to arrive in the United States in October 1939.

In the meantime, since the beginning of 1939, difficulties multiplied both for those who intended to leave Italy and for those who already lived abroad. In fact, the Fanos could not access the assets they had moved to Switzerland and deposited at the Union de Banques Suisses. Thus, at the suggestion of Giulio Racah, who paid a visit to the SPSL offices, Ugo wrote to the Society proposing a sort of agreement: an anonymous Swiss person (alias his father himself) would have issued a bank transfer to the organization, which the SPSL would pay to him in the form of grants in order to finance his research.[59] The SPSL accepted and sent Gino Fano the questionnaire to be filled in so that his case could fall under the administrative laws concerning asylum seekers.[60] Compiling these documents, although it was a mere formality, was unpleasant, even humiliating for a scholar who had prepared the last curriculum vitae for a competition 32 years earlier and for which fields such as "the name of the religion to which you belong: Jewish Orthodox or Jewish Reformed?" were incomprehensible (Fig. 5).[61] On March 6, 1939, the Union de

---

[59] SPSL, *GF*, fol. 303, 304, 305: U. Fano to SPSL, Paris 15 February 1939; SPSL to U. Fano, London, 17 February 1939; U. Fano to SPSL, Paris 19 February 1939.

[60] SPSL, *GF*, fol. 306, 307: SPSL to U. Fano, London, 20 February 1939, G. Fano to SPSL, Lausanne, 14 March 1939.

[61] SPSL, *GF*, fol. 296–298, 302: General & Confidential Information, Curriculum Vitae.

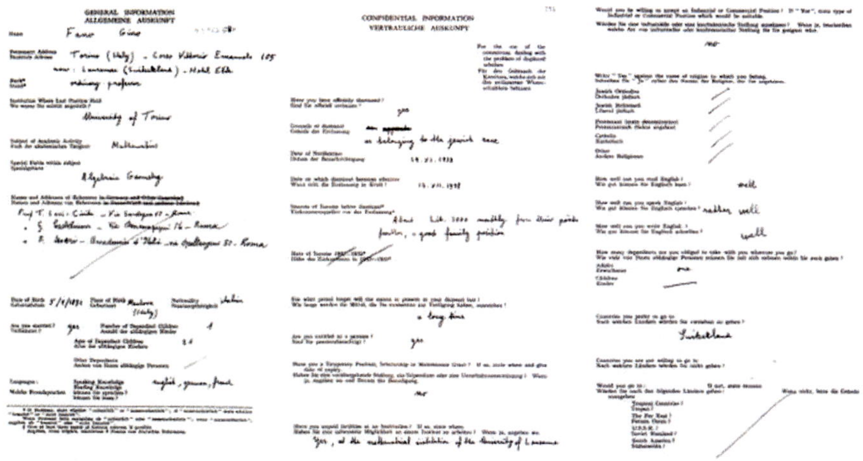
GENERAL INFORMATION
ALLGEMEINE AUSKUNFT

CONFIDENTIAL INFORMATION
VERTRAULICHE AUSKUNFT

Fig. 5 Fano's questionnaire submitted to SPSL

Banques Suisses informed the SPSL that one of its clients who did not want to be named had allocated 15,000 francs to Gino Fano to enable him to continue his scientific activity in Lausanne.[62] So, a practice began of monthly scholarships which, albeit with various obstacles due to Foreign Exchange Control, would allow the Fanos to pass through the war years with relative peace of mind.[63]

3.3 The Latest Studies on Varieties

Although it was a fictitious grant, the SPSL financed Fano to continue his geometric research and he was required to report annually on his outputs. As a matter of fact, despite his age, Fano was productive: between 1939 and 1945, he published 12 works in *Commentarii Mathematici Helvetici*, *Revista de la Universidad Nacional de Tucumán*, and *Atti della Pontificia Academia Scientiarum.*[64] The themes are those typical of Fano's scientific portfolio: algebraic curves, non-rationality and birational geometry in dimension three, cubic threefolds, Fano threefolds, Fano-Enriques threefolds, Enriques surfaces and their automorphisms. However, the fact that production renews in the wake of continuity was normal for senior mathematicians exiled from Italy, such as Fubini and B. Levi, and in some ways, it was mandatory,

62 SPSL, *GF*, fol. 308: Union Bank of Switzerland to SPSL, Lausanne, 6 March 1939.

63 Fano's dossier kept in the SPSL Archive includes extensive correspondence between the Foreign Exchange Control, the SPSL, and the Union de Banques Suisses concerning the installment of the grant, which had not gone unnoticed by the checks on foreign deposits in 1940 and 1941.

64 References are listed in [4], pp. 135–136.

since these scholars had been forced to separate themselves from their libraries, collections, and manuscripts. From this point of view, Fano was advantaged by the fact that up to the summer of 1939, his son Roberto forwarded to him the books and offprints received in Turin. After his departure for the United States and the outbreak of the war, this possibility ceased to exist and he had to "make do with" what he could find in Swiss libraries.[65]

Although there is no solution of continuity between the works issued before and after 1938, nor any trace of Fano's insertion into the new Swiss mathematical scene, which counted leading names such as G. De Rham, P. Finsler, A. Speiser, M. Plancherel, and R. Fueter, those of the Swiss period are not merely remakes or translations of publications that appeared before Fano's emigration. For example, the note *Sulle curve ovunque tangenti a una quintica piana generale* ([65], submitted to the *Commentarii Mathematici Helvetici* on October 10, 1939, and here published in vol. 12, 1940, pp. 172–190) gives evidence of a web of relationships between Fano and the Cambridge geometric School (J.S. Milne, H.F. Baker, ...) that followed his departure from Turin.[66]

But it was mainly in the field of threefolds that Fano achieved the most significant successes, arriving in Lausanne at some important results on classification and rationality problems for cubic varieties. The paper, ready since 1942, was presented by Severi to the Pontifical Academy in February 1943, but it would only come out in 1947. Its contents, however, were known both at national and international levels. For example, Beniamino Segre, John A. Todd, Lucien Godeaux, and Guido Castelnuovo discussed them in their letters:

> I was very glad to receive your letter, and I thank you for your friendly appreciation of my two notes. These will be followed in the Journal by other two. In one of them I prove the result quoted by Mordell in his note just appeared (but presented *after* mine!), while in the remaining one I put the final touch to theorem 1 of my note I, by solving parametrically the cubic Diophantine equation $z^2 = f(x, y)$, where f is any rational cubic polynomial in x, y which cannot be written as a polynomial in a single linear function of x and y. I have already obtained several additional results on cubic surfaces. One of them, by means of which theorem VIII of note I follows at once from theorem VII of the same note, is that "a non-singular cubic surface contains no homaloid linear system of complete intersections." An extension of this result to the non-regular V_3^3, would obviously prove its irrationality. I was told by Fano that this irrationality has been very recently proved by him, on considering the linear system of surfaces of genera 1 lying on V_3^3, but I have not seen the proof. I feel that one should be able to obtain the result also by my methods, but I have not yet had time of thinking seriously about this. (Caltech Archives, B. Segre Papers: B. Segre to J.A. Todd, Manchester 8 October 1943)

> During the war, I had some relations with M. Fano, a refugee in Lausanne; he managed to demonstrate the irrationality of the cubic variety of four-dimensional space, but I don't know his proof yet.[67] (Caltech Archives, B. Segre Papers: L. Godeaux to B. Segre, Liège 13 August 1945)

---

[65] Lincei National Academy, Levi-Civita Archive: G. Fano to T. Levi-Civita, Lausanne, 9 February 1939.

[66] See Sect. 4.3 in this paper.

Prof. Fano was ill in Boston, but now is well; he wrote me a long letter, and I have frequent news of him from my daughter Gina. His address is 3510 Rodman str., Washington D.C. N.W. The Memoir on the cubic variety which must be published by the Pontifical Academy has not yet come out; you will have seen the short summary published in the Lincei Rendiconti. In these days a very intelligent young man working here communicated to me a very simple and brief demonstration of the irrationality of the cubic variety based on topological considerations. But I need to think again about the matter.[68] (Caltech Archives, B. Segre Papers: G. Castelnuovo to B. Segre, Rome 19 December 1946)

For Fano, the work on V_3^3 would substantially be the last original article, since he was to stop active research in the middle of the war, because he felt himself inefficient, as "too often he had to read papers for a second time" ([6], p. 3).

3.4 The Return to Teaching in the Italian University Camp in Lausanne

In less than a month, between 8 September and 1 October 1943, Switzerland saved about 20,000 Italians fleeing German occupation and the Italian Social Republic [66–69]. Among them were the engineers Gustavo Colonnetti and Franco Levi; the mathematicians Modesto Dedò, Bruno Tedeschi, and Bonaparte Colombo; and the mathematics teachers Nedda Friberti, Ernesto Carletti, and Bianca Ottolenghi [70].

Most refugees were shunted off first to transit camps (*campi di smistamento*), where civilians were separated from the military, then to quarantine centers, and finally to labor camps, where they remained until repatriation. The life of the inmates, although different in various realities, was generally miserable. Refugees, especially simple soldiers, were crammed into stables and sheepfolds and forced to engage in penal labor.

To improve this situation, in September 1943, the Fond Européen de Secours aux Etudiants involved the Eidgenossisches Kommissariat für Internierung und Hospitalisierung in the formulation of a support program, aimed at providing refugees with "that intellectual and moral help that represents an imperative

---

67 *Pendant la guerre, j'ai eu quelques relations avec M. Fano, réfugié à Lausanne; il a réussi à démontrer l'irrationalité de la variété cubique de l'espace à quatre dimensions, mais je ne connais pas encore sa démonstration.*

68 *Il Prof. Fano è stato malato a Boston, ma ora sta benino, mi ha scritto egli stesso una lunga lettera, e di lui ho frequenti notizie dalla mia figlia Gina. Il suo indirizzo è 3510 Rodman str., Washington D.C. N.W. La Memoria sulla varietà cubica che deve esser pubblicata dall'Ac. Pontificia non è ancora uscita; avrà visto il breve estratto pubblicato nei Rendiconti dei Lincei. In questi giorni un giovane di qua, molto intelligente, mi ha comunicato una dimostrazione molto semplice e breve della irrazionalità della varietà cubica fondata su considerazioni topologiche. Ma ho bisogno di pensare ancora alla cosa.*

necessity and constitutes an indispensable complement to material aid."[69] The secretary of the Fond Européen, A. de Blonay, circulated a questionnaire in all the about 150 internment camps located in Switzerland, to register university students. On the date of November 13, 1140 questionnaires duly filled in had already been received by FESE. Seeing the census results, a Comité d'aide aux universitaires italiens en Suisse was set up in Lausanne, chaired by Colonnetti and by P. Bolla, the vice president of the federal court, which selected 540 applications. Those admitted were divided into four camps (Lausanne, Freiburg, Neuchâtel, and Geneva) in which parallel university courses were organized, so as to allow interned soldiers to enroll in Swiss universities and finish their studies. In favor of the excluded applicants, University Studia (*Studi Universitari*) for officers and subofficers were created in Mürren and Huttwil afterwards [71, 72]. Lausanne was chosen as the seat of the main Italian University Campus (CUI), inaugurated on January 26, 1944, and operating until May 1945, which hosted about 200 students. Its professors were prominent scholars: Colonnetti (dean and teacher in the courses of construction science), L. Einaudi, A. Fanfani, M.G. Levi, F. Levi, and L. Szegö [70].

For mathematical courses, Colonnetti immediately recruited Fano (the two had been colleagues at the Polytechnic of Turin for almost 20 years, from 1919 to 1938), who immediately agreed to resume his teaching in the Campus of Lausanne and Huttwil, with such energy and genuine enthusiasm as to deserve a personal letter of thanks from Colonnetti at the end of the first semester:

> At the moment when the university camp in Lausanne is near to closure, I would like to express to you all my gratitude for the service you have performed as a teacher of Analytical, Projective and Descriptive Geometry, with such a high sense of patriotism and solidarity for young soldiers interned.[70]

In addition to lecturing in Lausanne, Fano also committed himself in the University Studium of Mürren and Huttwil where, thanks to him and to Dedò, many "valid disciples, such as Aldo Andreotti,[71] were attracted to geometric studies."[72] In Huttwil, he was both teacher and president of the examination boards for all mathematics courses, and in just a semester, he set and held 166 oral tests, helped by two colleagues only: Alessandro Levi (philosopher of law) and Paolo D'Ancona

[69] Swiss Federal Archives, Bern: *Les universitaires italiens internés en Suisse*, November 1943, memorandum signed by de Blonay, E5791, 1, 18/1, fol. 4.

[70] ACT: G. Colonnetti to G. Fano, Lausanne, 24 July 1944: *Nel momento in cui sta per chiudersi il campo universitario di Losanna io desidero esprimerle tutta la mia riconoscenza per l'opera da Lei prestata in qualità di docente di Geometria Analitica Proiettiva e Descrittiva, con così alto spirito di patriottismo e di solidarietà per i giovani militari internati.*

[71] After taking shelter in Lausanne in October 1942, to avoid deportation to labor camps in Germany, Andreotti attended the courses of De Rham and B. Eckmann at the University and those of Fano in the CUI. Returning to Italy at the end of the war, in 1951, he took over from Fano the chair of analytical and projective geometry.

[72] ACT: E. Carletto to G. Fano, Mürren, 15 November 1943 and M. Dedò to G. Colonnetti, Mürren-Münchenbuchsee, 8 January 1944: *molti discepoli valorosi, come Aldo Andreotti, sono attirati verso gli studi geometrici.*

(historian of art). Of the geometric teaching imparted by Fano, there remains an evocative textual trace in two volumes of handouts: the *Lezioni di Geometria descrittiva* written by Roberto Ballarati and Franco Brindisi, and the lecture notes in analytical and projective geometry compiled by an anonymous student [73, 74]. Fano's expertise in this field was enormous. Suffice it to notice that in Turin, he had been the only full professor of projective and descriptive geometry from 1901 to the merger of the two courses. At the Polytechnic, he had taught projective geometry from the foundation of the School of Engineering, in 1908, up to the time of the racial laws. Equally impressive was his output as a textbook author, which included many treatises on descriptive, projective, and analytical geometry, two of which were co-authored by Terracini.[73] The second edition of *Lezioni di geometria analitica e proiettiva* had been their last shared commitment before the departures of Fano for Lausanne and Terracini for Tucumán (Argentina).[74]

The handouts of Fano's lessons at CUI reflect his mastery in this sector. In those of analytic and projective geometry, a sub-discipline to which he had always attributed special value for its mirrors on the development of geometric studies in Italy, one can seize the clarity of his didactic style, his way of proceeding "from a few premises to the construction of large theories, [a way] that left the students full of admiration and almost amazed, especially after the more complicated apparatus of elementary geometry, learned in secondary schools."[75] Equally evident are the legacy of his classical texts in descriptive geometry, specially the third edition of the *Lezioni di geometria descrittiva date nel R. Politecnico di Torino*, lectures which he based on his own theory chapters, with the exception of the section on quadrics, and for the applications (instruments for construction machinery and equipment, photogrammetry, etc.).

3.4.1 Keynote Lectures at the Cercle Mathématique

Fano is credited as one of the main popularizers of Italian geometric culture abroad, a kind of mathematics communication exercise which he began at the Mathematische Gesellschaft in Göttingen in 1893, continued in Aberystwyth in

---

73 G. Fano, *Lezioni di geometria descrittiva*, Torino, litogr. 1903; *Lezioni di geometria descrittiva date nel R. Politecnico di Torino*, Turin, Paravia, 1909, 1914^2, 1926^3; *Geometria Proiettiva. Lezioni raccolte da D. Pastore e E. Ponzano*, Turin, litogr. 1907; G. Fano, A. Terracini, *Lezioni di geometria analitica e proiettiva*, Turin, litogr. 1926, then Turin, Paravia, 1930, 1940^2, 1948^3. See Sect. 2.3.2 in this paper.

74 Fano returned to Turin a few times, up to 1940, to make arrangements with Paravia in view of the second edition of *Lezioni*. Cfr. BSMT, A. Terracini Papers: G. Fano to G. Sacerdote Terracini, 16 December 1947; G. Fano to A. Terracini, New York, 6 February 1948; Mantua 7 April 1948, 9 May 1948, 20 May 1948.

75 [11], p. 325: *da poche premesse alla costruzione di teorie di larga portata che lasciava gli studenti ammirati e quasi meravigliati, soprattutto dopo l'apparato più macchinoso della geometria elementare, appresa nelle scuole secondarie.*

1923 and Kazan in 1929, and ended in Lausanne. Here, in the span of 2 years, from May 1942 to February 1944, at the Cercle Mathématique, Fano held five invited lectures dedicated to Italian algebraic geometry,[76] from a historical and "School" perspective, which met with success and gave rise to lively discussions with his new Swiss friends and colleagues De Rham, G. Dumas, P.G. Javet, and J. Marchand.[77] The history of the Italian geometric Risorgimento from Cremona to the Veronese-Segre-Bertini triad and Castelnuovo and Enriques' classification of algebraic surfaces and contributions on threefolds and on birational transformations (including the research carried out in the last period) allowed Fano to promote (and celebrate) the results of a tradition, of a School to which he strongly felt he belonged.

In the first conversation, *Quelques aperçus sur le développement de la géométrie algébrique en Italie pendant le dernier siècle* (evening of May 4, 1942), Fano traces the recent and contemporary history of algebraic geometry by L. Cremona, who set the foundations of the Italian School, up to the arrival of the great masters: G. Veronese, C. Segre, and E. Bertini (Fig. 6). The second, *Géométrie sur les surfaces algébriques* (May 11, 1942), focuses on one of the masterpieces of the Italian School: the theory of algebraic surfaces by Castelnuovo and Enriques. The third, *Aperçu général sur les surfaces du 3ème ordre* (February 2, 1943), deals with a classic topic: the existence of the 27 lines on a cubic surface. The fourth lecture, *Les surfaces du 4ème ordre* (May 13, 1943), published posthumously by Aldo Andreotti [75], represents the natural continuation of the previous one, focusing on the quartic surfaces. In the last one, *Transformations de contact birationelles dans le plan* (February 10, 1944), Fano outlines his contributions concerning the birational geometry in dimension three. On this occasion, he had the opportunity to recapitulate some of his past studies, presented at the International Congress of Mathematicians in Bologna in 1928, and to highlight two important issues, on which he had resumed working "in recent times": systems ∞^2 of curves that correspond, under a birational contact transformation, to points of the plane or to straight lines, and the determination of the simplest operations with which to obtain, as products, the totality of birational transformations.

There exists a common thread that runs through these five conferences: the desire to exhibit, to display in front of foreign colleagues the best achievements of his own school, a desire that is not the nostalgic recollection of an old Italian geometer at the end of his career, but on the contrary permeates all his experiences of dissemination, from the end of the nineteenth century onwards.[78]

In this perspective, the first conversation held by Fano at the Cercle Mathématique, *Quelques aperçus sur le développement de la géométrie algébrique en Italie pendant le dernier siècle*, particularly stands out, as it nicely expresses what the

[76] The manuscript drafts of the five lectures are kept in BSMT, Fondo Fano, and are digitized in [28].

[77] Minutes of the sittings of the Cercle register the presence of at least 20 participants in each conversation, both Cercle associates and external guests.

[78] See Sect. 2.3.1 in this paper.

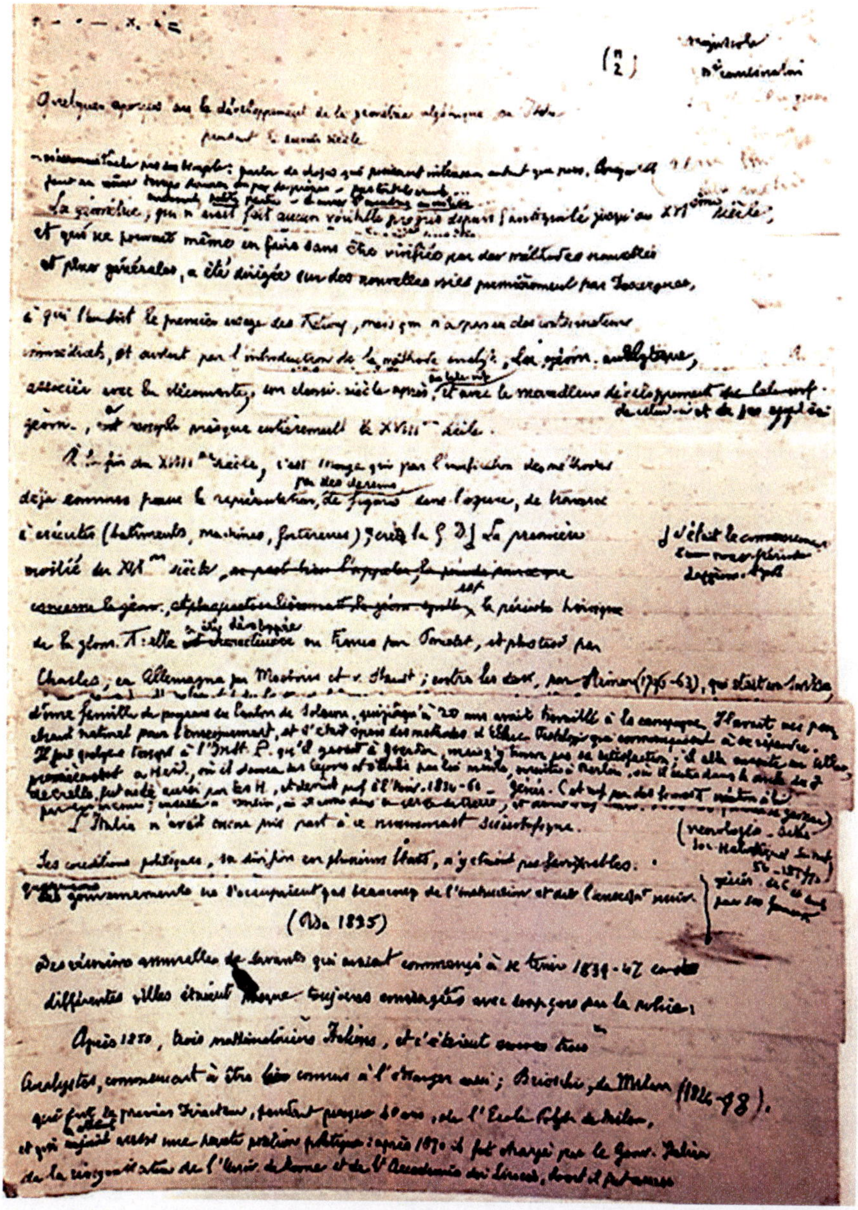

Fig. 6 Manuscript draft of the lecture *Quelques aperçus sur le développement de la géométrie algébrique en Italie pendant le dernier siècle* ([3], c. 53*r*)

Italian School of geometry was for its members, what it meant to be part of it, and how the birth, image, and evolution of this research group were declined abroad. The notion of School according to Fano and the main characters of the historical fresco offered by him in the *Aperçus* can be summarized in the following points:

- Continuity and stability of the development path of Italian geometry from Cremona to Severi;
- Importance of the social dimension for the flourishing, competitiveness, and attractiveness of research schools;
- Emergence of a national identity in mathematical studies only after the Risorgimento;
- Need for "grafts" (*innesti*) of foreign trends of studies (primarily from Germany) on the trunk of Italian geometric production;
- Scientific and moral role of the masters, both in the moment of creation of a school and in the successive phases of development, expression, and affirmation of the working group beyond geopolitical borders;[79]
- Added value of the collaboration: the school is not only a team of scholars who share a research project or a workplace, but is a network, in a way a family.

These are aspects which Fano had often thought about in the past and on which he would have continued to reflect until the last months of his life when, preparing the Lyncean commemoration of Castelnuovo, he

> lingered in thought on the schools that in the last decades of the 19th century had determined the revival of geometric studies in Italy, and on men, in the first place Luigi Cremona who had made it possible. One of the many clues to the continuity of Italian geometric thought,[80]

a continual tradition in the wake of which he had been able to harmoniously place his own contribution.

Pronounced at a terrible historical juncture (when Turin was being bombed and when mass deportations to concentration camps had already begun), the lecture *Quelques aperçus sur le développement de la géométrie algébrique en Italie* is characterized by some national markers which at first reading would appear singular, almost bizarre. They are definitely not: they similarly shape other series of seminars and conversations held by Italian refugees, for example the *Correrías en la logica matématica* by Levi (Tucumán, autumn 1942, then [77], pp. 13–78). As Levi "brought to Tucumán" the voices of Peano, Pieri, and Burali-Forti, Fano intended to revive in front of his colleagues at the Cercle Mathématique the studies and figures of those who had been his masters and friends (Cremona, Segre, Castelnuovo, Klein), in a word that golden season of Italian geometric research, which he had experienced first as an observer, and later as a player. The two contexts are

[79] On the notion and role of research Schools, according to Italian geometers (Segre, Castelnuovo, etc.), see [2, 20, 76].

[80] [12], pp. 702–703: *indugiato col pensiero sulle scuole che negli ultimi decenni del secolo scorso avevano determinato il rifiorire degli studi geometrici in Italia, e sugli uomini, in primo luogo Luigi Cremona* [. . .]. *Uno dei molteplici indizi della continuità del pensiero geometrico italiano.*

obviously very different: Lausanne is not Tucumán, the audience is different, but the underlying spirit and basic aim are analogous: to claim, even during the experience of exile, their own belonging to two mathematical Schools that had led Italy to achieve an international *Führende Stellung* (leading position).

3.5 *Between Turin and the United States*

Fano was both among the first to leave Italy and among the first to come back. That in Switzerland was a parenthesis for him, to be closed as soon as possible in view of the future, which Fano foresaw in Italy, where he returned immediately after the liberation. What Terracini defines "el dilema de la vuelta" did not touch him.[81]

The case of his sons was quite different. Italian Jews who arrived in America on the eve of the war can be somewhat schematically divided into two groups: those who had come as immigrants and who tried to integrate into the new world, and those who had come as refugees and were always ready to return to their homeland ([78], p. 357): the young Fanos fall into the former category. Both were now settled, married, and with children; they had obtained American citizenship in 1946 and had completed the path from denationalization to renationalization. The desire to establish themselves in their new homeland and to contribute to determining its technological-scientific primacy prevailed over the sense of being dispossessed that they had experienced after racial discrimination ([79], p. 363).

Faced with their determination to stay in America, this time it was the father who gave up, agreeing to spend the last few years of his life partly in Italy and partly in the United States, whence he came over for the first time after the war in August 1946. Thus, although reinstated in service, Fano only nominally resumed teaching from May to November 1946, when he retired.[82] Aware that his return to the chair was "fictitious," he immediately told his colleagues that the time had come to think about his succession, a succession that—as he confided to his friends Tricomi and Castelnuovo—he hoped would be ensured by Beniamino Segre or Alessandro Terracini, who "were meritorious of the country, having resumed their university positions."[83]

On the other hand, Fano continued to maintain his contacts with the Italian and American milieu and made a decisive contribution to the reconstruction of the heritage of the Special Mathematics Library of Turin, which had been partially

---

[81] SPSL Archive: SPSL to G. Fano, London, 15 May 1947.

[82] ASUT: Dean M. Allara to G. Fano, Turin, 5 November 1945; G. Fano to M. Allara, 8 December 1945.

[83] Caltech Archives, B. Segre Papers: G. Fano to B. Segre, New York, 21 March 1945: *si sono resi veramente benemeriti del Paese, riprendendo le loro cattedre universitarie.* See also G. Castelnuovo to C. Segre, Rome, 3 October 1946; F. Tricomi to C. Segre, Turin, 26 October 1946.

destroyed in the bombings of 1942–1943, incidentally by donating his entire collection of offprints (over 5000 items).[84] Declared an emeritus in the summer of 1948,[85] he dedicated the last years of his life to a project proposed to him by B. Segre: the monograph edition, under the auspices of the Italian Mathematical Union, of his writings concerning the demonstration of the irrationality of the V_3^3 of $\mathbb{P}^4$. His last talk, held at the Mathematical Seminar of the University and Polytechnic of Turin in February 1950 and published in his *Festschrift* volume [105], focused on this subject.[86]

4 On Fano's Material and Immaterial Heritage

In this part of the essay, we will focus on some Fano's mathematical contributions from the historiographical perspective of "mathematical heritages," referring with this term not only to material collections (libraries, miscellanies, museums, and archives) but also to the immaterial aspects of crystallization, transmission, and circulation of knowledge.[87] Such an approach can provide a new key to understanding a mathematical community, a well-characterized social group such as the Italian School of algebraic geometry. By encompassing both components, Fano's case study allows us to detect the close link between material and immaterial dimensions.

It is now well established that Fano's mathematical activity addressed different research topics, which are analyzed in ([4], pp. 115–122). Fano is best known for his pioneering work on three-dimensional algebraic varieties (or "threefolds")—to which his name is inextricably linked—and the problems of their classification and rationality. Within the literature in this field, two complementary and sometimes opposite trends emerge. On the historical side, it is agreed that Fano's research, especially that of the last period, was carried out in the declining phase of the Italian School, when results were "intuited" rather than "demonstrated." Fano's original papers are "very obscure and criticizable from the point of view of rigor," thus being indicative of the limits of Italian tradition, and "they were considered so even in the period of the full flowering of the School" ([86], p. 129). On the other hand, on the

[84] See [80, 81].

[85] ASUT: Dean M. Allara to the Ministry of Public Instruction, Turin, 18 June 1948; the Ministry of Public Instruction to M. Allara, Rome, 19 July 1948; M. Allara to G. Fano, Turin, 24 July 1948; G. Fano to M. Allara, Colognola ai Colli, 1 August 1948. The proposal to declare Fano Emeritus came from Terracini, who also drew up the report concerning "his merits as teacher and scientist" (*meriti come maestro e come scienziato*). The report, dated 7 June1948, is kept in ASUT, in Fano's personal file.

[86] ASUT, Corrispondence of BSMT: A. Terracini to G. Fano, Turin, 15 February 1950; G. Fano to A. Terracini, New York, 5 January 1951, 23 January 1951; A. Terracini to G. Fano, Turin, 29 January 1951.

[87] For some examples of such historiographical approach, see, inter alia, [14, 82–85].

mathematical side, the originality of Fano's ideas is widely recognized. As J. Murre pointed out, Fano's mathematical creativity enabled him

> to tackle these problems almost empty-handed because there was no foundation for higher dimensional algebraic varieties. The modern development has shown that Fano was essentially right and, once the foundations were available, his methods were correct and effective. ([18], p. 224)

A further aspect to be considered is the fact that Fano's research has been a source of inspiration for modern studies since the 1980s. Our aim is to reconsider this set of elements from a new perspective, adopting as investigation lens that of "patrimonialization," also in the light of some unpublished documents and manuscripts, through which the material and immaterial dimensions intertwine in a significant way. In the case of Fano's contributions on threefolds, this conception can be articulated on three different levels which will be investigated in the following paragraphs.

4.1 Fano's Work Within the Cultural Heritage of the Italian Geometric Tradition

Firstly, Fano's work is positioned within a specific cultural heritage, that of the Italian School of algebraic geometry [107]. From the point of view of research alone, it is characterized by the commonality not only of mathematical questions and research themes but also of the following:

- Method, which can be briefly described as a prevalently synthetic approach based on projective tools and a widespread use of so-called geometrical intuition;
- Sources where, alongside the works of Italian geometers, there are the German contributions of the late nineteenth and early twentieth centuries;
- Way of writing and presenting the mathematical findings, i.e., "in progress" and as a result of "experimental work".[88]

The starting point of Fano's research on threefolds is fully in line with the Italian tradition: as is well known, he started from Lüroth's problem in higher dimensions—asking whether every unirational variety is rational—with the aim of extending the results of rationality and classification of algebraic surfaces achieved by Castelnuovo and Enriques a few years earlier.

After a preliminary work published in 1904 [87] devoted to the cubic hypersurface of $\mathbb{P}^4$, in 1908 Fano began to study threefolds having the plurigenera equal to zero, stating that "for three-dimensional algebraic varieties the nulling of all

[88] For a detailed discussion of the latter aspect, see [76], pp. 196–197.

genera is not yet a sufficient condition for their biunivocal representation" on $\mathbb{P}^3$, i.e. the projective space of dimension three. His research then became directed towards showing "the existence—which occurs for the first time in the case of three-dimensional varieties—of birationally distinct types of varieties having all the genera equal to zero."[89]

From this moment on, Fano's studies unfolded in several directions, among which the introduction of the today called "Fano threefolds" definitely stands out: in modern language, they are three-dimensional varieties whose anticanonical bundle is ample. In order to understand Fano's printed and unpublished writings, it is necessary to introduce some fundamental elements of his notation. Denoting by p the genus of curve sections, Fano identified the families of threefolds M_3^{2p-2} of order $2p - 2$, embedded in the projective space $\mathbb{P}^{p+1}$. Such three-dimensional varieties contain the surfaces $F^{2p-2} \subseteq \mathbb{P}^p$ as hyperplane sections, having the same order of the starting varieties and all the plurigenera equal to one; from a modern point of view, they are K3 surfaces. Canonical curve sections $C_p^{2p-2} \subseteq \mathbb{P}^{p-1}$ of genus p and order $2p - 2$ are obtained from the intersection of two generic hyperplane sections. These particular varieties were not systematically introduced by Fano until 1928, during the International Congress of Mathematicians in Bologna. Here during the iconic plenary lecture on the Italian algebraic geometry by Castelnuovo, who had taken on the role of leader of the School after Segre's death, the centrality of the issues addressed by Fano was underlined with these words:

> How can we decide whether an assigned equation with four unknowns represents a rational or a semi-rational variety? We know nothing about this, not even for the lowest values of the degree, higher than 2. Actually, research that Fano has been carrying out for several years, and which he will present during his communication, shows how complex the question is. He examines the varieties that have all the genera and plurigenera equal to zero and he distributes them into a finite number of families: the first one is composed of rational varieties, the second of semi-rational varieties, and the others of varieties that become more detached from rationality. An accurate classification of these types would shed light on a question that needs to be solved for the future development of algebraic geometry.[90]

---

[89] [88], p. 973: [...] *per le varietà algebriche a tre dimensioni l'annullarsi di tutti i generi (analoghi ai precedenti) non è ancora condizione sufficiente perché esse possano rappresentarsi biunivocamente sullo spazio S_3; e scopo di questa breve Nota è appunto di assodare l'esistenza – che si presenta per la prima volta nel caso di varietà a tre dimensioni – di tipi birazionalmente distinti di varietà aventi tutti i generi nulli.*

[90] [89], p. 200: *Come decidere se una equazione assegnata a quattro incognite rappresenti una varietà razionale o semirazionale? Nulla sappiamo in proposito, nemmeno per i più bassi valori del grado, superiori a 2. Anzi, ricerche che il Fano prosegue da vari anni, e di cui vi parlerà in una sua comunicazione, fanno vedere quanto la questione sia complessa. Egli prende in esame le varietà che hanno nulli tutti i generi e i plurigeneri e le distribuisce in un numero finito di famiglie, di cui la prima si compone di varietà razionali, la seconda di varietà semirazionali e le altre di varietà che si staccano sempre più dalla razionalità. Una classificazione accurata di questi tipi getterebbe molta luce sopra una questione che è necessario risolvere per lo sviluppo futuro della geometria algebrica.*

However, some manuscript papers recently identified within the *Appunti vari* of the Fondo Fano (BSMT) show that Fano's research in this direction originated a few years earlier, starting not from this construction though, but from the idea of undertaking a complete analogy with the study of surfaces. This emerges from the drafts[91] of some lectures dedicated to the main differences encountered in the study of algebraic surfaces and in that of threefolds that Fano had planned to give during his cycle of lectures at the University College of Wales in Aberystwyth in 1923.[92] As he himself declared ([47], p. 3), these topics were not covered for reasons of time. The predominantly didactic approach adopted in this context further emphasizes Fano's position within the heritage of methods and approaches typical of the Italian School, among which proceeding by analogy plays a fundamental role.

At the beginning, Fano refers to Severi's research on higher dimensional varieties dating back to the years 1906–1909, introducing some fundamental notions: geometric and arithmetic genus (denoted by P_g and P_a, respectively), three-dimensional irregularity $q_1 = P_g - P_a$, surface irregularity q_2, linear connection $2q_2 + 1$, and sum of irregularities $q_1 + q_2$. At this juncture, especially in the introduction, Fano took almost verbatim some passages from Severi,[93] who in those years began to assume a central role in Italian mathematics. However, he also refers to the studies of Marino Pannelli, a mathematician specializing in algebraic geometry and a secondary figure in the Italian geometrical scenery, considered by Severi "one of the best of the Italian scholars of this discipline who were not appointed to a professorial chair."[94] In 1906, Pannelli determined some relations between the numerical characters of threefolds which are invariant under birational transformations, including Ω_2, the virtual arithmetic genus of the canonical surface, to which we will return in the next section.

This first manuscript already helps to shed light on Fano's positioning within the wide immaterial heritage of the Italian School of algebraic geometry, composed of the works by both great masters like Severi and minor authors like Pannelli. This is confirmed by an analysis of the citational network of Fano's papers on Fano threefolds:[95] indeed, almost all the cited works are publications in the "glorious" Italian geometric tradition (88% of citations), for a total of 22 Italian authors (Fig. 7). Among them, with at least ten references, besides Severi's (with 33

91 See [3], cc. 125–130.

92 See Sect. 2.3.3 in this chapter.

93 Compare [3], c. 125*r* with [90], p. 337.

94 [91], p. 83: *[Il Severi lo considerava come] uno dei migliori fra gli studiosi italiani di quella disciplina non pervenuti ad una cattedra universitaria.*

95 To perform this analysis, we considered the references contained in Fano's 16 publications on three-dimensional varieties, published between 1904 and 1950. Out of a total of 165 citations, only 19 papers are signed by foreign authors: 9 works were from Germany (F. Klein, M. Noether, T. Reye, G. Salmon, A. Voss, E. Weber), 6 from the United Kingdom (D.W. Babbage, A. Cayley, P. Du Val, L. Roth, J.A. Todd), 2 from Denmark (H.G. Zeuthen), and 1 each from Austria (K. Zindler) and Switzerland (L. Schläfli).

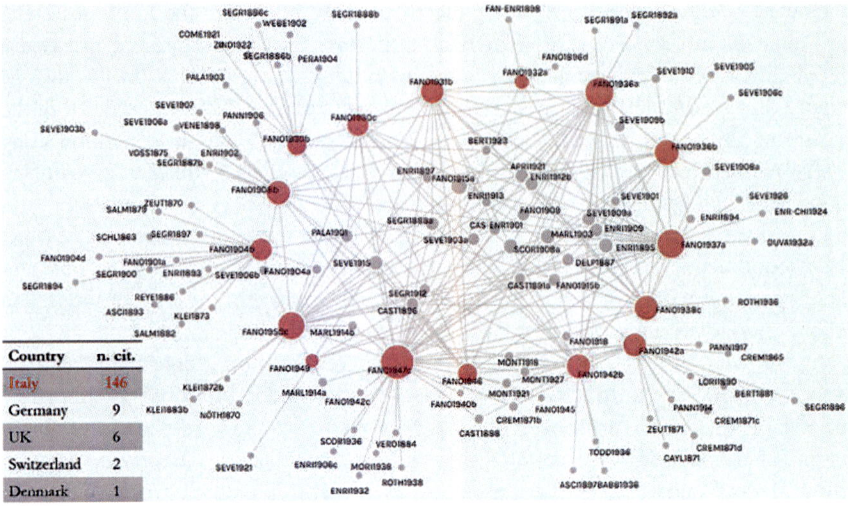

Fig. 7 Citational network of Fano's works on Fano threefolds

citations), the names of Enriques (30), C. Segre (18), Castelnuovo (12), and G. Marletta (10) stand out.

Besides the quoted works, at least a further aspect deserves to be considered in order to grasp Fano's commitment in the wake of the Italian geometric tradition. The opening of his talk in Bologna is representative in terms of both expository style and research methodology. Connecting up with Castelnuovo's talk, Fano introduced the subject as follows:

> The distinction, which seems traditional, between sciences of reasoning and experimental sciences is now outdated. In every science, experience and reasoning play a role; the distinction concerns only their reciprocal proportions. In mathematics, the part reserved for experience, small and limited to the stage of discovery, consists essentially in the careful examination of some particular cases. I propose to set forth here the result of a little experimental work, and of some further conjectures, regarding an arduous and important question, which has been waiting in vain for a solution for a long time.[96]

[96] [92], p. 115: *La distinzione, che pareva tradizionale, tra scienze di ragionamento e scienze sperimentali è ormai sorpassata. In ogni scienza hanno parte l'esperienza e il ragionamento; la distinzione concerne solo le reciproche proporzioni. In matematica la parte riservata all'esperienza, piccola e limitata alla fase di scoperta, consiste essenzialmente nell'esame accurato di qualche caso particolare. Io mi propongo appunto di esporre qui il risultato di un po' di lavoro sperimentale, e di qualche congettura ulteriore, riguardo a una questione ardua e importante, che da tempo attende invano la soluzione.*

4.2 A Specific Mathematical Heritage: Methods and Results

A second manuscript,[97] entitled *Appunti e vedute concernenti le varietà algebriche a tre dimensioni aventi tutti i generi nulli*, contains the preliminary studies of the communication that Fano presented at the International Congress in Bologna. It leads to a second level of investigation, that of immaterial heritage conceived as a specific set of contents, including the results obtained but also (and above all) the techniques and instruments developed to achieve them and the language used and/or coined for this purpose. In this sense, Fano's manuscripts and printed works constitute a true gold mine for historians of mathematics. While the results achieved or, at least, "suggested" by Fano were described in the aforementioned historical papers, the methods and tools of research deserve to be considered more carefully.

Returning to the invariant Ω_2, Fano's unpublished notes brought to light a first classification of Fano threefolds based on this numerical character (Fig. 8). Here, Fano addressed the issue starting from the need to identify an arithmetic character analogous to the Castelnuovo-Enriques invariant ω for surfaces: it is a relative invariant whose maximum value for a class of birationally equivalent surfaces is an absolute invariant, i.e. the virtual linear genus $p^{(1)}$ of the canonical curve. He thus identified the analogue of this character with the virtual arithmetic genus Ω_2 of the canonical surface of a threefold, whose maximum absolute value Ω turns out to be an absolute invariant. Fano then gave a first classification of these threefolds depending on such invariant, which essentially coincides with that subsequently presented at the ICM of Bologna, with the sole exception of the case $p = 8$. In the communication of 1928, Fano instead classified these varieties according to the value of p, the geometric genus of the curve-sections. However, in the paper published in the ICM proceedings, he again introduced Ω_2: differently from the manuscript, he added that for a variety M_3^{2p-2} this invariant is equal to $-(p+2)$ and coincides with the dimension of the systems of surfaces of genus one contained within such threefold increased by one unit. Taking into account both the handwritten notes and the 1931 printed work, Fano's first systematic classification of M_3^{2p-2} can be summarized as in Table 1.[98]

It seems natural to wonder why Fano reintroduced the invariant Ω_2, despite having already provided a classification of Fano threefolds based on the value of p. On one side, this invariant allowed him to classify not only these $M_3^{2p-2} \subseteq \mathbb{P}^{p+1}$, but also a second category of three-dimensional varieties. These are singular threefolds M_3^n of order n, immersed in $\mathbb{P}^4$ and containing a line of multiplicity $n-2$. On the other side, it seems emblematic of the author's desire to give the

[97] See [3], cc. 45–46 and 52.

[98] Here, the following notations are used: * = this threefold does not appear within the manuscript; $\Longleftrightarrow$ = birationally equivalent varieties; $V(d) \subseteq \mathbb{P}^N$ = hypersurface of degree d in $\mathbb{P}^N$; $V(d_1, \ldots, d_k) \subseteq \mathbb{P}^N$ = intersection of k hypersurfaces of degree $d_1, \ldots, d_k$ in $\mathbb{P}^N$; $Gr(n, k)$ = Grassmannian of $(k+1)$-dimensional vector subspaces of an $(n+1)$-dimensional vector space; H_i = hyperplane.

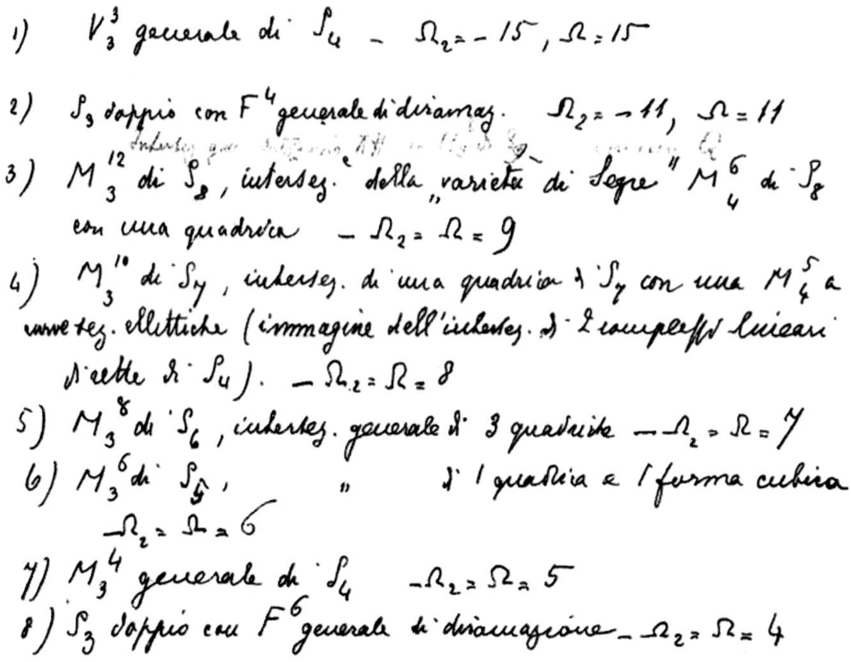
1) V_3^3 generale di S_4 – $\Omega_2 = -15$, $\Omega = 15$
2) S_3 doppio con F^4 generale di diramaz. $\Omega_2 = -11$, $\Omega = 11$
3) M_3^{12} di S_8, intersez.e della „varietà di Segre" M_4^6 di S_8 con una quadrica – $\Omega_2 = \Omega = 9$
4) M_3^{10} di S_7, intersez. di una quadrica di S_7 con una M_4^5 a curve sez. ellittiche (immagine dell'intersez. di 2 complessi lineari di rette di S_4). – $\Omega_2 = \Omega = 8$
5) M_3^8 di S_6, intersez. generale di 3 quadriche – $\Omega_2 = \Omega = 7$
6) M_3^6 di S_5, " di 1 quadrica e 1 forma cubica – $\Omega_2 = \Omega = 6$
7) M_3^4 generale di S_4 – $\Omega_2 = \Omega = 5$
8) S_3 doppio con F^6 generale di diramazione – $\Omega_2 = \Omega = 4$

Fig. 8 Manuscript draft of Fano's first classification of M_3^{2p-2} ([3], c. 52*v*)

Table 1 Fano's classification of Fano threefolds

p	$\|\Omega_2\|$	M_3^{2p-2}	Basic modern description
13	15	$V_3^3 \subseteq \mathbb{P}^4 \Longleftrightarrow M_3^{24} \subseteq \mathbb{P}^{14}$	Cubic threefold $V(3) \subseteq \mathbb{P}^4$
9	11	'double' $\mathbb{P}^3 \Longleftrightarrow M_3^{16} \subseteq \mathbb{P}^{10}$	Double cover of $\mathbb{P}^3$ with branched surface of degree 4
8	*	$M_3^{14} \subseteq \mathbb{P}^9$	$Gr(1,5) \bullet H_1 \bullet H_2 \bullet H_3 \bullet H_4 \bullet H_5 \subseteq \mathbb{P}^9$
7	9	$M_3^{12} \subseteq \mathbb{P}^8$	Intersection of a quadric hypersurface in $\mathbb{P}^8$ with the image of $\mathbb{P}^2 \times \mathbb{P}^2$ via the Segre embedding
6	8	$M_3^{10} \subseteq \mathbb{P}^7$	$Gr(1,4) \bullet V(2) \bullet H_1 \bullet H_2 \subseteq \mathbb{P}^7$
5	7	$M_3^8 \subseteq \mathbb{P}^6$	$V(2,2,2) \subseteq \mathbb{P}^6$
4	6	$M_3^6 \subseteq \mathbb{P}^5$	$V(3,2) \subseteq \mathbb{P}^5$
3	5	$V_3^4 = M_3^4 \subseteq \mathbb{P}^4$	Quartic threefold $V(4) \subseteq \mathbb{P}^4$
2	4	'double' $\mathbb{P}^3$	Double cover of $\mathbb{P}^3$ with branched surface of degree 6

"heritage status" to a harvest of discoveries and achievements in a new geometric field, where there was still much to explore and develop. This consideration is corroborated by Fano's introduction of new mathematical terminology. As noted in the margins of the manuscript, he defined varieties of the first type "semi-rational" (*semirazionali*) since they are "intermediate between rational entities and those having at least one of the genera and plurigenera greater than zero."[99] Threefolds of the second type, having "as an analogue, in the field of surfaces, something intermediate between rational surfaces and irrational ruled surfaces,"[100] are instead called "pseudo-rational" (*pseudorazionali*).

Although the work published in the ICM proceedings is the only printed work in which Fano followed an approach also based on Ω_2, the handwritten notes contain the germ of further ideas and techniques later extended and refined in various publications. Fano thus began to work out some fundamental notions that, developed and expanded during research activity of over 40 years, would become an integral part of the heritage of classical studies on three-dimensional varieties.

Regarding the problem of rationality, Fano claimed that "examining these different varieties, one gets the impression that, if they are not rational, they come closer to rationality as p increases, despite some restrictions."[101] Moreover, these threefolds carry an additional property: each of them can be projected from the lower order curves contained in them (and therefore from a straight line, if it exists) into a variety of the same type that corresponds to lower values of p (namely a $M_3^{2p-6} \subseteq \mathbb{P}^{p-1}$) and contains a ruled cubic surface as an image of the projection center. Lastly, Fano observed that "from the birational point of view (with some restrictions), each of the enumerated varieties includes the following threefolds as particular cases (corresponding to higher values of p) [...]; so that the increase of p implies, in principle, a progressive particularization of M_3."[102]

An interesting aspect is that in both manuscripts, unlike the publications, Fano explicitly stated his "work plan" for dealing with the issue of three-dimensional varieties. Indeed, he opened a window on his "intellectual laboratory,"[103] declaring that his research

> aimed at demonstrating, as far as possible, the irrationality of some of these varieties, was essentially directed at studying:
>
> (a) the linear systems at least ∞^2 of regular surfaces having all the genera $=1$;
> (b) the set (group) of possible birational transformations;

---

[99] [92], p. 120: *come intermedie fra gli enti razionali e quelli aventi almeno uno dei generi e plurigeneri maggiore di zero.*

[100] [92], p. 121: *come analogo, nel campo delle superficie, qualcosa di intermedio fra le superficie razionali e le rigate irrazionali.*

[101] [92], p. 118: *esaminando queste diverse varietà, si ha l'impressione che esse, qualora non siano razionali, tuttavia, al crescere di p, pur con qualche restrizione, vadano gradatamente accostandosi alla razionalità.*

[102] [92], p. 119: *dal punto di vista birazionale (con qualche limitazione), ciascuna delle varietà enumerate comprende come casi particolari le successive (corrispondenti a valori più elevati di p) [...]; sicché il crescere di p implica, in massima, una progressiva particolarizzazione della* M_3.

[103] See Sect. 2.3.1 in this chapter.

> and trying to find in the systems a) and in the transformations b)—in turn, naturally related to each other—some properties that are different from those of the space S_3, so that we can conclude that they are birat[iona]lly distinct entities.[104]

However, the apparatus of tools developed by Fano to address the problem of threefolds is not limited to the study of the relative invariant Ω_2, to the analysis of linear systems of K3 surfaces contained in M_3^{2p-2} (corresponding to the point *a)* of the manuscript), or to the comparison between the group of birational transformations on threefolds and that of $\mathbb{P}^3$ (point *b)*) with the aim of showing that $Bir\left(M_3^{2p-2}\right) \neq Bir\left(\mathbb{P}^3\right)$. Noteworthy is the study of homaloidic systems of surfaces to prove the irrationality of the quartic hypersurface of $\mathbb{P}^4$ and of the threefold obtained as the complete intersection of a quadric and a cubic hypersurface in $\mathbb{P}^5$. To examine the systems of surfaces contained in a certain threefold, Fano exploited the properties of canonical curves obtained as prime sections of such surfaces which are transmitted to the variety under investigation. After 1928, he began to extend the method of projection of a given threefold from a line, exploring the effects of projecting the variety from different vertices. In this way, the projection of M_3^{2p-2} from a conic of itself is a M_3^{2p-8} of canonical curve sections that contains a rational quartic ruled surface, the image of the starting conic; again, M_3^{2p-2} projects from the tangent space at a general point into a M_3^{2p-10} of canonical curve sections, on which the neighborhood of the vertex is mapped by a Veronese surface. But Fano did not extend the classical tool of projection from a line only in this direction, through the choice of an appropriate vertex. In the extensive memoir of 1937 [93], he introduced the method of so-called double projection that makes it possible birationally to refer each Fano threefold to a "simpler" (but not more general) variety of the same type, immersed in the projective space of dimension $p-6$. Indeed, after having projected $M_3^{2p-2} \subseteq \mathbb{P}^{p+1}$ from a line of itself into a $M_3^{2p-6} \subseteq \mathbb{P}^{p-1}$ which contains a ruled cubic surface spanning $\mathbb{P}^4$, it is possible to project M_3^{2p-6} from this $\mathbb{P}^4$ into a Fano threefold $M_3^{2p-18} \subseteq \mathbb{P}^{p-6}$. Within this vast immaterial heritage of techniques and tools, it is finally worth mentioning the analysis of involutions on threefolds, an area in which Fano's research intertwined with the contemporary studies by Enriques and G. Aprile.

In addition, Fano's way of writing and presenting his achievements is paradigmatic of the process of patrimonialization of mathematical knowledge, as emerges from his 1950 work. Invited at the Turin Mathematical Seminar to deliver a lecture

---

104 [3], c. 45*r*: *[Le mie ricerche], intese a dimostrare, per quanto possibile, la irrazionalità di alcune fra queste varietà, sono state essenzialmente dirette a studiare:*

(a) *i sistemi lineari almeno* ∞^2 *di superficie regolari aventi tutti i generi* $=1$;
(b) *l'insieme (gruppo) delle eventuali trasformazioni birazionali;*

e a cercare di trovare nei sistemi a) e nelle trasformazioni b) – naturalmente, a loro volta, legati fra loro – qualche proprietà che sia diversa da quelle dello spazio S_3, *in modo da poterne concludere che si tratta di enti bir[azionalmen]te distinti.*

on the occasion of his appointment as an emeritus professor, Fano did not merely describe his main scientific contributions during 40 years of academic activity but paid close attention to retracing the main stages of his research path.

4.3 Fano's Legacy in the Short and Long Terms

The third aspect that emerges from adoption of the heritage investigation lens is linked to the concepts of legacy and influence of a certain line of research, over a long or short period. Beyond the ideas and insights provided to modern algebraic geometry, Fano's studies gave an important impulse to research in different contexts but in close dialogue with the Italian School, even in the years of its decline, when at the international level different traditions and directions—such as those outlined by topology and abstract algebra—were becoming widespread. This is the case of the English School of geometry[105] on which the research of Italian geometers exerted a guiding action, at least until the 1930s. In particular, Fano's heritage was positively welcomed and continued in Cambridge. He maintained a regular exchange with this mathematical community, as evidenced by the correspondence with H.F. Baker who, in December 1931, wrote to Fano:

> Dear Sir,
>
> I was very honoured by, and very grateful to you for, your letter of 2 Dec., telling me that you had written further about my little Note of the Del Pezzo ψ^5. I shall look forward to the privilege of an offprint, when the paper is published. And, as soon as possible, I shall study your letter in detail, which I have not been able to do as yet.
>
> Our students in Cambridge read many of your published papers, and find them very helpful—so that I am particularly grateful to you for writing to me. (BSMT, Fondo Fano, letter no. 22: H.F. Baker to G. Fano, Cambridge, 14 December 1931)[106]

It is therefore not surprising that Fano's classical line of research on threefolds was taken up by Leonard Roth, who published it in an organic form in the treatise *Algebraic Threefolds: With Special Regard to Problems of Rationality* [95]. It is relevant that Roth had spent a year in Rome with Severi as the winner of a Rockefeller fellowship in 1930–1931. Fano's legacy was hence taken up by Roth, whose profound knowledge of Fano's studies on three-dimensional varieties shines through this volume. The scientific correspondence between the two mathematicians, which continued at least until Fano's relocation to Switzerland, shows far-reaching sharing both in terms of research themes and of geometric tools adopted. Favorite topics are certainly the results of rationality and unirationality, but there is also a focus on some specific results of Fano on threefolds such as those that led Roth to state: "I must say how satisfying it is to know that the series of the V_3^{2p-2} [should read

[105] For an insight into the development of geometry at Cambridge in these years, see [94], pp. 340–349.

[106] This letter can be accessed at https://www.corradosegre.unito.it/fondofano/lettera22.pdf.

M_3^{2p-2} in our notation] ends for $p = 37$ and if $p > 10$ they are rational."[107] As far as methods are concerned, Roth skillfully handled the tools pioneered by Fano, such as the study of homaloidic systems of surfaces and the analysis of the complete surface sections of a given threefold, but he also provided new ideas for advancing research. He combined in an original way the results obtained by English geometers with classical Italian techniques, like successive projections, for instance in the following terms:

> In a recent note—not yet published—I determined that a quartic form of S_4 [should read $\mathbb{P}^4$ in our notation] cannot have more than 45 isolated nodes and, after all, it is well known that this limit is reached, since a rational V_3^4 of this kind exists (see Todd, Quarterly Journal, Oxford 1936). Perhaps we might use this result to show, via subsequent projections, that the V_3^{2p-2} of the first species do not exist for $p > 23$.[108]

However, it should not be thought of as a "one-way" interaction, from the Italian to the English School. A broad conception of patrimonialization, conceived as a process of recognition and appropriation of an articulated set of elements, includes exchange and dialogue with other traditions with which there is a pooling of principles and fundamental aspects. The group of five writings cited by Fano in his works on threefolds, signed by the English geometers P. Du Val, D. Babbage, J. Todd, and Roth and published between 1932 and 1938, thus appears particularly significant. From these publications, Fano drew both specific results, such as those on the quartic hypersurface of $\mathbb{P}^4$, and certain procedures, like those adopted by Roth for the study of M_3^{14}. The fact that all the English geometers mentioned by Fano were students of Baker in Cambridge is not accidental. Besides Roth, Du Val too, thanks to a Trinity fellowship, visited Rome (1930–1932) where he worked with Enriques, specializing in the theory of algebraic surfaces according to the Italian study orientation. The outcome of this period of study abroad is the two works cited by Fano [33, 96], which represent Du Val's first two papers, written in Italian, devoted to the classification of surfaces.

If we look at Fano's contributions on another family of three-dimensional varieties, the so-called Fano-Enriques threefolds (i.e., special Fano threefolds whose general hyperplane section is an Enriques surface), a parallel argument can be made for the work of Lucien Godeaux—who, unsurprisingly, is the most represented

[107] BSMT, Fondo Fano, letter no. 23: L. Roth to G. Fano, London, 18 February 1937: [. . .] *ma devo dire quanto è soddisfacente sapere che la serie delle* V_3^{2p-2} *termina per* $p = 37$ *e che per* $p > 10$ *esse sono razionali.* This letter can be accessed at https://www.corradosegre.unito.it/fondofano/lettera23.pdf.

[108] BSMT, Fondo Fano, letter no. 23: L. Roth to G. Fano, London, 18 February 1937: *[. . .] in una Nota recente –non ancora pubblicata– ho stabilita che una forma quartica di* S_4 *non può aver più di 45 nodi isolati, e del resto si sa che tale limite è raggiunto, perché esiste una* V_3^4 *razionale di questa natura (ved. Todd, Quarterly Journal, Oxford 1936). Forse si potrebbe usare questo risultato per dimostrare, mediante proiezioni successive, che le* V_3^{2p-2} *della prima specie non esistono per* $p > 23$.

author within Fano's miscellany[109]—and the Belgian School of geometry. It must be remembered that Godeaux, in turn, spent a period of advanced training in algebraic geometry in Bologna with Enriques[110] between 1912 and 1914; he was greatly influenced by the work of the Italian School, and his future research directions were to a great extent laid down at this time. Godeaux's 1933 result about linear systems of surfaces over special threefolds ([98], p. 134) represents the starting point for Fano's introduction and classification of this completely new class of three-dimensional varieties 5 years later [99]. Looking at the citational network of Fano's papers on this theme, Godeaux's works make up 25% of the citations (excluding Fano's self-citations).[111] If on the one hand Fano drew directly from Godeaux's research, on the other hand the legacy of Fano's studies on Godeaux is manifest. Indeed, even though Fano's mathematical activity in this direction was not so successful and had a very limited reception, Godeaux's last works on the topic, dating back to the 1960s, were still within the classical research path traced by Fano.

5 Conclusive Remarks

The significant use of archival sources has made it possible to trace out a more "nuanced" vision of Fano's scientific biography than the one provided by the existing historiography (Fig. 9).

Manuscripts preserved in Turin, Göttingen, and Liverpool have contributed to placing him in the context of the Italian School of algebraic geometry, conceived as a scholarly community characterized by specific fields of research, a geometrical style, and some epistemological and linguistic patterns. Fano's research contributions were deeply rooted in the works of Italian masters and in Klein's ideas, as were many of Fano's assumptions in the methodological and teaching fields (use of intuition, dynamic teaching, role of *Approximationsmathematik*, emphasis on the concepts of function and group of transformation, importance of teacher training, ...). His dissemination activity too testifies to his looking back to his cultural roots. This action of promotion, developed in several contexts, reached its peak in the Aberystwyth lectures and ended in Lausanne talks at the Cercle Mathématique where, even from the pain of exile, Fano continued to proudly declare his belonging to the Italian geometric tradition.

[109] For a complete list of this library heritage, see [97] which can be accessed at https://www.corradosegre.unito.it/doc/fano_miscellanea.pdf. For an analysis of such patrimony, see [81].

[110] Also a pupil of Godeaux, Pol Burniat (1902–1975) went to Italy in 1929–1930 to study with Enriques, who had moved to Rome in the meantime.

[111] Excluding citations of his own work, in the two papers on Fano-Enriques threefolds, Fano referred mainly to the publications of the Italian School (15 citations), followed by those by Godeaux (9) and by three other foreign mathematicians (G. Darboux, S. Junski, T. Reye).

Fig. 9 Gino Fano in the 1930s

Documents preserved in the SPSL and Caltech archives, in addition to unpublished correspondence, have made it possible to document a short parenthesis in Fano's scientific life, generally dismissed as a sad epilogue of a successful career. By way of contrast, the contours of his experience in Switzerland, as reconstructed through such sources, are so peculiar that we can reasonably state that no other refugee from racist Italy had an analogous experience.

Fano's handwritten notes on threefolds have shed new light on his research activity. While perfectly appropriating the cultural heritage of the Italian School of algebraic geometry (a certain way of doing geometry and a set of readings and cultural references), Fano's contributions represented a heritage, in terms of questions dealt with, tools and techniques developed, and language used. In this sense, his manuscripts and publications on threefolds are paradigmatic of the process of patrimonialization of mathematical knowledge, which started at the end of the nineteenth century within the Italian context, continued in some contemporary geometric Schools (such as the English and Belgian ones), and is still living nowadays in some trends of contemporary algebraic geometry.[112]

[112] See [100] for an example regarding the Fano-Enriques threefolds.

In conclusion, this chapter, strongly grounded in archival research, can be considered as a first, but not provisional, contribution to the biography of an Italian geometer belonging to the School of Segre.

References

1. Casnati, G., Conte, A., Gatto, L., Giacardi, L., Marchisio, M., Verra, A.: From Classical to Modern Algebraic Geometry. Corrado Segre's Mastership and Legacy. Birkhäuser (Springer), Cham (2016)
2. Conte, A., Giacardi, L.: Segre's university courses and the blossoming of the Italian school of algebraic geometry. In: Casnati, G., et al. (eds.) From Classical to Modern Algebraic Geometry. Corrado Segre's Mastership and Legacy, pp. 3–91. Birkhäuser (Springer), Cham (2016)
3. Fano, G.: Appunti vari, Fondo Fano, BSMT in: Giacardi, L. (ed.) (2013–2021)
4. Collino, A., Conte, A., Verra, A.: On the life and scientific work of Gino Fano. La Matematica nella Società e nella Cultura. Rivista dell'UMI. **7**(1), 99–137 (2014)
5. Conte, A., Giacardi, L.: Gino Fano. In: Roero, C.S. (ed.) La Facoltà di Scienze Matematiche Fisiche Naturali di Torino, 1848–1998, vol. II, pp. 548–554. Deputazione subalpina di storia patria, Torino (1999)
6. Fano, R.: In loving memory of my father Gino Fano. In: Collino, A., Conte, A., Marchisio, M. (eds.) The Fano Conference, Turin 29 September–5 October 2002. Proceedings, pp. 1–4. Dipartimento di Matematica, Università di Torino, Turin (2004)
7. Fano, U.: The memories of an atomic physicist for my children and grandchildren. Phys. Essays. **13**(2/3), 176–197 (2000)
8. Janovitz, A., Mercanti, F.: Gino Fano. In: Sull'apporto evolutivo dei matematici ebrei mantovani nella nascente nazione italiana. Monografie di EIRIS (Epistemologia dell'Informatica e Ricerca Sociale), pp. 43–61. https://diazilla.com/doc/811027/sull-apporto-evolutivo-dei-matematici-ebrei-mantovani-nella.%20Accessed%205%20January%202022. Accessed 28 Feb 2022
9. Segre, B.: Gino Fano. Archimede. **4**, 262–263 (1952)
10. Terracini, A.: Gino Fano. Bollettino dell'UMI. **7**(3), 485–490 (1952)
11. Terracini, A.: Gino Fano. Annuario dell'Università di Torino, 325–328 (1952–53)
12. Terracini, A.: Commemorazione del Socio Gino Fano. Rendiconti Accademia Nazionale dei Lincei. **14**(8), 702–715 (1953)
13. Fano, G.: Sopra le curve di dato ordine e dei massimi generi in uno spazio qualunque. Memorie R. Accademia delle Scienze di Torino. **44**(2), 335–382 (1893)
14. Bruneau, O., d'Enfert, R., Ehrhard, C.: Patrimonialisation en mathématiques (18e–21e siècles). Cahier thématique de Philosophia Scientiæ. **26**(2) (2022)
15. Gario, P. (ed.): Lettere e Quaderni dell'Archivio di Guido Castelnuovo. http://operedigitali.lincei.it/Castelnuovo/Lettere_E_Quaderni/menu.htm. Accessed 28 Feb 2022
16. Hawkins, T.: Lie groups and geometry: the Italian connection. In: Brigaglia, A., Ciliberto, C., Sernesi, E. (eds.) Algebra e geometria (1860–1940): il contributo italiano, Supplemento ai Rendiconti del Circolo matematico di Palermo, vol. 36, pp. 185–206 (1994)
17. Hawkins, T.: Emergence of the Theory of Lie Groups. Springer, New York (2000)
18. Murre, J.: On the work of Gino Fano on tree–dimensional algebraic varieties. In: Brigaglia, A., Ciliberto, C., Sernesi, E. (eds.) Algebra e geometria (1860–1940): il contributo italiano, Supplemento ai Rendiconti del Circolo matematico di Palermo, vol. 36, pp. 219–229 (1994)
19. Segre, C.: Considerazioni intorno alla geometria delle coniche di un piano e alla sua rappresentazione sulla geometria dei complessi lineari di rette. Atti della R. Accademia delle Scienze di Torino. **20**, 487–504 (1884–1885)

20. Luciano, E., Roero, C.S.: From Turin to Göttingen: dialogues and correspondence (1879–1923). Bollettino di Storia delle Scienze Matematiche. **32**, 7–232 (2012)
21. Klein, F. (transl. by Fano, G.): Considerazioni comparative intorno a ricerche geometriche recenti. Annali di matematica pura ed applicata. **17**(2), 307–343 (1890)
22. Segre, C.: Introduzione alla geometria sugli enti algebrici semplicemente infiniti. Quaderni. **3**. In [28]
23. Fano, G.: Sui postulati fondamentali della geometria proiettiva in uno spazio lineare a un numero qualunque di dimensioni. Giornale di Matematiche. **30**, 106–132 (1892)
24. Amodeo, F.: Quali possono essere i postulati fondamentali della Geometria proiettiva di uno S_r. Atti dell'Accademia delle Scienze di Torino. **26**, 741–770 (1890–91)
25. Enriques, F.: Sui fondamenti della geometria proiettiva. Rendiconti dell'Istituto Lombardo di scienze e lettere. **27**(2), 550–567 (1894)
26. Enriques, F.: Sui postulati fondamentali della geometria proiettiva. Rendiconti del Circolo matematico di Palermo. **9**, 79–85 (1895)
27. Fano, G.: Sui postulati fondamentali della geometria proiettiva (Due lettere al Prof. Enriques). Rendiconti del Circolo matematico di Palermo. **9**, 79–82, 84–85 (1895)
28. Giacardi, L. (ed.): Corrado Segre e la Scuola Italiana di Geometria Algebrica. http://www.corradosegre.unito.it/ (2013–2022). Accessed 28 Feb 2022
29. Avellone, M., Brigaglia, A., Zappulla, C.: The foundations of projective geometry in Italy from De Paolis to Pieri. Arch. Hist. Exact Sci. **56**(5), 363–425 (2002)
30. Fano, G.: Osservazioni su alcune "geometrie finite" I. Rendiconti Accademia Nazionale dei Lincei. **26**, 55–60 (1936)
31. Fano, G.: Osservazioni su alcune "geometrie finite" II. Rendiconti Accademia Nazionale dei Lincei. **26**, 129–134 (1936)
32. Fano, G.: Sull'insegnamento della matematica nelle Università tedesche e in particolare nell'Università di Gottinga. Rivista di Matematica. **4**, 170–188 (1894)
33. Du Val, P.: Osservazioni sulle superficie di genere uno che non sono base per un sistema di quadriche. Rendiconti Accademia Nazionale dei Lincei. **15**(6), 345–347 (1932)
34. Rowe, D.: Segre, Klein, and the theory of quadratic line complexes. In: Casnati, G., et al. (eds.) From Classical to Modern Algebraic Geometry. Corrado Segre's Mastership and Legacy, pp. 243–263. Birkhäuser (Springer), Cham (2016)
35. Fano, G.: Sopra alcune considerazioni geometriche che si collegano alla teoria delle equazioni differenziali lineari. Rendiconti Accademia Nazionale dei Lincei. **4**, 18–25 (1895)
36. Enriques, F.: Le superficie con infinite trasformazioni proiettive in se stesse. Atti R. Istituto Veneto di scienze, lettere ed arti. **51**(7), 1590–1635 (1893)
37. Fano, G.: Sulle varietà algebriche dello spazio a quattro dimensioni con un gruppo continuo integrabile di trasformazioni proiettive in sé. Atti R. Istituto Veneto di scienze, lettere ed arti. **28**(7), 1069–1103 (1895–96)
38. Fano, G.: Sulle varietà algebriche con un gruppo continuo non integrabile di trasformazioni proiettive in sé. Memorie della R. Accademia delle scienze di Torino. **46**, 187–218 (1896)
39. Enriques, F., Fano, G.: Sui gruppi continui di trasformazioni cremoniane dello spazio. Annali di Matematica. **26**, 59–98 (1897)
40. Fano, G.: I gruppi di Jonquières generalizzati. Memorie della R. Accademia delle scienze di Torino. **48**, 221–278 (1898)
41. Fano, G.: Kontinuierliche geometrische Gruppen. Die Gruppentheorie als geometrisches Einteilungsprinzip. In: Enzyklopädie der mathematischen Wissenschaften, III, 4a, pp. 289–388. Teubner, Leipzig (1907)
42. Giacardi, L.: The Italian school of algebraic geometry and the teaching of mathematics in secondary schools: motivations, assumptions and strategies. In: Marchisio, M., Verra, A. (eds.) Geometry of Algebraic Varieties in Honor of Alberto Conte. Rendiconti del seminario matematico, vol. 71.3/4, pp. 421–461. Università e Politecnico di Torino (2013)
43. Fano, G.: Intenti, carattere, valore formativo della matematica: conferenza tenuta alla Scuola di guerra il 15 marzo 1924. Alere flammam. Bollettino del Gabinetto di cultura della Scuola di guerra. **2**(7), 9–32 (1924)

44. Fano, G.: Matematica esatta e matematica approssimata. Bollettino della Mathesis. **3**, 106–126 (1911)
45. Klein, F.: On the mathematical character of space–intuition and the relation of pure mathematics to the applied sciences. In: Lectures on Mathematics Delivered from Aug. 28 to Sept. 9, 1893 ... at Northwestern University Evanston, Ill. By F. Klein, Reported by A. Ziwet, pp. 41–50. Macmillan, New York (1894)
46. Fano, G.: Scuola operaia serale femminile. Relazione 1909–1910, Unione Femminile Nazionale. Sezione di Torino, pp. 1–24. G. Derossi, Torino (1910)
47. Fano, G.: A Series of Special Lectures on 'Italian Geometry'. Walker, Shrewsbury (1923)
48. Fano, G.: Vedute matematiche su fenomeni e leggi naturali. Discorso letto nella R. Università di Torino per l'inaugurazione dell'anno accademico 1922–23. In: Annuario R. Università di Torino 1922–23, pp. 15–45. Schioppo, Torino (1923)
49. Fano, G.: Un po' di matematica per i non matematici. Rivista d'Italia. **8**(9), 366–377 (1905)
50. Fano, G.: Promemoria. In: Lorey, W. (ed.) Zur Schulreform. Unterrichtsblätter für Mathematik und Naturwissenschaften, vol. 32.1, pp. 23–25 (1926)
51. Luciano, E., Scalambro, E.: Il dovere e il piacere di insegnare: l'impegno di Gino Fano nell'educazione matematica. Studi Piemontesi. (2022)
52. Giacardi, L.: From Euclid as textbook to the Giovanni Gentile reform (1867–1923). Problems, methods and debates in mathematics teaching in Italy. Paedagogica Historica. Int. J. Hist. Educ. **17**, 587–613 (2006)
53. Schubring, G.: Pure and applied mathematics in divergent institutional settings in Germany: the role and impact of Felix Klein. In: Rowe, D., McCleary, J. (eds.) The History of Modern Mathematics, vol. 2, pp. 170–220. Academic Press, London (1989)
54. Fano, G.: Le Scuole di Magistero, Relazione al Congresso della Società Italiana Mathesis. Periodico di Matematiche. **2**(4), 102–110 (1921)
55. Giacardi, L.: I matematici e la formazione degli insegnanti in Italia nel primo Novecento. In: Ghione, F. (ed.) La formazione degli insegnanti di matematica. L'esperienza italiana a confronto con alcune esperienze europee, PRISTEM STORIA, pp. 61–106. Università Bocconi, Milano (2013)
56. Fano, G., Terracini, A.: Lezioni di geometria analitica e proiettiva. Paravia, Torino (1929)
57. Fano, G.: Studio di alcuni sistemi di rette considerati come superficie dello spazio a cinque dimensioni. Annali di Matematica. **21**, 141–192 (1893)
58. Luciano, E.: Gino Fano in Svizzera (1939–1945). Bollettino di Storia delle Scienze Matematiche. **31**, 1–30 (2022)
59. Sutton, G.: The centenary of the birth of W. H. Young (20th October 1863). Math. Gaz. **49**(367), 16–21 (1965)
60. Luciano, E.: Looking for a Space of Intellectual Survival. The Jewish Mathematical Emigration from Fascist Italy. Series Science Networks. Historical Studies. Springer, Basel (2022)
61. Fano, G.: L'opera del Comitato regionale di mobilitazione industriale per il Piemonte: settembre 1915 - marzo 1919. Giani, Torino (1919)
62. Fano, G.: Il confine del Trentino e le trattative dello scorso aprile con la monarchia austro-ungarica. Conferenza tenuta alla Società di Cultura di Torino il giorno 11 giugno 1915. Armani e Stein, Roma (1915)
63. Broggini, R.: Terra d'asilo. I rifugiati italiani in Svizzera 1943–1945. Il Mulino, Bologna (1993)
64. Signori, E.: La Svizzera e i fuoriusciti italiani. Aspetti e problemi dell'emigrazione politica 1943–1945. Angeli, Milano (1983)
65. Fano, G.: Sulle curve ovunque tangenti a una quintica piana generale. Commentari Mathematici Helvetici. **12**, 172–190 (1940)
66. Broggini, R.: La frontiera della speranza. Gli ebrei dall'Italia verso la Svizzera, 1943–1945. Mondadori, Milano (1998)
67. Lasserre, A.: Frontières et camps: le refuge en Suisse de 1933 à 1945. Payot, Lausanne (1995)
68. Sarfatti, M.: Dopo l'8 settembre: gli ebrei e la rete confinaria italo-svizzera. La Rassegna Mensile di Israel. **47**, 150–173 (1981)

69. Wisard, F.: L'université vaudoise d'une guerre à l'autre. Payot, Lausanne (1998)
70. Luciano, E. (ed.): Scienza in esilio. Gustavo Colonnetti e i campi universitari in Svizzera (1943–1945). PRISTEM STORIA. Note di Matematica, Storia, Cultura 41–42. Egea, Milano (2017)
71. Colonnetti, G.: Pensieri e fatti dall'esilio. Accademia Nazionale dei Lincei, Roma (1973)
72. Levi, A.: I campi universitari italiani in Svizzera (1944–1945). Svizzera italiana. **62**(7), 93–101 (1947)
73. Fano, G.: Lezioni di geometria descrittiva tenute dal prof. Gino Fano, raccolte dagli studenti Roberto Ballarati e Franco Brindisi. Losanna, CUI Lezioni, vol. 37. FESE (1944)
74. Fano, G.: Lezioni di geometria analitica. Losanna, CUI Lezioni, vol. 6. FESE (1944)
75. Andreotti, A. (ed.), †Fano, G.: Les surfaces du quatrième ordre. Rendiconti del Seminario Matematico Università e Politecnico di Torino. **12**, 301–313 (1953–54)
76. Brigaglia, A.: The creation and persistence of national schools: the case of Italian algebraic geometry. In: Bottazzini, U., Dahan-Dalmédico, A. (eds.) Changing Images of Mathematics, pp. 187–206. Routledge, London (2001)
77. Levi, B.: Correría en la Logica matematica. Revista de matematicas y fisica teorica Univ. Nacional de Tucumán. **3**, 13–78 (1942)
78. Pontecorboli, G.: America nuova terra promessa. Storie di italiani in fuga dal fascismo. Brioschi, Milano (2013)
79. Campanile, B.: Robert Fano e il coraggio di vivere il "non luogo". Viaggiatori. Circolazioni, scambi ed esilio. **1**(2), 353–386 (2018)
80. Giacardi, L., Roero, C.S.: La Biblioteca speciale di matematica 'Giuseppe Peano'. In: Roero, C.S. (ed.) La Facoltà di Scienze Matematiche Fisiche Naturali di Torino, 1848–1998, vol. I, pp. 437–458. Deputazione Subalpina di Storia Patria, Torino (1999)
81. Luciano, E., Scalambro, E.: On Gino Fano's patrimony: library and miscellany. Rivista di Storia dell'Università di Torino. **10.1**, 45–73 (2021)
82. Beretta, M.: Storia materiale della scienza. Dal libro ai laboratori, Mondadori, Milano (2002)
83. Jovanovic, F., Rebolledo-Dhuin, V., Verdier, N. (eds.): Science(s) et édition(s), des années 1780 à l'entre-deux-guerres. Philos. Sci. **22**(1) (2018)
84. Luciano, E.: Constructing an international library: the collections of journals in Turin's special mathematics library (1883–1964). Hist. Math., 433–449 (2018)
85. Nabonnand, P., Peiffer, J., Gispert, H. (eds.): Circulation et échanges mathématiques. Études de cas, Philosophia Scientiæ. **19**(2) (2015)
86. Brigaglia, A., Ciliberto, C.: Italian Algebraic Geometry Between the Two World Wars. Queen's University, Kingston (1995)
87. Fano, G.: Ricerche sulla varietà cubica generale dello spazio a quattro dimensioni e sopra i suoi spazi pluritangenti. Annali di Matematica. **10**, 251–285 (1904)
88. Fano, G.: Sopra alcune varietà algebriche a tre dimensioni aventi tutti i generi nulli. Atti R. Acc. Sci. Torino. **43**, 973–984 (1908)
89. Castelnuovo, G.: La geometria algebrica e la scuola italiana. Atti del Congresso Internazionale dei Matematici, 3–10 Settembre 1928, vol. 1, pp. 191–201. Zanichelli, Bologna (1929)
90. Severi, F.: Alcune proposizioni fondamentali per la geometria sulle varietà algebriche. Rendiconti Accademia Nazionale dei Lincei. **16**(5), 337–344 (1907)
91. Tricomi, F.G.: Matematici italiani del primo secolo dello stato unitario. Memorie dell'Accademia delle Scienze di Torino. **1**(4), 1–120 (1962)
92. Fano, G.: Sulle varietà algebriche a tre dimensioni aventi tutti i generi nulli. In: Atti del Congresso Internazionale dei Matematici, 3–10 Settembre 1928, vol. 4, pp. 115–121. Zanichelli, Bologna (1931)
93. Fano, G.: Sulle varietà algebriche a tre dimensioni a curve–sezioni canoniche. Memorie della R. Accademia d'Italia. **8**, 23–64 (1937)
94. Barrow-Green, J., Gray, J.: Geometry at Cambridge. Hist. Math. **33**, 315–356 (2006)
95. Roth, L.: Algebraic Threefolds. With Special Regard to Problems of Rationality. Springer, Berlin (1955)

96. Du Val, P.: Superficie di genere uno che non sono base per un sistema di quadriche. Rendiconti Accademia Nazionale dei Lincei. **15**(6), 276–279 (1932)
97. Scalambro, E.: La miscellanea di Gino Fano. Schedatura. https://www.corradosegre.unito.it/fondo_fano_m.php (2021). Accessed 28 Feb 2022
98. Godeaux, L.: Sur les variétés algébriques à trois dimensions dont les sections hyperplanes sont des surfaces de genre zéro et de bigenre un. Bulletin de la Classe des sciences (Académie royale de Belgique). **19**, 134–140 (1933)
99. Fano, G.: Sulle varietà algebriche a tre dimensioni le cui sezioni iperpiane sono superficie di genere zero e bigenere uno. Memorie della Società Italiana delle Scienze (detta dei XL). **24**, 41–66 (1938)
100. Conte, A., Murre, J.: Algebraic varieties of dimension three whose hyperplane sections are Enriques surfaces. Annali della Scuola Normale Superiore di Pisa Classe di Scienze. **12**, 43–80 (1985)
101. Chislenko, E., Tschinkel, Y.: The Felix Klein protocols. AMS Notice. **54**(8), 960–970 (2007)
102. Fano, G.: Uno sguardo alla storia della matematica. Atti e Memorie R. Accademia Virgiliana. **15–16**, 3–34 (1895)
103. Fano, G.: Über Gruppen, insbesondere kontinuierliche Gruppen von Cremona-Transformationen der Ebene und des Raumes. In: Verhandlungen des ersten Internationalen Mathematiker-Kongresses in Zürich vom 9 bis 11 August 1897, pp. 254–255. Teubner, Leipzig (1898)
104. Fano, G.: Gegensatz von synthetischer und analytischer Geometrie in seiner historischen Entwicklung im XIX Jahrhundert. In: Enzyklopädie der mathematischen Wissenschaften, III, 4a, pp. 221–288. Teubner, Leipzig (1907)
105. Fano, G.: Irrazionalità della forma cubica generale dello spazio a quattro dimensioni. Rendiconti del Seminario Matematico Università e Politecnico di Torino. **9**, 21–45 (1950)
106. Giacardi, L.: Testimonianze sulla Scuola italiana di geometria algebrica nei fondi manoscritti della Biblioteca "Giuseppe Peano" di Torino. In: Montaldo, S., Novaria, P. (eds.) Gli Archivi della scienza. L'Università di Torino e altri casi italiani, pp. 105–119. Angeli, Milano (2011)
107. Severi, F.: La géométrie algébrique. In: Fields, J.C. (ed.) Proceedings of the International Mathematical Congress Held in Toronto, August 11–16, vol. 1, pp. 149–154. University Press, Toronto (1924)

From Enriques Surface to Artin-Mumford Counterexample

Alessandro Verra

Abstract After an introduction to the themes of Enriques surfaces and rationality questions, the Artin-Mumford counterexample to Lüroth problem is revisited. Its realization is explicitly connected to Enriques surfaces, more precisely to the special family of Reye congruences and their classical geometry.

1 Perspectives on Enriques Surface and Rationality[1]

This section serves as a non-technical introduction to this paper and to a wider theme of investigation, in algebraic geometry and its history, we can well summarize by the keywords *Enriques surface* and *rationality of algebraic varieties.* These words certainly represent central and related issues, typical of the Italian contribution to algebraic geometry: since the golden age of the classification of complex algebraic surfaces, by Castelnuovo and Enriques, to the present times. Their presence is therefore natural in the proceedings of an INdAM workshop bearing the title *The Italian contribution to Algebraic Geometry between tradition and future.*

In particular, the names of Castelnuovo and Enriques are definitely related to *Castelnuovo criterion of rationality* and to the discovery of *Enriques surfaces.* As is well known, these achievements, in the theory of complex algebraic surfaces and their birational classification, are strictly related and represent one of the culminating points, in the history of algebraic geometry and of the Italian contribution, during some fortunate years at the juncture of the nineteenth and twentieth centuries.

Without entering in technical issues, the discovery of Enriques surfaces was motivated, independently from the work of birational classification, by the search of counterexamples to a quite natural conjecture, concerning the rationality of an

[1] We work over the complex field.

A. Verra (✉)
Dipartimento di Matematica, Universitá degli Studi di Roma TRE, Roma, Italy

G. Bini (ed.), *Algebraic Geometry between Tradition and Future*,
Springer INdAM Series 53, https://doi.org/10.1007/978-981-19-8281-1_8

algebraic surface S. More precisely, it was somehow natural to expect that S is rational if and only if its geometric genus $p_g(S)$ and its irregularity $q(S)$ are zero.

This led Enriques, in the long days of a summer vacation of 1894, to consider the family of sextic surfaces, in the complex projective space $\mathbb{P}^3$, having multiplicity 2 along the edges of a tetrahedron. He proved that such a general sextic is non-rational, though its geometric genus and irregularity are zero. This is a celebrated episode; see the correspondence Castelnuovo-Enriques [7] p. x and 111.

In this way, an entire class of surfaces was discovered, and nowadays, these bear the name of Enriques surfaces. Actually, any Enriques surface turns out to be birational to a sextic as above. Castelnuovo's criterion was proven in the same period: let $P_2(S)$ be the bigenus of S, and then S is rational iff

$$P_2(S) = q(S) = 0.$$

A general motivation for this paper is to stress that, along their more than centennial history, Enriques surfaces certainly did not stay confined in the classification of algebraic surfaces as a kind of special or exotic plant. Instead, their original interaction with rationality problems in dimension 2 impressively started to extend to higher dimension, implicitly or explicitly, touching in particular the famous Lüroth problem.

Lüroth's problem is the question of deciding whether an algebraic variety V admitting rational parametric equations $f : \mathbb{C}^N \to V$ is also rational, that is, it admits birational parametric equations $g : \mathbb{C}^d \to V, d = \dim V$.

Lüroth's theorem and Castelnuovo's criterion imply this property for curves and surfaces, but the problem was staying open for years in dimension ≥ 3. This until the crucial 1971–1972, when three counterexamples came out in dimension 3: by Artin-Mumford, Clemens-Griffiths, and Manin-Iskovskikh. This famous triple episode can be considered as the beginning of a "modern era" in the domain of rationality problems; see [5]-1.3 and Section 4.1.

Whatever it is, starting from Castelnuovo and Enriques and considering all the decades until the present days of modern era, the overlapping of perspectives, on the mentioned themes of Enriques surfaces and rationality problems, is often visible and always very interesting.

As an example, let us mention the debate on Lüroth's problem, originated in the 1950s from Roth's book [25], and the related Serre's theorem that unirational implies simply connected [28]. The threefold considered is a unirational sextic in $\mathbb{P}^4$ whose hyperplane sections are sextic Enriques surfaces. The discussion was about, possibly, proving its irrationality via some features of irrationality of its hyperplane sections; see [4] and [24].

We cannot add more views and perspectives in this introduction but mention the extraordinary wave of breakthrough results from the very last years. This is true in particular for stable rationality of some threefolds and cohomological decomposition of their diagonal; see [30–32].

Coming to the contents of this paper, we revisit the counterexample of Artin-Mumford, putting in evidence the presence, sometimes behind the scene in the literature, of an Enriques surface S. S is embedded in the Grassmannian $\mathbb{G}$ of lines of $\mathbb{P}^3$ and well known as a Reye congruence.

S brings us in the middle of classical algebraic geometry: S is obtained from a general three-dimensional linear system W, a web, of quadric surfaces. Moreover the threefold to be considered is the double covering of W,

$$f : \tilde{W} \to W,$$

parametrizing the rulings of lines of the quadrics of W. It is easy to see that $\tilde{W}$ is unirational; see 4.7 and [4]. The branch surface of f is a Cayley quartic symmetroid $\tilde{S}_+$, parametrizing the singular quadrics of W. A natural desingularization $\tilde{W}'$ of W is the blow-up of $\tilde{W}$ at its ten singular points. Artin and Mumford prove that, for a smooth variety V, the torsion of $H^3(V, \mathbb{Z})$ is a birational invariant. Moreover, they prove

$$H^3(\tilde{W}', \mathbb{Z}) \cong \mathbb{Z}/2\mathbb{Z},$$

which implies the irrationality of $\tilde{W}'$, since $H^3(\mathbb{P}^3, \mathbb{Z}) = 0$. As is well known, the Enriques surface S has the same feature of irrationality:

$$H^3(S, \mathbb{Z}) \cong H_1(S, \mathbb{Z}) \cong \mathbb{Z}/2\mathbb{Z}.$$

We describe, and use, the nice geometry offered by S and by the Fano threefold $\tilde{W}$, to reconstruct explicitly, from the nonzero class of $H_1(S, \mathbb{Z})$, the nonzero class of $H^3(\tilde{W}', \mathbb{Z})$. In particular, we profit of the special feature of the Fano surface of lines of $\tilde{W}$. Indeed this is split into two irreducible components birational to S; see 4.4 and [17].

Remark 1.1 For $t \geq 3$, let W_t be a general linear system of dimension $\binom{t}{2}$ of quadrics of $\mathbb{P}^t$. Let $W_t^4 \subset W_t$ be the locus of quadrics of rank ≤ 4. Then $\dim W_t^4 = 2t - 3$, and we have a double covering $f : \tilde{W}_t^4 \to W_t^4$, parametrizing the rulings of subspaces of maximal dimension of the quadrics of W_t^4. A special family of these linear systems is associated with Reye congruences as in [11]. In this case, W_t defines an embedding of a Reye congruence S as a surface of class $(t, 3t - 2)$ in the Grassmannian $\mathbb{G}_t$ of lines of $\mathbb{P}^t$. It seems that, like for $t = 3$, S determines nonzero 2-torsion in $H^3(\tilde{W}_t', \mathbb{Z})$, $\tilde{W}_t'$ being a desingularization of $\tilde{W}_t$. This will be possibly reconsidered elsewhere.

2 Webs of Quadrics and Enriques Surfaces

Now we introduce the family of those Enriques surfaces bearing the name of *Reye congruences.*[2] Let $\mathbb{G}$ be the Grassmannian of lines of $\mathbb{P}^3$, and a general member S of this family is a smooth Enriques surface embedded in $\mathbb{G}$, such that its rational equivalence class in the Chow ring $\mathrm{CH}^*(\mathbb{G})$ is $(\sigma_{1,1}, (\sigma_{2,0})$, is the class of the set of lines in a plane (through a point).)

$$7\sigma_{2,0} + 3\sigma_{1,1}. \tag{1}$$

Then S has degree 10 and sectional genus 6 in the Plücker embedding of $\mathbb{G}$. Moreover, $\mathcal{O}_S(1)$ is an example of Fano polarization of an Enriques surface [14] 3.5. Notice that, in the period space of Enriques surfaces, the locus of Reye congruences is an irreducible divisor and coincides with the locus of periods of Enriques surfaces containing a smooth rational curve [23]. The construction of these surfaces was given by Reye [25]. It relies on projective methods and the beautiful geometry of webs of quadric surfaces. We concentrate on these related topics, recovering a general picture.

2.1 *Webs of Quadrics*

Let E be a four-dimensional vector space; we fix the notation

$$\mathbb{P}^3 := \mathbb{P}(E) \tag{2}$$

and define the Plücker embedding of the Grassmannian of lines of $\mathbb{P}^3$ by

$$\mathbb{G} \subset \mathbb{P}^{5-}, \tag{3}$$

where $\mathbb{P}^{5-} := \mathbb{P}(\wedge^2 E)$. In particular, $\mathbb{G}$ is a smooth quadric hypersurface. Then we consider $E \otimes E$ and its standard direct sum decomposition

$$E \otimes E = \wedge^2 E \oplus \mathrm{Sym}^2 E, \tag{4}$$

via the eigenspaces of the involution exchanging the factors of $E \otimes E$. The induced involution on $\mathbb{P}^{15} := \mathbb{P}(E \otimes E)$ will be denoted by

$$\iota : \mathbb{P}^{15} \to \mathbb{P}^{15}. \tag{5}$$

[2] See [16] 3.7 for a historical note.

We also put $\mathbb{P}^{9+} := \mathbb{P}(\mathrm{Sym}^2 E)$ and observe that the set of fixed points of ι is $\mathbb{P}^{5-} \cup \mathbb{P}^{9+}$. Then we define the commutative diagram

$$\begin{array}{ccccc} \mathbb{G} & \longleftarrow & \mathbb{P}^3 \times \mathbb{P}^3 & \longrightarrow & \mathbb{S} \\ \downarrow & & \downarrow & & \downarrow \\ \mathbb{P}^{5-} & \xleftarrow{\lambda_-} & \mathbb{P}^{15} & \xrightarrow{\lambda_+} & \mathbb{P}^{9+}. \end{array} \tag{6}$$

Here $\mathbb{S}$ is the set of points defined by symmetric tensor $a \otimes b + b \otimes a$ having rank ≤ 2. Notice that $\mathbb{S}$ is biregular to the quotient $\mathbb{P}^3 \times \mathbb{P}^3 / \langle \iota \rangle$. The vertical arrows are the natural inclusions. Moreover, λ_+ and λ_- are the natural linear projections onto the spaces $\mathbb{P}^{9+}$ and $\mathbb{P}^{5-}$. The restrictions of λ_- and λ_+ to $\mathbb{P}^3 \times \mathbb{P}^3$ admit an elementary description. Let $(x, y) \in \mathbb{P}^3 \times \mathbb{P}^3$ and let $\ell \in \mathbb{G}$, where ℓ is the line through x, y if $x \neq y$; then we have

$$\lambda_-(x, y) = \ell \in \mathbb{G} \text{ and } \lambda_+(x, y) = x + y \in \mathbb{S}. \tag{7}$$

Definition 2.1 $\mathbb{T}^r$ is the set of points in $\mathbb{P}^{15}$ defined by tensors $t \in E \otimes E$ having rank $\leq r$. In particular, $\mathbb{T}^2$ is $\mathbb{P}^3 \times \mathbb{P}^3$ and $\mathbb{T}^1$ is its diagonal.

From the loci $\mathbb{T}_r$, we have of course the rank stratification

$$\mathbb{T}^1 \subset \mathbb{T}^2 \subset \mathbb{T}^3 \subset \mathbb{P}^{15}. \tag{8}$$

Now we pass to the dual space $\mathbb{P}^{15*}$ of bilinear forms, to be denoted by

$$\mathbb{B} := \mathbb{P}(E^* \otimes E^*). \tag{9}$$

Let $b \in \mathbb{B}$, and then $b^\perp \subset \mathbb{P}^{15}$ will denote its orthogonal hyperplane. Moreover let $W \subset \mathbb{B}$, and then the orthogonal subspace of W is by definition

$$W^\perp := \bigcap_{b \in W} b^\perp. \tag{10}$$

Let W be a subspace of dimension c, and then codim $W^\perp = c + 1$, and we have

$$|\mathcal{I}_{W^\perp}(1)| = W, \tag{11}$$

where we reserve the notation $\mathcal{I}_{W^\perp}$ to the ideal sheaf of $W^\perp$ in $\mathbb{P}^{15}$. Notice that $\mathbb{B}$ contains the subspace $\mathbb{Q} := \mathbb{P}(\mathrm{Sym}^2 E^*)$ and that this is just the space of quadrics of $\mathbb{P}^3$. We denote its rank stratification by

$$\mathbb{Q}^1 \subset \mathbb{Q}^2 \subset \mathbb{Q}^3 \subset \mathbb{Q}. \tag{12}$$

Finally we come to web of quadrics of $\mathbb{P}^3$. We use the traditional word *web* for a three-dimensional linear system of divisors of a variety. For short, we will use the word *web of quadrics* for a web of quadric surfaces of $\mathbb{P}^3$. Let

$$W \subset \mathbb{Q} \subset \mathbb{B} \tag{13}$$

be a general web of quadrics, and then W naturally defines an Enriques surface which is a Reye congruence [16] 7. We construct it as follows. We have

$$W = \mathbb{P}(V), \tag{14}$$

where $V \subset \mathrm{Sym}^2 E^*$ is a general four-dimensional space. Notice that ι^* is the identity on $V = H^0(\mathcal{I}_{W^\perp}(1))$; therefore, the five-dimensional subspace

$$W^\perp \subset \mathbb{P}^{15} \tag{15}$$

satisfies $\iota(W^\perp) = W^\perp$. Now observe that the set of fixed points of the involution $\iota|W^\perp$ is the disjoint union of the space $\mathbb{P}^{5-} \subset W^\perp$ and of

$$W^{\perp +} := W^\perp \cap \mathbb{P}^{9+}. \tag{16}$$

Therefore, $\mathbb{P}^{5-} \cup W^{\perp +}$ generates $W^\perp$ and codim $W^{\perp +} = 4$ in $\mathbb{P}^{9+}$, so that

$$\lambda_+^*(W^{\perp +}) = W^\perp. \tag{17}$$

This implies that the family of spaces $W^{\perp +}$ coincides with the Grassmannian of codimension 4 spaces in $\mathbb{P}^{9+}$. Notice also that the family of spaces $W^\perp$ is the family of codimension 4 spaces of $\mathbb{P}^{15}$ containing $\mathbb{P}^{5-}$. Now consider

$$\lambda_+|\mathbb{T}_2 : \mathbb{T}_2 \to \mathbb{S} \tag{18}$$

as in (73). This is the quotient map of $\iota|\mathbb{T}_2$ and a finite double cover branched on Sing $\mathbb{S}$, that is, on the image of the diagonal $\mathbb{T}_1$ of $\mathbb{T}_2 = \mathbb{P}^3 \times \mathbb{P}^3$.

2.2 *Reye Congruences of Lines*

Definition 2.2 Given a web W, let us fix the notation

$$S_+ := \mathbb{S} \cdot W^{\perp +} \;, \quad \tilde{S} := \mathbb{T}_2 \cdot W^\perp. \tag{19}$$

We assume that W is general, so that $W^{\perp +}$ is disjoint from Sing $\mathbb{S}$ and transversal to the map λ_+. By Bertini theorem, $\tilde{S}_+$ and S are smooth, irreducible surfaces and linear sections of $\mathbb{T}_2$ and $\mathbb{S}$. It is also clear that

$$\iota|\tilde{S} : \tilde{S} \to \tilde{S}. \tag{20}$$

is a fixed-point-free involution. Indeed ι^* is the identity on V and $\tilde{S}$ is disjoint from the set $\mathbb{T}_1$ of fixed points of ι. Hence $\lambda_+ : \mathbb{T}_2 \to \mathbb{S}$ restricts to an étale double covering $\lambda_+|\tilde{S} : \tilde{S} \to S_+$. To complete the picture, we consider the rational map $\lambda_- : \mathbb{P}^{15} \to \mathbb{P}^{5-}$ and its restriction $\lambda_-|\tilde{S}$, and we fix the notation $S := \lambda_-(\tilde{S})$. Of course, we have

$$S \subset \mathbb{G} \subset \mathbb{P}^{5-}. \tag{21}$$

Theorem 2.1 *The surfaces S and S_+ are embeddings, of degree* 10 *and sectional genus* 6*, of the same Enriques surface $\tilde{S}/\langle\iota|\tilde{S}\rangle$. Moreover one has*

$$\omega_{S_+}(1) \cong \mathcal{O}_S(1). \tag{22}$$

Proof $\tilde{S}$ is a K3 surface, endowed with the fixed-point-free involution $\iota|\tilde{S}$ and embedded in $\mathbb{T}_2 = \mathbb{P}^3 \times \mathbb{P}^3$ as a complete intersection of four elements of $|\mathcal{O}_{\mathbb{P}^3\times\mathbb{P}^3}(1,1)|$. This follows by its construction and adjunction formula. Now S and S_+ are embeddings of the same smooth surface $\tilde{S}/\langle\iota|\tilde{S}\rangle$. Since $\iota|\tilde{S}$ is fixed-point-free and $\tilde{S}$ is a smooth K3 surface, then $\tilde{S}/\langle\iota\rangle$ is a smooth Enriques surface. On the other hand, the K3 surface $\tilde{S}$ is embedded in $W^\perp$ by $\mathcal{O}_{\tilde{S}}(1) := \mathcal{O}_{\mathbb{P}^3\times\mathbb{P}^3}(1,1) \otimes \mathcal{O}_{\tilde{S}}$, a polarization of genus 11 and degree 20. Moreover the involution $\iota|\tilde{S}$ acts on $H^0(\mathcal{O}_{\tilde{S}}(1))$, and its quotient map $\lambda_+|\tilde{S} : \tilde{S} \to S_+$ is the double cover defined by the canonical sheaf ω_{S_+} and $\lambda_+^*\mathcal{O}_{S_+}(1) \cong \mathcal{O}_{\tilde{S}}(1)$; it follows that we have the decomposition

$$H^0(\mathcal{O}_{\tilde{S}}(1)) = \lambda_+^* H^0(\mathcal{O}_{S_+}(1)) \oplus \lambda_+^* H^0(\omega_{S_+}(1)). \tag{23}$$

The summands, respectively, are the six-dimensional $+1$ and -1 eigenspaces of $(\iota|\tilde{S})^*$. This implies the last part of the statement; we omit some details. □

Clearly $\lambda_-|\tilde{S} : \tilde{S} \to \mathbb{P}^{5-}$ factors through $\iota|\tilde{S}$ and defines the embedding

$$S \subset \mathbb{G} \subset \mathbb{P}^{5-}. \tag{24}$$

Both S and S_+ are examples of Fano models of an Enriques surface. S is known as *Reye congruences of lines*. A general Fano model is projectively normal; hence, it is not contained in a quadric, cfr. [14], 3.5. Instead S is contained in the smooth quadric $\mathbb{G}$. This is the special feature of this family. Each S is endowed with a special rank 2 vector bundle, namely, the restriction of the universal bundle of $\mathbb{G}$. We will use it to introduce the classical construction of S; see [15, 25], and [16]-7.

2.3 Further Notation

The universal bundle over $\mathbb{G}$ is $p : U \to \mathbb{G}$, and then $U_\ell \subset E$, and we define

$$\mathbb{U}_\ell := \mathbb{P}(U_\ell) \subset \mathbb{P}^3, \tag{25}$$

for each $\ell \in \mathbb{G}$. This is the line in $\mathbb{P}^3$ defining the point $\ell \in \mathbb{G}$. After some tradition, we say that a ℓ is *a ray* of $\mathbb{G}$. We will also use the $\mathbb{P}^3$ bundle

$$\pi : \mathbb{P}(U \otimes U) \to \mathbb{G}, \tag{26}$$

whose fiber at ℓ is $\mathbb{P}^3_\ell := \mathbb{P}(U_\ell \otimes U_\ell)$. Let $\tau : \mathbb{P}(U \otimes U) \to \mathbb{P}^{15}$ be its tautological map, and then $\tau_\ell : \mathbb{P}^3_\ell \to \mathbb{P}^{15}$ is the linear embedding defined by the inclusion of $U_\ell \otimes U_\ell \subset E \otimes E$. Moreover τ is a morphism birational onto its image. We identify $\mathbb{P}^3_\ell$ to its image by τ_ℓ so that $\mathbb{P}^3_\ell \subset \mathbb{P}^{15}$. Since we have

$$U_\ell \otimes U_\ell = \wedge^2 U_\ell \oplus \mathrm{Sym}^2 U, \tag{27}$$

then $\mathbb{P}^3_\ell$ is ι invariant. Its point $\ell = \mathbb{P}(\wedge^2 U_\ell)$ and its plane $Q_\ell := \mathbb{P}(\mathrm{Sym}^2 U_\ell)$ are the set of fixed points of $\iota|\mathbb{P}^3_\ell$. We observe that $\mathbb{T}_2 \cap \mathbb{P}^3_\ell$ is the quadric $\mathbb{U}_\ell \times \mathbb{U}_\ell$ and $\mathbb{T}_1 \cap \mathbb{P}^3_\ell$ its diagonal. Finally, the next diagram will be useful:

$$\begin{array}{ccccc} \mathbb{P}^{15} & \xleftarrow{\tau} & \mathbb{P}(U \otimes U) & \xrightarrow{\pi} & \mathbb{G}. \\ \uparrow & & \uparrow & & \uparrow \\ \mathbb{T}_2 & \xleftarrow{\delta} & \mathbb{D} & \xrightarrow{\pi} & \mathbb{G} \end{array} \tag{28}$$

Its vertical arrows are the natural inclusions, and $\mathbb{D}$ is the projectivized set of points defined in $\mathbb{P}(U \otimes U)$ by the decomposable tensors in $U \otimes U$.

3 Reye Congruences and Symmetroids

3.1 The Classical Construction

It is now useful to revisit in modern terms Reye's classical construction of the surface S, cfr. [13, 15, 16, 25]. Let W be a general web of quadrics, defining as above a smooth $K3$ surface $\tilde{S}$ and the embeddings $S \subset \mathbb{G}$ and $S_+ = \mathbb{S} \cdot W$ of S. We assume $W = \mathbb{P}(V)$, and then V is a space of quadratic forms on E, and we have the natural inclusions

$$V \subset H^0(\mathcal{O}_{\mathbb{P}^3}(2)) \subset H^0(\mathrm{Sym}^2 U^*). \tag{29}$$

The evaluation of global sections of $\mathrm{Sym}^2 U^*$ defines a morphism

$$e : \mathcal{O}_{\mathbb{G}} \otimes V \to \mathrm{Sym}^2 U^*, \tag{30}$$

of vector bundles of ranks 4 and 3. Counting dimensions, the degeneracy scheme of e is a surface, provided it is proper and non-empty.

Theorem 3.1 *The degeneracy scheme of e is the Enriques surface S, and its rational equivalence class is $7\sigma_{1,1} + 3\sigma_{2,0}$ in $\mathrm{CH}^*(\mathbb{G})$.*

Proof Assume the degeneracy scheme S_e of e is proper. Then, computing its class in the Chow ring of $\mathbb{G}$, we obtain that $[S_e] = 7\sigma_{1,1} + 3\sigma_{2,0}$ in $\mathrm{CH}^*(\mathbb{G})$. Hence $\deg S_e = \deg S = 10$, and the equality $\mathrm{Supp}\, S_e = S$ implies the statement. To show the equality, consider $\ell \in \mathbb{G}$ and the fiberwise map

$$e_\ell : V \to H^0(\mathcal{O}_{\mathbb{U}_\ell}(2)). \tag{31}$$

This is the restriction map to the line $\mathbb{U}_\ell \subset \mathbb{P}^3$, defined as above. Equivalently, keeping our previous identifications, we can assume that the curve

$$\Delta_\ell \subset \mathbb{U}_\ell \times \mathbb{U}_\ell \subset \mathbb{P}^3_\ell \subset \mathbb{P}^{15} \tag{32}$$

is the diagonal embedding of $\mathbb{U}_\ell$ and that $V \subset H^0(\mathcal{O}_{\mathbb{P}^{15}}(1))$. Then e_ℓ is the restriction map $V \to H^0(\mathcal{O}_{\Delta_\ell}(1))$. Moreover we know that V has a basis a_1, a_2, a_3, s so that $\iota^* a_j = -a_j$ and $\iota^* s = s$, where s is zero on Δ_ℓ. Hence e_ℓ degenerates iff its rank is two. Equivalently $V \to H^0(\mathcal{O}_{\mathbb{P}^3_\ell}(1))$ defines a pencil of planes in $\mathbb{P}^3_\ell$, and its base line intersects $\mathbb{U}_\ell \times \mathbb{U}_\ell$ in two distinct points $x, y \in \tilde{S}$ such that $y = \iota(x)$. This implies $\mathrm{Supp}\, S_e = \lambda_-(\tilde{S}) = S$. □

The proof reveals the main geometric feature of S, and we have

$$S = \{\ell \in \mathbb{G} \mid \dim(V \cap H^0(\mathcal{I}_\ell(2)) = 2\}. \tag{33}$$

In other words, $\ell \in S$ iff two quadrics of W contain ℓ, that is, ℓ is in the base locus of a pencil of quadrics of W. We can conclude as follows.

Definition 3.1 The Reye congruence of W is the degeneracy scheme S of the previous morphism e, provided S is proper.

Theorem 3.2 *The Reye congruence of W is the family of lines of $\mathbb{P}^3$ which are in the base locus of a pencil of quadrics contained in W.*

3.2 Order and Class of S

Following the classical language, the order of a surface $Y \subset \mathbb{G}$ is the number a of rays of Y passing through a general point, and the class is the number b of rays of Y in a general plane. Then one has $[Y] = a\sigma_{1,1} + b\sigma_{2,0}$ in $CH^*(\mathbb{G})$. Of course, these notions naturally extend for surfaces in any Grassmannian of lines. Let us motivate geometrically the equality

$$[S] = 7\sigma_{1,1} + 3\sigma_{2,0} \in \mathrm{CH}^*(\mathbb{G}).$$

The coefficient 7 means that exactly seven rays of S contain a general point $o \in \mathbb{P}^3$. Consider the net $W_o \subset W$ of quadrics through o, and then its base locus is a set of eight distinct points $\{o\ o_1 \dots o_7\}$. It is easy to see that the lines $\overline{oo_1} \dots \overline{oo_7}$ are the seven rays of the family passing through o.

The coefficient 3 means that exactly three rays of S are contained in a general plane P. The web W restricts on P to a web W_P of conics. How many lines in P are fixed component of a pencil of conics of W_P? The answer is classical: W_P is generated by four *double* lines, supported on lines in general position. These define a complete quadrilateral. One can check that its three diagonals are precisely the lines with the required property.

3.3 The Quartic Symmetroid

Now we concentrate further on W, considering the quartic surface

$$\tilde{S}_+ = \mathbb{Q}^3 \cdot W \tag{34}$$

parametrizing the singular quadrics of a general W. Clearly such a surface is defined, in the projective space W, by the determinant of a symmetric 4×4 matrix of linear forms. For this reason, it bears the following name.

Definition 3.2 A quartic symmetroid is a surface $\tilde{S}_+$, constructed as above from a general web of quadrics W.

This classical surface is well known [16], 7.4, p. 93. Let $W \subset \mathbb{Q}$ be transversal to the quartic hypersurface $\mathbb{Q}^3 \subset \mathbb{Q}$ and to its singular locus $\mathbb{Q}^2$. Then, counting dimensions and degrees, we have $\mathbb{Q}^1 \cap W = \emptyset$, and, moreover, the set

$$\mathrm{Sing}\ \tilde{S}_+ = \mathbb{Q}^2 \cap \tilde{S}_+ \tag{35}$$

consists of ten ordinary modes. Actually $\tilde{S}_+$ is a birational projective model of $\tilde{S}$, as we are going to see. To this purpose, let us fix on $\mathbb{P}^3 \times \mathbb{P}^3$ coordinates

$$(x, y) := (x_1 : x_2 : x_3 : x_4) \times (y_1 : y_2 : y_3 : y_4), \tag{36}$$

so that $\tilde{S}$ is defined by the four symmetric bilinear equations

$$\sum_{1 \le i, j \le 4} q_{ij}^{[k]} x_i y_j = 0, \quad k = 1 \dots 4. \tag{37}$$

The set of quadratic forms $q^{[k]} := \sum q_{ij}^{[k]} x_i x_j$ is a basis for the vector space V such that $W = \mathbb{P}(V)$. From now on, we set $z := (z_1 : z_2 : z_3 : z_4)$ and

$$q_z := z_1 q^{[1]} + z_2 q^{[2]} + z_3 q^{[3]} + z_4 q^{[4]}, \tag{38}$$

denoting by Q_z the quadric in $\mathbb{P}^3$ defined by q_z. Then we compute that

$$\frac{\partial q_z}{\partial x_j} = z_1 q_j^{[1]} + z_2 q_j^{[2]} + z_3 q_j^{[3]} + z_4 q_j^{[3]}, \quad j = 1 \dots 4, \tag{39}$$

where we put $q_j^{[k]} := \sum_{1 \le i \le 4} q_{ij}^{[k]} x_i$. Clearly the four bilinear equations

$$\frac{\partial q_z}{\partial x_j} = 0, \quad j = 1 \dots 4, \tag{40}$$

define in the product $\mathbb{P}^3 \times W$ the incidence correspondence

$$\Xi := \{(x, z) \in \mathbb{P}^3 \mid x \in \operatorname{Sing} Q_z\}. \tag{41}$$

This is the universal singular locus over the family of quadrics W. Let

$$p_x : \Xi \to \mathbb{P}^3 \text{ and } p_z : \Xi \to W, \tag{42}$$

be the projections of Ξ to the factors of $\mathbb{P}^3 \times W$. Obviously $p_z(\Xi)$ is the quartic symmetroid $\tilde{S}_+$, defined by the symmetric determinant

$$\det(z_1 q_{ij}^{[1]} + \dots + z_4 q_{ij}^{[4]}). \tag{43}$$

Let $\tilde{S}_x = p_x(\Xi)$; then, eliminating $(z_1 : z_2 : z_3 : z_4)$ from the equations of Ξ, the equation of $\tilde{S}_x$ is a quartic form in $(x_1 : x_2 : x_3 : x_4)$, namely,

$$\tilde{S}_x = \{\det (q_j^{[k]}) = 0\}. \tag{44}$$

Now consider $\tilde{S} \subset \mathbb{P}^3 \times \mathbb{P}^3$ and its equations in (x, y). Let $\tilde{p}_x : \tilde{S} \to \mathbb{P}^3$ be the first projection and $\tilde{S}_v = \tilde{p}_x(\tilde{S})$. Eliminating y, one computes that

$$\tilde{S}_v = \{\det (q_j^{[k]}) = 0\}. \tag{45}$$

Then $\tilde{S}_x = \tilde{S}_v$, and hence, the surfaces $\tilde{S}_v$, $\tilde{S}$, and $\tilde{S}_+$ are birational projective models of the same symmetroid $\tilde{S}_+$. Moreover let $\tilde{p}_y : \tilde{S} \to \mathbb{P}^3$ be the second projection and $\tilde{S}_y = p_y(\tilde{S})$, and then $\tilde{S}_x$ and $\tilde{S}_y$ are projectively isomorphic. This follows because $\iota(\tilde{S}) = \tilde{S}$ and hence $\tilde{p}_y = \tilde{p}_x \circ \iota$. We keep the notation $\tilde{S}_v$ for $\tilde{S}_x$ and $\tilde{S}_y$. In particular, $\tilde{S}_v$ is the birational image of $p_z : \tilde{S}_+ \to \mathbb{P}^3$. The next theorem summarizes our discussion and implements the picture.

Theorem 3.3 *The $K3$ surface $\tilde{S}$ in $\mathbb{P}^3 \times \mathbb{P}^3$ and the quartic symmetroid $\tilde{S}_+$ in W are birational to the quartic surface $\tilde{S}_v$. Moreover this surface is the locus in $\mathbb{P}^3$ of the singular points of the singular quadrics of W.*

Occasionally $\tilde{S}_v$ is said to be the *Steinerian* of $\tilde{S}_+$, after some uncertain tradition. It is now the time to recall some facts on quartic double solids.

3.4 *Quartic Double Solids*

To begin, we recall that a *quartic double solid* is a finite double cover

$$f : X \to \mathbb{P}^3 \tag{46}$$

whose branch scheme is a quartic surface $B \subset \mathbb{P}^3$. See [9, 29], and [8] for new results and update. *We assume that no line is in B and that* Sing B *is a finite set of ordinary double points.* Let us consider the blowing up

$$\sigma : \mathbb{P}^{3'} \to \mathbb{P}^3, \tag{47}$$

of Sing B; then a desingularization of X is provided by the base change

$$\begin{array}{ccc} X' & \xrightarrow{f'} & \mathbb{P}^{3'} \\ {\scriptstyle\sigma'}\downarrow & & {\scriptstyle\sigma}\downarrow \\ X & \xrightarrow{f} & \mathbb{P}^3. \end{array} \tag{48}$$

Then f' is the finite double cover branched on the strict transform of B by σ. This is a smooth, minimal model of B embedded in $\mathbb{P}^{3'}$, which we denote by

$$B' \subset \mathbb{P}^{3'}. \tag{49}$$

The line geometry of $\mathbb{P}^3$ strongly influences the geometry of X. Consider indeed the universal line $\mathbb{U} = \{(x, \ell) \in \mathbb{P}^3 \times \mathbb{G} \mid x \in \mathbb{U}_\ell\}$ and its projections

$$\mathbb{P}^3 \xleftarrow{\;t\;} \mathbb{U} \xrightarrow{\;u\;} \mathbb{G}. \tag{50}$$

Then u is the projective universal bundle and t its tautological morphism. The quartic surface B clearly defines a rational section

$$s : \mathbb{G} \to \mathbb{P}(\mathrm{Sym}^4 U^*), \tag{51}$$

sending the ray $\ell \in \mathbb{G}$ to the intersection divisor $\mathbb{U}_\ell \cdot B \in |\mathcal{O}_{\mathbb{U}_\ell}(4)|$.

Definition 3.3 $s : \mathbb{G} \to \mathbb{P}(\mathrm{Sym}^4 U^*)$ is the section defined by B.

Since B does not contain lines, the rational map s is a morphism. Let

$$\mathbb{P}(\mathrm{Sym}^2 U^*) \subset \mathbb{P}(\mathrm{Sym}^4 U^*).$$

be the embedding defined by the squaring map. By definition, this means that any point $d \in |\mathcal{O}_{\mathbb{U}_\ell}(2)| = \mathbb{P}(\mathrm{Sym}^2 U^*)_\ell$ is embedded as the point

$$2d \in |\mathcal{O}_{\mathbb{U}_\ell}(4)| = \mathbb{P}(\mathrm{Sym}^4 U^*)_\ell.$$

We use this embedding to define the family of bitangent lines to B.

Definition 3.4 $\mathbb{F}(B)$ is the pull-back of $\mathbb{P}(\mathrm{Sym}^2 U^*)$ by s.

Clearly, Supp $\mathbb{F}(B)$ is the set of bitangent lines to B. The structure of $\mathbb{F}(B)$ is known; see [29] 3 and [12] 3.4. We summarize as follows.

Theorem 3.4 *If B is general, $\mathbb{F}(B)$ is a smooth integral surface and*

$$[\mathbb{F}(B)] = 12\sigma_{1,1} + 28\sigma_{2,0} \in CH^*(\mathbb{G}).$$

In the classical language, $\mathbb{F}(B)$ has *order* 12 and *class* 28. These numbers are easily explained: 28 is the number of bitangent lines to a general plane section of B, which is a smooth plane quartic. Instead 12 is the number of ordinary nodes of the branch curve of a general projection $B \to \mathbb{P}^2$.

Definition 3.5 $\mathbb{F}(B)$ is the congruence of bitangent lines of B.

The surprising case of $\mathbb{F}(B)$, B a quartic symmetroid, will be discussed in detail. Let us fix our notation for the natural involutions of X and X'.

Definition 3.6 We, respectively, denote by $j' : X' \to X'$ and by $j : X \to X$ the biregular involutions induced by $f' : X' \to \mathbb{P}^{3'}$ and by $f : X \to \mathbb{P}^3$.

Finally we introduce *the Fano surface of lines* of X'. Let $\ell \in \mathbb{F}(B)$, and then the line $\mathbb{U}_\ell$ is bitangent to B. Moreover, for ℓ general in $\mathbb{F}(B)$, it is also true that $\mathbb{U}_\ell \cap$

Sing $B = \emptyset$. Assuming this, the curve $f'^{*}\mathbb{U}_\ell$ splits as follows:

$$f'^{*}\mathbb{U}_\ell = R_{\ell,+} + R_{\ell,-} \tag{52}$$

where $R_{\ell,+}$, $R_{\ell,-}$ are biregular to $\mathbb{P}^1$ and exchanged by the involution j'. They belong to the Hilbert scheme of curves of arithmetic genus 0 and of degree 1 for $(f' \circ \sigma)^*\mathcal{O}_{\mathbb{P}^3}(1)$. This is a well-known connected surface, and the family of curves $R_{\ell,+}$, $R_{\ell,-}$ is open and dense in it. We denote it by $\mathbb{F}(X')$.

Definition 3.7 $\mathbb{F}(X')$ is the Fano surface of lines of X'.

The surface $\mathbb{F}(X')$ is *smooth and irreducible* for a general B [9, 12, 29]. We point out that $j' : X' \to X'$ defines a biregular involution

$$j'^{*} : \mathbb{F}(X') \to \mathbb{F}(X'). \tag{53}$$

As above, let $\sigma : \mathbb{P}^{3'} \to \mathbb{P}^3$ be the blow-up of Sing B and let $\mathbb{G}'$ the Hilbert scheme of the pull-back by σ of a general line of $\mathbb{P}^3$. Then the push-forward of cycles by the blowing up σ defines a natural birational morphism

$$\sigma_* : \mathbb{G}' \to \mathbb{G}. \tag{54}$$

Let $\mathbb{F}(B) \to \mathbb{G}$ be the inclusion map, and then the base change by σ_*

$$\begin{array}{ccc} \mathbb{F}(B') & \longrightarrow & \mathbb{G}' \\ \downarrow & & \downarrow{\scriptstyle \sigma_*} \\ \mathbb{F}(B) & \longrightarrow & \mathbb{G} \end{array} \tag{55}$$

defines the surface $\mathbb{F}(B')$ in $\mathbb{G}'$. If Sing $B = \emptyset$, this is $\mathbb{F}(B)$, and we only mention that $\mathbb{F}(B') \to \mathbb{F}(B)$ is the normalization map if B is general with Sing $B \neq \emptyset$. Moreover the push-forward of cycles defines the next commutative diagram, where f'_* generically coincides with the quotient map of j'^{*}:

$$\begin{array}{ccc} \mathbb{F}(X') & \xrightarrow{f'_*} & \mathbb{F}(B') \\ \downarrow & & \downarrow{\scriptstyle \sigma_*} \\ \mathbb{F}(X) & \xrightarrow{f_*} & \mathbb{F}(B). \end{array} \tag{56}$$

To conclude, we recall the following theorem.

Theorem 3.5 *Let X be a general quartic double solid, and then the map*

$$f'_* : \mathbb{F}(X') \to \mathbb{F}(B')$$

is an étale double covering of smooth regular surfaces of general type.

See the seminal papers [9, 29], and [12] for some recent new results.

Remark 3.1 Clearly, for a general B, the morphism is defined by a nonzero 2-torsion element of Pic $\mathbb{F}(B')$. Then $\mathbb{F}(B')$ is not simply connected. However, it is regular and of general type. Coming to a *quartic symmetroid* $\tilde{S}_+$, the surface $\mathbb{F}(\tilde{S}_+)$ shows up as an unexpected limit of a general $\mathbb{F}(B)$. We will see that it is singular and normalizes to the Reye congruence S.

4 The Artin-Mumford Counterexample Revisited

4.1 Artin-Mumford Double Solids

The study of the rationality problem for quartic double solids, with its related issues, plays a very important role, historically and not only. The unirationality of a quartic double solid is known since longtime, cfr. [3]-4, [25]-10, 11. Its irrationality, when it is branched on a quartic symmetroid, was proven in 1972 by Artin and Mumford [2]. This result is one of the three first counterexamples to Lüroth problem in dimension 3, appearing simultaneously in 1971–1972 and relying on different methods. The other examples also rely on very famous results: the proof of the irrationality of a smooth cubic threefold, by Clemens and Griffiths, and the irrationality of a smooth quartic threefold, proven by Manin and Iskovskikh [10, 22]. Nowadays the rationality problem for quartic double solids is settled, by application of similar methods and further work, at least as follows, cfr. [8].

Theorem 4.1 *Let $f : X \to \mathbb{P}^3$ be a quartic double solid and let $m = |\mathrm{Sing}\, B|$. Then X is irrational if $0 \leq m \leq 6$ and rational if $m \geq 11$.*

Nevertheless, this matter is a not exhausted field of top interest for several reasons. Though there is no space for further digression, let us mention once more the results on *non-stable rationality* and *non-decomposition* of the diagonal in $H^*(X, \mathbb{Z})$, for a very general quartic double solid [31, 32].

Definition 4.1 We say that a quartic double solid is an Artin-Mumford double solid if its branch surface is a general quartic symmetroid $\tilde{S}_+$.

To treat Artin-Mumford double solids, *let us fix our notation as follows:* as above, $W \subset \mathbb{Q}$ is a *general* web in the space $\mathbb{Q}$ of quadric surfaces of $\mathbb{P}^3$. Then W is a three-dimensional subspace and $\tilde{S}_+ = \mathbb{W} \cdot \mathbb{Q}^3$. We denote by

$$f : \tilde{W} \to W \tag{57}$$

the finite double covering of W branched on $\tilde{S}_+$. The next diagram provides, exactly as in (48) of the previous section, a desingularization $\tilde{W}'$ of $\tilde{W}$:

$$\begin{array}{ccc} \tilde{W}' & \xrightarrow{f'} & W' \\ \sigma' \downarrow & & \sigma \downarrow \\ \tilde{W} & \xrightarrow{f} & W. \end{array} \tag{58}$$

We will say, with a slight abuse, that the morphism $f' : \tilde{W}' \to W'$ is the desingularization of $f : \tilde{W} \to W$ and that $\tilde{W}'$ is the smooth model of $\tilde{W}$.

In the above diagram, constructing an Artin-Mumford double solid, the implicit presence of an Enriques surface is clear. $\tilde{S}$ is indeed the minimal desingularization of $\tilde{S}_+$ and an étale double covering of the Reye congruence S. The existence of S is not mentioned in Artin-Mumford paper [2]. However, the irrationality of $\tilde{W}$ follows there from the same irrationality feature of S, namely, the presence of nonzero torsion in the third cohomology group. At first, it is shown in [2] that the torsion subgroup of it is birationally invariant, for any smooth projective variety. Then the next theorem is proven.

Theorem 4.2 *The torsion of $H^3(\tilde{W}', \mathbb{Z})$ is non-trivial.*

The proof relies on the notion of Brauer group and on some Severi-Brauer varieties, conic bundles in this case, related to $\tilde{W}$. Since $H^3(\mathbb{P}^3, \mathbb{Z}) = 0$, the irrationality of W follows. Moreover, as we will see, $\tilde{W}$ is unirational. Then $\tilde{W}$ is a counterexample to Lüroth problem. Since the above torsion group is a stably rational invariant, $\tilde{W}$ is not stably rational as well [31].

Actually both S and $\tilde{W}'$ have a nonzero torsion subgroup in the third cohomology, and this is, in both cases, $\mathbb{Z}/2\mathbb{Z}$. The existence of S is variably used and considered in the literature on Artin-Mumford double solids. The same is even more true for an explicit and geometric description of the relation between $H^3(S, \mathbb{Z})$ and $H^3(\tilde{W}', \mathbb{Z})$. As far as we know, the surface S was used at first by Beauville in [3]-9, to this respect, as follows.

Let $\tilde{\mathbb{G}}$ be the blow-up of $\mathbb{G}$ at S; one constructs very geometrically a dominant morphism $\tilde{\phi} : \tilde{\mathbb{G}} \to \tilde{W}$. Since $\mathbb{G}$ is rational, $\tilde{W}$ is unirational. Moreover $H^*(S, \mathbb{Z})$ is a summand of $H^*(\tilde{\mathbb{G}}, \mathbb{Z})$, and hence $\tilde{\phi}$ naturally defines a homomorphism $h : H^*(\tilde{W}', \mathbb{Z}) \to H^*(S, \mathbb{Z})$. Using it, a proof of the previous theorem is given by application of cohomological methods; see [3] p. 30. In particular, h restricts to an isomorphism between the torsion subgroup of $H^3(\tilde{W}', \mathbb{Z})$ and $H^3(S, \mathbb{Z}) \cong \mathbb{Z}/2\mathbb{Z}$. See also [21]-4.

Next we introduce some geometry linking S and $\tilde{W}'$. From it, a quite explicit description of the torsion of $H_3(\tilde{W}', \mathbb{Z})$, via $H_1(S, \mathbb{Z})$, will follow. Let $\mathbb{U}|S$ be the universal line $\mathbb{U} \to \mathbb{G}$ restricted over S. The description essentially relies on a rational map $\upsilon : \mathbb{U}|S \to \tilde{W}'$ embedding a general fiber of $\mathbb{U}|S$ as a line of $\tilde{W}'$, that is, an element of the Fano surface $\mathbb{F}(\tilde{W}')$. Then a "cylinder map" $c_\upsilon : H_1(S, \mathbb{Z}) \to$

$H_3(\tilde{W}', \mathbb{Z})$ is induced by υ. One can show that c_υ is injective. Hence $H_1(S, \mathbb{Z}) = \mathbb{Z}/2\mathbb{Z}$ injects in $H_3(\tilde{W}', \mathbb{Z})$ and, by Poincaré duality, in $H^3(\tilde{W}', \mathbb{Z})$. Actually c_υ is an isomorphism.

Let us also point out that $\mathbb{U}|S$ defines a morphism onto the surface of bitangent lines to $\tilde{S}_+$, sending $\ell \in S$ to the pencil $\mathbb{I}_\ell$ of all quadrics of W through $\mathbb{U}_\ell$. This is a bitangent line to $\tilde{S}_+$. We denote this morphism by

$$\nu : S \to \mathbb{F}(\tilde{S}_+). \tag{59}$$

This is peculiar to $\mathbb{F}(\tilde{S}_+)$ and determined by W. Different from a general quartic, $\mathbb{F}(\tilde{S}_+)$ is non-normal and birational to the Reye congruence S. We outline here a description of $\mathbb{F}(\tilde{S}_+)$. However, during our work, we became aware of the beautiful description already performed by Ferretti [17, 18]-3.

Remark 4.1 The latter one relates $\mathbb{F}(\tilde{S}_+)$ to the theory of EPW sextics, when the corresponding hyperkahler fourfold is the Hilbert scheme $\tilde{S}^{[2]}$ of two points of $\tilde{S}$. The surface $\mathbb{F}(\tilde{S}_+)$ is birational to the locus of fixed points of the natural involution induced on $\tilde{S}^{[2]}$ by the quartic $\tilde{S}_+$ [17]-3.1.1.

4.2 *The Congruence of Bitangent Lines* $\mathbb{F}(\tilde{S}_+)$

As in (12), let $\mathbb{Q}$ be the space of quadrics of $\mathbb{P}^3$, and we assume that $W \subset \mathbb{Q}$ is a general web. Then $\tilde{S}_+$ is a general quartic symmetroid. Since now

$$\mathbb{G}_\mathbb{Q} \subset \mathbb{P}^{44} \tag{60}$$

is the Plücker embedding of the Grassmannian of lines of $\mathbb{Q}$, and then $\mathbb{G}_Q$ is the family of all pencils of quadrics of $\mathbb{P}^3$. The space of its orbits, under the action of Aut $\mathbb{P}^3$, classifies these pencils up to projective equivalence. The classification goes back at least to Corrado Segre [26, 27]. See [20] and [1]-7 for some recent revisiting. It follows from the classification that the locus

$$\mathbb{F}(\mathbb{Q}^3) \subset \mathbb{G}_\mathbb{Q}, \tag{61}$$

of pencils which are bitangent lines to the quartic discriminant $\mathbb{Q}^3 \subset \mathbb{Q}$, is an irreducible subvariety of codimension 2. Its description is classical and very well known: a *general* element P of $\mathbb{F}(\mathbb{Q}^3)$ is a pencil of quadrics whose base scheme is a complete intersection of two quadrics

$$L \cup C \subset \mathbb{P}^3,$$

where L is a line and C is a smooth, rational normal cubic curve. Moreover

$$L \cap C = \{v_1, v_2\} = \text{Sing } L \cup C$$

where v_1, v_2 are ordinary nodes of $L \cup C$. Notice that P is generated by two quadrics Q_1 and Q_2 of rank 3, respectively, singular at v_1 and v_2. These are the tangency points of P to $\mathbb{Q}^3$ and the two singular quadrics of the pencil.

More globally $\mathbb{F}(\mathbb{Q}^3)$ is parametrized by the smooth correspondence

$$\mathcal{G} := \{(\ell, P) \in \mathbb{G} \times \mathbb{G}_{\mathbb{Q}} \mid \mathbb{U}_\ell \text{ is in the base scheme of } P\}. \tag{62}$$

Indeed consider the natural projections

$$\mathbb{G} \xleftarrow{u_{\mathcal{G}}} \mathcal{G} \xrightarrow{t_{\mathcal{G}}} \mathbb{F}(\mathbb{Q}^3), \tag{63}$$

and then $u_{\mathcal{G}} : \mathcal{G} \to \mathbb{G}$ is smooth and a Grassmann bundle whose fiber at ℓ is the Grassmannian of the pencils of quadrics containing $\mathbb{U}_\ell$. Moreover it is easy to see that $t_{\mathcal{G}} : \mathcal{G} \to \mathbb{G}_{\mathbb{Q}}$ is a birational morphism onto its image $\mathbb{F}(\mathbb{Q}^3)$.

Clearly $t_{\mathcal{G}}$ is biregular over each $P \in \mathbb{F}(\mathbb{Q}^3)$ such that P is a general pencil as above. Notice also that, for any $P \in \mathbb{F}(\mathbb{Q}^3)$, the fiber of $t_{\mathcal{G}}$ at P is a scheme supported on the points $(\ell, P) \in \mathbb{G} \times \{P\}$ such that $\mathbb{U}_\ell$ is in the base scheme of P. This remark is the starting point to describe Sing $\mathbb{F}(Q^3)$.

Theorem 4.3 Sing $\mathbb{F}(\mathbb{Q}^3)$ *is irreducible of codimension* 3 *in* $\mathbb{G}_{\mathbb{Q}^3}$. *Let P be general in* Sing $\mathbb{F}(\mathbb{Q}^3)$, *then P is a non-normal ordinary double point, and the base scheme B_P of P contains a conic $D = L \cup L'$ of rank* 2.

Actually the pairs $(L, P), (L', P) \in \mathcal{G}$ correspond to the branches of the node $P \in \mathbb{F}(\mathbb{Q}^3)$. Then such a general singular point P is a pencil whose base scheme is a complete intersection $D \cup D', '$ being a smooth conic. This implies that P contains a quadric Q of rank 2, namely, $Q \in \mathbb{Q}^2$ is the union of the planes supporting D and D'. In particular, let us consider

$$\Sigma_{\mathbb{Q}^3} \subset \mathbb{G}_{\mathbb{Q}}, \tag{64}$$

the locus of all pencils P intersecting $\mathbb{Q}^2$; then the next theorem follows.

Theorem 4.4 Sing $\mathbb{F}(\mathbb{Q}^3) = \mathbb{F}(\mathbb{Q}^3) \cdot \Sigma_{\mathbb{Q}^2}$.

We use these properties for a basic description of the surface $\mathbb{F}(W)$. Let

$$\mathbb{G}_W \subset \mathbb{G}_{\mathbb{Q}} \tag{65}$$

be the Grassmannian of lines of W, and then $\mathbb{G}_W$ is a Schubert variety and a smooth four-dimensional quadric in $\mathbb{G}_{\mathbb{Q}}$. According to the classification of pencils of quadrics, the space $\mathbb{F}(\mathbb{Q}^3)$ is quasi-homogeneous for the action of Aut $\mathbb{P}^3$, the union of finitely many orbits Γ. By transversality of general translate, we can assume that $\mathbb{G}_W$ is transversal to each Γ. Let

$$\mathsf{O} \subset \mathbb{F}(\mathbb{Q}^3) \tag{66}$$

be the family of pencils $P \in \mathbb{F}(\mathbb{Q}^3)$ which are general as above, and it is easy to see that O is irreducible and the unique orbit which is open in $\mathbb{F}(\mathbb{Q}^3)$. By transversality of general translate again, $\mathsf{O} \cap \mathbb{G}_W$ is a smooth, irreducible open set of the surface $\mathbb{F}(\tilde{S}_+)$. From Segre classification for pencils P of $\mathbb{F}(\mathbb{Q}^3)$, one can see the complement of $\mathsf{O} \cap \mathbb{F}(\tilde{S}_+)$ in $\mathbb{F}(\tilde{S}_+)$. This is a union of curves in $\mathbb{F}(\tilde{S}_+)$. For it we fix our notation and description as follows:

$$\mathbb{F}(\tilde{S}_+) - (\mathsf{O} \cap \mathbb{F}(\tilde{S}_+)) = N \cup D. \tag{67}$$

(1) **The singular curve** $N :=$ Sing $\mathbb{F}(\tilde{S}^+)$. The above properties imply that N is the family of bitangent lines intersecting Sing $\tilde{S}_+$. Therefore, we have $N = \cup_{o \in \text{Sing } \tilde{S}_+} N_o$, where the curve N_o is

$$N_o := \{P \in \mathbb{F}(\tilde{S}_+) \mid o \in P\}. \tag{68}$$

The curve N_o lies in the plane $\mathbb{P}^2 \subset \mathbb{G}_W$ parametrizing all rays passing through o. As is well known, N_o is the union of two smooth cubics intersecting at nine points, three of which are on a smooth conic. Moreover, N_o is the discriminant curve of the conic bundle structure naturally defined by

$$\pi_o \circ f' : \tilde{W}' \dashrightarrow \mathbb{P}^2,$$

where $\pi_o : W \to \mathbb{P}^2$ is the linear projection of center o, cfr. [16]-7, [21]-4.

(2) **The curve** D **of hyperflex bitangent lines.** We say that P is a hyperflex tangent line if $P \cdot \tilde{S}_+$ is a unique point of P, with multiplicity 4. We omit more details on the well-known curve D, since we will not use it.

In what follows, we will concentrate on two important and elementary peculiarities of Artin-Mumford double solids, which are determinant for the existence of a $\mathbb{Z}/2\mathbb{Z}$ torsion subgroup of $H^3(\tilde{W}', \mathbb{Z})$. These are:

- A rational map $\psi : \mathbb{G} \dashrightarrow W$ factorizing as follows:

$$\begin{array}{ccc} \tilde{\mathbb{G}} & \xrightarrow{\tilde{\phi}} & \tilde{W} \\ {\scriptstyle\beta}\downarrow & & \downarrow{\scriptstyle f} \\ \mathbb{G} & \xrightarrow{\psi} & W, \end{array} \tag{69}$$

where $\beta : \tilde{G} \to \mathbb{G}$ is the blowing up of S and $\psi \circ \beta$ is a morphism. A remarkable fact is that f is the Stein factorization of $\psi \circ \beta$.

- A birational morphism $\nu : S \to \mathbb{F}(\tilde{S}_+)$ and its birational lifting

$$\tilde{\nu} : S \to \mathbb{F}(\tilde{W}'),$$

onto its image $\mathbb{F}^+ \subset \mathbb{F}(\tilde{W}')$. Remarkably the surface $\mathbb{F}(\tilde{W}')$ splits as

$$\mathbb{F}(\tilde{W}) = \mathbb{F}^+ \cup \mathbb{F}^-, \tag{70}$$

with $\mathbb{F}^+, \mathbb{F}^-$ birational to S and exchanged by the involution j'^*.

4.3 The Rational Map $\psi : \mathbb{G} \dashrightarrow W$

Let W be a general web as above; we consider the rational map

$$\psi : \mathbb{G} \dashrightarrow W, \tag{71}$$

sending a general $\ell \in \mathbb{G}$ to the unique quadric $Q \in W$ through $\mathbb{U}_\ell$. We recall that the Reye congruence S of W is by definition the degeneracy scheme of the map of vector bundles in (30). This easily implies that S is the indeterminacy scheme of ψ. To have a resolution of it, we study the graph

$$\tilde{\mathbb{G}} := \{(\ell, Q) \in \mathbb{G} \times W \mid \mathbb{U}_\ell \subset Q\}, \tag{72}$$

of ψ. We also consider its two natural projections

$$\mathbb{G} \xleftarrow{\ \beta\ } \tilde{\mathbb{G}} \xrightarrow{\ \phi\ } W. \tag{73}$$

From the description of S in (3.2), it follows that β is the contraction to $\mathbb{G}$ of a $\mathbb{P}^1$ bundle over S. Then β is the blowing up of $\mathbb{G}$ at S. Notice also that the linear system of the quadrics of W through $\mathbb{U}_\ell$ is the fiber $\beta^*(\ell)$ of β. For the exceptional divisor of β, we fix the notation

$$\mathbb{I} := \{(\ell, Q) \in \mathbb{G} \times W \mid \dim \beta^*(\ell) = 1\}. \tag{74}$$

Clearly the birational morphism β restricts on $\mathbb{I}$ to the $\mathbb{P}^1$ bundle

$$\beta|\mathbb{I} : \mathbb{I} \to S$$

and, at $\ell \in S$, its fiber $\mathbb{I}_\ell$ is the pencil of quadrics of W through $\mathbb{U}_\ell$. Let $w \in W$, and then its corresponding quadric embedded in $\mathbb{P}^3$ will be denoted by

$$Q_w \subset \mathbb{P}^3. \tag{75}$$

Now we consider the projection map $\phi : \tilde{\mathbb{G}} \to W$. This is a morphism whose general fiber is not connected. Indeed let $w \in W$, and then we have

$$\phi^*(w) = \tilde{\mathbb{G}} \cdot (\mathbb{G} \times \{w\}). \tag{76}$$

The equality just says that $\phi^*(w)$ is the Hilbert scheme of lines of the quadric surface Q_w. Then the proof of the next theorem is elementary.

Theorem 4.5 *For a general $w \in W$, the two rulings of lines of Q_w are the connected components of $\phi^*(w)$. For any fiber $\phi^*(w)$ can be as follows:*

(i) rank $Q_w = 4$*: disjoint union of two smooth conics,*
(ii) rank $Q_w = 3$*: a smooth conic having multiplicity* 2*,*
(iii) rank $Q_w = 2$*: union of two planes P_1, P_2. $P_1 \cap P_2 = \{$one point$\}$.*

Clearly $\phi^*(w)$ has type (iii) iff $w \in \text{Sing } \tilde{S}_+$ and (ii) iff $w \in \tilde{S}_+ - \text{Sing } \tilde{S}_+$. The type is (i) and the fiber is not connected iff $w \in \mathbb{P}^3 - \tilde{S}_+$.

Passing to the *Stein factorization of* ϕ, we have the diagram

$$\tilde{\mathbb{G}} \xrightarrow{\tilde{\phi}} \tilde{W} \xrightarrow{f} W, \tag{77}$$

where $\phi = \tilde{\phi} \circ f$ and f is a finite double covering. It is clear that

$$f : \tilde{W} \to W \tag{78}$$

is branched on the symmetroid $\tilde{S}_+$ of W and defines the Artin-Mumford double solid $\tilde{W}$. Let $w \in W$, and then the fiber $f^*(w)$ is finite of length two. To denote the points of Supp $f^*(w)$, we fix the following convention:

$$\text{Supp } f^*(w) := \{w^+, w^-\}, \tag{79}$$

moreover $w^+ = w^-$ iff $w \in \tilde{S}_+$. Clearly w^+ and w^- label the rulings of lines of the quadric Q_w if its rank is ≥ 3. Let us roughly summarize as follows.

Remark 4.2 $\tilde{W}$ is a parameter space for pairs (w, w^+), (w, w^-), where $w \in W$ and w^+, w^- are the rulings of Q_w. f is the natural forgetful map.

Finally we see in (77) that a rational variety, namely, $\tilde{\mathbb{G}}$, dominates $\tilde{W}$. As is well known, this implies that $\tilde{W}$ is unirational. Then the same is true for its birational model $\tilde{W}'$, and the next theorem follows, cfr. [5]-6.

Theorem 4.6 *$\tilde{W}'$ is unirational.*

Remark 4.3 Notably, the properties of a blowing up imply

$$H^*(\tilde{\mathbb{G}}, \mathbb{Z}) \cong H^*(\mathbb{G}, \mathbb{Z}) \oplus \sigma^* H^*(S, \mathbb{Z}), \tag{80}$$

where $\sigma : \tilde{\mathbb{G}} \to \mathbb{G}$ is the blowing up of $\mathbb{G}$ at S. That has its importance.

4.4 *S and the Fano Surface $\mathbb{F}(\tilde{W}')$*

Now we want to see that S interacts so much with the Fano surface $\mathbb{F}(\tilde{W}')$. Actually this is the union of two irreducible components, birational to S. We begin with the Grassmannian $\mathbb{G}_W \subset \mathbb{G}_{\mathbb{Q}}$ of lines of W and fix the notation

$$\delta : S \to \mathbb{G}_W \tag{81}$$

for the following rational map. For $\ell \in S$, let $\mathbb{I}_\ell \subset W$ be the linear system of quadrics through $\mathbb{U}_\ell$. For each $\ell \in S$, this is a pencil, defining a point of $\mathbb{G}_W$. By definition, this is $\delta(\ell)$ and, moreover, $\delta : S \to \mathbb{G}_W$ is a morphism.

Theorem 4.7 *$\delta : S \to \mathbb{G}_W$ is birational onto its image; moreover, we have*

$$\delta(S) = \mathbb{F}(\tilde{S}_+).$$

Proof Recall that W is always general, so that $\mathbb{G}_W$ is transversal to $\mathbb{F}(\mathbb{Q}^3)$. Now consider the orbit $\mathsf{O} \subset \mathbb{F}(\mathbb{Q}^3)$ as in (66). As observed, this is the unique irreducible open orbit under the action of Aut $\mathbb{P}^3$ on $\mathbb{F}(\mathbb{Q}^3)$. Moreover, $\mathsf{O} \cap \mathbb{F}(\tilde{S}_+)$ is a smooth, irreducible open set of the surface $\mathbb{F}(\tilde{S}_+)$. Finally a point P of it is a pencil whose base scheme B_P is $L \cup C \subset \mathbb{P}^3$, where C is a smooth rational normal cubic and L is a line. Let $\ell \in \mathbb{G}$ be the point defined by L; then we have $P = \mathbb{I}_\ell$ and $\delta(\ell) = P$. Since the only line in B_P is L, $\delta^{-1}(P) = \{\ell\} \subset S$. Moving P along the mentioned open set of $\mathbb{F}(\tilde{S}_+)$, it follows that $\delta : S \to \mathbb{F}(\tilde{S}_+)$ is invertible and dominant. □

4.5 *The Counterexample of Artin-Mumford*

Finally we want to show that $H^3(\tilde{W}', \mathbb{Z})$ has a nonzero 2-torsion element, determined by the nonzero element of $H_1(S, \mathbb{Z})$. Let $\pi_1(S)$ be the fundamental group of S and $\mathring{S} \subset S$ a non-empty Zariski open set. We recall that the inclusion $i : \mathring{S} \to S$ defines a surjective homomorphism

$$i_* : \pi_1(\mathring{S}) \to \pi_1(S).$$

This is well known for any complex, irreducible algebraic variety, [19] 0.3. Since we have $H_1(S, \mathbb{Z}) = \pi_1(S) = \mathbb{Z}/2\mathbb{Z}$, the next property follows.

Proposition 4.1 *For the generator* $[\gamma]$ *of* $H_1(S, \mathbb{Z})$ *one can choose* γ *so that its image is in* $\mathring{S}$. *Then* γ *defines a nonzero 2-torsion element of* $H_1(\mathring{S}, \mathbb{Z})$.

After this remark we consider the exceptional divisor $\mathbb{I}$ of the blowing up of $\mathbb{G}$ at S. As in (74) this is a $\mathbb{P}^1$-bundle $\beta : \mathbb{I} \to S$. Let $\mathbb{I}_\ell$ be its fiber at $\ell \in S$, then $\mathbb{I}_\ell$ sits in $\{\ell\} \times W = W$ as the pencil of quadrics

$$\mathbb{I}_\ell = \{Q \in W \mid Q \text{ contains the line of } \mathbb{P}^3 \text{defined by } \ell\}. \tag{82}$$

We already know that the surface $\mathbb{F}(\tilde{S}_+)$, of bitangent lines to the quartic symmetroid $\tilde{S}_+$, is the family of pencils $\{\mathbb{I}_\ell \subset W, \ \ell \in S\}$. Moreover, in the Grassmannian $\mathbb{G}_W$ of lines of W, $\mathbb{F}(\tilde{S}_+)$ is the birational image of

$$\delta : S \to \mathbb{G}_W, \tag{83}$$

the morphism sending ℓ to $\mathbb{I}_\ell$. Let $f : \tilde{W} \to W$ be the finite double cover branched on $\tilde{S}_+$. We also know that the fiber of f at $Q \in W$ is bijective to the set of connected components of the family of lines of Q.

Proposition 4.2 *Let* $\ell \in S$, *then its inverse image* $f^{-1}(\mathbb{I}_\ell)$ *is a union of two or one irreducible components, biregular to* $\mathbb{I}_\ell$ *via* f.

Proof Let $Q \in \mathbb{I}_\ell$, then the connected components of its family of lines are distinguished by the property of either containing the point ℓ or not. This easily implies that $f^{-1}(\mathbb{I}_\ell)$ is the union prescribed by the statement. □

Let $w \in W$, we keep our notation Q_w for the corresponding quadric in $\mathbb{P}^3$ and denote by R^+_ℓ the family of ℓ in Q_w, (by R^-_ℓ the other family). We have

$$f^{-1}(\mathbb{I}_\ell) = R^+_\ell \cup R^-_\ell. \tag{84}$$

The equality implies that S parametrizes an irreducible component of the Fano surface $\mathbb{F}(\tilde{W})$ of lines of $\tilde{W}$. Let $j : \tilde{W} \to \tilde{W}$ be the involution induced by f, and let $\mathbb{F}^\pm := \{R^\pm_\ell, \ \ell \in S\}$, then the next property easily follows.

Proposition 4.3 $\mathbb{F}(\tilde{W}) = \mathbb{F}^+ \cup \mathbb{F}^- = \delta(S) \cup j^*\delta(S)$.

Remark 4.4 (A Wirtinger construction) Notice that $F^+ \cup \mathbb{F}^-$ normalizes to the disjoint union $S' \vee S''$ of two copies of S and that j^* lifts to the involution $j^\vee : S' \vee S'' \to S' \vee S''$, exchanging S' and S'' via the identity map of S. Then $\mathbb{F}(\tilde{W})$ is a suitable quotient of $S' \vee S''$. Now let $\{f_t : X_t \to \mathbb{P}^3, t \in T\}$ be an integral, flat family of quartic double solids, where X_t is general and f_o is $f : \tilde{W} \to W$ for some $o \in T$. Then a general member of the family of involutions $\{j^*_t : \mathbb{F}(X_t) \to \mathbb{F}(X_t), t \in T\}$ is fixed point free and the family is a deformation of

$j^* : \mathbb{F}^+ \cup \mathbb{F}^- \to \mathbb{F}^+ \cup \mathbb{F}^-$. A similar deformation, for curves with a fixed-point-free involution, is known as *a Wirtinger construction.* □

We can think of a point of $\mathbb{I}$ as a pair $(\ell, w) \in S \times W$ such that Q_w contains the line $\mathbb{U}_\ell$. Moreover we can think of a point of $\tilde{W}$ as a pair $(w, r) \in W \times \tilde{W}$, where r is a connected component of the family of lines in Q_w.

Definition 4.2 The fundamental morphisms are the maps

$$\upsilon^+ : \mathbb{I} \to \tilde{W} \ , \ \upsilon^- : \mathbb{I} \to \tilde{W} \tag{85}$$

defined as follows: $\upsilon^+(\ell, w) = (w, w^+)$, w^+ being the connected component of ℓ in the family of lines of Q_w. Moreover the map w^- is $j^* \circ \upsilon^+$.

Theorem 4.8 *The fundamental morphisms have degree* 6.

Proof The degree of υ^+ is the number of pairs (ℓ', w') such that $w = w'$ and $w^+ = r$, where $(w, r) = \upsilon^+(\ell, w)$ and (ℓ, w) is general in $\tilde{W}$. Then we can assume rank $Q_w = 4$ and ℓ general in S. To compute $\deg \upsilon^+$ we consider the restriction of W to Q_w: this is a plane N in $|\mathcal{O}_{Q_w}(2H)| := \mathbb{P}^8$, where $H \in |\mathcal{O}_{Q_w}(1)|$. Hence $\deg \upsilon^+$ is the number of reducible elements of N, containing a line of the ruling of ℓ. Let $|L| := \mathbb{P}^1$ be such a ruling, then $|2H - L| := \mathbb{P}^5$ is a five-dimensional linear system of rational cubics. Let

$$s : \mathbb{P}^1 \times \mathbb{P}^5 \to \mathbb{P}^8$$

be the sum map, sending $(L, C) \in \mathbb{P}^1 \times \mathbb{P}^5$ to $L + C \in |2H| = \mathbb{P}^8$. We claim that s^*N is finite. Then $\deg \upsilon^+$ is its length and coincides with the degree of the Segre embedding of $\mathbb{P}^1 \times \mathbb{P}^5$, which is six. To prove our claim, assume s^*N is not finite. Then each element of $|L|$ defines a point of $\mathbb{G}$ which is in S and $|L|$ derfines a conic in S. Since ℓ is general in S, then S is uniruled: a contradiction. Finally $\upsilon^- = j^* \circ \upsilon^+$ implies $\deg \upsilon^- = \deg \upsilon^+ = 6$. □

Finally let $\sigma' : \tilde{W}' \to \tilde{W}$ be the natural desingularization obtained, as in (58), by blowing up the set of ten nodes Sing $\tilde{W}$. Then the diagram

$$S \xleftarrow{\ \beta\ } \mathbb{I} \xrightarrow{\ \upsilon^+\ } \tilde{W} \xleftarrow{\ \sigma'\ } \tilde{W}' \tag{86}$$

defines a "cylinder map"

$$\mathsf{c} : H_1(S, \mathbb{Z}) \to H_3(\tilde{W}', \mathbb{Z}), \tag{87}$$

where $\mathsf{c} := \sigma'^* \circ \upsilon^+_* \circ \beta^*$. Now let us consider the non-empty Zariski open set $\mathring{S} \subset S$ of points $\ell \in S$ satisfying the following conditions:

(a) $\upsilon^+ : \mathbb{I} \to \tilde{W}$ *is finite over* $\upsilon^+(\mathbb{I}_\ell)$ *and generically unramified at* $\mathbb{I}_\ell$,
(b) $\upsilon^+(\mathbb{I}_\ell) \cap \operatorname{Sing} \tilde{W} = \emptyset$, *that is,* rank $Q_w \geq 3$, $\forall (w, r) \in \upsilon^+(\mathbb{I}_\ell)$,
(c) $\delta(\ell) \notin \operatorname{Sing} \delta(S)$, *where the map* $\delta : S \to \mathbb{G}_W$ *sends* ℓ *to* $\mathbb{I}_\ell$.

Let $[\gamma] \in H_1(S, \mathbb{Z}) = \mathbb{Z}/2\mathbb{Z}$ be the generator of this group. By proposition 4.1 we can choose γ so that its image is in the Zariski open set $\overset{\circ}{S}$. Now, in the euclidean topology, $\mathbb{I}$ is homeomorphic to $S \times \mathbb{P}^1$ and we have

$$H_3(\mathbb{I}, \mathbb{Z}) \cong H_1(S, \mathbb{Z}) \otimes H_2(\mathbb{P}^1, \mathbb{Z}) \cong \mathbb{Z}/2\mathbb{Z}. \tag{88}$$

Indeed, one has $H_1(\mathbb{P}^1, \mathbb{Z}) = H_3(\mathbb{P}^1, \mathbb{Z}) = 0$. Moreover, by Poincaré duality, one has $H_3(S, \mathbb{Z}) \cong H^1(S, \mathbb{Z})$. Furthermore $H^1(V, \mathbb{Z})$ has no torsion for any smooth, projective variety V: see [31] proof of lemma 2.9 or deduce this property from the exponential sequence of V . Since the first Betti number of S is zero, it follows $H^1(S, \mathbb{Z}) = 0$. Then, applying Künneth formulae to $\mathbb{I}$, the above isomorphism follows. Now let us consider on $\mathbb{I}$ the 3-cycle

$$\beta^* \gamma := \mathbb{I}_\gamma.$$

Clearly $\mathbb{I}_\gamma$ defines the 2-torsion class generating $H_3(\mathbb{I}, \mathbb{Z})$ and the 3-cycle

$$T_\gamma := \upsilon^+_*(\mathbb{I}_\gamma).$$

By (b) T_γ is a 3-cycle of $\tilde{W} - \operatorname{Sing} \tilde{W}$, moreover σ' is biregular over the set $\tilde{W} - \operatorname{Sing} \tilde{W}$. Then consider the 3-cycle $T'_\gamma := \sigma'^*(T_\gamma)$ of $\tilde{W}'$ and its class

$$\tau' = \mathsf{c}([\gamma]) \in H_3(\tilde{W}', \mathbb{Z}). \tag{89}$$

This is a 2-torsion element, we can now conclude via the following result.

Theorem 4.9 *τ' is a nonzero element.*

Proof It suffices to show that the image of τ' by $(\upsilon^{+*} \circ \sigma'_*)$ is nonzero in $H_3(\mathbb{I}, \mathbb{Z})$. This is equivalent to show that

$$(\upsilon^{+*} \circ \sigma'_* \circ \sigma'^*)(\tau') = \upsilon^{+*}(\tau') = [\mathbb{I}_\gamma].$$

Since σ' is biregular over $\tilde{W} - \operatorname{Sing} \tilde{W}$ the first equality is immediate. Let us prove the second one. For the class $[\gamma]$ of $H_1(\overset{\circ}{S}, \mathbb{Z})$ we can even assume that the map $\gamma : (0, 1) \to \overset{\circ}{S}$ is a real analytic embedding. We have

$$\upsilon^{+*} T'_\gamma = m\mathbb{I}_\gamma + Z,$$

where Z is the image by i_* of a cycle of $\mathbb{I} - \mathbb{I}_\gamma$ and $i : (\mathbb{I} - \mathbb{I}_\gamma) \to \mathbb{I}$ is the inclusion. Let $\Gamma := \gamma([0, 1])$ and $\ell \in \Gamma$, then, by assumption (a) on γ, the morphism υ^+ is finite over $\upsilon^+(\mathbb{I}_\ell)$ and generically unramified along $\mathbb{I}_\ell$. This implies $m = 1$ and that Z is a 3-cycle. Now let $t \in S - \Gamma$, we claim that $\upsilon^+(\mathbb{I}_t)$ is not in $\upsilon^+(\mathbb{I}_\gamma)$. Hence no summand of Z is a pull-back of a cycle by $\beta : \mathbb{I} \to S$. Let $H_3(\mathbb{I}, \mathbb{Z}) \cong \bigoplus_{a+b=3} H_a(S, \mathbb{Z}) \otimes H_b(\mathbb{P}^1, \mathbb{Z})$ be the Künneth isomorphism, then no summand of

Z defines a class in $H_1(S, \mathbb{Z}) \otimes H_2(\mathbb{P}^1, \mathbb{Z})$. This implies that the class of Z is zero and hence the theorem follows. To complete the proof we show the above claim. Let $t \in S$, in what follows $W_t \subset W$ is the pencil of quadrics defined by t, that is, the image of $\upsilon^+(\mathbb{I}_t)$ by $f : \tilde{W} \to W$. Now let $t \in S - \Gamma$, then, by our assumption (c), Γ is in $\delta(S) - \operatorname{Sing}\delta(S)$, hence W_t is not in the family $\{W_\ell,\ \ell \in \Gamma\}$. Since W_t, W_ℓ are lines, they intersect in at most one point and the same is true for $\upsilon^+(\mathbb{I}_t)$, $\upsilon^+(\mathbb{I}_\ell)$. Let $W_\gamma = \bigcup_{\ell \in \Gamma} W_\ell$, this implies by (a) that the intersection of $\mathbb{I}_t$ and ${\upsilon^+}^{-1}(W_\gamma)$ has real dimension ≤ 1. Hence $\upsilon^+(\mathbb{I}_t)$ is not in $\upsilon^+(\mathbb{I}_\gamma)$. □

Since $H^3(\tilde{W}', \mathbb{Z})$ admits nonzero torsion and this is a birational invariant, $\tilde{W}$ is not rational. Since it is unirational, it is a counterexample to Lüroth problem, see [2]. This offers a partially new proof of Artin-Mumford counterexample to Lüroth problem, based on some geometry of Enriques surfaces, more precisely of Reye congruences of lines.

Acknowledgments We are grateful to Claire Voisin and to Bert van Geemen for a useful conversation and advice on the subject of this paper. Let us also thank the Scientific Committee of this very interesting conference for its realization.

This work was partially supported by INdAM-GNSAGA.

References

1. Abdallah, N., Emsalem, J., Iarrobino, A.: Nets of conics and associated Artinian algebras of length 7 Translation and update of 1977 version, preprint (2021)
2. Artin, M., Mumford, D.: Some elementary examples of unirational varieties which are not rational. Proc. London Math. Soc. **25**, 75–95 (1972)
3. Beauville, A.: Variétés rationnelles et unirationnelles. In: Algebraic Geometry-Open Problems. Lecture Notes in Mathematics, vol. 997, pp. 16–33. Springer, Berlin (1983)
4. Beauville, A.: Castelnuovo and the Lüroth problem, Conference In memoria di Guido Castelnuovo slides (2015)
5. Beauville, A.: The Lüroth Problem. In: Rationality Problems in Algebraic Geometry. Lecture Notes in Mathematics, vol. 2172, pp. 16–33. Springer I.P., Switzerland (2016)
6. Beauville, A.: Recent progress on rationality problems, slides from Duke Math. J. Conference (2018)
7. Castelnuovo, G., Enriques, F.: Riposte Armonie, Lettere di Federigo Enriques a Guido Castelnuovo, Bottazzini, U., Conte, A., Gario, P. (eds.). Bollati Boringhieri, Torino (1996)
8. Cheltsov, I., Przyjalkowski, V., Shramov, C.: Which quartic double solids are rational? J. Algebraic Geom. **28**, 201–243 (2019)
9. Clemens, H.: Double solids. Adv. Math. **47**, 107–230 (1983)
10. Clemens, H., Griffiths, Ph.: The intermediate Jacobian of the cubic threefold. Ann. Math. **95**, 281–356 (1972)
11. Conte, A., Verra, A.: Reye constructions for nodal Enriques surfaces. Transactions AMS **336**, 79–100 (1993)
12. Corvaja, A., Zucconi, F.: Bitangents to a quartic surface and infinitesimal deformations (2021). ArXiv 191001365v.2
13. Cossec, C.F.: Reye congruences. Transactions of Amer. Math. Soc. **280**, 737–751 (1983)
14. Cossec, C.F., Dolgachev, I., Liedtke, C.: Enriques Surfaces I, Preprint, http://www.math.lsa.umich.edu/~idolga/EnriquesOne.pdf (2022)

15. Darboux, G.: Sur systemes linèaires de coniques et de surfaces du seconde ordre. Bull. Sci. Math. Astr. **1**, 348–358 (1870)
16. Dolgachev, I., Kondo, S.: Enriques Surfaces II, Preprint (2022). http://www.math.lsa.umich.edu/~idolga/EnriquesTwo.pdf
17. Ferretti, A.: The Chow ring of double EPW sextics. Rend. Mat. Appl. **31**, 69–217 (2011)
18. Ferretti, A.: Special subvarieties of EPW sextics. Math. Z. **272**, 1137–1164 (2012)
19. Fulton, W., Lazarsfeld, R.: Connectivity and its applications in algebraic geometry in Algebraic Geometry Lecture Notes in Math. **862**, 26–92 (1980)
20. Fevola, C., Mandelshtam, Y., Sturmfels, B.: Pencils of quadrics: old and new. Le Matematiche **76**, 319–335 (2021)
21. Iliev, A., Katzarkov, L.,Przyjalkowski, V.: Double solids, categories and non-rationality Proc. Edinb. Math. Soc. **57**, 145–173 (2014); **361**, 107–133 (2015)
22. Iskovskikh, V., Manin, Y.: Three-dimensional quartics and counterexamples to the Lüroth problem. Math. USSR Sbornik **15**, 141–166 (1971)
23. Namikawa, Y.: Periods of enriques surfaces. Math. Ann. **270**, 201–222 (1985)
24. Picco Botta, L., Verra, A.: The non-rationality of generic enriques threefold. Compositio Math. **48**, 167–184 (1983)
25. Reye, T.: Die Geometrie der Lage, Reprint of 1866 1st edn. Salzwasser-Verlag Reprint, Paderborn (2021)
26. Segre, C.: Studio sulle quadriche in uno spazio lineare ad un numero qualunque di dimensioni. Memorie Acc. Sci. Torino **36**, 3–86 (1883)
27. Segre, C.: Sulla geometria della retta e delle sue serie quadriche. Memorie Acc. Sci. Torino **36**, 87–157 (1883)
28. Serre, J.P.: On the fundamental group of a unirational variety. J. London Math. Soc. **34**, 481–484 (1959)
29. Welters, G.E.: Abel–Jacobi Isogenies for Certain Types of Fano Threefolds. Mathematical Centre Tracts, vol. 141. Mathematisch Centrum, Amsterdam (1981)
30. Voisin, C.: Unirational threefolds with no universal codimension 2 cycle. Invent. Math. **201**, 207–237 (2015)
31. Voisin, C.: Stable birational invariants and the Lüroth problem. In: Surveys in Differential Geometry, vol. 21, pp. 313–342. International Press, Boston (2016)
32. Voisin, C.: Birational invariants and decomposition of the diagonal. In: Birational Geometry of Hypersurfaces. Lectures Notes UMI, vol. 26, pp. 3–71. Springer, Cham (2019)

The Theorem of Completeness of the Characteristic Series: Enriques' Contribution

Ciro Ciliberto

Abstract In this paper I will recall the content of the so-called theorem of completeness of the characteristic series, whose algebro-geometric proof has been one of the unarrived targets of the algebraic geometers of the classical Italian school. After having recalled also one of its main consequences, namely, the fundamental theorem of irregular surfaces, I will focus on Enriques' attempts to prove the theorem of completeness of the characteristic series. As Mumford shows in Mumford (Notices AMS 58, 250–260, 2011), some of Enriques' intuitions were perfectly sound and, had he had at his disposal the right algebro-geometric tools, they could have led to a proof of the theorem.

Keywords Surfaces · Characteristic series · Irregular surfaces

1991 Mathematics Subject Classification 14B10, 14C20, 14C22, 01A60

1 Introduction

The so-called theorem of completeness of the characteristic series is basically an assertion about the smoothness, under suitable conditions, of the Hilbert scheme of curves on a surface (see Sect. 2 below). A first result in this direction was stated by Enriques in [6]: in a modern language, this says that the Hilbert scheme of curves on a complex surface is smooth at points corresponding to *sufficiently big* curves. For a curve C on a surface S to be *sufficiently big* meant, in Enriques' view, what we may express today by saying that C is *regular* and *non-special*, i.e., that $h^i(S, \mathcal{O}_S(C)) = 0$ for $i = 1, 2$.

At that time, and for a long while, Enriques' proof was considered to be valid, and, building on it, various geometers, like Castelnuovo and Severi, drew important

C. Ciliberto (✉)
Dipartimento di Matematica, Università di Roma Tor Vergata, Roma, Italy
e-mail: cilibert@mat.uniroma2.it

G. Bini (ed.), *Algebraic Geometry between Tradition and Future*,
Springer INdAM Series 53, https://doi.org/10.1007/978-981-19-8281-1_9

consequences from it. One of these is the so-called fundamental theorem of irregular surfaces, which says that the various definitions of irregularity that can be given for a surface in the attempt of extending to surfaces the concept of genus of a curve were all equivalent (see Sect. 4).

In 1910 another proof of Enriques' statement was given by Poincaré in [22] with transcendental techniques (see Sect. 3). In the 1921 paper [27], Severi simplified Poincaré's proof and pointed out that in Enriques' original argument there was a serious gap. From that point on, Enriques, Severi, and other Italian algebraic geometers of that time, like B. Segre, tried to fix Enriques' original proof using only algebro-geometric techniques, avoiding transcendental methods. In their view the result was too important for the Italian algebraic geometers to admit that it was not possible to prove it with the purely algebro-geometric techniques typical of their approach to the classification of surfaces. However none of them succeeded in providing a correct complete proof, though they made various attempts in the course of the years, and Enriques and Severi started blameworthy controversies between them on who had the most convincing arguments.

The first complete proof under the appropriate hypotheses of semiregularity of the curves (see Definition 6) is due to Mumford who exposed it in the text [18]. On the other hand, the case of Enriques–Poincaré's statement for regular curves follows easily from the great machinery (Hilbert schemes, Picard schemes, etc.) that Grothendieck created in the 1960s and is exposed at the end of Sect. 2. In the last Sect. 5, following Mumford (see [19]), I focus on some of Enriques' ideas for the proof of the theorem of completeness of the characteristic series. As Mumford shows some of Enriques' intuitions are interesting and sound and, with little effort, can be used for providing a correct proof of the theorem, at least for regular curves.

The present paper is purely expository and contains neither new result nor any information that cannot be found already in the current literature. It is just a slightly expanded version of the talk I gave at the INdAM Workshop "Italian algebraic geometry between tradition and future" held in Rome, December 6–8, 2021. Its only merit, if any, is to collect in one place a few informations that are scattered in the literature.

In this paper I will work over the complex numbers. For information about some of the topics of this paper in positive characteristic, see [5].

2 The Theorem of Completeness of the Characteristic Series

Throughout this paper, S will be a smooth, irreducible, projective, complex surface, and C will be a curve on S. Though not strictly necessary, I will assume that C is irreducible. Consider the *normal (line) bundle* N_C of C in S, which fits in the exact sequence

$$0 \longrightarrow \mathcal{O}_S \longrightarrow \mathcal{O}_S(C) \longrightarrow N_C \longrightarrow 0. \tag{1}$$

I will denote by ν_C the complete linear series determined by N_C, i.e.,

$$\nu_C = \mathbb{P}(H^0(C, N_C)).$$

Consider a flat family $F = \{\mathcal{C} \longrightarrow W\}$ of deformations of C on S, where W is a variety (i.e., a reduced scheme) and $\mathcal{C}$ is an effective divisor on $S \times W$ flat over W, whose fiber under the map to W over a point $w \in W$ is equal to C. Let $T_w(W)$ be the Zariski tangent space to W at w. In the above setting, one has the *characteristic (linear) map* at C

$$\rho_{F,C} : T_w(W) \longrightarrow H^0(C, N_C),$$

often simply denoted by ρ if there is no danger of confusion. The projectivized image $\mathbb{P}(\text{Im}(\rho))$ is a linear subseries of ν_C, which is called the *characteristic series* of the family F at C, and I will denote it by $\nu_{F,C}$.

Definition 1 In the above setting, the family F is said to enjoy the property of the *completeness of the characteristic series* at C if ρ is surjective, i.e., if and only if $\nu_{F,C}$ is complete.

Remark 2 Let $\mathbf{t} \in T_w(W)$ be a vector. This corresponds to an *infinitesimal first-order deformation* of C inside F that is called an *infinitely near curve* to C at the first order in F. It is obtained by base change from F via the morphism $\text{Spec}(\mathbb{C}[t]/(t^2)) \longrightarrow W$ corresponding to $\mathbf{t}$, in which the closed point of $\text{Spec}(\mathbb{C}[t]/(t^2))$ maps to w.

Suppose that, given $\mathbf{t} \in T_w(W)$, there is a morphism $\tau : \mathbb{D} \to W$, where $\mathbb{D}$ is a disc, such that $\tau(0) = w$ and the differential of τ maps the tangent vector d/dt to $\mathbb{D}$ at 0 to $\mathbf{t}$. This is certainly the case if W is smooth at w. Then by base change from F, there is a flat subfamily of curves $F_{\mathbf{t}} = \{C_t\}_{t \in \mathbb{D}}$ of F, with $C_0 = C$. Suppose that $F_{\mathbf{t}}$ is not constant equal to C, so I can assume $C_t \neq C$ for all $t \in \mathbb{D} \setminus \{0\}$. Then the divisor D_t cut out by C_t on C for $t \in \mathbb{D} \setminus \{0\}$ is well defined, and one can consider

$$D = \lim_{t \to 0} D_t.$$

Then D is the zero divisor of the section $\rho(\mathbf{t}) \in H^0(C, N_C)$ so that D is in $\nu_{F,C}$. If W is smooth at w, in this way we get all divisors of $\nu_{F,C}$.

Given the family F as above, there is a unique morphism $f : W \to \mathcal{H}$, where $\mathcal{H}$ is the Hilbert scheme of curves on S, such that the pullback via f of the universal family over $\mathcal{H}$ is F. Let $c = f(w)$ be the point of $\mathcal{H}$ corresponding to the curve C. Then we have the diagram

$$\begin{array}{ccc} T_w(W) & \xrightarrow{\ \text{id}\ } & T_w(W) \\ {\scriptstyle df}\downarrow & & \downarrow{\scriptstyle \rho} \\ T_c(\mathcal{H}) & \xrightarrow{\ \sim\ } & H^0(C, N_C) \end{array}$$

One has:

Lemma 3 (Artin-Kleiman, see [17]) *If in the above setting W is smooth at w and F enjoys the property of the completeness of the characteristic series at C, then $\mathcal{H}$ is smooth at c.*

The theorem of completeness of the characteristic series deals with the problem of finding reasonable conditions under which there is a flat family $F = \{\mathcal{C} \longrightarrow W\}$ of deformations of a curve C on S, whose characteristic series is complete at C and the parameter space W is smooth at the point w corresponding to C. By Artin–Kleiman's Lemma, this is the same as giving conditions for the Hilbert scheme to be smooth at the point c corresponding to C.

Example 4 If $C^2 < 0$, then the trivial family parametrized by a point, consisting of the only curve C, enjoys the property of the completeness of the characteristic series, because ν_C is empty, being of negative degree. So by Artin–Kleiman's Lemma, the Hilbert scheme is smooth at any point corresponding to an irreducible curve with negative self-intersection.

Example 5 A few examples of singular points of the Hilbert scheme of curves on a surface have been classically known. Such examples have been given, for example, by Albanese, Enriques, Rosenblatt, and Severi (see [29, §30–32]). In the paper [30] of 1943, Zappa gave an interesting example of singular point on a positive dimensional, generically smooth, component of the Hilbert scheme.

However often one may be interested in having examples of entire components of the Hilbert scheme of curves on a surface that are singular, i.e., non-reduced. It has been believed for a long time by the geometers of the Italian school that no example of this sort should exist. By contrast, such example has been produced by Zappa [31] and Severi (see l.c. §33, where also Zappa's examples are exposed). For a modern exposition of Zappa's example in [31] and generalizations, see [24].

Severi's example (sketched by Severi in [29, §33], implicit in the paper [23, §15] by Corrado Segre and worked out by Mumford in [18, p. 155]) is very simple and worth to be recalled here. Let E be an elliptic curve. Consider a non-split extension

$$0 \longrightarrow \mathcal{O}_E \longrightarrow \mathcal{E} \longrightarrow \mathcal{O}_E \longrightarrow 0$$

corresponding to a non-zero element of $\mathrm{Ext}^1_{\mathcal{O}_E}(\mathcal{O}_E, \mathcal{O}_E) \simeq H^1(E, \mathcal{O}_E) = \mathbb{C}$. Consider the ruled surface $S = \mathbb{P}(\mathcal{E})$, that is, a $\mathbb{P}^1$-bundle over E. The quotient $\mathcal{E} \longrightarrow \mathcal{O}_E$ corresponds to a section $C \simeq E$ of $S \longrightarrow E$ and one has $N_C = \mathcal{O}_C$. The point c corresponding to C is an isolated point of the Hilbert scheme; otherwise, $\mathcal{E}$ would split. Since $H^0(C, N_C) = \mathbb{C}$, the Zariski tangent space to the Hilbert scheme at c is one-dimensional, whereas the Hilbert scheme is zero-dimensional. Hence one has a non-reduced component of the Hilbert scheme.

Severi found in [28] the right concept for giving somehow minimal sufficient conditions for the validity of the completeness of the characteristic series.

Definition 6 (Severi, see [28] and [29], §21) A curve C on the surface S is said to be *semiregular* if the map

$$u : H^1(S, \mathcal{O}_S(C)) \longrightarrow H^1(C, N_C)$$

arising from the exact sequence (1) is the zero map, or equivalently, if

$$h^0(C, N_C) = h^1(S, \mathcal{O}_S) + \dim(|C|) - h^1(S, \mathcal{O}_S(C)).$$

Note that the concept of semiregularity has been extended by S. Bloch in [2] and used by him and, after him, by various mathematicians.

Remark 7 There are two obvious sufficient conditions for a curve C to be semiregular. One is that $h^1(S, \mathcal{O}_S(C)) = 0$, in which case the curve is said to be *regular*. Another is $h^1(C, N_C) = 0$ which is well-known to be a sufficient condition for the Hilbert scheme to be smooth at the point corresponding to C.

Note also that C is semiregular if and only if the map

$$v : H^1(C, N_C) \longrightarrow H^2(S, \mathcal{O}_S)$$

is injective, or dually, if the map

$$v^* : H^0(S, \Omega_S^2) \longrightarrow H^0(C, \omega_C \otimes N_C^*) = H^0(C, \Omega^2_{S|C})$$

is surjective, which means that the canonical system $|K_S|$ of S cuts out on C a complete linear series. This was the original definition of Severi's.

Theorem 8 (Theorem of Completeness of the Characteristic Series) *Let C be a semiregular curve on the surface S. Then the Hilbert scheme is smooth at the point corresponding to the curve C.*

Proof in the Regular Case (Following Grothendieck) A great part of Mumford's book [18] is devoted to the proof of this theorem. The general proof is not easy. However the case in which C is regular (which, as we will see, is sufficient for the crucial application to Enriques–Poincaré's Theorem 9 below) is quite simple, once we accept all Grothendieck's general machinery consisting in the introduction of Hilbert schemes and Picard schemes (see [13, 14]).

First of all, given the surface S, we can consider its *Picard group* $\mathrm{Pic}(S) = H^1(S, \mathcal{O}_S^*)$ whose points correspond to line bundles on S. If $\mathrm{Pic}^\tau(S)$ is the subgroup of $\mathrm{Pic}(S)$ of line bundles $\mathcal{L}$ numerically equivalent to zero (i.e., for any curve C on S, $\deg(\mathcal{L}_{|C}) = 0$), one can define the *Néron–Severi group* $\mathrm{Num}(S) = \mathrm{Pic}(S)/\mathrm{Pic}^\tau(S)$ of S, whose points correspond to line bundles on S up to numerical equivalence.

Grothendieck proved that, given any class $\xi \in \mathrm{Num}(S)$, one can consider the *Hilbert scheme* $H(\xi)$, a projective scheme that parametrizes all curves C on S such that the class of $\mathcal{O}_S(C)$ in $\mathrm{Num}(S)$ is ξ. He also proved that one can consider the *Picard scheme* $P(\xi)$, a projective scheme that parametrizes all line bundles on S

whose class in Num(S) is ξ. Actually $H(\xi)$ and $P(\xi)$ represent functors, the functor of curves, and the functor of line bundles of class ξ. In particular one can consider $P(0)$ that is a group scheme, and, since we work over $\mathbb{C}$ it is smooth. It is easy to see that the isomorphism class of $P(\xi)$ does not depend on ξ, so that for all ξ, $P(\xi) \simeq P(0)$); hence, $P(\xi)$ is also smooth.

There is an obvious morphism $p : H(\xi) \longrightarrow P(\xi)$ that maps a curve C to the line bundle $\mathcal{O}_S(C)$. The fiber of this map over a point $\mathcal{L}$ in $P(\xi)$ is the complete linear system $|\mathcal{L}|$ of all curves C such that $\mathcal{O}_S(C) \simeq \mathcal{L}$.

Now suppose we have a regular curve C on S with class ξ in Num(S). Set $\ell = \dim(|C|)$. Let $c \in H(\xi)$ be the point corresponding to C and $\mathcal{O}_S(C) = p(c) \in P(\xi)$. There is then an open neighborhood U of $\mathcal{O}_S(C)$ in $P(\xi)$ such that for all $\mathcal{L}$ in U one has $h^1(S, \mathcal{L}) = 0$. For any such $\mathcal{L}$, one has

$$h^0(\mathcal{L}) + h^2(\mathcal{L}) = \chi(S, \mathcal{L}) = \chi(\mathcal{O}_S) + \frac{\xi(\xi - K_S)}{2}.$$

Both $h^0(\mathcal{L})$ and $h^2(\mathcal{L})$ are upper semicontinuous in U, so they are constant in U, i.e., $\dim(|\mathcal{L}|) = \ell$ for all $\mathcal{L} \in U$. Then all fibers of $p^{-1}(U) \to U$ are projective spaces of dimension ℓ. More precisely, one finds a vector bundle $\mathcal{E}$ of rank $\ell + 1$ on U such that $p^{-1}(U) \simeq \mathbb{P}(\mathcal{E})$, and this shows that $p^{-1}(U)$, being a $\mathbb{P}^\ell$ bundle on U is smooth, proving the assertion. □

3 Enriques–Poincaré Theorem

Let S be a surface as above. Define $h^1(S, \mathcal{O}_S) = h^1(S, \Omega^2_S)$ to be the *arithmetic irregularity* of S, denoting it by $q_a(S)$, or simply by q_a if S is intended. This is a birational invariant of S.

Let now H be an irreducible component of the Hilbert scheme of curves on S, with its induced scheme structure (at its general point). Let C be the curve corresponding to the general point of H. Set $\ell_H := \dim(|C|)$, and note that this depends only on H and not on C. By the sequence (1), one has

$$0 \longrightarrow H^0(S, \mathcal{O}_S) \longrightarrow H^0(S, \mathcal{O}_S(C)) \longrightarrow H^0(C, N_C) \longrightarrow H^1(S, \mathcal{O}_S)$$

and therefore we get

$$\dim(H) \leqslant h^0(C, N_C) \leqslant q_a + \ell_H.$$

One defines the *geometric irregularity* of S, denoted by $q_g(S)$, or simply by q_g, the maximum integer q such that for a component H of the Hilbert scheme one has

$$\dim(H) = q + \ell_H.$$

This is again a birational invariant of S. By the above, one has

$$q_g \leqslant q_a. \tag{2}$$

The following theorem is crucial.

Theorem 9 (Enriques-Poincaré) *One has* $q_g = q_a$.

Proof The proof is an easy consequence of the theorem of completeness of the characteristic series in its mildest form, i.e., for regular curves.

Indeed, take D to be a regular curve on S (e.g., by Serre's vanishing theorem [15, Chapt. III, Thm. 5.2], D can be taken to be linearly equivalent to a suitably large multiple of a very ample curve on S). Then the Hilbert scheme is smooth at the point corresponding to D. If H is the unique component of the Hilbert scheme passing through the point corresponding to D, the curve C corresponding to the general point of H is also regular; therefore, $\ell_H = \dim(|D|) = \dim(|C|)$, and one has

$$\dim(H) = h^0(D, N_D) = h^1(S, \mathcal{O}_S) + \dim(|D|) = q_a + \ell_H$$

and the assertion is proved. □

Enriques stated this theorem in [6]. Enriques' proof contained a serious gap, pointed out by Severi only in 1921 [27]. This was the beginning of a long and harsh dispute between Enriques and Severi (see [12, Chapt. IX, §6], [29, §45] and [3]).

Enriques' idea for the proof of the theorem relied on Castelnuovo's theorem on the *defect of completeness* of the characteristic series. This theorem implies that if C is a regular curve on a surface S, the corank of the *restriction map*

$$\sigma : H^0(C, \mathcal{O}_S(C)) \longrightarrow H^0(C, N_C)$$

is the arithmetic irregularity q_a of S (see sequence (1)). Namely, the *characteristic series* of the linear system $|C|$ on C has defect of completeness q_a. Enriques' intuition was that this defect should be accounted by the fact that $|C|$ is contained in a *complete flat family* F whose characteristic series on C is complete. In modern terms, Enriques looks at the infinitesimal deformations of C on S, which are given by the sections of the normal bundle N_C. Some infinitesimal deformations, those in $\mathrm{Im}(\sigma)$, are accounted by moving C inside the linear system $|C|$. The others, the ones which contribute to $\mathrm{cork}(\sigma)$, should be accounted, in Enriques' mind, by deforming the curve off $|C|$ in some continuous, non-linear, way. Enriques' argument fails since nobody ensures that the infinitesimal deformations contributing to $\mathrm{cork}(\sigma)$ are not *obstructed*, i.e., that they correspond to some *true* deformation.

Theorem 9 was proved by Poincaré in 1910. His proof, simplified by Severi in [27], is based on an ingenuous application of Abel's theorem and Jacobi inversion theorem. What Poincaré does is to consider a general pencil $\mathcal{P}$ of hyperplane sections of genus g of the surface S in a suitable projective embedding (a *Lefschetz pencil*, in the modern terminology). Then he proves that one can *rationally determine* on the curves C of $\mathcal{P}$ a basis $\omega_1, \ldots, \omega_g$ of the space of the holomorphic differentials, i.e., one may take a basis of $H^0(C, \Omega^1_C)$ which is defined over the base field of $\mathcal{P}$. This enables Poincaré to also rationally determine the Abel–Jacobi map $\alpha : C(g) \to J(C)$ (where $C(g)$ is the symmetric product g times of the curve C), which is a birational map by Jacobi's inversion theorem. Now, if one takes a sufficiently general point $\xi \in J(C)$ which is also rational on the base field of $\mathcal{P}$, then $\alpha^{-1}(\xi)$ is an effective divisor of degree g on C which, as C varies in $\mathcal{P}$, describes a curve Ξ on S. One may hope that Ξ is an algebraic curve which varies in a continuous, non-linear, system on S. As a matter of fact, Poincaré proves that this is not the case if ξ is general in $J(C)$, but it is so if and only if ξ is general in an appropriate rationally determined abelian subvariety A of dimension q_a of $J(C)$. In this way one constructs a flat family of dimension q_a of non-linearly equivalent algebraic curves on S, and this proves the theorem.

Poincaré's proof was obtained by *trascendental methods*, i.e., using complex analysis and topology. Enriques, and other Italians, like Severi and B. Segre, went on for years fighting with the difficulty of the subject and between them in order to find an *algebro-geometric proof* of it which would not use transcendental techniques (see, for a, rather partisan, account [12, chapt. IX, §6]). All these attempts have been frustrated by insuccess; however, as we will see later, Enriques had right ideas and intuitions, though not the appropriate technical tools, for solving the problem.

4 The Fundamental Theorem of Irregular Surfaces

The irregularity of a surface S is one of the possible extensions to surfaces of the concept of genus of a curve. Besides the two concepts of irregularity, both birational invariants, the arithmetic and the geometric one, that as we saw are equal, one can define other two concepts of irregularity, the *analytic irregularity* $q_{an}(S) = h^0(S, \Omega^1_S)$ (or simply q_{an}) and the *topological irregularity* $q_t(S)$ (or simply q_t), with the property that its double is the first Betti number of the surface S, which is always even. The analytic irregularity q_{an} equals the maximum number of linearly independent holomorphic 1-forms on the surface. It is well-known that any such differential ω is closed (see [1, p. 137–138]), so it is locally integrable, by Poincaré's Lemma. Hence

$$\int \omega$$

is a (transcendental) holomorphic function on S that is multivalued, due to the presence of the periods of ω along the 1-cycles of S. There is another integer related to q_{an}. One defines a *differential of the second kind* on S as a closed, meromorphic 1-form on S such that its integral is a multivalued meromorphic function on S, i.e., this integral does not have logarithmic singularities. In particular, the differentials of rational functions on S are differentials of the second kind. Hence the vector space

$\Omega(S)$ of differentials of the second kind on S is infinite dimensional, containing the subspace $d\mathbb{C}(S)$ of all differentials of rational functions. However, modulo this subspace, it is finite dimensional, and the dimension $r = r(S)$ of the vector space $\Omega(S)/d\mathbb{C}(S)$ is another birational invariant of the surface S. If we fix a basis $\gamma_1, \dots, \gamma_{2q_t}$ of 1-cycles on S, one has the *period map*

$$\pi : \omega \in \Omega(S) \to \left(\int_{\gamma_1} \omega, \dots, \int_{\gamma_{2q_t}} \omega\right) \in \mathbb{C}^{2q_t}$$

whose kernel is immediately seen to be $d\mathbb{C}(S)$. So one has $r \leqslant 2q_t$. It is a classical result of Picard (see [21, p. 150]) that the period map π is also surjective; hence,

$$r = 2q_t. \tag{3}$$

The surjectivity of the period map π also implies with an easy argument that

$$r \geqslant 2q_{an}. \tag{4}$$

Finally, a result by Severi [26] and Picard [20] states that

$$r - q_{an} = q_a. \tag{5}$$

In any event one has the following result.

Theorem 10 (Fundamental Theorem of Irregular Surfaces) *One has* $q_a = q_g = q_{an} = q_t$.

So one sets $q = q_a = q_g = q_{an} = q_t$ and calls it the *irregularity* of the surface. Hence $r = 2q$.

This theorem has been proved, as a consequence of Theorem 9, at the same time by Castelnuovo [4] and Severi [25] in 1905. Castelnuovo's proof is very illuminating, so I sketch it here.

Castelnuovo's Proof of Theorem 10 Castelnuovo starts from Enriques' idea. He takes an irreducible, reduced component H of the Hilbert scheme whose point corresponds to a regular curve C on the surface S. One has $\dim(H) = q_a + \ell$, where $\ell = \dim(|C|)$. Castelnuovo constructs a variety X of dimension q_a with a surjective morphism $p : H \longrightarrow X$ whose fibers are the linear systems (in general of dimension ℓ) contained in H. He proves that X is an abelian variety that he calls the *Picard variety* $\mathrm{Pic}^0(S)$ of the surface S. He shows that there are infinitely many subvarieties V of H such that the restriction $p_{|V} : V \to \mathrm{Pic}^0(S)$ is birational. Let Γ be a sufficiently general (smooth) curve in $\mathrm{Pic}^0(S)$. By taking its proper transform via $p_{|V}$, it can be seen as a curve γ in V and then in H. The curve γ determines a one-dimensional, flat family $\mathcal{G}$ of curves in H, and if $x \in S$ is a general point, there are ν curves of $\mathcal{G}$ passing through x (ν is called the *index* of $\mathcal{G}$). In this way one has a rational map $S \dashrightarrow \Gamma(\nu)$ that to a general point $x \in S$

associates the divisor of degree ν of Γ whose points correspond to the ν curves of $\mathcal{G}$ passing through x. On the other hand, the sum in $\mathrm{Pic}^0(S)$ induces a morphism $\Gamma(\nu) \to \mathrm{Pic}^0(S)$, and by composition, one has a map $f : S \longrightarrow \mathrm{Pic}^0(S)$ that is a morphism because $\mathrm{Pic}^0(S)$ is an abelian variety; hence, it does not contain any rational curve. One has $q_a = h^0(\mathrm{Pic}^0(S), \Omega^1_{\mathrm{Pic}^0(S)})$, and by pulling back via f the 1-forms in $H^0(\mathrm{Pic}^0(S), \Omega^1_{\mathrm{Pic}^0(S)})$, one sees that $q_{an} \geqslant q_a$. But it cannot be the case that $q_{an} > q_a$. In fact, if this where the case, since $r \geqslant 2q_{an}$ by (4), one would have $r - q_{an} \geqslant q_{an} > q_a$ contrary to (5). This proves that $q_a = q_{an}$ and that $r = 2q_{an}$. Finally, by (3), we have $q_a = q_{an} = q_t$. □

Remark 11 From a modern viewpoint, using the powerful machinery of Hodge theory, the proof that $q_a = q_{an} = q_t$ is very easy. In fact Hodge theory tells us that $H^1(S, \mathbb{C}) = H^{0,1}(S) \oplus H^{1,0}(S)$, with $H^{0,1}(S) \simeq H^{1,0}(S)$. Moreover $H^{0,1}(S) \simeq H^1(S, \mathcal{O}_S)$ and $H^{1,0}(S) = H^0(S, \Omega^1_S)$. From this the assertion immediately follows. The equality $q_g = q_a$ is the non-trivial part of the fundamental theorem.

5 Enriques' Attempts for the Proof of Theorem 9

As I explained, the first proof by Enriques of Theorem 9 in [6] was wrong. After Severi pointed out the gap in [27], Enriques came back several times on this subject in various papers [8–11] and in the books [7, 12], outlining basically two different approaches to the problem. In this section, following Mumford [19], I will explain how, though Enriques did not have at his disposal the right algebraic tools, he had however some right ideas to attack the proof of Theorem 9. As Mumford puts it, Enriques "anticipated Grothendieck in understanding that the key to unlocking the Fundamental Theorem was understanding and manipulating geometrically higher order deformations."

Enriques focused on curves C on S such that $h^i(S, \mathcal{O}_S(C)) = 0$, for $i = 1, 2$. To prove Theorem 9, Enriques tried to show that, given a curve C as above, and given a non-zero section $s \in H^0(C, N_C)$, i.e., an infinitesimal deformation to first order of C in S, namely, an abstract infinitely near curve to C on S, this can be prolonged to an infinitesimal deformation of C of indefinitely higher order, i.e., to an infinitely near curve to C of indefinitely higher order. By invoking the existence of the Hilbert scheme, this implies the existence of a component H of this scheme containing a point corresponding to the curve C, such that

$$\dim(H) \geqslant q_a + \ell_H,$$

so that $q_g \geqslant q_a$. On the other hand, we have the opposite inequality (2), and this proves $q_g = q_a$ as wanted.

Enriques did not have a precise definition of higher-order infinitesimal deformations (or, equivalently, of infinitely near curve of higher order), and so he was not confident about the arguments that could be used with this concept. For attacking the problem, he had two different arguments. One was to try to construct infinitely near curves of higher order by an "addition argument" that I'll review later. The other one was based on the fact that higher-order infinitesimal divisors (and linear series) on a curve are easy to define rigorously using symmetric products (and the Jacobian). Then he tried to use this and construct higher-order infinitesimal curves by lifting to the surface higher-order infinitesimal divisors on curves of a Lefschetz pencil of curves on S. This has some interesting similarity with Poincaré's argument for the proof of Theorem 9. This latter approach however has some serious gaps which are not easy to fill up, so I will not review it here (the interested reader may look at Mumford's beautiful paper [19]). The former approach instead, which is much easier by the way, has, as we will see, the possibility of being made rigorous in modern terms. Therefore I will focus on this in this section.

First of all, Enriques' proof requires the construction that Enriques did not have at hand, but he implicitly assumed, of Hilbert schemes and Picard schemes, both due, as we saw above, to Grothendieck.

Secondly, Enriques did not know that $H^0(C, N_C)$ coincides with the space of first-order infinitesimal deformations of C on the surface S. Then he essentially proves this, with a rather complicated but correct argument (see [19, p. 254]) that I skip here.

Thirdly, as I said, Enriques did not have a formal definition of higher-order infinitely near curve to a given curve C on the surface S. From our viewpoint, an infinitely near curve to C of order n is a flat family of curves on S parametrized by the scheme $\mathrm{Spec}\,\mathbb{C}[t]/(t^{n+1})$, such that the curve corresponding to the unique closed point of $\mathrm{Spec}\,\mathbb{C}[t]/(t^{n+1})$ is C. Note that the obvious surjective maps

$$\mathbb{C}[t]/(t^{n+1}) \longrightarrow \mathbb{C}[t](t^{m+1})$$

for $m < n$ determine injective morphisms

$$\mathrm{Spec}(\mathbb{C}[t](t^{m+1})) \longrightarrow \mathrm{Spec}(\mathbb{C}[t]/(t^{n+1}))$$

hence, by base change, an infinitely near curve to C of order n determines (by truncation) an infinitely near curve to C of order m, with $m < n$, and the former is called a *prolongation* of the latter.

More generally, if we have a flat family of curves parametrized by a disc $\mathbb{D} = \mathrm{Spec}(\mathbb{C}\{t\})$, with $\mathbb{C}\{t\}$ the ring of converging power series at 0, with C corresponding to 0, similar as above we have injective morphisms

$$\mathrm{Spec}(\mathbb{C}[t]/(t^{n+1})) \longrightarrow \mathbb{D}$$

for every positive integer n, hence truncations of the flat family parametrized by $\mathbb{D}$ to infinitely near curves to C of any order, and the former is said to *prolong* or *integrate* the latters.

Conversely, if we have a sequence of infinitely near curves to C of any order, each prolongation of the other, there is a flat family parametrized by $\mathbb{D}$ that integrates all of them. This follows by the existence of the Hilbert scheme.

Next notice that sections of $H^0(C, N_C)$ in the image of the restriction map

$$\sigma : H^0(S, \mathcal{O}_S(C)) \longrightarrow H^0(C, N_C)$$

can clearly be integrated to flat families parametrized by $\mathbb{D}$ inside the linear system $|C|$. So the real problem is the integration of infinitely near curves of order one to C corresponding to sections in $H^0(C, N_C)/\mathrm{Im}(\sigma)$.

Let $\xi \in \mathrm{Num}(S)$ be the point corresponding to $\mathcal{O}_S(C)$. As in the proof of Theorem 8, we have the morphism $p : H(\xi) \to P(\xi)$, whose fibers are complete linear systems. If $c \in H(\xi)$ is the point corresponding to the curve C, the sections of $\mathrm{Im}(\sigma)$ are in the kernel of $dp_{|c}$. Let $\ell = \dim(|C|)$. Let $p_1, \ldots, p_\ell$ be sufficiently general points on C such that C is the unique curve in $|C|$ containing them. Then consider the subscheme H of $H(\xi)$ consisting of all curves in $H(\xi)$ containing $p_1, \ldots, p_\ell$. This is the image of a section of $p : H(\xi) \to P(\xi)$ in a neighborhood of c, and as such it is isomorphic to $P(\xi)$ in a neighborhood of $p(c)$. The space $H^0(C, N_C)/\mathrm{Im}(\sigma)$ can be interpreted as the tangent space to $P(\xi)$ at $p(c)$. Now we know that

$$H^0(C, N_C)/\mathrm{Im}(\sigma) \simeq H^1(S, \mathcal{O}_S)$$

but Enriques did not know it. However, $P(\xi)$ is isomorphic to $\mathrm{Pic}^0(S)$, which is an abelian variety. Since we are over $\mathbb{C}$, $\mathrm{Pic}^0(S)$ is smooth, and all its tangent spaces are naturally isomorphic via translations. So we may assume that $p(c)$ coincides with the 0 in $\mathrm{Pic}^0(S)$. So ultimately higher-order infinitely near curves to C prolonging sections in $H^0(C, N_C)/\mathrm{Im}(\sigma)$ can be seen as higher-order infinitely near points to 0 in $\mathrm{Pic}^0(S)$.

Let us quote from [8]:

> Re-examining the same question in my "Lessons on the classification of surface", edited by L. Campedelli [see [7]], I observed, however, that the conclusions of my treatment would mantain their validity if one admitted that curves infinitely near to a given curve on a surface had an effective existence and that one could operate on them as on finite curves, by adding and subtracting. Thus, letting C_1 be a curve infinitely near to C and inequivalent to it [i.e., corresponding to a section in $H^0(C, N_C)/\mathrm{Im}(\sigma)$], the operation $+C_1 - C$, successively repeated, serves to define, in the neighborhood of any curve K whatsoever, a series of infinitely near curves $K_1, K_2, K_3, \ldots$ belonging to a suitably high order neighborhood and this leads to the conclusion that this K should belong to a continuous nonlinear series in which that K_1 would be close to K.
>
> It remained however to justify the intuitive truth: that one can effectively operate on infinitely near curves on a surface by addition and subtraction. And this is precisely the aim of the present note.

This passage contains the central idea in Enriques' argument for the proof of Theorem 9 using infinitely near curves of higher order. Given a section in $H^0(C, N_C)/\mathrm{Im}(\sigma)$, this provides, as indicated above, a tangent vector $\mathbf{t}$ to $\mathrm{Pic}^0(S)$

at the origin, i.e., a map $\mathrm{Spec}(\mathbb{C}[t]/(t^2)) \longrightarrow \mathrm{Pic}^0(S)$, with the closed point of $\mathrm{Spec}(\mathbb{C}[t]/(t^2))$ mapping to the origin. Roughly speaking, what Enriques had in mind was to act with the addition in the group scheme $\mathrm{Pic}^0(S)$, in order to deduce from $\mathrm{Spec}(\mathbb{C}[t]/(t^2)) \longrightarrow \mathrm{Pic}^0(S)$, maps $\mathrm{Spec}(\mathbb{C}[t]/(t^{n+1})) \longrightarrow \mathrm{Pic}^0(S)$ prolonging it for any integer $n \geqslant 1$. Enriques did not have an appropriate argument for this, but in fact it is possible to fill up this gap.

This is based on the following:

Lemma 12 (Mumford's Lemma, see [24]) *The subring of*

$$\mathbb{C}[t_1, \dots, t_n]/(t_1^2, \dots, t_n^2)$$

invariant under the action of the symmetric group $\mathfrak{S}_n$ on $t_1, \dots, t_n$, is isomorphic to $\mathbb{C}[t]/(t^{n+1})$ where $t = t_1 + \cdots + t_n$.

Proof The invariant subring is generated by the elementary symmetric polynomials in $t_1, \dots, t_n$. For all positive integer k, one has

$$(t_1 + \cdots + t_n)^k \equiv k!\phi_k(t_1, \dots, t_n), \text{ (modulo } (t_1^2, \dots, t_n^2)),$$

where $\phi_k(t_1, \dots, t_n)$ is the kth symmetric polynomial in $t_1, \dots, t_n$. In particular

$$\phi_k(t_1, \dots, t_n) \equiv \frac{1}{k!}(t_1 + \cdots + t_n)^k, \quad \text{(modulo } (t_1^2, \dots, t_n^2)),$$

for all $k = 1, \dots, n$ (we can divide by $k!$ because we are in characteristic zero) and

$$(t_1 + \cdots + t_n)^k \equiv 0, \quad \text{(modulo } (t_1^2, \dots, t_n^2)),$$

for all $k > n$. The assertion follows. □

This lemma can be applied to our situation as follows. Given a tangent vector $\mathbf{t}$ to $\mathrm{Pic}^0(S)$ at the origin, i.e., a map $\phi : \mathrm{Spec}(\mathbb{C}[t]/(t^2)) \to \mathrm{Pic}^0(S)$, with the image of the closed point of $\mathrm{Spec}(\mathbb{C}[t]/(t^2))$ equal to the origin, we get the following n-fold summation by adding via the group law on $\mathrm{Pic}^0(S)$

$$\phi \circ p_1 + \cdots + \phi \circ p_n : \mathrm{Spec}(\mathbb{C}[t_1, \dots, t_n]/(t_1^2, \dots, t_n^2)) \longrightarrow \mathrm{Pic}^0(S)$$

where $p_1, \dots, p_n$ are the obvious projections. By commutativity of the sum in $\mathrm{Pic}^0(S)$, the pullback of all functions on $\mathrm{Pic}^0(S)$ is permutation invariant. By Lemma 12, the map factors through $\mathrm{Spec}(\mathbb{C}[t]/(t^{n+1}))$, and this gives the required infinitely near point of order n.

To make the above computation more explicit, let us argue (with Mumford) as follows. Suppose that the tangent vector $\mathbf{t}$ to $\mathrm{Pic}^0(S)$ at the origin corresponds to a 1-cocycle $\{1 + tf_{ij}\}_{(i,j)\in I^2}$ (with $t^2 = 0$). The cocycle condition reads

$$f_{ij} + f_{jk} = f_{ik}.$$

This is a first-order infinitesimal divisor class on S. We want to add it n times to get an nth order infinitesimal divisor class on S. To do this we consider

$$\prod_{h=1}^{n}(1+t_h f_{ij}) = \sum_{k=0}^{n} \phi_k(t_1,\ldots,t_n) f_{ij}^k \equiv \sum_{k=0}^{n} \frac{t^k}{k!} f_{ij}^k, \quad (\text{modulo } (t_1^2,\ldots,t_n^2)),$$

and this is the cocycle defining the nth order infinitesimal divisor class on S obtained from the sum, n times, of $\mathbf{t}$ with itself. By the property of the (truncated) exponential series, it is easy to see that this is in fact a cocycle.

As Mumford points out "Enriques certainly did not know such an argument, but this at least confirms that his intuition was completely sound." Moreover this argument shows that working in characteristic zero (that allows to divide by $k!$ as we did above) is crucial in this problem. Indeed, Enriques–Poincaré's Theorem does not hold in positive characteristic, as shown by a counterexample by Igusa [16].

Acknowledgement The author is a member of GNSAGA of the Istituto Nazionale di Alta Matematica "F. Severi."

References

1. Barth, W., Hulek, K., Peters, C., van de Ven, A.: Compact Complex Surfaces, 2nd Enlarged Edition, Ergebnisse der Mathematik und ihrer Grenzgebiete, vol. 4. Springer, Berlin (2003)
2. Bloch, S.: Semi-regularity and de Rham cohomology. Invent. Math. **17**, 51–66 (1972)
3. Brigaglia, A., Ciliberto, C.: Italian algebraic geometry between the two world wars. Queen's Papers in Pure and Applied Mathematics, Kingston (1995)
4. Castelnuovo, G.: Sugli integrali semplici appartenenti ad una superficie irregolare. Rend. R. Accad. Lincei, V **14**, 545–556, 593–598, 655–663 (1905)
5. Ciliberto, C., Fontanari, C.: Two letters by Guido Castelnuovo. In: G. Bini (ed.) Algebraic Geometry between Tradition and Future: An Italian Perspective. Springer INdAM Series, vol. 53. Springer Nature Singapore, Singapore (2023)
6. Enriques, F.: Sulla proprietà caratteristica delle superficie algebriche irregolari. Rend. Accad. Sci. Bologna **9**, 5–13 (1904)
7. Enriques, F.: Lezioni sulla teoria delle superficie algebriche, Campedelli, L. (ed.) CEDAM, Padova (1932)
8. Enriques, F.: Curve infinitamente vicine sopra una superficie algebrica. Rend. del Seminario Mat. dell'Univ. di Roma **1**(4), 1–9 (1936)
9. Enriques, F.: Addizione alla memoria: "Curve infinitamente vicine sopra una superficie algebrica". Rend. del Seminario Mat. dell'Univ. di Roma **1**(4) , 119 (1936)
10. Enriques, F.: La proprietà caratteristica delle superficie algebriche irregolari e le curve infinitamente vicine. Rend. della R. Acc. Nazionale dei Lincei **23**(6), 459–462 (1936)
11. Enriques, F.: Curve infinitamente vicine sopra una superficie algebrica. Rend. della R. Acc. Nazionale dei Lincei **26**(6), 193–197 (1937)
12. Enriques, F.: Le superficie algebriche. Zanichelli, Bologna (1949)
13. Grothendieck, A.: Techniques de construction et théoremes d'existence en géométrie algébrique IV: Les schémas de Hilbert. Séminaire Bourbaki **6**(221), 249–276 (1960)
14. Grothendieck, A.: Technique de descente et théoremes d'existence en géométrie algébrique V. Les schémas de Picard: théoremes d'existence. Séminaire Bourbaki **7**(232), 143–161 (1961)

15. Hartshorne, R.: Algebraic Geometry. Graduate Texts in Mathematics. Springer, New York (1977)
16. Igusa, J.: On some problems in abstract algebraic geometry. Proc. Natl. Acad. Sci. U. S. A. **41**(11), 964–967 (1955)
17. Kleiman, S.: Completeness of the characteristic series. Adv. Math. **11**, 304–310 (1973)
18. Mumford, D.: Lectures on Curves on an Algebraic Surface. Annals of Mathematics Studies, vol. 59. Princeton University Press, Princeton (1966)
19. Mumford, D.: Intuition and Rigor and Enriques's quest. Notices AMS **58**(2), 250–260 (2011)
20. Picard, É.: Sur quelques thèoremes relatifs aux surfaces algébriques de connexion linéaire supérieure à l'unitè. Comptes Rendus de Séances de l'Acad. des Sc. de Paris **140**, 117–122 (1905)
21. Picard, É., Simart, G.: Thèorie des fonctions algèbriques de deux variables indèpendantes, vol. I, Paris (1897)
22. Poincaré, H.: Sur les courbes tracées sur les surfaces algébriques. In: Annales scientifiques de l'É.N.S. 3e série, vol. 27, pp. 55–108 (1910)
23. Segre, C.: Ricerche sulle rigate ellittiche di qualunque ordine. Atti della R. Acc. delle Scienze di Torino **21**, 868–891 (1885–1886)
24. Sernesi, E.: Severi, Zappa and the characteristic system. In: G. Bini (ed.) Algebraic Geometry between Tradition and Future: An Italian Perspective. Springer INdAM Series, vol. 53. Springer Nature Singapore, Singapore (2023)
25. Severi, F.: Il teorema d'Abel sulle superficie algebriche. Annali di Mat. **12**(3), 55–79 (1905)
26. Severi, F.: Sulla differenza fra i numeri degli integrali di Picard della prima e della seconda specie appartenenti ad una superficie irregolare. Atti della R. Acc. delle Scienze di Torino **40**, 254–262 (1905)
27. Severi, F.: Sulla teoria degli integrali semplici di 1^a specie appartenenti ad una superficie algebrica (7 notes). Rend. R. Accad. Lincei **30**(5), 163–167, 204–208, 231–235, 276–280, 328–332, 365–367 (1921)
28. Severi, F.: Sul teorema fondamentale dei sistemi continui di curve sopra una superficie algebrica. Annali di Mat. **23**(4), 149–181 (1944)
29. Severi, F.: Geometria dei sistemi algebrici sopra una superficie e sopra una varietà algebrica, vol. 2. Cremonese, Roma (1958)
30. Zappa, G.: Sull'esistenza di curve algebricamente non isolate, a serie caratteristica non completa, sopra una rigata algebrica. Acta Pont. Accad. Sci. **7**, 4–8 (1943)
31. Zappa, G.: Sull'esistenza, sopra le superficie algebriche, di sistemi continui completi infiniti, la cui curva generica è a serie caratteristica incompleta. Acta Pont. Accad. Sci. **9**, 91–914 (1945)

Severi, Zappa, and the Characteristic System

Edoardo Sernesi

Abstract Two examples of obstructed curves on an algebraic surface, due to G. Zappa, are given in modern language, after a short description of the historical context of Zappa's work.

Keywords Vector bundle · Characteristic series · Obstructed curve · Hilbert scheme

1 Introduction

This is a partial report of joint work with G. Ottaviani. In Algebraic Geometry the name of Guido Zappa is associated with his discovery of an important example, namely, of a positively dimensional family of smooth curves on an algebraic surface all of whose members are singular points of the Hilbert scheme, i.e., they are *obstructed curves*. In classical language, this is expressed by saying that all members of the family have "incomplete characteristic linear system." This example, published in [24], has some peculiar, not widely known, historical features. Its publication was preceded by the paper [23], where another interesting similar example appeared. In this note, I will briefly describe the historical context of Zappa's work, focusing on the late attempts to give a geometric proof of the theorem of completeness of the characteristic linear system of a complete continuous system of curves on an algebraic surface. In the final part, I will describe both examples in detail.

E. Sernesi (✉)
Dipartimento di Matematica e Fisica, Università Roma Tre, Roma, Italia

G. Bini (ed.), *Algebraic Geometry between Tradition and Future*,
Springer INdAM Series 53, https://doi.org/10.1007/978-981-19-8281-1_10

2 The Fundamental Problem

At the turn of the twentieth century, the main general problem, called "fundamental problem," in algebraic surface theory was to prove that the irregularity $h^{01}(S) := h^1(S, \mathcal{O}_S)$ of a (complex projective nonsingular) surface S is equal to $h^{10}(S) := h^0(S, \Omega^1_S)$, the dimension of the Picard variety. This theorem, called fundamental theorem, was proved by Poincaré [14] in 1910 using transcendental methods. There still remained the problem of giving a purely algebro-geometric proof of it. Severi, in 1904 [18], introduced the notion of *characteristic linear series (or system)* on the curves of a continuous system of curves. He proved that the fundamental theorem is equivalent to proving the completeness of the characteristic linear system on the curves of a sufficiently good complete continuous system. The meaning of "sufficiently good" remained vague for a long time and was clarified only at the very end of the story (see below). The completeness statement is equivalent, in modern language, to the unobstructedness of the curves of the system, i.e., to the nonsingularity of the corresponding points of the Hilbert scheme. This equivalence is not difficult to prove and very clearly explained in [11], Lecture 2.

Enriques, B. Segre, and Severi tried to prove the fundamental theorem without success for long time. The most significant highlights of the entire story are [4, 15, 19], but I will skip them, being mostly spoiled by errors and sterile controversies.[1] In this volume, C. Ciliberto gives an accurate explanation of the fundamental problem as well as historical details [3]. I also recommend the papers [1, 12], for an extended historical discussion. So I will point directly to what happened at the very end, between 1941 and 1945.

In 1941, Severi published [20] where he criticized Enriques' work on the problem, and Enriques harshly responded in 1942 in [5], published on *Commentarii Mathematici Helvetici*. Severi was informed by the editors about this paper and was given the possibility of answering with another paper [21] which appeared on the same issue of the journal.

Here Zappa, who was assistant of Severi, comes on stage. In the last mentioned paper, Severi claimed to prove that the general curve of any positive dimensional complete continuous system is unobstructed. Zappa, following the suggestion of his master to look for examples on ruled surfaces, published a paper in the same year [23] with the purpose of showing the sharpness of Severi's criterion. In fact it contains an example of a positive dimensional system of curves on a ruled surface of genus 2 whose general curve is unobstructed but containing a special obstructed curve. Shortly after, Zappa discovered his second example [24], the important one, which showed that Severi's criterion was incorrect: it consisted of a positive dimensional, everywhere obstructed, system of curves on another ruled surface of genus 2. It is likely that Severi was inspired by this example and pushed by his self-esteem: he then published another paper [22] where he succeeded in

[1] Mumford in [11], p. 7, calls them "depressing."

giving the celebrated notion of *semiregularity* (called by him *emiregolarità*) which is a sufficient condition to guarantee the completeness of the characteristic system. This is the notion studied in modern language by Kodaira and Spencer [9] and subsequently by S. Bloch [2] in higher codimension. So, finally, Severi was able to give a substantial contribution to the proof of the fundamental theorem by giving a satisfactory definition of "sufficiently good" curve.

It is interesting to observe that of the two Zappa's papers, only [24] has been quoted and acknowledged in the modern literature, but none of the examples has been explained in modern language in any published paper or book, as far as I know. More precisely:

(i) In [8], Kodaira quotes Zappa by saying that [24] contains an example of a positive dimensional complete continuous system whose general member is obstructed. But he does not give the example.
(ii) On p. 271 of [6], Grothendieck quotes [24] along the same lines as Kodaira does. Again, the example is not given explicitly.
(iii) In [11], p. 155, Mumford gives an example of an obstructed isolated curve inside an elliptic ruled surface, and he credits Severi and Zappa for it. Actually the example he gives is not given by Zappa neither in [23] nor in [24], even though his construction does not differ much from Zappa's. It is interesting that this example can be traced back to C. Segre [16] §15, where of course no mention is made of the characteristic system nor of the fundamental theorem. The same example is reproduced in my book [17] and attributed to Zappa.

3 The Examples

In this section, I will construct two classes of ruled surfaces X of genus 2 endowed with a section $C_0 \subset X$. The properties of the Hilbert scheme of X around $\{C_0\}$ in each class correspond to those of the examples appearing in [23] and in [24].

I work over $\mathbb{C}$. Let C be a projective nonsingular connected curve of genus 2, and consider a non-split exact sequence of the following form on C:

$$0 \longrightarrow \mathcal{O}_C \longrightarrow E \longrightarrow \omega_C \longrightarrow 0. \tag{1}$$

It corresponds to a non-zero element e of

$$\mathrm{Ext}^1(\omega_C, \mathcal{O}_C) \cong H^1(C, \omega_C^{-1}) = H^0(C, \omega_C^2)^\vee.$$

Therefore, e defines a point of $\mathbb{P}H^0(C, \omega_C^2)^\vee \cong \mathbb{P}^2$, the target space of the bicanonical map of C. Note that $\varphi_{2K}(C) \subset \mathbb{P}^2$ is a conic q.

Lemma 3.1 $h^0(C, E) = 2$, *resp.* $= 1$, *according to whether* $e \in q$ *or* $e \notin q$.

Proof The multiplication map $S^2H^0(C,\omega_C) \longrightarrow H^0(C,\omega_C^2)$ is easily seen to be surjective. Therefore, the coboundary map ∂_e in (1) cannot be zero. Moreover, by [10], Lemma 2, ∂_e has rank one precisely when $e \in q$. □

We let $X = \mathbb{P}(E)$ and we denote by $\pi : X \longrightarrow C$ the projection. Let $C_0 \subset X$ be the image of the section of π corresponding to the quotient $E \longrightarrow \omega_C$ in (1). We want to study the Hilbert scheme of X nearby the point $\{C_0\}$. We have $\mathcal{O}_{C_0}(C_0) = \omega_{C_0}$ by Sernesi [17, Cor. 4.6.3]. Thus, $C_0^2 = 2$ and

$$T_{\{C_0\}}\mathrm{Hilb}_X = H^0(C_0, \omega_{C_0})$$

has dimension 2. Moreover, by Hartshorne [7, Prop. V.2.6 p. 371], we have $\mathcal{O}_X(C_0) \cong \mathcal{O}_X(1)$ and therefore

$$\dim(|C_0|) = h^0(X, \mathcal{O}_X(C_0)) - 1 = h^0(C, E) - 1. \tag{2}$$

A final ingredient in this analysis is the remark that all deformations of C_0 inside X are still sections of self-intersection 2 and therefore correspond to quotients of E of the form

$$0 \longrightarrow \zeta^{-1} \longrightarrow E \longrightarrow \omega_C(\zeta) \longrightarrow 0 \tag{3}$$

for some $\zeta \in \mathrm{Pic}^0(C)$. Let's consider the two possibilities in the statement of Lemma 3.1.

First Possibility $e \in q$, i.e., $h^0(C, E) = 2$. The identity (2) implies that the linear system $|C_0|$ on X is a pencil. If $\zeta \neq \mathcal{O}_C$, then the exact sequence (3) implies that $h^0(C, E) \leq 1$, contradicting our assumptions. Therefore, all quotients of E are of the form $E \longrightarrow \omega_C$, and therefore, C_0 can only deform inside $|C_0|$.

The tangent space to $|C_0|$ at $\{C_0\}$ is

$$T_{\{C_0\}}|C_0| = H^0(X, \mathcal{O}_X(C_0))/\langle C_0\rangle$$

(where $\langle - \rangle$ denotes linear span), and the characteristic map of this family

$$T_{\{C_0\}}|C_0| \longrightarrow T_{\{C_0\}}\mathrm{Hilb}_X = H^0(C_0, \omega_{C_0})$$

is induced by the restriction

$$H^0(X, \mathcal{O}_X(C_0)) \longrightarrow H^0(C_0, \omega_{C_0}).$$

Since

$$\dim[T_{\{C_0\}}|C_0|] = 1 < 2 = \dim[T_{\{C_0\}}\mathrm{Hilb}_X]$$

we see that the characteristic map has corank 1; in particular, it is not surjective. Therefore, $\{C_0\}$ is a singular point of Hilb_X, and the same clearly holds for every $\{C'\} \in |C_0|$. More precisely Hilb_X is isomorphic to a nonreduced scheme of dimension 1 supported on $|C_0| \cong \mathbb{P}^1$. So we have constructed an example having the same properties of the example constructed in [24].

Remark 3.2 A similar construction can also be made in genus $g = 1$; the output is the example given by Mumford in [11].

Second Possibility $e \notin q$, i.e., $h^0(C, E) = 1$. Arguing exactly as above, we find this time that

$$h^0(X, \mathcal{O}_X(C_0)) = h^0(C, E) = 1$$

and therefore the linear system $|C_0|$ is zero-dimensional. The exact sequences (3) are in 1–1 correspondence with the set

$$Z := \{\xi \in \mathrm{Pic}^0(C) : H^0(C, E \otimes \xi) \neq 0\}.$$

Assume for a moment that E is stable. Then, by [13], Theorem 2, Z is a curve,[2] and therefore Hilb_X is one-dimensional around $\{C_0\}$. Since $T_{\{C_0\}}\mathrm{Hilb}_X$ is two-dimensional, C_0 is obstructed. On the other hand, at a point $\xi \in Z$, the corresponding section $C_\xi \subset X$ satisfies

$$T_{\{C_\xi\}}\mathrm{Hilb}_X = H^0(C_\xi, \mathcal{O}_{C_\xi}(C_\xi)) \cong H^0(C, \omega_C \xi^2)$$

which has dimension 1 except at the finitely many points where $\xi^2 = \mathcal{O}_C$ and the dimension is 2. Therefore, Hilb_X is unobstructed at the general $\{C_\xi\}$. We then see that we have the same situation as that of the example in [23]. It remains to prove the following:

Lemma 3.3 *E is stable.*

Proof By contradiction, assume that there exists $\eta \subset E$, $\deg(\eta) \geq 1$, destabilizing E. Then we have a commutative diagram:

[2] In [13], it is assumed that $\det(E) = \mathcal{O}_C$, but one can easily reduce to the case $\det(E) = \omega_C$.

$$
\begin{array}{ccccccccc}
 & & & & 0 & & & & \\
 & & & & \downarrow & & & & \\
 & & & & \eta & \overset{a}{\searrow} & & & \\
 & & & & \downarrow & & & & \\
0 & \longrightarrow & \mathcal{O}_C & \longrightarrow & E & \longrightarrow & \omega_C & \longrightarrow & 0 \\
 & & & & \downarrow & & & & \\
 & & & & \omega_C \eta^{-1} & & & & \\
 & & & & \downarrow & & & & \\
 & & & & 0 & & & &
\end{array}
$$

where $a \neq 0$. If $\deg(\eta) = 2$, then a is an isomorphism and (1) splits, a contradiction. Therefore, $\deg(\eta) = 1$. Since $\chi(\eta) = 0$, it must be $h^0(C, \eta) = 1$, because otherwise the vertical exact sequence would imply that $h^0(C, E) = 0$, a contradiction. But since $h^0(C, E) = 1$, the above diagram shows that $\mathcal{O}_C \subset \eta \subset E$, which is clearly impossible because it implies that the torsion sheaf $\eta/\mathcal{O}_C$ is contained in ω_C. Therefore, E is stable. □

Remark 3.4 In the first case considered ($e \in q$), the subset $Z \subset \mathrm{Pic}^0(C)$ described above consists solely of the point $\{\mathcal{O}_C\}$. Remembering Theorem 2 of [13], one deduces that E is not stable. In fact, it is easy to show that E is strictly semistable and sits in an exact sequence of the form:

$$0 \longrightarrow \mathcal{O}_C(Q_1) \longrightarrow E \longrightarrow \mathcal{O}_C(Q_2) \longrightarrow 0$$

for some $Q_1 + Q_2 \in |\omega_C|$. Details will appear in a work in preparation in collaboration with G. Ottaviani, where generalizations to higher genera will also be discussed.

References

1. Babbitt, D., Goodstein, J.: Federigo Enriques's quest to prove the "Completeness Theorem". Notices AMS **58**, 240–249 (2011)
2. Bloch, S.: Semiregularity and De Rham cohomology. Invent. Math. **17**, 51–66 (1972)
3. Ciliberto, C.: The theorem of completeness of the characteristic series: Enriques' contribution. In: G. Bini (ed.) Algebraic Geometry between Tradition and Future: An Italian Perspective. Springer INdAM Series, vol. 53. Springer Nature Singapore, Singapore (2023)
4. Enriques, F.: Sulla proprietà caratteristica delle superficie algebriche irregolari. Rend. R. Acc. Ist. Scienze Bologna **9**, 5–13 (1905)
5. Enriques, F.: Sui sistemi continui di curve appartenenti ad una superficie algebrica. Comment. Math. Helvetici **15**, 227–237 (1942–1943)

6. Grothendieck, A.: Les schemas de Hilbert. *Seminaire Bourbaki* exp. 221 (1960). In: Seminaire Bourbaki, Année 1960/61, Exposés 205-222, Société Math. de France, pp. 249–276 (1995)
7. Hartshorne, R.: Algebraic Geometry, GTM, vol. 52. Springer, Berlin (1977)
8. Kodaira, K.: Characteristic linear systems of complete continuous systems. Amer. J. Math. **78**, 716–744 (1956)
9. Kodaira, K., Spencer, D.C.: A theorem of completeness of characteristic systems of complete continuous systems. Amer. J. Math. **81**, 477–500 (1959)
10. Mukai, S.: Curves and K3 Surfaces of Genus 11. Lecture Notes in Mathematics, vol. 172, pp. 189–197. Dekker, New York (1996)
11. Mumford, D.: Lectures on Curves on an Algebraic Surface. Annals of Mathematics Studies, vol. 59. Princeton University Press, Princeton (1966)
12. Mumford, D.: Intuition and rigor and Enriques's quest. Notices AMS **58**, 250–260 (2011)
13. Narasimhan, M.S., Ramanan, S.: Moduli of vector bundles on a compact Riemann surface. Ann. Math. **89**, 14–51 (1969)
14. Poincaré, H.: Sur les courbes tracées sur les surfaces algébriques. Ann. École Norm. Sup. **27**(3), 55–108 (1910)
15. Segre, B.: Un teorema fondamentale della geometria sulle superficie algebriche ed il principio di spezzamento. Ann. Mat. Pura Appl. **17**(4), 107–126 (1938)
16. Segre, C.: Ricerche sulle rigate ellittiche di qualunque ordine. Atti R. Accad. Sc. Torino, XXI, 1885–86, 628–651. Reprinted in "Opere", vol. 1, Cremonese (1957)
17. Sernesi, E.: Deformations of Algebraic Schemes. Grundlehren der mathematischen Wissenschaften, vol. 334. Springer, Berlin (2006)
18. Severi, F.: Osservazioni sui sistemi continui di curve appartenenti ad una superficie algebrica. Atti R. Acc. Scienze Torino **39**, 371–392 (1904)
19. Severi, F.: Sulla teoria degli integrali semplici di 1^a specie appartenenti ad una superficie algebrica. Nota V. Rend. R. Acc. Naz. dei Lincei **30**(5), 296–301 (1921)
20. Severi, F.: La teoria generale dei sistemi continui di curve sopra una superficie algebrica. Memorie R. Acc. d'Italia **12**, 337–430 (1941).
21. Severi, F.: Intorno ai sistemi continui di curve sopra una superficie algebrica. Comment. Math. Helvetici **15**, 238–248 (1942–1943)
22. Severi, F.: Sul teorema fondamentale dei sistemi continui di curve sopra una superficie algebrica. Annali di Matematica **23**(4), 149–181 (1944)
23. Zappa, G.: Sull'esistenza di curve algebricamente non isolate, a serie caratteristica non completa, sopra una rigata algebrica. Acta Pontif. Acad. Sci. **v.VII**(2), 1–5 (1943)
24. Zappa, G.: Sull'esistenza, sopra le superficie algebriche, di sistemi continui infiniti, la cui curva generica è a serie caratteristica incompleta. Acta Pontif. Acad. Scient. **v.IX**(9), 91–93 (1945)

Two Letters by Guido Castelnuovo

Ciro Ciliberto and Claudio Fontanari

Abstract In this expository paper, we transcribe two letters by Guido Castelnuovo, one to Francesco Severi and the other to Beniamino Segre, and explain the contents of both, which basically focus on the quest for an algebraic proof of the equality between the analytic and the arithmetic irregularity and of the closedness of regular 1-forms on a complex, projective, algebraic surface. Such an algebraic proof has been found only in the 1980s by Deligne and Illusie.

Keywords Guido Castelnuovo · Beniamino Segre · Francesco Severi · Holomorphic forms · Hodge theory · Frölicher spectral sequence

1 Introduction

As it is well-known, the treatise by Federigo Enriques epitomizing the celebrated classification of algebraic surfaces by the Italian school of algebraic geometry has been published posthumously in 1949, a few years after the sudden death of the author in 1946. As pointed out by Guido Castelnuovo in the preface (see [8]),

> (...) dove il terreno è meno solido l'Autore mette sull'avviso lo studioso. Di questi punti ancora fluidi quello che presenta la difficoltà più ardua ed il maggiore interesse *è la teoria* dei sistemi continui di curve algebriche (...) che esistono sopra ogni superficie irregolare. (...) tutti i tentativi compiuti (...) per dimostrarla mediante considerazioni algebrico-geometriche si sono urtati contro difficoltà sinora insuperate. (...) l'Autore dà anche suggerimenti sopra una via da tentare per giungere alla meta. Debbo confessare che non vedo come quella via possa tradursi in un procedimento irreprensibile.

C. Ciliberto
Dipartimento di Matematica, Università di Roma Tor Vergata, Roma, Italia
e-mail: cilibert@axp.mat.uniroma2.it

C. Fontanari (✉)
Dipartimento di Matematica, Università degli Studi di Trento, Povo, Trento
e-mail: claudio.fontanari@unitn.it

G. Bini (ed.), *Algebraic Geometry between Tradition and Future*,
Springer INdAM Series 53, https://doi.org/10.1007/978-981-19-8281-1_11

> (...) where the ground is less solid the Author warns the reader. Among these still unsteady points the most difficult and interesting one *is the theory* of the continuous systems of algebraic curves (...) existing on any irregular surface. (...) all attempts (...) towards an algebro-geometric approach have been frustrated by still insurmountable difficulties. (...) the Author provides some hints about a strategy to reach the goal. I should confess I cannot see how that strategy may be translated into a fully rigorous argument.

The two letters by Guido Castelnuovo that we transcribe and translate into English in Sect. 2, the first one addressed to Francesco Severi and dated 1947 and the second one addressed to Beniamino Segre and dated 1950, provide firsthand witness of Castelnuovo's attempts to a purely algebro-geometric understanding of irregular surfaces.

In Sect. 3 we explain the background of Castelnuovo's letters, using modern terminology. In particular we explain a classical method, very familiar to Castelnuovo and due to Picard and Severi, of constructing regular 1-forms on a surface. As explained in Sect. 3.3, one of the crucial points of Castelnuovo's approach is the attempt of proving the closedness of global regular 1-forms, a fact that today we are aware to strongly rely on the characteristic zero assumption. Indeed, Castelnuovo's remarks in his letters turn out to be quite inconclusive and sometimes even unprecise, as we discuss in Sect. 4 that is devoted to explaining most of the issues raised by Castelnuovo in his two letters. The algebro-geometric proof of the results that Castelnuovo sought (i.e., closedness of global regular 1-forms and equality of different definitions of irregularity, in characteristic zero) is now available; it is due to work by Deligne and Illusie in the 1980s and turns out to be completely out of reach of Castelnuovo's classical tools, since (somehow paradoxically) it involves a tricky reduction to the case of positive characteristic. Section 5 is devoted to give a brief account on how these algebraic proofs can be obtained using modern tools.

We stress that the present note does not contain any original result, but in our opinion the contents of Castelnuovo's letters are worthy of careful consideration from both a historical and a mathematical viewpoint. This paper is addressed to readers who are well aware of rather advanced concepts in algebraic geometry, so we do not dwell on explaining standard technical details when they occur.

2 The Letters

In this section we transcribe two letters of Castelnuovo, the first one of November 26, 1947, to Francesco Severi and the second one of January 15, 1950, to Beniamino Segre. The first letter belongs to the "Fondo Guido Castelnuovo" of the Accademia Nazionale dei Lincei that has been edited by Paola Gario, has been digitalized, and can be found on the web page

http://operedigitali.lincei.it/Castelnuovo/Lettere_E_Quaderni/menuL.htm

The second letter comes from the collection of documents of Beniamino Segre kept at the University of Caltech.

2.1 Guido Castelnuovo to Francesco Severi

Roma, 26 novembre 1947

Caro Severi,

aderendo al tuo desiderio ti comunico alcuni risultati sulle superficie irregolari; parecchi si ottengono senza difficoltà e possono servire come esercizio per i tuoi discepoli.

Lo scopo remoto ed ambizioso che mi proponevo era di costruire una teoria delle dette superficie indipendente dalla nozione di sistema continuo di curve, teoria in cui si ritrovassero il teorema sul numero $(p_g - p_a)$ dei differenziali totali indipendenti di prima specie, il teorema di Hodge, ecc.. Il programma è appena iniziato; ma si deve raggiungere la meta, a meno che la teoria delle superficie irregolari non riservi delle sorprese che non saprei nemmeno immaginare.

I. Indico con $|C|$ un sistema regolare di grado n e genere π; in molti casi occorre supporre che $|C|$ sia *abbastanza ampio*, contenga entro di sè il sistema canonico $|K|$ od anche un suo multiplo; ricercando caratteri invarianti, tutto ciò non ha importanza. Indico con χ il tuo invariante q', cioè il numero delle curve indipendenti di $|2C + K|$ che passano per il gruppo jacobiano G_δ di un fascio $|C|$ e in conseguenza per il gruppo base G_n del fascio. Indico con G_k il gruppo dei k punti cuspidali di una superficie, d'ordine n, a singolarità ordinarie, le cui sezioni piane appartengano al sistema $|C|$.

   Ecco un significato di χ che si raggiunge subito:

   (1) *E' χ la sovrabbondanza del sistema $|4C+2K|$ rispetto al gruppo dei punti cuspidali G_k* (cioè G_k impone $k - \chi$ condizioni al detto sistema).

   Invece G_k presenta condizioni indipendenti ai sistemi $|mC+K|$, $|mC+2K|, \ldots$, per $m \geqslant 5$.

   Per $m = 4$ vi è un risultato di Enriques ottenuto indirettamente attraverso il computo dei moduli di una superficie, risultato che converrebbe dimostrare direttamente; lo ricordo perché interviene tra poco: "La sovrabbondanza del sistema $|4C + K|$ rispetto al gruppo G_k dei punti cuspidali è un invariante", che indicherò con Q' e di cui sotto darò l'espressione.

   (2) *La serie completa g_k determinata dal gruppo G_k sopra una curva di $|4C+K|$ passante per esso ha la dimensione χ.*

   (3) *La serie completa g_k determinata dal gruppo G_k sopra una curva di $|3C+K|$ passante per esso (ad es.: sulla $f = f'_x = 0$) ha la dimensione $2n-\pi+2p_g+p_a-(I+4)+\theta$ dove $0 \leqslant \theta \leqslant p_g - p_a$* (si suppone $|C|$ abbastanza ampio). E' θ un invariante?

II. Il procedimento che ti ha condotto a stabilire l'invarianza di $q' = \chi$ fa vedere subito che:

(4) *E' invariante il numero delle curve linearmente indipendenti di* $|2C + 2K|$ *che passano per il gruppo jacobiano* G_δ *di un fascio* $|C|$ *ed anche per il gruppo base* G_n; indicherò questo invariante con Q.

Si vede poi (se è esatto il risultato di Enriques sopra citato) che:

(5) *E' pure invariante il numero delle curve linearmente indipendenti di* $|2C + 2K|$ *che passano per il gruppo* G_δ *senza esser costrette a passare per* G_n; questo nuovo invariante uguaglia l'invariante di Enriques Q'.

E' quindi invariante il numero delle condizioni che una curva di $|2C + 2K|$ passante per il gruppo jacobiano G_δ di un fascio $|C|$ deve soddisfare per contenere il gruppo base. Si dimostra che questo invariante soddisfa alla diseguaglianza $Q' - Q \leqslant p_g$.

Quanto alle espressioni di Q e Q' posso dir questo.

Se le ∞^{Q-1} curve di $|2C + 2K|$ passanti per $G_\delta + G_n$ segano sopra una curva di $|2C + K|$ passante per lo stesso gruppo una serie *completa* (residua di $G_\delta + G_n$ rispetto alla serie canonica) allora:

$$Q - 1 = p_a + p_g + p^{(1)} - (I - 4) + \omega$$

dove ω ($\leqslant p_g - p_a$) è un nuovo invariante che ha un significato molto semplice: $I + 4 - \omega - 1$ è il numero delle condizioni che un gruppo G_{I+4} della tua serie d'equivalenza (in senso stretto) presenta alle curve *bicanoniche* costrette a contenerlo.

Se la serie lineare nominata non è completa, dall'espressione di $Q - 1$ va tolta la deficienza $\leqslant p_g - p_a$ della serie stessa.

Nello stesso ordine d'idee ti comunico ancora questo risultato:

(6) *La sovrabbondanza del sistema* $|3C + K|$ *rispetto al gruppo jacobiano* G_δ *di un fascio* $|C|$ *è un invariante e vale precisamente* $2p_g$ (se il sistema completo $|C|$ cui il fascio appartiene è abbastanza ampio).

III. Altre questioni. Come sai il teorema fondamentale da dimostrare è questo: una curva Γ di $|2C + K|$ passante per il gruppo jacobiano G_δ e il gruppo base G_n di un fascio $|C|$ sega sopra la curva generica C del fascio (fuori di G_n) un gruppo canonico *che non appartiene ad una curva aggiunta* C'. Ho cercato di trasformare la condizione in altre equivalenti. Tale è ad esempio questa: il gruppo G_n su quella curva deve presentare condizioni indipendenti alla serie caratteristica di Γ *resa completa*. Alla serie caratteristica in senso stretto, G_n presenta solo $n - 1$ condizioni.

Altra forma: Scriviamo la identità di Picard in coordinate omogenee:

$$Xf'_x + Yf'_y + Zf'_z + Tf'_t = 0$$

dove $X = 0, \ldots$ sono superficie d'ordine $n - 3$. Occorre aggiungere la condizione (non detta esplicitamente) che le superficie $yX - xY = 0, \ldots, tZ -$

$zT = 0$ siano aggiunte alla $f = 0$ d'ordine n. Segue che le $X = 0, \dots, T = 0$ passano semplicemente per i t punti tripli di f ed hanno inoltre in comune un gruppo di $(n-4)d - 3t$ punti sulla curva doppia d'ordine d di f; esse segano inoltre rispettivamente i piani $x = 0, \dots, t = 0$ in curve aggiunte d'ordine $n-3$. Da ciò segue che quel gruppo di punti della curva doppia appartiene alla serie segata su questa dalle superficie d'ordine $n-4$ passanti semplicemente per i t punti tripli, purch'e questa serie venga resa completa, mentre essa ha la deficienza $p_g - p_a$ per la definizione stessa di irregolarità. Orbene il teorema fondamentale equivale al seguente: *Per quel gruppo di* $(n-4)d-3t$ *punti della curva doppia non passa nessuna superficie d'ordine* $n-4$ *che contenga i* t *punti tripli di* f. Questo enunciato si traduce in questo altro, molto elegante dal punto di vista analitico: *Non è possibile soddisfare una identità del tipo:*

$$\overline{X} f'_x + \overline{Y} f'_y + \overline{Z} f'_Z + \overline{T} f'_t \equiv Q f$$

ove $\overline{X} = 0, \dots, \overline{T} = 0$ *sono superficie aggiunte d'ordine* $n-3$ *e* $Q = 0$ *una superficie d'ordine* $n-4$ *se non nel caso banale* $\overline{X} = \frac{1}{n} x Q, \dots, \overline{T} = \frac{1}{n} t Q$.

Ritornando all'identità di Picard scritta sopra, ti consiglio di far studiare da qualche discepolo la omografia tra il sistema di superficie non aggiunte di ordine $n-3$ $\lambda X + \mu Y + \nu Z + \rho T = 0$ e il sistema di piani $\lambda x + \mu y + \nu z + \rho t = 0$, ognuno dei quali taglia la superficie corrispondente in una curva aggiunta. Nel caso delle rigate irrazionali dei primi ordini si trovano proprietà elegantissime.

IV. Finalmente alcune osservazioni che ti potranno servire se esporrai in lezione la tua Nota sugli integrali semiesatti.

Tu dimostri che ad ogni curva di $|2C + K|$ passante per il gruppo jacobiano e per il gruppo base di un fascio $|C|$ (*curva covariante del fascio*, come io la chiamo) si può associare una determinata curva covariante di ogni altro fascio $|D|$. Due curve associate segano sullo stesso gruppo di punti la curva di contatto di due fasci. Esse inoltre si segano in un gruppo G_{I+4} della tua serie. Si vede facilmente che questo gruppo è comune a tutta la famiglia di curve covarianti associate relative agli infiniti fasci esistenti sulla superficie. Ogni curva C di $f = 0$ è segata dalla curva covariante della famiglia in un gruppo canonico che dirò *gruppo traccia.*

Preso un punto P della superficie, esistono infinite curve per P per le quali P appartiene al gruppo traccia. *Tutte queste curve si toccano in* P. Vuol dire che ad ogni punto P di $f = 0$ è collegata una direzione tangente, o un elemento lineare uscente da P (indeterminato solo se P appartiene al gruppo G_{I+4}). Connettendo tutti questi elementi si viene a ricoprire la superficie con un fascio di curve (trascendenti) che risultano esser le curve integrali dell'equazione $Bdx - Ady = 0$, dove $A = 0$ e $B = 0$ sono due superficie aggiunte d'ordine $n-2$ secanti su f le curve covarianti dei fasci $x =$ cost., $y =$ cost. Resterebbe naturalmente da far vedere che $1/f'_z$ è fattore integrante dell'espressione differenziale.

Al variare di P su f quella tangente in P descrive una congruenza algebrica di classe $2\pi - 2$ e di ordine $k - \nu = 6\pi - 6 + p^{(1)} - 1 - (I + 4)$ ove k è il numero di punti cuspidali e ν è l'ordine della curva $f = f'_x = 0$.

E qui termino questa lunghissima lettera che vorrei potesse spingere a colmare nella teoria delle superficie quella lacuna che tutti avvertiamo.

Cordiali saluti dal tuo aff.mo

GUIDO CASTELNUOVO

Rome, November 26, 1947

Dear Severi,

following your wishes I am going to tell you some results about irregular surfaces; many of them are easily obtained and may be useful as exercises for your students. My ambitious and ultimate purpose was to build a theory of such surfaces independent of the notion of continuous system of curves, a theory embracing the theorem on the number $(p_g - p_a)$ of independent global differentials of the first kind, the theorem of Hodge, etc.. This program has just started; but the goal should be achieved, unless the theory of irregular surfaces hide amazing things I could not even imagine.

I. Let $|C|$ be a regular system of degree n and genus π; in many cases we need to assume that $|C|$ is *sufficiently ample*, containing the canonical system $|K|$ or even one of its multiples; since we are looking for invariant characters, this is immaterial. I denote by χ your invariant q', namely, the number of independent curves of $|2C+K|$ passing through the jacobian group G_δ of a pencil $|C|$, hence through the base locus G_n of the pencil. Let G_k be the group of the k cuspidal points[1] of a surface, of degree n, with only ordinary singularities, and whose plane sections belong to the system $|C|$.

   Here is a meaning of χ which is immediate:

   (1) *The invariant χ is the superabundance of the system $|4C+2K|$ with respect to the cuspidal points G_k* (i.e. G_k imposes $k-\chi$ conditions to such system).

   On the other hand, G_k gives independent conditions to the systems $|mC + K|$, $|mC + 2K|$, ..., per $m \geqslant 5$.

   For $m = 4$ there is a result of Enriques, indirectly obtained by a moduli computation for a surface, but which should be directly proven; I recall it because it is coming into play shortly later: "The superabundance of the system $|4C + K|$ with respect to the group G_k of cuspidal points is an invariant", which I will denote Q' and whose expression I am going to give below.

   (2) *The complete series g_k determined by the group G_k on a curve of $|4C+K|$ passing through it has dimension χ.*

[1] The usual English term for *cuspidal points* is *pinch points*.

(3) *The complete series* g_k *determined by the group* G_k *on a curve of* $|3C+K|$ *passing through it (for instance: on* $f = f'_x = 0$*) has dimension* $2n - \pi + 2p_g + p_a - (I+4) + \theta$ *where* $0 \leqslant \theta \leqslant p_g - p_a$ (assume $|C|$ sufficiently ample). Is θ an invariant?

II. The same argument which led you to establish the invariance of $q' = \chi$ immediately shows:

(4) *It is invariant the number of linearly independent curves of* $|2C + 2K|$ *passing through the jacobian group* G_δ *of a pencil* $|C|$ *and also through the base group of* G_n; I will denote this invariant by Q.

Then one sees (if the aforementioned result of Enriques is correct) that:

(5) *It is also invariant the number of linearly independent curves of* $|2C+2K|$ *passing through the group* G_δ *without having to pass through* G_n; this new invariant is equal to Enriques invariant Q'.

It is therefore invariant the number of conditions that a curve of $|2C + 2K|$ passing through the jacobian group G_δ of a pencil $|C|$ has to satisfy in order to contain the base group. One proves that this invariant satisfies the inequality $Q' - Q \leqslant p_g$.

Regarding the expressions of Q e Q' I can state the following.

If the ∞^{Q-1} curves of $|2C+2K|$ passing through $G_\delta + G_n$ cut on a curve of $|2C + K|$ passing through the same group a *complete* series (residual of $G_\delta + G_n$ with respect to the canonical series) then:

$$Q - 1 = p_a + p_g + p^{(1)} - (I - 4) + \omega$$

where ω ($\leqslant p_g - p_a$) is a new invariant which has a very simple meaning: $I+4-\omega-1$ is the number of conditions that a group G_{I+4} of your series of equivalence (in the strict sense) prescribes to the *bicanonical* curves forced to contain it.

If such a linear series is not complete, from the expression of $Q - 1$ one has to subtract the deficiency $\leqslant p_g - p_a$ of the series.

In the same circle of ideas I also tell you the following result:

(6) *The superabundance of the system* $|3C + K|$ *with respect to the jacobian group* G_δ *of a pencil* $|C|$ *is an invariant and its value is precisely* $2p_g$ (if the complete system $|C|$ to which the pencil belong is sufficiently ample).

III. Other issues. As you know, the fundamental theorem to be proven is the following: a curve Γ of $|2C + K|$ passing through the jacobian group G_δ and the base group G_n of a pencil $|C|$ cuts on the generic curve C of the pencil (off G_n) a canonical group *which does not belong to an adjoint curve* C'. I tried to translate this condition into other equivalent formulations. Such is for instance the following one: the group G_n on that curve has to impose independent conditions to the characteristic series of Γ *made complete*. To the characteristic series in the strict sense, G_n imposes only $n - 1$ conditions.

Other formulation: Let us write Picard's identities in homogeneous coordinates:

$$Xf'_x + Yf'_y + Zf'_z + Tf'_t = 0$$

where $X = 0, \ldots$ are surfaces of degree $n - 3$. We have to add the condition (not explicitly stated) that the surfaces $yX - xY = 0, \ldots, tZ - zT = 0$ are adjoint to $f = 0$ of degree n.[2] It follows that $X = 0, \ldots, T = 0$ pass simply through the t triple points of f and moreover share a group of $(n - 4)d - 3t$ points on the double curve of degree d of f; furthermore, they cut the planes $x = 0, \ldots, t = 0$, respectively, in adjoint curves of degree $n - 3$. Hence it follows that such group of points of the double curve belongs to the series cut on this curve by the surfaces of degree $n - 4$ passing simply through the t triple points, provided this series has been made complete, while it has deficiency $p_g - p_a$ by the very definition of irregularity. Now, the fundamental theorem is equivalent to the following: *Through such groups of* $(n - 4)d - 3t$ *points of the double curve it does not pass any surface of degree* $n - 4$ *containing the* t *triple points of* f.[3] This statement translates into the following one, which is quite elegant from the analytic viewpoint: *It is impossible to verify an identity of the form:*

$$\overline{X}f'_x + \overline{Y}f'_y + \overline{Z}f'_Z + \overline{T}f'_t \equiv Qf$$

where $\overline{X} = 0, \ldots, \overline{T} = 0$ *are adjoint surfaces of degree* $n - 3$ *and* $Q = 0$ *is a surface of degree* $n - 4$ *except in the trivial case* $\overline{X} = \frac{1}{n}xQ, \ldots, \overline{T} = \frac{1}{n}tQ$.

Going back to Picard's identity as written above, I suggest to you to propose to some student to investigate the homography between the system of non–adjoint degree $n - 3$ surfaces $\lambda X + \mu Y + \nu Z + \rho T = 0$ and the system of planes $\lambda x + \mu y + \nu z + \rho t = 0$, each cutting the corresponding surface in an adjoint curve. In the case of irrational ruled surfaces of low degree one finds very elegant properties.

IV. Finally, a few remarks you may find useful if you will present in a course your note about semiexact integrals.

You prove that to every curve of $|2C+K|$ passing through the jacobian group and the base group of a pencil $|C|$ (*covariant curve of the pencil*, as I call it) one can associate a unique covariant curve of every other pencil $|D|$. Two associated curves cut on the same group of points the contact curve of two pencils. They moreover cut each other in a group G_{I+4} of your series. One easily checks that this group is common to the whole family of associated covariant curves with

[2] This is clearly an error, Castelnuovo means adjoint of degree $n - 2$.

[3] The right statement here would be: *Through such groups of* $(n - 4)d - 3t$ *points of the double curve it does not pass any surface of degree* $n - 4$ *containing the* t *triple points of* f *and not containing the double curve.*

respect to the infinitely many pencils on the surface. Every curve C of $f = 0$ is cut by the covariant curve of the family in a canonical group which I will call *trace group*.

Taken a point P of the surface, there exist infinitely many curves through P such that P belongs to the trace group. *All these curves intersect in* P. It means that to every point P of $f = 0$ is associated a tangential direction, or a linear element (not defined only if P belongs to the group G_{I+4}). By connecting all these elements the surface is covered by a pencil of (transcendental) curves which turn out to be the integral curves of the equation $Bdx - Ady = 0$, where $A = 0$ and $B = 0$ are two adjoint surfaces of degree $n - 2$ cutting on f the covariant curves of the pencils $x = \text{const.}$, $y = \text{const.}$ Of course one should show that $1/f'_z$ is an integral factor of the differential expression.

Varying P on f the tangent in P describes an algebraic congruence of class $2\pi - 2$ and degree $k - \nu = 6\pi - 6 + p^{(1)} - 1 - (I + 4)$ where k is the number of cuspidal points and ν is the degree of the curve $f = f'_x = 0$.

Here I stop this quite long letter I wish it could stimulate to fill in the theory of surfaces that gap we all perceive.

Best regards, yours friendly

GUIDO CASTELNUOVO

2.2 *Guido Castelnuovo to Beniamino Segre*

Roma, 15 genn. 50

Caro Professore,

In relazione alla nostra conversazione di venerdì scorso e al programma di ricerche di cui Le parlavo, penso di sottoporle una questione, risolta la quale si sarebbe compiuto un passo notevole verso la meta cui Le accennavo. Si tratta di una questione di geometria algebrica, la quale, ove si possano togliere alcune restrizioni forse non necessarie, si muta in una questione relativa alle equazioni alle derivate parziali con condizioni al contorno. Con i mezzi svariati e potenti di cui Ella dispone potrà affrontarla e pervenire alla risposta desiderata.

Sia $f(x, y, z, t)$ una superficie (in coord. omog.) d'ordine n, irriducibile, con singolarità ordinarie; e siano $X = 0$, $Y = 0$, $Z = 0$, $T = 0$ quattro superficie aggiunte d'ordine $n - 3$. Si tratta di dimostrare che un'identità del tipo

$$Xf'_x + Yf'_y + Zf'_z + Tf'_t = Qf, \tag{1}$$

con Q polinomio di grado $n - 4$, non può sussistere salvo nel caso banale (identità di Eulero) $X = \frac{1}{n}xQ, \dots, T = \frac{1}{n}tQ$. Per farle vedere l'interesse della questione Le dirò che se si toglie la condizione che le sup. $X, \dots, T$ siano aggiunte, e si sostituisce con la condizione meno stretta che siano aggiunte le sei superficie d'ordine $n - 2$ $yX - xY = 0, \ldots$, allora la identità può sussistere con Q identicamente nulla; anzi di identità di quel tipo ve ne sono $p_g - p_a$ indipendenti per una superficie irregolare (Picard).

Ritornando alla (1), supposto che essa possa aver luogo, si vedrebbe che la superficie $Q = 0$ incontra la curva doppia di $f = 0$ nei punti tripli e nei punti ove $X'_x + Y'_y + Z'_z + T'_t = 0$, donde si concluderebbe che $Q \equiv X'_x + Y'_y + Z'_z + T'_t + \overline{Q}$ (salvo un fattore costante), essendo $\overline{Q} = 0$ una superficie aggiunta d'ordine $n - 4$ che darebbe luogo a un integrale doppio senza periodi; e di qua l'assurdo (Hodge). Ma io richiedo evidentemente una dimostrazione più diretta e più elementare di quella qui abbozzata.

Ci pensi quando ha tempo, perché mi pare ne valga la pena. Cordiali saluti; aff.mo

G. Castelnuovo

Rome, January 15, 1950

Dear Professor,

Concerning our conversation of last Friday and the research program I exposed to you, I am going to propose to you a question, whose solution would provide a remarkable step towards the goal I mentioned. It is a question in algebraic geometry, which, up to removing some maybe unnecessary restrictions, translates into a question in partial differential equations with boundary conditions. By applying the many and poweful tools you have at your disposal you could address it and obtain the desired answer.

Let $f(x, y, z, t)$ be a surfaces (in homogeneous coordinates) of degree n, irreducible, with ordinary singularities; let $X = 0$, $Y = 0$, $Z = 0$, $T = 0$ be four adjoint surfaces of degree $n - 3$. The point is to show that an identity of the form

$$Xf'_x + Yf'_y + Zf'_z + Tf'_t = Qf, \tag{2}$$

with Q polynomial of degree $n - 4$, is not satisfied unless in the trivial case (Euler identity) $X = \frac{1}{n}xQ, \ldots, T = \frac{1}{n}tQ$. In order to show you the interest of the question I will tell you that if one drops the condition that the surfaces $X, \ldots, T$ are adjoint, and one replaces it by the less strict condition that the six degree $n-2$ surfaces $yX - xY = 0, \ldots$ are adjoint, then the identity may hold with Q identically zero; indeed, there are $p_g - p_a$ independent such identities for an irregular surface (Picard).

Coming back to (2), assuming it may hold, one would see that the surface $Q = 0$ meets the double curve of $f = 0$ in the triple points and in the points where $X'_x + Y'_y + Z'_z + T'_t = 0$, whence one would conclude that $Q \equiv X'_x + Y'_y + Z'_z + T'_t + \overline{Q}$ (up to a constant factor), where $\overline{Q} = 0$ would be an adjoint surface of degree $n - 4$ which would give rise to a double integral without periods, hence a contradiction (Hodge). But of course I am looking for a more direct and more elementary proof than the one sketched here.

Please think about that when you have time, because I believe it is worth the trouble. Best regards; yours friendly

G. Castelnuovo

3 Regular 1-Forms on a Surface

If X is a smooth, irreducible, projective surface over an algebraically closed field $\mathbb{K}$, the elements of $H^0(X, \Omega^1_X)$ are called *regular 1-forms* on X. We will denote the dimension of $H^0(X, \Omega^1_X)$ by $q_{\mathrm{an}}(X)$ (or simply by q_{an} if there is no danger of confusion), and we will call it the *analytic irregularity* of X (see [6]).

In this section we want to explain the background of Castelnuovo's letters, using modern terminology. In particular we want to explain a classical method, very familiar to Castelnuovo and due to Picard and Severi, of constructing regular 1-forms on a surface. In Sect. 3.3, we explain Castelnuovo's viewpoint on the attempts of proving closedness of regular 1-forms on a surface.

3.1 *The General Set Up*

Let X be a smooth, irreducible, projective surface over an algebraically closed field $\mathbb{K}$. We may assume X to be linearly normally embedded as a surface of degree d in a projective space $\mathbb{P}^r$, with $r \geqslant 5$, in such a way that the following happens. If we consider a general projection π of S to $\mathbb{P}^3$, whose image is a surface S of degree d, then S has *ordinary singularities* (see [20, Thm. 2]), i.e., it has:

- an irreducible *nodal* double curve Γ, i.e., S has normal crossings at the general point of Γ,
- a finite number of triple points for both Γ and S; the triple points for Γ are *ordinary*, i.e., the tangent cone there to Γ consists of the union of three non-coplanar lines, and the tangent cone there to S consists of the union of three distinct planes,
- finitely many *pinch points* on Γ; we will denote by G_c the *pinch points scheme*, i.e., the reduced zero-dimensional scheme on X where the differential of π drops rank, so that G_c is mapped by π to the set of pinch points of S on Γ. We will set $\gamma = \mathrm{length}(G_c)$.

The map $\pi : X \to S$ is the normalization map.

We will introduce homogeneous coordinates $[x_1, x_3, x_3, x_4]$ in $\mathbb{P}^3$ and related affine coordinates (x, y, z), with

$$x = \frac{x_1}{x_4}, \quad y = \frac{x_2}{x_4}, \quad z = \frac{x_3}{x_4}$$

so that $x_4 = 0$ is the *plane at infinity*. We assume that the coordinates (i.e., the corresponding fundamental points) are general with respect to S. The homogeneous equation of S is of the form $F(x_1, x_2, x_3, x_4) = 0$, with F an irreducible homogeneous polynomial of degree d, and the affine equation of S is $f(x, y, z) = 0$, with $f(x, y, z) = F(x, y, z, 1)$. We will denote by f_x, f_y, f_z the partial derivatives of f with respect to x, y, z and by F_i the partial derivative of F with respect to x_i, for $1 \leqslant i \leqslant 4$ (we will use similar notations for other polynomials). Note that

$$f_x(x, y, z) = F_1(x, y, z, 1) \tag{3}$$

and similarly for the other derivatives. Therefore, by Euler's identity, we have

$$d \cdot f(x, y, z) = xF_1(x, y, z, 1) + yF_2(x, y, z, 1) + zF_3(x, y, z, 1) + F_4(x, y, z, 1). \tag{4}$$

By the generality assumption of the coordinates with respect to S, we have that:

- the plane at infinity is not *tangent* to S, i.e., it cuts out on S a curve whose pullback on X via π is smooth;
- each of the pencils $\mathcal{P}_i$ of planes with homogeneous equations $hx_i = kx_4$, with $(h, k) \in \mathbb{K} \setminus \{(0, 0)\}$, pulls back via π to a *Lefschetz pencil* $\mathcal{X}_i$ on X, with $1 \leqslant i \leqslant 3$;
- the pullback Γ_i on X of the curve γ_i cut out on S off the double curve Γ by the *polar* surfaces $F_i = 0$ is smooth for $1 \leqslant i \leqslant 3$.

Remark 1 By the genericity of the position of S with respect to the coordinate system, one sees that the curves Γ_i, for $1 \leqslant i \leqslant 3$, contain the pinch points scheme G_c and intersect pairwise transversely there.

The singular points of the finitely many curves in the pencil $\mathcal{X}_i$ are nodes and form a reduced zero-dimensional scheme $\mathcal{J}_i$ on X, which is called the *Jacobian scheme* of $\mathcal{X}_i$, for $1 \leqslant i \leqslant 3$. We will assume that, for all $i \in \{1, 2, 3\}$, the image J_i of this scheme on S, called the *Jacobian scheme* of $\mathcal{P}_i$, has no intersection with the double curve Γ.

It is also easy to check that the curve Γ_i cuts out on Γ_j the divisor $G_c + \mathcal{J}_k$, where $\{i, j, k\} = \{1, 2, 3\}$. So, in particular, taking into account that $\Gamma_i \in |3C + K_X|$, for $1 \leqslant i \leqslant 3$ (we denote by C a hyperplane section of X), one has

$$\mathcal{O}_{\Gamma_3}(G_c + \mathcal{J}_1) = \mathcal{O}_{\Gamma_3}(G_c + \mathcal{J}_2) = \mathcal{O}_{\Gamma_3}(3C + K_X), \tag{5}$$

hence

$$\mathcal{O}_{\Gamma_3}(\mathcal{J}_1) = \mathcal{O}_{\Gamma_3}(\mathcal{J}_2). \tag{6}$$

Similar relations hold on Γ_2 and Γ_3. Note that (5) implies that $|3C + K_X|$ has no fixed component and $(3C + K_X)^2 > 0$; hence, $3C + K_X$ is big and nef.

Let $e := e(X)$ be the *Euler–Poincaré characteristic* of X (i.e., the second *Chern class* of the tangent bundle of X) and g the arithmetic genus of the hyperplane sections of X. By the *Zeuthen–Segre formula* (see [10, p. 301]), the length δ of J_i is

$$\delta = e + 4(g-1) + d.$$

3.2 The Expression of 1-Forms on a Surface

It is a result by Picard (see [19, p. 116]; Picard works over $\mathbb{C}$, but it is easy to check that his argument works on any algebraically closed field $\mathbb{K}$) that if ω is a regular 1-form on X, then it is the pullback on X of a rational 1-form of the type

$$\frac{Ady - Bdx}{f_z} \tag{7}$$

where $A = 0$, $B = 0$ are affine equations of two *adjoint surfaces* of degree $d-2$ to S. Recall that a surface is said to be adjoint to S if it contains the double curve Γ of S.

In the 1-form (7), we can make a change of variables passing from x, y to x, z. From the relation

$$f_x dx + f_y dy + f_z dz = 0$$

we deduce

$$dy = -\frac{f_x dx + f_z dz}{f_y}.$$

Substituting into (7) we find

$$\frac{-\frac{Af_x + Bf_y}{f_z}dx - Adz}{f_y}$$

and this has to be of the same form as (7) with respect to the variables x, z. This implies that there must be a polynomial C of degree $d-2$ such that $C = 0$ is the affine equation of an adjoint surface to S, such that

$$-\frac{Af_x + Bf_y}{f_z} = C, \quad \text{modulo} \quad f = 0.$$

This yields the *Picard's relation*

$$Af_x + Bf_y + Cf_z = Nf \tag{8}$$

where N is a suitable polynomial of degree $d-3$. The Picard relation has some remarkable consequences, pointed out by Severi (see [22, §9]). Before stating Severi's result, we recall the following:

Lemma 2 (Castelnuovo's Lemma) *Let $g(x_1, x_2, x_3) = 0$ be the equation of an irreducible plane curve of degree n with no singular points except nodes. Then there is no non-trivial syzygy of degree $l \leqslant d-2$ of the triple (g_1, g_2, g_3) of derivatives of g.*

For the proof see [22, §7] or [14, p. 34]. Next we can prove Severi's result:

Proposition 3 *If A, B, C are non-zero polynomials verifying (8), then the (projective closure of the) surface with equation $A = 0$ [resp. $B = 0$, $C = 0$] contains the base line of the pencil of planes $\mathcal{P}_1$ [resp. of $\mathcal{P}_2$, of $\mathcal{P}_3$] and also the Jacobian scheme J_1 [resp. J_2, J_3] of this pencil. Moreover the (projective closures of the) surfaces $A = 0, B = 0, C = 0$ cut out on the plane at infinity the same curve off the aforementioned lines.*

Proof First we prove that the surface with equation $A = 0$ contains the scheme J_1. Let P be a point of J_1. Then f, f_y, f_z vanish at P. Hence by (8), also Af_x vanishes at P. However f_x does not vanish at P because P does not belong to the double curve Γ of S. Hence A vanishes at P. Similarly for the surface with equation $B = 0$ [resp. $C = 0$] containing the scheme J_2 [resp. J_3].

Next, homogenize (8). By (3) (and the similar for the other derivatives), we get a relation of the form

$$\bar{A}F_1 + \bar{B}F_2 + \bar{C}F_3 = \bar{N}F$$

where we denote by the bars the homogenization of the corresponding polynomials. By taking into account the Euler identity, this relation takes the form

$$(d\bar{A} - x_1\bar{N})F_1 + (d\bar{B} - x_2\bar{N})F_2 + (d\bar{C} - x_3\bar{N})F_3 = x_4\bar{N}F_4.$$

Setting $x_4 = 0$ and taking into account Castelnuovo's Lemma 2, we have identically

$$d\bar{A} - x_1\bar{N} \equiv 0, \quad d\bar{B} - x_2\bar{N} \equiv 0, \quad d\bar{C} - x_3\bar{N} \equiv 0$$

under the condition $x_4 = 0$. This implies that

$$dA_0 - x_1N_0 \equiv 0, \quad dB_0 - x_2N_0 \equiv 0, \quad dC_0 - x_3N_0 \equiv 0$$

where A_0, B_0, C_0 are the homogeneous components of A, B, C in degree $d-2$ and N_0 is the homogeneous component of N of degree $d-3$. The assertion follows right away. □

Remark 4 Note that the surfaces with equations $A = 0, B = 0, C = 0$ in Proposition 3 are not necessarily adjoint. Keeping the notation of the proof of

Proposition 3, set $N_0 = \vartheta$. Then we have identities of the form

$$A = x\vartheta + A_1, \quad B = y\vartheta + B_1, \quad C = z\vartheta + C_1, \quad N = d\vartheta + N_1 \tag{9}$$

where A_1, B_1, C_1 are (non-homogeneous) polynomials of degree at most $d-3$ and N_1 has degree at most $d-4$.

Severi next proved the following proposition (see [22, §9]):

Proposition 5 *Let $A = 0$ be the affine equation of an adjoint surface of degree $d-2$ to S containing the scheme J_1. Then there are uniquely determined adjoint surfaces of degree $d-2$ to S with affine equations $B = 0$ and $C = 0$, containing the schemes J_2 and J_3, respectively, such that (8) holds. Each of the polynomials A, B, C uniquely determines the other two.*

Proof Consider A as in the statement. The complete linear system $|2C + K_X|$ is the pullback to X of the curves cut out on S, off the double curve Γ, by the adjoint surfaces of degree $d-2$. Looking at the exact sequence

$$0 \longrightarrow \mathcal{O}_X(-C) \longrightarrow \mathcal{O}_X(2C + K_X) \longrightarrow \mathcal{O}_{\Gamma_3}(2C + K_X) \longrightarrow 0$$

we see that $|2C + K_X|$ cuts out on Γ_3 a complete linear series ξ, because $h^1(X, \mathcal{O}_X(-C)) = 0$ (by the Kodaira vanishing theorem, see [11, p. 154]). Moreover, since $h^0(X, \mathcal{O}_X(-C)) = 0$, the restriction map

$$H^0(X, \mathcal{O}_X(2C + K_X)) \longrightarrow H^0(\Gamma_3, \mathcal{O}_{\Gamma_3}(2C + K_X))$$

is injective.

Let us abuse notation and denote by $A \in |2C + K_X|$ the pullback on X of the curve cut out on S by the (projective closure of the) surface $A = 0$ off Γ. Then A cuts out on Γ_3 a divisor of the form $\mathcal{J}_1 + Z \in \xi$. Since $\mathcal{J}_2 + Z \in \xi$ by (6), there is a unique curve $B \in |2C + K_X|$ that cuts out $\mathcal{J}_2 + Z$ on Γ_3. By abusing notation, we denote by B a non-zero polynomial, uniquely defined up to a constant, such that $B = 0$ is the adjoint surface cutting out on S off Γ the curve whose pullback on X is B. The surfaces Af_x and Bf_y cut out on the curve γ_3 the same divisor; hence, there is a non-zero constant b such that $Af_x - bBf_y = 0$ on γ_3. By substituting B with $-bB$, we may assume that $Af_x + Bf_y = 0$ on γ_3.

Consider now the complete intersection scheme Y, whose ideal is generated by f and f_z, which consists of two components given by γ_3 and by Γ with a double structure. Since $Af_x + Bf_y$ vanishes on γ_3 and vanishes with multiplicity 2 on Γ, then $Af_x + Bf_y$ vanishes on Y, and therefore $Af_x + Bf_y$ is a combination of f and f_z, i.e., there are polynomials C and N, of degrees $d-2$ and $d-3$, respectively, such that (8) holds. Note that C cannot be identically zero. Otherwise we would have an identity of the sort

$$Af_x + Bf_y = Nf.$$

This is impossible, because then Bf_y would vanish along the curve γ_1, but neither f_y nor B can vanish along this curve. Since Af_x, Bf_y, and f vanish doubly along Γ, then C vanishes along Γ so that $C = 0$ is adjoint to S. Moreover C is uniquely determined. In fact, from another identity of the form

$$Af_x + Bf_y + C' f_z = N' f,$$

subtracting memberwise from (8), we deduce

$$(C - C') f_z = (N - N') f$$

and f would divide the left-hand side, which is impossible because both factors there have degree smaller than f. The assertion follows. □

By taking into account Proposition 3, one has the:

Corollary 6 *Every adjoint surface to S of degree $d - 2$ containing the scheme J_1 contains also the base line of the pencil $\mathcal{P}_1$.*

We can state this corollary in an intrinsic form:

Corollary 7 *Let X be a smooth, irreducible, projective surface, C a very ample effective divisor on X, and $\mathcal{P}$ a Lefschetz pencil in $|C|$. Then any curve in $|2C + K_X|$ containing the* Jacobian scheme *of the pencil $\mathcal{P}$ (i.e., the scheme of double points of the singular curves in $\mathcal{P}$) also contains the* base locus scheme *of $\mathcal{P}$.*

Now, given an adjoint surface of degree $d-2$ to S containing the scheme J_1, with affine equation $A = 0$, consider the other two adjoint surfaces $B = 0$ and $C = 0$ existing by Proposition 5. We can consider the three regular 1-form pullbacks on X of the forms

$$\frac{Ady - Bdx}{f_z}, \quad \frac{Bdz - Cdy}{f_x}, \quad \frac{Cdx - Adz}{f_y}.$$

By the very proof of Proposition 3, we see that these forms are equal. In conclusion, if we consider the vector space $\mathrm{Adj}_{d-2}(S)$ of (non-homogeneous) polynomials of degree (at most) $d - 2$ defining adjoint surfaces to S passing through J_1, this determines an isomorphism

$$\varphi : \mathrm{Adj}_{d-2}(S) \to H^0(X, \Omega^1_X). \tag{10}$$

The map φ sends a polynomial A to the 1-form pullback of the form (7) to X, where $B = 0$ is the adjoint surface of degree $d - 2$ described in Proposition 5. The same by exchanging J_1 with J_2 or J_3.

3.3 *Closedness of 1-Forms*

The following result is well-known:

Proposition 8 *If X is a complex, smooth, compact surface, any regular 1-form on X is closed.*

Proof This proof is extracted from [2, p. 137–138].

Let $\omega \in H^0(X, \Omega^1_X)$ be a non-zero regular form. By Stokes' Theorem one has

$$\int_X d\omega \wedge d\bar{\omega} = \int_X d(\omega \wedge d\bar{\omega}) = 0. \tag{11}$$

Write down locally $d\omega = f dz_1 \wedge dz_2$. Then

$$d\omega \wedge d\bar{\omega} = -|f|^2 dz_1 \wedge d\bar{z}_1 \wedge dz_2 \wedge d\bar{z}_2 = 4|f|^2 dx_1 \wedge dy_1 \wedge dx_2 \wedge dy_2$$

where $z_j = x_j + iy_j$, for $1 \leqslant j \leqslant 2$, so that by (11) one gets $f = 0$, i.e., $d\omega = 0$. □

The proof of this proposition is analytic and does not hold in positive characteristic. In fact in positive characteristic there are counterexamples to Proposition 8 (see [16, Corollary]). There is then the problem, which was classically well-known (see [24, p. 185]) and considered also in the two letters by Castelnuovo, of finding a purely algebraic proof of Proposition 8. It is useful for us to review the classical viewpoint on this subject.

Let us keep the notation introduced above. Let ω be a regular 1-form on the surface X, which is the pullback on X of the rational 1-form (7). Then we have $d\omega = \phi dx \wedge dy$, with

$$\phi = \frac{\partial}{\partial x}\left(\frac{A}{f_z}\right) + \frac{\partial}{\partial y}\left(\frac{B}{f_z}\right)$$

where it is intended that the differentiations take place on the surface X, so that z is function of x, y implicitly defined by $f(x, y, z) = 0$. So, for instance,

$$\frac{\partial z}{\partial x} = -\frac{f_x}{f_z}$$

and

$$\begin{aligned}\frac{\partial}{\partial x}\left(\frac{A}{f_z}\right) &= \frac{\left(A_x + A_z \frac{\partial z}{\partial x}\right) f_z - A\left(f_{zx} + f_{zz}\frac{\partial z}{\partial x}\right)}{f_z^2} = \\ &= \frac{\left(A_x - A_z \frac{f_x}{f_z}\right) f_z - A\left(f_{zx} - f_{zz}\frac{f_x}{f_z}\right)}{f_z^2} = \\ &= \frac{f_z^2 A_x - f_z(A f_{zx} + A_z f_x) + f_{zz} A f_x}{f_z^3}\end{aligned}$$

and similarly

$$\frac{\partial}{\partial y}\left(\frac{B}{f_z}\right) = \frac{f_z^2 B_y - f_z(Bf_{zy} + B_z f_y) + f_{zz}Bf_y}{f_z^3}$$

so that

$$\phi = \frac{f_z^2(A_x + B_y) - f_z(Af_{zx} + A_z f_x + Bf_{zy} + B_z f_y) + f_{zz}(Af_x + Bf_y)}{f_z^3}. \tag{12}$$

Taking into account (8) and the identity

$$\frac{\partial(Af_x + Bf_y)}{\partial z} = A_z f_x + Af_{xz} + B_z f_y + Bf_{yz},$$

Equation (12) becomes

$$\phi = \frac{1}{f_z^3}\Big[f_z^2(A_x + B_y + C_z - N) + f(Nf_{zz} - f_z N_z)\Big]$$

so that

$$\phi = \frac{A_x + B_y + C_z - N}{f_z}, \quad \text{modulo} \quad f$$

and this is regular on X. Hence if we set

$$Q = A_x + B_y + C_z - N$$

the polynomial Q has to vanish on the double curve Γ of S, because it has to vanish where f_z vanishes.

A priori Q is a polynomial of degree $d - 3$, but one has actually:

Lemma 9 *In the above setting, Q has degree $d - 4$.*

Proof By taking into account the identities (9) in Remark 4, we have

$$A_x = \theta + x\theta_x + \frac{\partial A_1}{\partial x}, \quad B_y = \theta + y\theta_y + \frac{\partial B_1}{\partial y}, \quad C_z = \theta + z\theta_z + \frac{\partial C_1}{\partial z}$$

where θ is a homogeneous polynomial of degree $d - 3$. Hence, by Euler's identity, we get

$$\begin{aligned} A_x + B_y + C_z - N &= d\theta + \tfrac{\partial A_1}{\partial x} + \tfrac{\partial B_1}{\partial y} + \tfrac{\partial C_1}{\partial z} - (d\theta + N_1) = \\ &= \tfrac{\partial A_1}{\partial x} + \tfrac{\partial B_1}{\partial y} + \tfrac{\partial C_1}{\partial z} - N_1 \end{aligned} \tag{13}$$

which proves the assertion. □

In conclusion, we have

$$\frac{\partial}{\partial x}\left(\frac{A}{f_z}\right)+\frac{\partial}{\partial y}\left(\frac{B}{f_z}\right)=\frac{Q}{f_z}$$

and with similar computations, one finds

$$\frac{\partial}{\partial y}\left(\frac{B}{f_x}\right)+\frac{\partial}{\partial z}\left(\frac{C}{f_x}\right)=\frac{Q}{f_x},\quad \frac{\partial}{\partial z}\left(\frac{C}{f_y}\right)+\frac{\partial}{\partial x}\left(\frac{A}{f_y}\right)=\frac{Q}{f_y}.$$

In any event, the form ω as above is closed if and only if $Q = 0$ modulo f. But, since Q has degree smaller than d, this is the case if and only if Q is identically zero. So, taking into account (13), the problem of giving an algebraic proof of Proposition 8 translates in the following:

Problem 10 Find an algebraic proof that (8) implies either one of the two equivalent relations

$$N = A_x + B_y + C_z,\quad N_1 = \frac{\partial A_1}{\partial x}+\frac{\partial B_1}{\partial y}+\frac{\partial C_1}{\partial z} \tag{14}$$

each of which is called the *integrability condition.*

We want to stress that any solution of Problem 10 must use the fact that the base field $\mathbb{K}$ has characteristic zero.

3.4 *Homogeneous Form of Picard's Relation*

It is useful to describe the homogeneous form of Picard's relation (8). This is contained in [19, p. 119], and we expose this here for the reader's convenience.

By (4), we can rewrite (8) as

$$\begin{aligned} d\cdot AF_1(x,y,z,1)+d\cdot BF_2(x,y,z,1)+d\cdot CF_3(x,y,z,1) &= \\ = N(xF_1(x,y,z,1)+yF_2(x,y,z,1)+zF_3(x,y,z,1)+F_4(x,y,z,1))& \end{aligned}$$

Set

$$X_1=\overline{dA-xN},\quad X_2=\overline{dB-yN},\quad X_3=\overline{dC-zN},\quad X_4=-\bar{N}$$

where, as usual, the bars stay for homogenization. By (9), we have

$$X_1=\overline{dA_1-xN_1},\quad X_2=\overline{dB_1-yN_1},\quad X_3=\overline{dC_1-zN_1},\quad X_4=-\bar{N}$$

and the polynomials X_i, with $1 \leqslant i \leqslant 4$, are of degree $d-3$. Then we have the relation

$$X_1F_1 + X_2F_2 + X_3F_3 + X_4F_4 = 0 \tag{15}$$

which is the *homogeneous Picard's relation*. If we consider the matrix

$$M = \begin{pmatrix} X_1 & X_2 & X_3 & X_4 \\ x_1 & x_2 & x_3 & x_4 \end{pmatrix}, \tag{16}$$

all minors of order 2 of M, after dehomogenization, are linear combinations of A, B, C and so are in $\mathrm{Adj}_{d-2}(S)$.

Suppose the homogeneous Picard's relation (15) holds. Taking into account the expressions of the polynomials X_i, for $1 \leqslant i \leqslant 4$, and the relations (9), the integrability relation in the form of the right-hand side of (14) becomes

$$\frac{\partial X_1}{\partial x_1} + \frac{\partial X_2}{\partial x_2} + \frac{\partial X_3}{\partial x_3} + \frac{\partial X_4}{\partial x_4} = 0 \tag{17}$$

which is the *homogeneous integrability condition*. Problem 10 can now be expressed in homogeneous form as:

Problem 11 Find an algebraic proof that (15) (with all minors of order 2 of the matrix M in (16), after dehomogenization, in $\mathrm{Adj}_{d-2}(S)$) implies the homogeneous integrability condition (17).

4 Comments on Castelnuovo's Letters

This section is devoted to explaining most of the issues raised by Castelnuovo in his two letters. Both letters focus on the understanding of the algebro-geometric meaning of the analytic irregularity q_{an} and on solving Problems 10 or 11.

In §§1 and 2 of the first letter, Castelnuovo suggests, with no proofs, various geometric interpretations of q_{an}. Analogous remarks have been partially included by Castelnuovo in the paper [4] published 2 years after this letter. Castelnuovo does not say it, but maybe he had in mind in the letter that the various geometric interpretations of q_{an} could have been useful to algebro-geometrically prove the equality between q_{an} and $q_a := h^1(X, \mathcal{O}_X)$ that we will call the *arithmetic irregularity*, an equality that Castelnuovo proved with analytic methods in the paper [3] of 40 years before (a different proof was given by Severi in [21]; see also [6]). Note that this equality does not hold in positive characteristic, as proved by Igusa in [12] (see also [17]).

Let us keep the notation introduced so far. The first result Castelnuovo states in his letter to Severi is the following:

Proposition 12 *Let $|C|$ be a very ample linear system on X. Then*

$$h^1(X, \mathcal{O}_X(4C + 2K_X) \otimes \mathcal{I}_{G_c|X}) = q_{\text{an}}.$$

Proof In Remark 1 we saw that $3C + K_X$ is big and nef. This implies that also $4C + K_X$ is big and nef.

Look at the exact sequence

$$0 \longrightarrow \mathcal{O}_X(C + K_X) \longrightarrow \mathcal{O}_X(4C + 2K_X) \longrightarrow \mathcal{O}_{\Gamma_3}(4C + 2K_X) \longrightarrow 0.$$

We have $h^i(X, \mathcal{O}_X(C + K_X)) = 0$ for $1 \leqslant i \leqslant 2$ (by the Kodaira vanishing theorem), and $h^1(X, \mathcal{O}_X(4C + 2K_X)) = 0$, because $4C + K_X$ is big and nef (by Mumford's theorem, see [18, §II]). This implies that $|4C + 2K_X|$ cuts out on Γ_3 a complete, non-special linear series g^r_n, where

$$r = n - p_a(\Gamma_3)$$

(recall the definition of the curves Γ_i, $i = 1, 2, 3$, from the beginning of Sect. 3.1).

Set now

$$h^1(X, \mathcal{O}_X(4C + 2K_X) \otimes \mathcal{I}_{G_c|X}) = h.$$

The linear system $|\mathcal{O}_X(4C + 2K_X) \otimes \mathcal{I}_{G_c|X}|$ cuts out on Γ_3, off G_c, a complete linear series $\xi = g^{r-\gamma+h}_{n-\gamma}$ (recall that $\gamma = \text{length}(G_c)$), so that h is the index of speciality of ξ.

Let G be a general divisor of ξ, so that

$$\mathcal{O}_{\Gamma_3}(G + G_c) = \mathcal{O}_{\Gamma_3}(4C + 2K_X).$$

Let G' be a divisor on Γ_3 such that

$$\mathcal{O}_{\Gamma_3}(G' + \mathcal{J}_1) = \mathcal{O}_{\Gamma_3}(2C + K_X).$$

Adding up these two relations and subtracting (5), we get

$$\mathcal{O}_{\Gamma_3}(G + G') = \mathcal{O}_{\Gamma_3}(3C + 2K_X) = \omega_{\Gamma_3}.$$

So we get

$$h = h^0(\Gamma_3, \mathcal{O}_{\Gamma_3}(G')).$$

On the other hand, by looking at the exact sequence

$$0 \longrightarrow \mathcal{O}_X(-C) \longrightarrow \mathcal{O}_X(2C+K_X) \longrightarrow \mathcal{O}_{\Gamma_3}(2C+K_X)) \longrightarrow 0,$$

since $h^1(X, \mathcal{O}_X(-C)) = 0$ (by the Kodaira vanishing theorem), we see that $|2C + K_X|$ cuts out on Γ_3 a complete linear series; hence

$$h = h^0(\Gamma_3, \mathcal{O}_{\Gamma_3}(G')) = h^0(X, \mathcal{O}_X(2C+K_X) \otimes \mathcal{I}_{\mathcal{J}_1|X})$$

and the assertion follows by the isomorphism φ in (10). □

After this Castelnuovo claims that

$$h^1(X, \mathcal{O}_X(nC+mK_X) \otimes \mathcal{I}_{G_c|X}) = 0$$

for $n \geqslant 5$ and $m \geqslant 1$. We have not been able to prove (or disprove) this assertion.

Another geometric interpretation of $q_{\rm an}$ that Castelnuovo suggests in the letter to Severi is the following: let D be a curve in $|4C + K_X|$ that contains G_c and it is smooth there; then

$$h^0(D, \mathcal{O}_D(G_c)) = q_{\rm an} + 1. \tag{18}$$

Also for this statement, we could not come up with a proof (or a counterexample).

Remark 13 It looks rather difficult that (18) could hold. In fact, consider again the curve Γ_3. Then D cuts out on Γ_3 a divisor $G_c + G$, where, by (5), one has

$$\mathcal{O}_{\Gamma_3}(G) = \mathcal{O}_{\Gamma_3}(\mathcal{J}_1 + H) \tag{19}$$

where H is a divisor cut out on Γ_3 by a hyperplane. By looking at the exact sequence

$$0 \longrightarrow \mathcal{O}_X(-C) \longrightarrow \mathcal{O}_X(3C+K_X) \longrightarrow \mathcal{O}_D(3C+K_X) \longrightarrow 0$$

and since $h^1(X, \mathcal{O}_X(-C)) = 0$, we see that $|3C + K_X|$ cuts out on D a complete linear series. Hence the linear series $|\mathcal{O}_D(G_c)|$ is cut out on D, off G, by the linear system $|\mathcal{O}_X(3C+K_X) \otimes \mathcal{I}_{G|X}|$, and therefore

$$h^0(D, \mathcal{O}_D(G_c)) = h^0(X, \mathcal{O}_X(3C+K_X) \otimes \mathcal{I}_{G|X}). \tag{20}$$

From the exact sequence

$$0 \longrightarrow \mathcal{O}_X \longrightarrow \mathcal{O}_X(3C+K_X) \otimes \mathcal{I}_{G|X} \longrightarrow \mathcal{O}_{\Gamma_3}(3C+K_X) \otimes \mathcal{I}_{G|X} \longrightarrow 0$$

we have

$$h^0(X, \mathcal{O}_X(3C+K_X) \otimes \mathcal{I}_{G|X}) \leqslant h^0(\Gamma_3, \mathcal{O}_{\Gamma_3}(3C+K_X) \otimes \mathcal{I}_{G|X}) + 1. \tag{21}$$

By (19), we have

$$\mathcal{O}_{\Gamma_3}(3C + K_X) \otimes \mathcal{I}_{G|X} = \mathcal{O}_{\Gamma_3}(2C + K_X) \otimes \mathcal{I}_{\mathcal{J}_1|X}.$$

Since, as we saw in the proof of Proposition 12, $|2C+K_X|$ cuts out on Γ_3 a complete linear series, we have

$$h^0(\Gamma_3, \mathcal{O}_{\Gamma_3}(2C + K_X) \otimes \mathcal{I}_{\mathcal{J}_1|X}) = h^0(X, \mathcal{O}_X(2C + K_X) \otimes \mathcal{I}_{\mathcal{J}_1|X}) = q_{\mathrm{an}}.$$

Putting together this, (20) and (21), one gets

$$h^0(D, \mathcal{O}_D(G_c)) \leqslant q_{\mathrm{an}} + 1.$$

Now the equality holds if and only if the restriction map

$$H^0(X, \mathcal{O}_X(3C + K_X) \otimes \mathcal{I}_{G|X}) \longrightarrow H^0(\Gamma_3, \mathcal{O}_{\Gamma_3}(3C + K_X) \otimes \mathcal{I}_{G|X})$$

is surjective. This looks difficult because the map

$$H^0(X, \mathcal{O}_X(3C + K_X)) \longrightarrow H^0(\Gamma_3, \mathcal{O}_{\Gamma_3}(3C + K_X))$$

is not surjective (it has corank q_a).

At the end of the first section of his letter to Severi, Castelnuovo claims that: if D is a curve in $|3C + K_X|$ containing G_c and smooth there, then

$$h^0(D, \mathcal{O}_D(G_c)) = 2d - g + 2p_g + \chi - e - 1 + \theta$$

with $0 \leqslant \theta \leqslant q_a$, and, as usual, $\chi = \chi(\mathcal{O}_X)$ and $p_g = h^0(X, \mathcal{O}_X(K_X))$. It is easy to check that this is equivalent to

$$2p_g + \chi - 1 \leqslant h^1(D, \mathcal{O}_D(G_c)) \leqslant 3p_g.$$

However we have not been able to prove this.

In the second section of the letter to Severi, Castelnuovo states the:

Proposition 14 *Let $\mathcal{P}$ be a Lefschetz pencil in $|C|$, with Jacobian scheme $\mathcal{J}$. Then*

$$h^1(X, \mathcal{O}_X(4C + K_X) \otimes \mathcal{I}_{G_c|X}) = h^0(X, \mathcal{O}_X(2C + 2K_X) \otimes \mathcal{I}_{\mathcal{J}|X}).$$

The proof of this, not so different from the one of Proposition 12, is contained in [4, §3], and we do not reproduce it here. Let B be the base locus scheme of the Lefschetz pencil $\mathcal{P}$. Castelnuovo compares $h^0(X, \mathcal{O}_X(2C + 2K_X) \otimes \mathcal{I}_{\mathcal{J}|X})$ with $h^0(X, \mathcal{O}_X(2C + 2K_X) \otimes \mathcal{I}_{\mathcal{J}+B|X})$. This does not look particularly interesting, and we do not dwell on it here. Castelnuovo also claims that if $\mathcal{P}$ is a Lefschetz pencil in $|C|$ with Jacobian scheme $\mathcal{J}$, then

$$h^1(X, \mathcal{O}_X(3C + K_X) \otimes \mathcal{I}_{\mathcal{J}|X}) = 2p_g$$

but we have not been able to prove it.

Let us jump for a moment to section 4 of the letter to Severi. In this part, as well as in the paper [4], Castelnuovo takes for granted the existence of the so-called Severi equivalence series. Severi claimed in [23] that, unless the surface has an irrational pencil, there exists the rational equivalence series, of dimension $q_{\rm an}-1$, of the zero-dimensional schemes of length e that are zeros of non-zero 1-forms in $H^0(X,\Omega^1_X)$. In addition Severi claimed that Ω^1_X is generated by global sections. These claims are false in general, as shown by F. Catanese in [5, §6]. Hence the contents of section 4 of the letter and of the paper [4] have biases because of this.

Let us now go back to section 3 of the letter to Severi. The focus of this section is on Problems 10 or 11. Castelnuovo proposes a few equivalent formulations of these problems, the most interesting of which, in our opinion, is the following, which is also the topic of the letter to B. Segre.

Problem 15 Suppose there is a (homogeneous) relation of the form

$$Y_1F_1+Y_2F_2+Y_3F_3+Y_4F_4=QF \tag{22}$$

where $Y_i=0$ are adjoint surfaces of degree $d-3$ to S and $Q=0$ is a surface of degree $d-4$. Prove (algebraically) that Q is an adjoint surface and that

$$Y_i=\frac{1}{d}x_iQ.$$

The solution of this problem implies the solution of Problem 11. Indeed, we can rewrite (22) as

$$\sum_{i=1}^{4}\left(Y_i-\frac{1}{d}Qx_i\right)F_i=0$$

and this is a homogeneous Picard's relation of the type (15), with

$$X_i=Y_i-\frac{1}{d}Qx_i,\quad 1\leqslant i\leqslant 4.$$

Problem 11 asks to prove that (17) holds, whereas Problem 15 asks to prove much more, i.e., that $X_i=0$, for $1\leqslant i\leqslant 4$. So Problem 15 does not look equivalent to Problem 11, and it is not at all clear if it has a solution or not.

5 Algebraic Proofs via the Hodge–Frölicher Spectral Sequence

This section is devoted to giving a brief account on how algebraic proofs of the closedness of regular 1-forms and of the equality between algebraic and analytic irregularity (both in characteristic zero) can be obtained using modern tools.

5.1 *Global Regular 1-Forms Are Closed in Characteristic Zero*

Let X be a smooth, irreducible, and projective variety of arbitrary dimension over an algebraically closed field $\mathbb{K}$. Let Ω^i_X be the sheaf of algebraic differential i-forms on X. The exterior derivative $d : \Omega^i_X \to \Omega^{i+1}_X$ allows to define a complex (the so-called algebraic de Rham complex) and a spectral sequence (the so-called Hodge–Frölicher spectral sequence):

$$E_1 = \bigoplus_{i,j \geqslant 0} E_1^{i,j}$$

where

$$E_1^{i,j} := H^j(X, \Omega^i_X)$$

and

$$d_1 : E_1^{i,j} \to E_1^{i+1,j}$$

is given by

$$d : H^j(X, \Omega^i_X) \to H^j(X, \Omega^{i+1}_X).$$

If this spectral sequence degenerates at E_1, then in particular we have $E_2^{1,0} = E_1^{1,0}$, i.e.,

$$\frac{\mathrm{Ker}(H^0(X, \Omega^1_X) \to H^0(X, \Omega^2_X))}{\mathrm{Im}(H^0(X, \mathcal{O}_X) \to H^0(X, \Omega^1_X))} = \mathrm{Ker}(H^0(X, \Omega^1_X) \to H^0(X, \Omega^2_X)) = H^0(X, \Omega^1_X).$$

Hence we see that if the Hodge–Frölicher spectral sequence degenerates at E_1, then all global regular 1-forms are closed.

An algebraic proof of the degeneration at E_1 of the Hodge–Frölicher spectral sequence in characteristic zero has been obtained by Deligne and Illusie in the paper [7] published in 1987 (see also [9] and [13] for more detailed and self-contained expositions). The strategy involves two steps: first, the result is proven under suitable assumptions in positive characteristic; then, by applying standard "spreading out" techniques, it is extended to characteristic zero.

Theorem 16 ([13], Corollary 5.6) *Let k be a perfect field of characteristic p, and let X be a smooth and proper k-scheme of dimension $< p$. If X satisfies a technical assumption (namely, X can be lifted over the ring $W_2(k)$ of Witt vectors of length 2 over k), then the Hodge–Frölicher spectral sequence of X over k degenerates at E_1.*

Corollary 17 ([13], Theorem 6.9) *Let $\mathbb{K}$ be a field of characteristic zero, and let X be a smooth and proper $\mathbb{K}$-scheme of arbitrary dimension. Then the Hodge–Frölicher spectral sequence of X over $\mathbb{K}$ degenerates at E_1.*

For a friendly introduction to this circle of ideas, we refer the interested reader to the informal survey [15] (see also [17], which explains the role of Witt vectors in studying the irregularity in positive characteristic). Unluckily, it seems that in order to address the case of surfaces, one needs to apply the whole machinery developed for the general case.

5.2 Analytic Irregularity and Arithmetic Irregularity Coincide

Let X be a smooth and projective surface over the complex field $\mathbb{C}$. As already realized (at least implicitly) by Castelnuovo, the fact that the *analytic irregularity* $q_{\mathrm{an}}(X) = h^0(X, \Omega^1_X)$ is equal to the *arithmetic irregularity* $q_a(X) = h^1(X, \mathcal{O}_X)$ (which holds in general only in characteristic zero) is strictly related to the closedness of global regular 1-forms.

A crucial additional ingredient for proving algebraically that $q_{\mathrm{an}}(X) = q_a(X)$ is the following equality, which admits a purely algebraic proof (see, for instance, [24, Mumford's remarks i) and iii) on p. 200], and [1, Theorem 5.1]):

$$h^1(X, \mathbb{C}) = 2h^1(X, \mathcal{O}_X) = 2q_a(X). \tag{23}$$

As in [2], proof of Lemma (2.6) on p. 139, there is a natural exact sequence

$$0 \to \mathbb{C} \to \mathcal{O}_X \to \mathcal{S} \to 0,$$

where $\mathcal{S}$ denotes the sheaf of closed regular 1-forms on X. Since all global regular 1-forms are closed, we get an exact sequence

$$0 \to H^0(X, \Omega^1_X) \to H^1(X, \mathbb{C}) \to H^1(X, \mathcal{O}_X).$$

It follows that $h^1(X, \mathbb{C}) \leqslant h^0(X, \Omega^1_X) + h^1(X, \mathcal{O}_X)$ and together with (23) we may deduce

$$h^1(X, \mathcal{O}_X) \leqslant h^0(X, \Omega^1_X).$$

On the other hand, the opposite inequality turns out to be much subtler and seems to require the full strength of the Hodge–Frölicher spectral sequence. Indeed, if one defines the *algebraic de Rham cohomology* $H^*_{\mathrm{dR}}(X/\mathbb{K})$ as the hypercohomology of the algebraic de Rham complex, then the equality

$$\dim(H^1_{\mathrm{dR}}(X/\mathbb{K})) = q_{\mathrm{an}}(X) + q_a(X)$$

is a formal consequence of the degeneration at E_1 of the Hodge–Frölicher spectral sequence (see for instance [15], Lemma 3.4). In particular, for $\mathbb{K} = \mathbb{C}$ we have

$$q_{\mathrm{an}}(X) + q_a(X) = \dim(H^1_{\mathrm{dR}}(X/\mathbb{K})) = h^1(X, \mathbb{C}) = 2q_a(X)$$

by (23); hence, we obtain $q_{\mathrm{an}}(X) = q_a(X)$.

Acknowledgments The authors are members of GNSAGA of the Istituto Nazionale di Alta Matematica "F. Severi." This research project was partially supported by PRIN 2017 "Moduli Theory and Birational Classification."

References

1. Badescu, L.: Algebraic Surfaces. Springer, New York (2001)
2. Barth, W., Hulek, K., Peters, C., van de Ven, A.: Compact Complex Surfaces, 2nd Enlarged Edition, Ergebnisse der Mathematik und ihrer Grenzgebiete, vol. 4. Springer, Berlin (2003)
3. Castelnuovo, G.: Sugli integrali semplici appartenenti ad una superficie irregolare. Rend. R. Accad. Lincei, V **14**, 545–556, 593–598, 655–663 (1905)
4. Castelnuovo, G.: Sul numero dei moduli di una superficie irregolare. Rend. della R. Acc. Nazionale dei Lincei **7** (8), 3–7 and 8–11 (1949)
5. Catanese, F.: On the moduli space of surfaces of general type. J. Diff. Geom. **19**, 483–515 (1984)
6. Ciliberto, C.: The theorem of completeness of the characteristic series: Enriques' contribution. In: G. Bini (ed.) Algebraic Geometry between Tradition and Future: An Italian Perspective. Springer INdAM Series, vol. 53. Springer Nature Singapore, Singapore (2023)
7. Deligne, P., Illusie, L.: Relèvements modulo p^2 et décomposition du complexe de de Rham. Invent. Math. **89**(2), 247–270 (1987)
8. Enriques, F.: Le Superficie Algebriche. Nicola Zanichelli, Bologna (1949)
9. Esnault, H., Viehweg, E.: Lectures on Vanishing Theorems. DMV Seminar, vol. 20. Birkhäuser Verlag, Basel (1992)
10. Fulton, W.: Intersection Theory, Ergebnisse der Mathematik und ihrer Grenzgebiete, vol. 2. Springer, Berlin (1984)
11. Griffiths, Ph., Harris, J.: Principles of Algebraic Geometry. Wiley, Hoboken (2014)
12. Igusa, J.-I.: A fundamental inequality in the theory of Picard varieties. Proc. Nat. Acad. Sci. U.S.A. **41**, 317–320 (1955)
13. Illusie, L.: Frobenius et dégénérescence de Hodge. Introduction à la théorie de Hodge, pp. 113–168. Panor. Synthèses, 3, Soc. Math. France, Paris (1996)
14. Kawahara, Y.: On the differential forms on algebraic surfaces. Nagoya Math. J. **4**, 73–78 (1952)
15. Martin, I.: Algebraic de Rham cohomology and the Hodge spectral sequence (2020). http://math.uchicago.edu/~may/REU2020/REUPapers/Martin.pdf.

16. Mumford, D.: Pathologies of modular algebraic geometry. Amer. J. Math. **83**(2), 339–342 (1961)
17. Mumford, D.: Lectures on Curves on an Algebraic Surface. Annals of Mathematics Studies, vol. 59. Princeton University Press, Princeton (1966)
18. Mumford, D.: Some Footnotes to the Work of C.P. Ramanujam, Ramanujam, C.P., A tribute. Springer, Berlin (1978)
19. Picard, E., Simart, G.: Théorie des Fonctions Algébriques de deux variables indépentandes, vol. I. Gauthier-Villars et Fils, Paris (1897)
20. Roberts, J.: Generic projections of algebraic varieties. Amer. J. Math. **93**(1), 191–214 (1971)
21. Severi, F.: Il teorema d'Abel sulle superficie algebriche. Annali di Mat. **12**(3), 55–79 (1905)
22. Severi, F.: Sugl'integrali algebrici semplici e doppi, (4 papers). Rend. della R. Acc. Nazionale dei Lincei **7**(6), 3–8, 9–14, 101–108, 161–169 (1928)
23. Severi, F.: La serie canonica e la teoria delle serie principali di gruppi di punti sopra una superficie algebrica. Comment. Math. Helv. **4**, 268–326 (1932)
24. Zariski, O.: Algebraic Surfaces, 2nd Supplemented Edition, Ergebnisse der Mathematik und ihrer Grenzgebiete, vol. 61. Springer, Berlin (1971)

Guido Castelnuovo and His Heritage: Geometry, Combinatorics, and Teaching

Claudio Fontanari

Abstract We approach Guido Castelnuovo's intellectual world by focusing on a trilogy of papers published in 1889 and by drawing a few remarks about Castelnuovo's scientific interests and attitudes.

Keywords Guido Castelnuovo · Algebraic geometry · Enumerative combinatorics · Teaching of mathematics

1 Introduction

For an account of the life and works of Guido Castelnuovo (1865–1952), we refer to [2] and [9], which also provide a fascinating insight into his familiar milieu, and to the biographical sketch [7]. Here instead we are going to focus our attention on three early papers, published by Castelnuovo in 1889, which are specifically devoted to enumerative geometry and algebraic curves, but shed light more generally on his intellectual world.

Indeed, Castelnuovo's contribution to the development of Brill-Noether theory is well-known; a terse presentation in modern terms is provided, for instance, in [8], p. 242:

> Let's now examine this history a bit more closely. To begin with, Brill and Noether asserted the truth of the theorem based, apparently, on a naive dimension count bolstered by the calculation of examples in low genus.
>
> Exactly how the desired variational element might enter into the proofs was first suggested by Castelnuovo. His goal was not to establish any of the present theorems. Rather, he assumed the statement of the Brill-Noether theorem and applied it to compute the *number* of g^r_d's on a general curve in the case $\rho = 0$, when we expect it to be finite.
>
> To do this, Castelnuovo looked not at any smooth curve of genus g, but at a g-nodal curve C_0: that is, a rational curve with g nodes $r_1, \ldots, r_g$ obtained by identifying g pairs of points (p_i, q_i) on $\mathbb{P}^1$.

C. Fontanari (✉)
Dipartimento di Matematica, Università degli Studi di Trento, Trento, Italy
e-mail: claudio.fontanari@unitn.it

G. Bini (ed.), *Algebraic Geometry between Tradition and Future*,
Springer INdAM Series 53, https://doi.org/10.1007/978-981-19-8281-1_12

> Any g^r_d on C_0, Castelnuovo reasoned, would pull back to a g^r_d on $\mathbb{P}^1$, which could then be represented as the linear series cut out on a rational normal curve $C \cong \mathbb{P}^1 \hookrightarrow \mathbb{P}^d$ of degree d by those hyperplanes containing a fixed $(d-r-1)$-plane $\Lambda \subset \mathbb{P}^d$.
>
> The condition that the g^r_d on $\mathbb{P}^1$ be the pullback of one on C_0 was simply that every divisor of the g^r_d containing p_i should contain q_i as well; in other words, Λ should meet each of the chords $\overline{p_i q_i}$ to C in $\mathbb{P}^d$.
>
> The number of g^r_d's on a general curve of genus g was thus, according to Castelnuovo, the number of $(d-r-1)$-planes in $\mathbb{P}^d$ meeting each of g lines, a problem in Schubert calculus that Castelnuovo went on to solve (to obtain the correct value for the number of g^r_d's on a general curve).
>
> It was Severi who first pointed out, some twenty years later, that Castelnuovo's computation might serve as the basis of a proof of the Brill-Noether statement.

Our plan is to give a closer look at the original papers, by providing in Sect. 2 an English translation of some key passages and by drawing in Sect. 3 a series of remarks based on Castelnuovo's own statements.

This material has been presented at the conference TiME 2019 (Levico Terme, September 2–6, 2019). The author is grateful to the organizers for their kind invitation and to Enrico Arbarello for enlightening conversations about Guido Castelnuovo and the genuine meaning of his heritage.

2 A Trilogy of Papers

G. Castelnuovo, *Una applicazione della geometria enumerativa alle curve algebriche*, Rendiconti del Circolo Matematico di Palermo, t. III, 1889 ([5], III, pp. 45–53).

> In questo lavoro ci proponiamo due fini: esporre un metodo utile in molte ricerche della teoria delle curve; presentare alcune formole che ci sembrano notevoli e in se stesse, e per le loro conseguenze. A queste formole noi siamo giunti applicando il principio della conservazione del numero a curve degeneri.

> In this paper we have a twofold goal: to discuss a method useful in several researches in curve theory; to present some formulas we believe remarkable both in themselves and for their consequences. We came to these formulas by applying the principle of conservation of the number to degenerate curves.

G. Castelnuovo, *Numero degli spazi che segano più rette in uno spazio ad n dimensioni*, Rendiconti della R. Accademia dei Lincei, s. IV, vol. V, 4 agosto 1889 ([5], IV, pp. 55–64).

> Fra le questioni che appartengono alla Geometria Enumerativa, va notata per la sua importanza algebrica e geometrica la seguente: Quanti sono gli spazi ad s dimensioni che soddisfanno a più condizioni fondamentali date in uno spazio ad n dimensioni? Naturalmente le condizioni si suppongono tali da rendere determinato il problema.
>
> (...)

Un caso particolare interessante è il seguente:

Date hk rette in uno spazio a $\{(k+1)(h-1)\}$ *dimensioni, il numero degli spazi a* $\{k(h-1)-1\}$ *dimensioni che segano in punti queste rette è uguale a*

$$\frac{1!2!3!\ldots(h-1)!1!2!3!\ldots(k-1)!(hk)!}{1!2!3!\ldots(h+k-1)!}.$$

Among the questions pertaining to Enumerative Geometry, we point out the following one, due to its both algebraic and geometric relevance: How many are the s-dimensional spaces satisfying several fundamental conditions given in an n-dimensional space? Obviously, we suppose these conditions are such that the problem is made determinate.

(...)

A particular interesting case is the following:

Given hk lines in a $\{(k+1)(h-1)\}$*-dimensional space, the number of* $\{k(h-1)-1\}$*-dimensional spaces cutting these lines in points is equal to*

$$\frac{1!2!3!\ldots(h-1)!1!2!3!\ldots(k-1)!(hk)!}{1!2!3!\ldots(h+k-1)!}.$$

G. Castelnuovo, *Numero delle involuzioni razionali giacenti sopra una curva di dato genere*, Rendiconti R. Accademia dei Lincei, s. IV, vol. V, 1 settembre 1889 ([5], V, pp. 65–68).

È noto che sopra una curva di genere p con moduli generali esistono delle *serie lineari* $g_m^{(q)}$ (involuzioni razionali di ∞^q gruppi di m punti) in numero finito, quando sia

$$m-q=(p-m+q)q; \tag{1}$$

quante sono queste $g_m^{(q)}$? (...) Noi ci proponiamo di risolvere il problema in tutta la sua generalità approfittando del seguente concetto di geometria enumerativa, che ci servì in altra occasione:[1]

"il numero (supposto finito) degli $[r]$ (spazi ad r dimensioni) che segano in σ punti una curva C_p^n (di ordine n e genere p) appartenente ad un $[s]$ non muta, o diventa infinito, quando alla curva data si sostituisca l'insieme di più curve, purché l'ordine ed il genere della curva composta siano ancora risp. n e p".

Noi useremo soltanto curve costituite da una curva semplice con più corde, e precisamente in luogo di una curva di genere p, considereremo una curva razionale insieme a p delle sue corde.

(...)

[1] V. la memoria III, *Una applicazione della geometria enumerativa.* Dobbiamo riconoscere che nello stabilire questo concetto ci fondiamo più sulla intuizione (e su varie verificazioni), che sopra un vero ragionamento matematico. Alla dimostrazione si potrà forse arrivare considerando la curva in uno spazio superiore come intersezione parziale di più varietà e trattando algebricamente il problema degli spazi secanti; si troverebbe che il numero delle soluzioni è indipendente dalla posizione particolare delle varietà. Ma un ragionamento di tal natura non potrà farsi che quando la teoria delle curve negli spazi superiori sarà più completa. Ci permettiamo però di approfittare di un principio non ancora dimostrato per risolvere un difficile problema, perché crediamo che anche con simili tentativi si possa giovare alla scienza, quando si dichiari esplicitamente ciò che si ammette e ciò che si dimostra.

La C_p^{m+p} si scinda in una curva razionale C_0^m appartenente ad $[m]$ ed in p delle sue corde scelte ad arbitrio.

(...)

Se è soddisfatta la (1), il numero delle serie $g_m^{(q)}$ *esistenti sopra una curva di genere* p *uguaglia il numero degli spazi* $[m-q-1]$ *che segano* p *rette di* $[m]$.

Ora quest'ultimo numero fu già determinato;[2] se per semplicità poniamo

$$p-1-(m-q)=Q$$

ossia:

$$m=p-1-(Q-q)$$

e quindi per la (1)

$$p=(q+1)(Q+1),$$

si trova che il numero di cui si parla è dato da

$$\frac{1!2!3!\dots q!1!2!3!\dots Q!p!}{1!2!3!\dots(q+Q+1)!}.$$

It is known that on a curve of genus p with general moduli there exist a finite number of *linear series* $g_m^{(q)}$ (rational involutions of ∞^q groups of m points) whenever

$$m-q=(p-m+q)q; \tag{2}$$

how many are these $g_m^{(q)}$? (...) Our goal is to solve this problem in complete generality by exploiting the following notion of enumerative geometry, which we applied in another occasion:[3]

"the number (assumed to be finite) of the $[r]$ (r-dimensional spaces) cutting σ points on a curve C_p^n (of degree n and genus p) in a $[s]$ does not change, nor becomes infinite, when the given curve is replaced by the union of several curves, provided that the degree and the genus of the reducible curve are still n e p, respectively".

We are going to use only curves made by an irreducible curve together with several chords, more precisely, instead of a curve of genus p we will consider a rational curve together with p of its chords.

(...)

[2] V. la nostra nota IV: *Numero degli spazi che segano più rette in uno spazio ad n dimensioni*, §10.

[3] See the paper III, *Una applicazione della geometria enumerativa.* We have to admit that in establishing this notion we rely more on intuition (an on several checks) than on a complete mathematical argument. Maybe the proof will be reached by considering the curve in a higher space as the partial intersection of several varieties and by treating algebraically the problem of secant spaces; we would find that the number of solutions is independent on the special position of the varieties. But such an argument could be carried out only when the theory of curves in higher spaces will be more developed. On the other hand, we are going to apply a still unproven principle in order to solve a difficult problem since we believe that such attempts may be useful to the progress of science, provided one explicitly declare what is admitted and what is proven.

Let the C_p^{m+p} be split into a rational curve C_0^m in $[m]$ and into p of its chords, arbitrarily chosen.

(...)

If (2) is satisfied, the number of series $g_m^{(q)}$ *laying on a curve of genus* p *is equal to the number of space* $[m-q-1]$ *cutting* p *lines of* $[m]$.

Now, this last number was already computed;[4] if for simplicity we set

$$p - 1 - (m - q) = Q$$

that is to say:

$$m = p - 1 - (Q - q)$$

hence by (2)

$$p = (q+1)(Q+1),$$

we find that the number we are looking for is given by

$$\frac{1!2!3!\ldots q!1!2!3!\ldots Q!p!}{1!2!3!\ldots (q+Q+1)!}.$$

Guido Castelnuovo: *Memorie scelte*, Aggiunta alle Memorie III, IV e V, p. 69.

Nell'autunno 1888 C. Segre, che allora studiava le superficie rigate algebriche degli iperspazi, aveva ricondotto la determinazione del numero delle direttrici (curve unisecanti le generatrici) d'ordine minimo di una rigata al calcolo del numero degli spazi S_{s-1} che contengono s generatrici di una rigata di S_s, e mi propose il problema di stabilire questo numero. Di qua l'origine della Mem. III, che contiene al n. 4 il risultato richiesto, del quale ha tenuto conto il Segre ("Mathematische Annalen", vol. 34, 1889). L'idea che mi ha permesso di raggiungere rapidamente questo e altri risultati consiste nel sostituire ad una curva irriducibile d'ordine n e genere p di un iperspazio, una curva composta di una curva d'ordine $n-1$ e di una retta unisecante o bisecante, secondo che quest'ultima curva ha genere p o $p-1$.

Questo *principio di degenerazione*, che serve anche nella Mem. V, è semplicemente ammesso; la prima dimostrazione che lo spezzamento non altera i numeri richiesti fu data per via topologica (ricorrendo alle superficie di Riemann) da F. Klein in un suo corso del 2° semestre 1892 (*Riemannsche Flächen*, II, pag. 110 e segg., lezioni litografate 1892). Per via algebrica occorre far vedere che la curva spezzata può esser riguardata come limite di una curva irriducibile variante entro un sistema continuo, ciò che, sotto ipotesi assai larghe, ha dimostrato F. Severi nelle *Vorlesungen über algebraische Geometrie*, Anhang G. (B. G. Teubner, Leipzig-Berlin, 1921); v. in particolare pag. 392.

In autumn 1888 C. Segre, who was then investigating ruled algebraic surfaces in hyperspaces, had reduced the determination of the number of directrices (cubes unisecant the generatrices) of minimal degree of a ruled surface to the computation of the number of spaces S_{s-1} containing s generatrices of a ruled surface in S_s, and he posed to me the problem of establishing this number. Hence the origin of paper III, containing at n. 4 the requested result, taken into account by Segre ("Mathematische Annalen", vol. 34, 1889).

[4] See our note IV: *Numero degli spazi che segano più rette in uno spazio ad n dimensioni*, §10.

The idea that allowed me to quickly reach this and other results consists in substituting to an irreducible curve of degree n and genus p in a hyperspace, a reducible curve union of a curve of degree $n - 1$ and of a line, either unisecant or bisecant according to the fact that this last curve has genus either p or $p - 1$.

This *principle of degeneration*, applied also in the paper V, is simply admitted; the first proof that the splitting does not change the required numbers was given through topology (by exploiting Riemann surfaces) by F. Klein in his course delivered on the second semester of 1892 (*Riemannsche Flächen*, II, p. 110 and ff., lithographed lectures 1892). To follow an algebraic way one needs to show that the reducible curve may be obtained as a limit of an irreducible curve varying in a continuous system, that is, under very mild assumptions, F. Severi proved in *Vorlesungen über algebraische Geometrie*, Anhang G. (B. G. Teubner, Leipzig-Berlin, 1921); see in particular p. 392.

3 A Few Remarks

3.1 Guido Castelnuovo and Corrado Segre

We have seen that the starting point of Castelnuovo's research is an enumerative question posed to him by Corrado Segre. The role of the intense scientific dialogue between Segre and Castelnuovo in the flourishing of the Italian school of algebraic geometry should not be underestimated: in a letter to Amodeo sent on February 6, 1893, Castelnuovo mentions with longing the *orge geometriche torinesi* (*geometric orgies in Turin*). As pointed out by Enrico Arbarello (private communication):

> When thinking of Castenuovo, and of his walks with Corrado Segre through the streets of Turin, during which they rapidly absorbed Riemann's point of view through the prism of their Italian taste, I can't help but think of how unfamiliar, abstract, and foreign it must have felt to them. I think that without being exposed to those revolutionary ideas, which freed him from the extrinsic world of Cremona transformations, Castelnuovo might not have arrived at his contraction principle. Indeed, what I would celebrate in Castelnuovo is his absolute open-mindedness, his taste for adventure, and his eagerness to create and follow new paths, no matter how faint their trace.

The mail correspondence between Segre and Castelnuovo witnesses their both scientific and personal deep fellowship. Here Segre comments Castelnuovo's results presented above:

C. Segre a G. Castelnuovo, 20 IX 1888 ([1])

Torino, 20 IX 88

Carissimo Castelnuovo,

Alcuni dei teoremi che mi comunichi mi paiono veramente importanti. Importante l'idea di servirsi di curve di genere p degeneri. (...)

C. Segre to G. Castelnuovo, 20 IX 1888

Torino, 20 IX 88

My dearest Castelnuovo,

Some of the theorems you told me seem to me really important. Important the idea of exploiting degenerations of curves of genus p. (...)

Here instead Segre addresses touching words to Castelnuovo just after his move from Turin to Rome:

C. Segre a G. Castelnuovo, 12 XI 1891 ([1])

Torino, 12 XI 91

Mio carissimo,

Ricevo la tua affettuosa lettera, e te ne ringrazio. Da Lunedì tu mi manchi ed io sento vivamente questa lacuna. Tu accenni a quel po' di giovamento che hai potuto trarre in questi quattro anni dalla mia compagnia. Se ciò è vero, è pur vero che da te io ho avuto un completo ricambio, e che il tuo ingegno acuto, come la tua bontà di cuore m'han reso continuamente utili e piacevoli le tante ore che passavamo insieme (...) Tu m'hai fatto del bene, lo ripeto, non solo intellettualmente ma anche moralmente. Ed ora che tu mi manchi sento realmente un vuoto, che non sarà colmato da nessuno. (...) Conservami sempre il tuo affetto. (...)

E ancora una volta un abbraccio affettuosissimo dal

Tuo aff.mo C. Segre

C. Segre to G. Castelnuovo, 12 XI 1891 (English version by Enrico Arbarello in [2], p. 21)

Torino, 12 XI 91

My Dearest,

I am in receipt of your affectionate letter and I thank you for it. Ever since Monday I miss you, and I feel this void very deeply. You mention the bit of benefit that you might have been able to gain from the past four years in my company. If that is indeed true, it is also true that it was a completely even exchange, and that your acute insight, as well as the goodness of your heart, have continuously made the many hours I spent with you useful and pleasant (...) You did me good, I repeat, not only intellectually, but also morally. And now that you are missing, I really feel a void which cannot be filled by anyone. (...) Keep me forever in your affection. (...) And once again, a very big hug, from

Your affectionate C. Segre

3.2 *Geometry and Probability*

We have seen that the number computed by Castelnuovo:

$$\frac{1!2!3!\dots q!1!2!3!\dots Q!p!}{1!2!3!\dots (q+Q+1)!}$$

presents a pronounced combinatorial flavour. It is well-known that Castelnuovo is the author of the first Italian treatise on the Calculus of Probability (1919), and this pioneering work [4] outside the realm of algebraic geometry is usually motivated by a gradual shift of his scientific interests towards applications of mathematics to natural and social phenomena (see, for instance, [7], p. 167). The above computation suggests a different (maybe complementary) explanation, by showing a much earlier Castelnuovo's taste for combinatorial manipulations analogous to the ones involved in discrete probability.

3.3 Guido ed Emma Castelnuovo

We have seen that Castelnuovo claims to apply *a still unproven principle in order to solve a difficult problem since... such attempts may be useful to the progress of science, provided one explicitly declares what is admitted and what is proven.* This intellectual habit, which is a very personal mixture of free open-mindedness and strict moral rigour, is typical of Castelnuovo's attitude to both mathematical research and teaching. His daughter, Emma Castelnuovo, consciously collects his heritage in her masterpiece volume [3] devoted to school teaching of mathematics. In particular, two explicit quotations in her book are devoted to Guido Castelnuovo. The first one is on p. 5:

> Guido Castelnuovo [espone] delle riflessioni fortemente indicative per un moderno insegnamento[5] e a cui, rilette a distanza di più di cinquanta anni, potrebbero ispirarsi oggi i compilatori dei programmi di matematica: *È questo il torto precipuo dello spirito dottrinario che invade la nostra scuola. Noi vi insegnamo a diffidare dell'approssimazione, che è realtà, per adorare l'idolo di una perfezione che è illusoria. Noi vi rappresentiamo l'universo come un edificio, le cui linee hanno una perfezione geometrica e ci sembrano sfigurate e annebbiate in causa del carattere grossolano dei nostri sensi, mentre dovremmo far comprendere che le forme incerte rivelateci dai sensi costituiscono la sola realtà accessibile, alla quale sostituiamo, per rispondere a certe esigenze del nostro spirito, una precisione ideale...*
>
> Guido Castelnuovo presents a few strongly suggestive remarks for a modern teaching,[6] from which school guidelines for mathematics could take inspiration today, more than fifty years later: *This is the main fault of the dogmatic spirit invading our school. There we teach to distrust approximation, which is reality, in order to worship the idol of a perfection, which is illusion. There we represent the universe as a building, whose lines have a geometric perfection and look deformed and obfuscated because of the rough character of our senses, while we should make clear that the confused forms disclose to us by our senses are the only achievable reality, to which we replace, in order to satisfy certain needs of our spirit, an ideal precision...*

The second, even more amazing, quotation of Guido Castelnuovo appears on p. 157 of [3]:

> Si dirà che è impossibile dare al bambino una nozione certa di funzione, che è pericoloso parlare del concetto di limite in termini vaghi, si dirà che quanto si insegna deve essere perfetto per non originare idee false che poi sarebbe difficile sradicare per sostituirle con appropriate definizioni. Mi torna alla mente quanto scriveva, nel lontano 1912, Guido Castelnuovo a questo proposito: *Ciò che si sa dal professore o dall'allievo – mi fu detto –, sia pur limitato, ma deve sapersi perfettamente. Orbene, io sono uno spirito mite e tollerante; ma tutte le volte che questa frase mi fu obiettata, un maligno pensiero mi ha attraversato come un lampo la mente. Oh, se potessi prendere in parola il mio interlocutore,*

---

[5] G. Castelnuovo, *La scuola nei rapporti con la vita e la scienza moderna*, conferenza tenuta a Genova nel 1912 in occasione del III Congresso della Mathesis, e riprodotta in *Archimede*, n. 2–3, 1962.

[6] G. Castelnuovo, *La scuola nei rapporti con la vita e la scienza moderna*, talk held in Genova in 1912 at the III Congress of Mathesis, and reproduced in *Archimede*, n. 2–3, 1962.

e con un magico potere riuscissi a spegnere per un istante nel suo cervello tutte le cognizioni vaghe per lasciar sussistere soltanto ciò che egli sa perfettamente! Voi non immaginate mai quale miserando spettacolo potrei presentarvi! Ammesso pure che dopo una così crudele mutilazione qualche barlume rimanesse ancor nel suo intelletto, e di ciò fortemente dubito, somiglierebbe questo ad un gioco di fuochi folletti sperduti in tenebre profonde e sconfinate. La verità è che noi nulla sappiamo perfettamente... [7]

One could object that it is impossible to provide a child with a rigorous notion of function, that it is dangerous to introduce the concept of limit in vague terms, one could object that what is taught has to be perfect in order to avoid misconceptions hard to eradicate and replace with appropriate definitions. It comes back to my mind what Guido Castelnuovo wrote about this, long ago in 1912: *What is known by the teacher or by the pupil – they say –, let it be limited, but it should be perfectly grasped. Well, I am a mild and indulgent spirit, but every time this statement has been objected to me, a malicious thinking has crossed my mind as a flash: Oh, if only I could take my discussant at his word, and with a magic power I could turn off in his brain for one moment every vague cognition, by letting there only what he knows perfectly! Provided after such a cruel mutilation some glimmer would survive in his intellect, which I strongly doubt, this would resemble a will o' wisp lost into a deep and immense darkness. The truth is that we perfectly know nothing...* [8]

For further links between the approach to mathematics of Guido and Emma Castelnuovo, we refer to [6].

Acknowledgments This research project was partially supported by GNSAGA of INdAM and by PRIN 2017 "Moduli Theory and Birational Classification".

References

1. Accademia Nazionale dei Lincei: Lettere e Quaderni dell'Archivio Guido Castelnuovo, a cura di Paola Gario. http://operedigitali.lincei.it/Castelnuovo/Lettere_E_Quaderni/menu.htm
2. Arbarello, E.: In honor of Guido Castelnuovo, Venezia 2015, slides available online at https://www1.mat.uniroma1.it/people/arbarello/G.Castelnuovo-Venezia.pdf
3. Castelnuovo, E.: Didattica della matematica. La Nuova Italia, Firenze (1963)
4. Castelnuovo, G.: Calcolo delle Probabilità. Società Dante Alighieri, Milano-Roma-Napoli (1919)
5. Castelnuovo, G.: Memorie scelte. Zanichelli, Bologna (1937)
6. Fontanari, C.: Guido Castelnuovo e la geometria delle curve algebriche: dalla ricerca alla didattica. Rend. Mat. Appl. **37**(7), 137–145 (2016)
7. Gario, P.: Guido Castelnuovo: l'uomo e lo scienziato. Rend. Mat. Appl. **37**(7), 147–183 (2016)
8. Harris, J., Morrison, I.: Moduli of Curves. Springer, New York (1998)
9. Rogora, E.: Guido Castelnuovo and his family. In: G. Bini (ed.) Algebraic Geometry between Tradition and Future, Springer INdAM Series, vol. 53. Springer Nature Singapore, Singapore (2023)

[7] G. Castelnuovo, *La scuola nei rapporti con la vita e la scienza moderna*, conferenza tenuta a Genova nel 1912 in occasione del III Congresso della Mathesis, e riprodotta in *Archimede*, n. 2–3, 1962.

[8] G. Castelnuovo, *La scuola nei rapporti con la vita e la scienza moderna*, talk held in Genova in 1912 at the III Congress of Mathesis, and reproduced in *Archimede*, n. 2–3, 1962.

Guido Castelnuovo and His Family

Enrico Rogora

Abstract In this paper the importance is discussed of studying the direct influence of Guido Caselnuovo's family environment in shaping his interests in Statistics and Probability Theory and on his thinking about education. The influences of his father Enrico, his uncle Luigi Luzzatti, and his grandmother Adele Levi Della Vida will be especially considered.

Keywords History of mathematics · Guido Castelnuovo · History of statistics · Teaching of mathematics

1 Guido Castelnuovo, His Times, and His Family

In 2015 I organized a conference at the Department of Mathematics of the University Sapienza in Rome, celebrating the 150th anniversary of the birth of Guido Castelnuovo, after whom the Department is named. Some of the conference contributions have been published in Volume **37** (3–4) 2016 of *Rendiconti di Matematica Pura e Applicata* and are freely available.[1] I recommend Paola Gario's contribution [22], a broad and articulated synthesis of Guido Castelnuovo's whole human and scientific itinerary. Also, I found particularly stimulating the contribution of Enrico Arbarello [1], highlighting the importance of his family context, to better appreciate the breadth of his thought. In this contribution I would like to deepen the knowledge of the family environment that surrounded Guido Castelnuovo, as it revealed particularly useful to clarify his opinions on teaching and to properly appreciate his manifold institutional activities.

[1] At https://www1.mat.uniroma1.it/ricerca/rendiconti/.

E. Rogora (✉)
Dipartimento di Matematica, Sapienza Università di Roma, Roma, Italy
e-mail: enrico.rogora@uniroma1.it

G. Bini (ed.), *Algebraic Geometry between Tradition and Future*,
Springer INdAM Series 53, https://doi.org/10.1007/978-981-19-8281-1_13

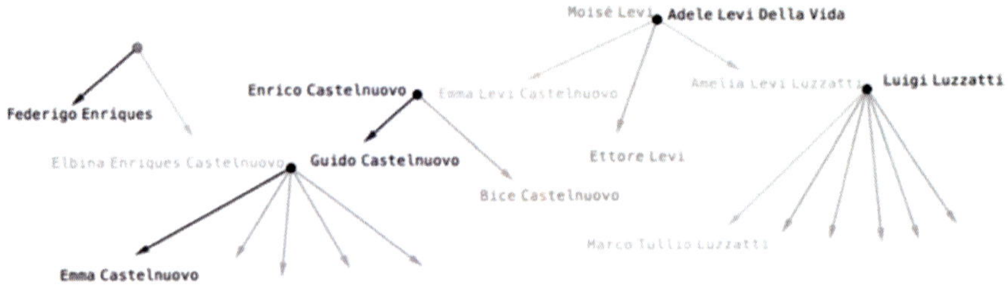

Fig. 1 Family tree of Guido Castelnuovo's family

1.1 Family Words

> A volte, mentre stava alla lavagna, con il gesso in mano, guardando i ragazzi nei banchi, fermava la lezione ed esclamava: "Evviva la libertà!" [16].[2]

I chose to begin with these words, which impressed me very much while reading the beautiful book [16], by Carla degli Esposti and Nicoletta on Emma Castelnuovo.

Freedom is one of those words whose sound echoes in all Guido Castelnuovo's family events, like *Solidarity*, *Science*, *Homeland*, *Culture*, and *Teaching*. They are "family words," the threads of the warp and the weft of his thought, which are also deeply rooted in the most genuine values of the Italian Risorgimento.

Guido Castelnuovo's family relationships are quite ramified. In the fragment family tree shown in Fig. 1, in black are marked the relatives on which attention will be focused in the following.

2 Enrico Castelnuovo (1839–1915): Passion and Duty

The first character in this story is Guido's father, Enrico Castelnuovo. Enrico's father, i.e., Guido's grandfather, left his family to seek fortune in Egypt when Enrico was one and a half years old. Then, his mother settled in Venice, relying on her relatives' solidarity. As for their son's education

> Non sembra che abbia ricevuto da bambino una vera educazione religiosa; comunque da adulto si mantenne sempre fedele ad un atteggiamento di positivista laico [35].[3]

After attending a technical school, not finished because of urging to contribute to his family's needs, Enrico found an employment at Della Vida commercial house,

---

[2] *Translation:* Sometimes, while she was at the blackboard, with chalk in her hand, looking at her pupils in the desks, she would stop the lesson and exclaim: " Long live freedom!". – *All translations are provided by the author.*

[3] *Translation:* As a child, there is no record of a true religious upbringing; however, as an adult, he always remained faithful to a positivist attitude.

owned by one of his uncles, and started studying French, English, and German. He devoted himself scrupulously to his work, at the same time cultivating his literary vocation, also through the reading of foreign novelists in original language, including Dickens and Zola. In 1865 he married his cousin Emma Levi and his son Guido was born the same year. In 1869, Emma died at her daughter Bice's birth. Della Vida commercial house was liquidated soon after.

Enrico personally took care of his children's education, with the illuminated support of his mother-in-law Adele Levi Della Vida, and became a collaborator of the newspaper *La stampa di Venezia*, press organ of the *azzurri*, a group of young liberals, who supported the leaders of the historical Right (Massimo d'Azeglio, Camillo Benso Count of Cavour, Vittorio Sella, Marco Minghetti). In his commemoration of Enrico Castelnuovo [21], Antonio Fradeletto gives a concise and incisive portrait of those young patriots and their activities:

> Durante gli ultimi anni della dominazione austriaca del Veneto, s'era formato un gruppo di giovani d'alto intelletto e di moderna coltura, i quali impresero un'efficace propaganda politica ed economica, con l'intento ideale e pratico di tenere sempre vivo lo spirito di libertà e d'italianità e di preparare [con il loro esempio e il loro insegnamento] i concittadini [...] al degno esercizio dei nuovi doveri. Erano tra essi Antonio Tolomei, Emilio Morpurgo, Alessandro Pascolato, Alberto Errera, e primeggiava un uomo destinato ad esercitare nella vita pubblica italiana un'opera infaticabile di elevazione e di solidarietà sociale, Luigi Luzzati [21].[4]

Continuing with Luzzatti's words

> Poi si associa al manipolo Enrico Castelnuovo, una felice tempra di sentimenti e di intelletto. (...) Durante i frequenti convegni nei caffè, nelle passeggiate in piazza San Marco, nelle pubbliche lezioni (...), nuove scintille si sprigionavano e maturavano poi in idee luminose. Spesso le riunioni si tenevano al Caffè Brigiacco, altre volte, di sera, si ritrovavano nella mia casa e lì si studiava, si filosofava, si inneggiava all'Italia [26].[5]

In his commemoration at Veneto Institute of Sciences, Letters and Arts, the mathematician Alesandro Bordiga recalled an episode that literally impressed the visible sign of freedom in the life of his good friend Enrico Castelnuovo

> [nel 1870] egli si era dato a dirigere il giornale *la Stampa*, nel quale portò contributo di serena fermezza e di rettitudine esemplare. In quel breve tempo, aspre com'erano per tutta Italia le lotte politiche, difendendo uomini più temperati di lui da violenza di più acri

[4] *Translation:* During the last years of the Austrian domination of Veneto, a group of young people of high intellect and modern culture had formed, who embarked on effective political and economic propaganda, with the ideal and practical intention of keeping the spirit of freedom and Italian patriotism alive and of preparing [with their example and their teaching] their fellow citizens [...] for the worthy exercise of their new duties. Among them there were Antonio Tolomei, Emilio Morpurgo, Alessandro Pascolato, Alberto Errera, and excelled a man whose fate would be to exercise a relentless work of elevation and social solidarity in Italian public life, Luigi Luzzati.

[5] *Translation:* Then, Enrico Castelnuovo joins the bunch, a happy temper of feelings and intellect. (...) During the frequent conferences in the cafes, in the walks in Piazza San Marco, in the public lessons (...), new sparks were released and then matured into bright ideas. Often, our meetings held at Caffè Brigiacco, other times, in the evening, we met at my home and there we studied, philosophized, and praised Italy.

> avversari, *sostenendo le ragioni della libertà e della civile tolleranza*, fu tratto a duello con un acceso soldato (...) della schiera dei Mille, e ne uscì offeso gravemente nella mano, la quale portò impresso per tutta la vita il segno ingiusto [5].[6]

The Sicilian economist Francesco Ferrara had been appointed director of the Higher School for Commerce, established in Venice in 1868, thanks to the strenuous efforts of Luigi Luzzatti. Towards the end of 1872, when "La Stampa di Venezia" ceased its publications, Ferrara called Enrico Castelnuovo to teach "commercial institutions".[7] A new period of his life began, relentlessly dedicated to his teaching duties and to increasing the role of this Institution for the good of Venice and Italy.[8]

In 1905 Enrico Castelnuovo become the third director of the Higher Institute for Commerce, after Francesco Ferrara, which served from 1868 to 1900, and Andrea Pascolato, who served from 1900 to 1905.

Enrico Castelnuovo was renowned for the clarity of his teaching, the reciprocal affection with his students, and his absolute dedication to the institution. On the occasion of his scientific jubilee,[9] his son Guido wrote:

> La verità è che mio padre, come io stesso, pur svolgendo materie tanto diverse, abbiano riguardato all'insegnamento come una missione, un'altissima missione. [33][10]

Enrico Castelnuovo's mission was teaching, but his passion was writing.

Mission and passion; duty and pleasure: dichotomies also evoked in the famous address delivered by Guido Castelnuovo before the International Commission for the Teaching of Mathematics in 1914:

> Ci domandiamo talvolta se il tempo che dedichiamo alle questioni d'insegnamento non sarebbe meglio impiegato nella ricerca scientifica. Ebbene, rispondiamo che è un dovere sociale che ci obbliga a trattare questi problemi [9].[11]

Behind his "rispondiamo" I like to hear the voice of his father too, who died just a few months after the address.

Enrico Castelnuovo's literary activity consisted of about 20 novels, numerous short stories, and various essays published in the proceedings of the Venetian Institute of Sciences and Letters, of which he was a corresponding member first

---

[6] *Translation:* [in 1870] he had set out to run the newspaper *la Stampa*, in which he brought a contribution of serene firmness and exemplary righteousness. In that short time, harsh as they were political struggles throughout Italy, while defending more temperate men than him from violence of several opponents, *supporting the reasons of freedom and civil tolerance*, he was dueled with a lit soldier (...) of Garibaldi's ranks of "I Mille" and went out gravely offended in the hand, which he carried imprinted throughout his life, as an unjust sign.

[7] In Italian, "Istituzioni commerciali." The exceptional nature of those times allowed talented young people being hired for teaching, even if they did not have proper academic qualifications.

[8] The institute later became the first Italian university dedicated to economic studies.

[9] Guido began his scientific career in 1887.

[10] *Translation:* The truth is that my father, like myself, despite being involved in the teaching of such different subjects, considered teaching as a mission, a most high mission.

[11] *Translation:* Sometimes we wonder whether the time we devote to teaching would not be better spent on scientific research. Well, we answer that it is a social duty that obliges us to deal with these problems.

(like his son) and an ordinary one later. His masterpiece is unanimously considered to be his last novel, *I Moncalvo*, where many readers have seen, in the figure of Professor Giacomo Moncalvo, a reference to his son Guido.

I believe it is not easy to tell apart, in the words of Giacomo Moncalvo, the father's from the son's voice. Perhaps, the following words may be those Enrico hoped he could convey to Guido:

> Ciò non toglie che io m'auguri prossimo il tempo in cui la morale possa reggersi da sola come un monumento che si regga senza l'armatura. Vedi, la religione è come il dizionario, ch'è sempre in arretrato quando lo si paragoni alla lingua viva [8].[12]

Benedetto Croce, writing sympathetically about Enrico Castelnuovo:

> notava com'egli sia stato "uno dei migliori, dei più stimabili autori italiani" di quelli che possono dirsi "romanzi ben fatti", cioè, non "grandi opere di bellezza", ma favole piacevoli ed oneste. Ed osservava altresì come, fra i libri del Nostro, meritino d'essere particolarmente apprezzati "i romanzi di costume", specchio sincero dei tempi [24].[13]

The pugnacious idealistic Italian philosopher had his say about two other members of Guido Castelnuovo's family, arguing polemically with Guido's brother-in-law Federigo Enriques and Luigi Luzzatti, his uncle.

Croce's opinion on Enrico Castelnuovo was benevolent, despite the fact that he:

> ostenti sovente, con ottocentesco ottimismo, fiducia nel progresso della civiltà e dell'umana conoscenza [24][14]

The explicit underlying positivistic philosophy of the author of these *favole piacevoli*[15] does not deserve the indignant invectives reserved for the unwelcome and, according to him, pretentious and amateurish philosophical encroachments of Enriques and Luzzatti.

3 Luigi Luzzatti (1841–1927): Solidarity and Institutions

Luigi Luzzatti, Guido's uncle, was a very good friend of Enrico Castelnuovo. They become relatives since both Luigi's wife Amelia and Enrico's wife Emma were sisters, offsprings of Adele Levi Della Vida. On Adele, Sect. 4 will be focused.

---

[12] *Translation:* Nevertheless, I hope the time is near when morality can stand on its own like a monument without armor. You see, religion is like the dictionary, always late when compared to living language.

[13] *Translation:...* noted that he was "one of the best, of the most respectable Italian authors" of novels which may be qualified as "well done," that is, not "great works of beauty," but pleasant and honest fables. Also, he observed that, among his books, his "romance of manners" deserve to be particularly appreciated, being a sincere mirror of his times.

[14] *Translation:* often flaunts with nineteenth-century optimism his confidence in progress of civilization and human knowledge.

[15] Pleasant fables.

He was one of the leading politicians of post-unification Italy, collaborator of Quintino Sella, Marco Minghetti, and Giovanni Giolitti. He held prestigious government positions, was appointed several times to the Minister of the Treasury and Agriculture, and became Prime Minister between 1910 and 1911.

He served as a professor of constitutional law, which he considered the *science of freedom*, at the University of Padua and Rome.

Before being called at university, he taught statistics at the technical institute of Milan (forerunner of Politecnico), run by Francesco Brioschi (with whom he had some tasty squabbles). His interest in statistics and his conviction of its fundamental usefulness for a good governance of the country dated back to his time as a student, when he was fascinated by Messedaglia's lecturing: indeed he always maintained relationships of high reciprocal esteem with this distinguished statistician and economist from Verona.

He shared the German economist Hermann Schulze-Delitzsch's theories on the social function of credit; thus, he became a staunch supporter of cooperative institutions, especially in the field of consumer credit, considered a useful tool to combat loan sharking and help both working and middle classes. He started many popular banks starting with the popular bank of Lodi in 1863. In agreement with Benedetto Cairoli, he promoted legislation for the protection of female and child labor.

> Per tutta la vita, Luzzatti [ricercò] sempre, sia in quanto uomo politico sia in quanto studioso, il punto di equilibrio tra solidarismo e scienza dell'economia, tra lavoro e capitale. Pur essendo ideologicamente vicino ai principi liberali, Luzzatti si discostava tuttavia dalla convinzione che il processo naturale e graduale di selezione tra gli uomini dovesse prescindere dal concetto di fraternità [32].[16]

3.1 The Higher School of Commerce of Venice

An initiative of Luzzatti which is important for our story was the establishment of the Commercial Institute of Venice, which he recalls in his Memories with words that, *mutatis mutandis*, could have been used by his nephew to describe the reasons behind the foundation of the school of statistical and actuarial sciences of the University of Rome, conceived to fulfill the urgent needs of a proper training for a new generation of state officials. [17]

> L'Italia mancava ancora di una scuola superiore di commercio e alla fine del 1866 io ne avevo lanciata l'idea nella mia città natale, insieme alla riforma degli studi professionali. Poi, nelle mie conversazioni con Frère–Orban e Léon Saymi mi ero persuaso che i due

[16] *Translation:* Throughout his life, Luzzatti always [pursued], as both a politician and scholar, a balance between solidarity and economics, and between labor and capital. Although ideologically close to liberal principles, yet Luzzatti diverged from the conviction that natural and gradual processes of selection among humans had to ignore the concept of fraternity.

[17] See [3].

grandi tipi di questa provvida istituzione erano l'Ecole Supérieure de Commerce di Anversa e quella di Mulhouse. Ne studiai sul luogo gli ordinamenti e gli organi con analitica precisione. In ambedue le città appresi la funzione di una cattedra di cui non c'era l'idea nel nostro paese, cioè l'insegnamento in atto della contabilità commerciale. Tornato in Italia mi presentai a Venezia con questi due doni scientifici, mostrai il valore politico ed economico di dare alla grande città, allora liberata, il benefizio di una scuola nuova educatrice dei grandi commercianti e di quella burocrazia commerciale competente sino allora desiderata invano nel nostro paese.

(...)

La scuola ebbe un effetto straordinario; attrasse a Venezia professori principali e discepoli che poi divennero grandi commercianti o diplomatici all'estero [26].[18]

3.2 *Literary Interlude*

In a letter of Enrico Castelnuovo to Luigi Luzzatti, written in February 1913 and kept in Luigi Luzzatti Archives at the Istituto Veneto di Scienze, Lettere ed Arti, we read:

Carissimo Luigi, oggi aprendo il Marzocco ho visti alcuni di quei versi scritti da te per le mie nozze la bellezza di cinquant'anni fa [2].[19]

Il Marzocco, a weekly literary magazine printed in Firenze, dedicated six articles to the "Poets of Montecitorio," written by the deputy Giovanni Rosadi about Leonida Bissolati (number **49**, 1912); Filippo Turati (n. **51**, 1912); Vittorio Cottafavi (n. **5**, 1912); Alfonso Lucifero e Emilio Pinchia (n. **2**, 1913); Ferdinando Martini (n. **5**, 1913); Luigi Luzzatti (n. **7**, 1913).

---

[18] *Translation:* Italy still lacked a high school of commerce, and at the end of 1866, I launched the idea in my hometown, along with a reform of professional studies. Then, in my conversations with Frère-Orban and Léon Saymi, I was persuaded that the two best types of this useful institutions were the École Supérieure de Commerce of Antwerp and that of Mulhouse. On the spot I studied the arrangements and the organs with analytical precision. In both cities I learned the function of a teaching whose idea was not yet conceived in our country, that is, the teaching of commercial accounting. Back to Italy, I brought with myself to Venice these two scientific gifts, and I showed the political and economic value of giving the great city, just liberated, the benefit of a new school for educating major traders and that competent commercial bureaucracy in vain wished for in our country, up to that moment.

(...)

The school had an extraordinary effect; it attracted top class professors and disciples to Venice, who later became major traders or diplomats abroad.

[19] *Translation:* Dear Luigi, today, opening the Marzocco, I saw some of those verses written by you for my wedding, 50 years ago.

Luzzatti's wedding poem was dedicated to Enrico Castelnuovo's bride, but does not deserve to be remembered; however, the article contains other fragments revealing a genuine passion behind Luzzatti's political engagement. For example, in a verse dedicated to the inauguration of an Asilo Mariuccia in December 1912,[20] he wrote:

Palpito nasce e poi diventa idea, La feconda il pensier, ma il cuor la crea [23][21]

With his caricatural and colorful writing, Giovanni Rosadi highlighted Luigi Luzzatti's human solidarity. He wrote *Alla fecondità del pensiero, alla magniloquenza della frase, il Luzzatti parve a qualcuno un uomo del seicento (...) i critici non si accorsero che il largo stile che a loro pareva retorico non era se non poetico, quello stile che il poeta del 63 non lasciò nel 908 alla porta di palazzo Braschi quando salì alla capitudine del governo. Infatti il suo gabinetto telegrafava a un prefetto che gli chiedeva istruzioni rispetto a certi operai scioperanti:* "li riceva con sorrisi lampeggianti di minacce"*; e a un altro prefetto che doveva prendere provvedimenti vessatori:* "distribuisca equamente il malcontento"*; e a un intendente di finanza che gli denunziava l'insubordinazione collettiva de' suoi impiegati:* "li punisca con il mio perdono". [22]

Enrico Castelnuovo gives us a glimpse of the relations between Guido Castelnuovo's family and Luigi Luzzatti's in Rome, at the end of his abovementioned letter:

Guido e l'Elbina mi scrivono d'aver assistito con molto piacere al dibattito cristiano – buddista fra te e il prof. Formichi [23].[23]

The Christian-Buddhist debate took place on the sidelines of a conference on Buddhism, held by the orientalist Carlo Formichi (later academic and vice-president of the Academy of Italy) at the Philosophical Circle, on February 8, 1913.

[20] The foundation "Asilo Mariuccia" was established in 1901 by Ersilia Bronzini Majno in memory of her daughter Maria with the goal of "addestrare all'emancipazione le fanciulle pericolanti" (help young women in danger to become emancipated).

[21] *Translation:* Heartbeat is born but then it becomes idea, the thought fecundates it, but the hearth builds it.

[22] *Translation:* Because of the fecundity of his thought, of the grandiloquence of the sentence, Luzzatti looked like a man of the seventeenth century to some of us (...) his critics did not realize that his broad style that seemed rhetorical was nothing but poetic, that style that the poet of 63 [when he wrote the verses for the wedding] did not drop in 908, at the door of Palazzo Braschi, when he raised to the head of the Government. In fact, his cabinet telegraphed to a prefect who asked him for instructions regarding striking workers: *"receive them with flashing smiles of threats"*; and to another prefect who was to take vexatious measures: *"distribute discontent with equanimity"*; and to a finance superintendent who denounced the collective insubordination of his employees: *"punish them with my forgiveness "*.

[23] *Translation:* Guido and Elbina wrote to me that they had been very pleased to witness the Christian-Buddhist debate between you and prof. Formichi.

3.3 *Freedom of Conscience and Science*

As I said before, Luigi Luzzatti too had his controversy with Benedetto Croce. It was about his book *The Freedom of Science and Conscience* which, as we read in a letter to his niece Maria, one of Guido's daughters,

> è il libro dove ho cercato di condensare le più alte idealità delle quali è capace l'anima mia [2].[24]

Croce wrote about Luzzatti's book in his Journal *La Critica*:

> Negli scritti messi in fondo al volume, il Luzzatti, si vanta di aver tenuto fede fin dal 1876 all'idealismo scientifico. Ed è un fatto che egli ha contrastato sempre il materialismo e il determinismo, e affermato, sempre, la libertà, i valori morali, la religiosità.
>
> Tutto ciò è nella sua mente, rimasto assai vago; e riesce impossibile stabilire con quali argomenti fondi la libertà, quale religione egli professi, o quale sia il suo sistema filosofico. Piuttosto che elaborare filosoficamente le sue idee, egli le ha asserite, aspettando fiducioso il pensiero, che un giorno dovrà giustificarle: "è lecito augurare che, quando siano maturi i tempi che noi non vedremo, sorga un maestro mirabile delle scienze naturali e filosofiche, capace di raccogliere in una sintesi luminosa il mondo spirituale e naturale". Ma sarebbe ingiusto negargli il merito di avere dato prova di sano istinto e di buone tendenze in tempi di grossolano naturalismo, imperversante nel campo economico e politico, non meno che in quello letterario e filosofico [15].[25]

Luzzatti answered to Croce during the third congress of the Italian Philosophical Society (organized by Federigo Enriques). The controversy widened with the Osservatore Romano on Croce's side and both Pompeo Molmenti and Francesco Ruffini on Luzzatti's. (see [27], vol. 1, pp.L-LIV).

Luzzatti, like Enriques, whose polemics with Croce is well-known and does not deserve to be recalled here (see, e.g.,[34]), did not sign the Croce manifesto, which Molmenti, Ruffini, and Guido Castelnuovo signed instead.

3.4 *Luigi Luzzatti's Judaism*

Luigi Luzzatti was born in Venice into a wealthy Jewish family.

---

[24] *Translation:* is the book where I tried to condense the highest ideals of which is capable my soul.

[25] *Tanslation:* In the writings placed at the end of his volume, Luzzatti boasts of having kept faith in scientific idealism since 1876. And it is a fact that he has always opposed materialism and determinism and has always affirmed freedom, moral values, religiosity.

All this has remained very vague in his mind; and it is impossible to establish on which arguments he bases freedom, what religion he professes, or what is his philosophical system. Rather than elaborating his ideas philosophically, he just asserts them, looking forward the philosophy which will justify them one day: "When times (that we will not see) will be ripe, it is legitimate to hope that an admirable master of natural and philosophical sciences will arise, capable of bringing together the spiritual and natural world in a luminous synthesis." But it would be unfair to deny him the merit of having given proof of healthy instinct and good tendencies in times of crude naturalism, raging in the economic and political fields, no less than in the literary and philosophical ones.

Around the age of 16 he distances himself from Judaism, but continues to be faithful, in a secular sense, to the fundamental values of the Jewish tradition, such as solidarity and freedom, which he had been educated to as a child. Like many other Italian patriots of Jewish origin, he struggles to place these values at the foundation of the new Italy.

What are the links between the values of freedom and solidarity proclaimed by Luzzatti and the traditional ones transmitted by the community in which he was born and educated? What turmoil and what tensions have shaped the form of their political sharing? I do not have the competence to deal with these aspects, which has already been considered by others,[26] However, I think it appropriate at least to point them out because of their relevance in our particular history and in the history of Italy in general.

It is worth to mention Luzzatti's answer to a letter of Geremia Bonomelli written on November 5, 1899, in which the bishop recalled Luzzatti's Jewish origins:[27]

> Io sono nato israelita e ci ritorno fieramente ogni volta che mi si rimprovera di esserlo e che l'esserlo mi espone ad un pericolo. Vi è una dignità a sostenere il peso della persecuzione e sarebbe vile il cansarlo. Ma fuori di questo, la mia educazione, le mie aspirazioni intendono a un largo cristianesimo, come traspare dai miei scritti [26][28]

3.5 Solidarity and Social Security

The value of solidarity was taken in great esteem by Luigi Luzzatti, as his Memories [26] clearly show. The pursue of this value explains his strong commitment for establishing a social security system in Italy[28]. Indeed, his approach to the problem is a scientific one. Moreover, it is crucial, according to him, a preliminary commitment for educating the masses in order to make them aware of the need to actively participate in social security and share its aims

His program is highlighted in two significant excerpts of his memories that I am going to report.

In the first, he asks for help of probability theory for establishing a social security system. In the second, he points out the need to clarify the mathematical mechanisms for those who do not have the necessary knowledge to appreciate them.

> lo Stato, offrendo di amministrare esso il risparmio popolare anche per le pensioni dei vecchi, ha la probabilità di raccogliere molti clienti e di poter porre ad effetto con minor

[26] See, for example, [4, 6].

[27] Bonomelli wrote: You are an Israelite, I am a Bishop.

[28] *Translation:* I was born an Israelite and I proudly return there every time I am reproached for being one, and that being one exposes me to danger. There is a dignity to bear the weight of persecution, and it would be coward to remove it. But out of this, my education, my aspirations point to a broad Christianity, as can be seen from my writings.

dispendio le leggi dell'assicurazione, le quali domandano l'aiuto *dei grandi numeri* per non fallire alla prova [25], p. 4.[29]

Nelle menti volgari la previdenza ha prospetti lontani, i cui effetti utili si maturano più tardi, s'intende meno di quella ad effetti immediati; e nell'economia è indizio di civiltà e di elevazione delle classi popolari la loro maggiore attitudine ad eleggersi i tipi di previdenza più elaborati.

La Cassa di Risparmio è intesa più e prima del mutuo soccorso, il mutuo soccorso per la malattia prima di quello per la vecchiaia, e le *forme razionali* dell'assicurazione si apprezzano con maggior difficoltà, perché, mancando la possibilità della esperienza immediata, è necessario un atto di ragionamento e un atto di fede, ardui ambedue negli animi semplici e diffidenti. Da queste difficoltà intrinseche pigliano qualità e modo i nuovi disegni tedeschi e italiani, gli uni e gli altri vogliono forzare a queste maniere di previdenza più sottili ed elaborate i lavoratori, moltiplicandone gli effetti utili coll'aiuto dello Stato. *Compelle intrare!* [25], p. 7.[30]

I like imagining that these reflections, as well as others of the same kind, left a deep impression on Guido Castelnuovo and they highly contributed to orient his commitment in promoting Probability Theory and Statistics, both as scientific disciplines and as substantial part of the background knowledge of officials of social security institutions and of Ministries of Economy and Finance.

3.6 *A Letter from Paolo Medolaghi to Friedrich Engel*

In the early twentieth century, Italy, and Rome in particular, were great social laboratories. The first economic miracle, which followed the great depression of the late nineteenth century, raised serious social tensions but also created new opportunities in the field of social legislation and workers protection, of which Luigi Luzzatti was one of the first to perceive the urgency. Different but equally great problems than those that arose at the time of the unification of Italy, capable

---

[29] *Translation:* The State, offering to administer popular savings even for old people's pensions, has the probability of gathering many customers and being able to implement insurance laws with less waste, and this calls for the help *of large numbers* to avoid failing the test.

[30] *Translation:* In vulgar minds, social security has distant prospects, whose useful effects mature later, of course, than for other forms of security with immediate effects; and in the economy a sign of civilization and the elevation of the popular classes is their greater aptitude for choosing the most elaborate types of social security. Saving Banks are understood better and before Mutual Aid, Mutual Aid for sickness before Mutual Aid seniority, and *rational forms* of insurance are appreciated with greater difficulty, because they cannot provide the possibility of immediate experience; an act of reasoning and an act of faith are necessary, both of which are difficult in simple and suspicious souls. The new German and Italian security plans are shaped, in quality and manner, according to these intrinsic difficulties; both of them aim at forcing workers to accept subtler and more elaborate pension schemes, multiplying their useful effects with the aid of the State. *Compelle intrare!*.

of mobilizing men (and women!) of good will and even many scientists, as in the time of the Risorgimento. Here is an excerpt of a letter of Paolo Medolaghi, a pupil of Castelnuovo, to Friedrich Engel to witness the ideal power of these challenges in shaping the interests of young intellectuals.[31]

> La nostra corrispondenza è diventata ora molto meno frequente: io ho interrotto da qualche tempo le mie ricerche sulla teoria dei gruppi e mi occupo di Geometria differenziale, E poi la maggior parte della mia giornata è occupata da affari che non hanno nulla a che fare con la Matematica. In Germania vi è tutto un sistema di leggi che proteggono gli operai: in Italia questa opera di protezione è cominciata assai più tardi, e si affida sopratutto alle iniziative individuali. La mia ambizione più alta è quella di contribuire con tutte le mie forze allo sviluppo ed al perfezionamento di questa opera, e, a dire il vero, sacrificherei per essa anche gli studi astratti se questo sacrificio fosse necessario [31].[32]

Medolaghi, after giving remarkable contributions Continuous Groups Theory, began a career as a top manager in social security institutions.

3.7 *Guido Castelnuovo's Commitment to Promoting Statistics and Probability Theory*

In 1914–1915 Guido Castelnuovo dedicated his Course lessons in Higher Geometry to the Theory of Probabilities, and in 1919 he publishes the first edition of his famous book [10]. This was the first step in his commitment to promoting the scientific knowledge necessary for planning an economic development that does not forget the protection of the weakest sections of the population.

In 1920 he got from Marco Besso Foundation funds for financing courses in Mathematical Statistics and Actuarial Mathematics at the Faculty of Sciences at the University of Rome.

Verbale di Facoltà: Seduta del 17 maggio 1920 (vol. 8)

> (...)
>
> Il Preside [Fano] legge una lettera del comm. Marco Besso che comunica che la Fondazione che da lui porta il nome ha stanziato £ 5000 annue per tre anni consecutivi allo scopo di istituire presso la Facoltà di Scienze gli insegnamenti di Statistica matematica superiore e di Matematica attuariale e dà la parola al prof. Castelnuovo cui spetta il merito di tale iniziativa. Il prof. Castelnuovo rende conto delle pratiche svolte col benemerito comm.

---

31 The letter, written on March 12, 1900.

32 *Translation:* Our correspondence has now become much less frequent: For some time I interrupted my research on Group Theory and dedicated myself to Differential Geometry. Moreover, most of my day is occupied with business having nothing to do with mathematics. In Germany there is a whole system of laws protecting workers: in Italy such commitment began much later, and, above all, it relies on individual initiatives. My highest ambition is to contribute with all my strength to the development and improvement of this work, and, to tell the truth, I would also sacrifice abstract studies for it if this sacrifice were necessary.

Besso il quale ha accolto con vero entusiasmo la proposta ed ha ottenuto dal Consiglio di amministrazione della sua Fondazione lo stanziamento dei fondi necessari per l'istituzione, per ora almeno in via provvisoria, dei due insegnamenti suddetti [36].[33]

Marco Besso, born in Trieste into a Jewish family, was President of "Assicurazioni Generali"[34] and of "Società Italiana contro gli Infortuni"[35] and a member of the Board of Directors of "Banca Commerciale".[36] On June 1st, 1918, he donated a substantial part of his assets to create a foundation, which is still alive[37] and having the purposes of:

- increasing national economy;
- improving moral and social conditions of working classes;
- spreading general knowledge.

As we said, the values of solidarity that inspire the social legislation on the protection of workers can also be traced in the family and in the non-religious Jewish environment in which Guido Castelnuovo grew up. These values are revealed in Castelnuovo's interest in Statistics and Probability Theory, which are linked to social problems in a way repeatedly indicated by his uncle Luigi in his Memories.

La statistica era per me la scienza mirabilmente adatta all'indole del secolo, spesso censurato con la taccia di positivista, perché, applicando alle scienze morali il processo logico del metodo naturale, proclamava candidamente di voler sostituire le pazienti analisi alle sterili ipotesi, la inesorabile ragione delle cifre alle intuizioni mistiche e dubbie. E mentre questa scienza sempre più aspirava con i suoi metodi elaborati e rigorosi a chiudere nell'orbita dei numeri la espressione delle condizioni dei paesi e dei popoli, si proponeva di non dimenticare che anche le cifre avevano i loro sofismi e che era d'uopo correggerle e dirigerle col genio del progresso, dell'amore e della libertà. Goethe, il quale col portentoso ingegno aveva saputo volare nei più sublimi campi della poesia e dell'arte, esprimeva una verità quando diceva: *Si afferma che i numeri governano il mondo; è certo però che i numeri mostrano come il mondo sia governato* [26], p. 152.[38]

---

[33] *Translation:* Session of May 17, 1920 (vol. 8). The Dean [Fano] reads a letter from comm. Marco Besso who announces that the Foundation bearing his name allocated £ 5000 per year for three consecutive years to establish the teaching of Higher Mathematical Statistics and Actuarial Mathematics at the Faculty of Sciences and gives floor to prof. Castelnuovo who deserves the credit for this initiative. Professor. Castelnuovo reports on the practices carried out with the meritorious comm. Besso who welcomed the proposal with great enthusiasm and obtained from the Board of Directors of his Foundation the allocation of the necessary funds for the institution, of the two aforementioned courses, at least provisionally for now.

[34] A private insurance company based at Trieste.

[35] The Italian Society against Accidents.

[36] The Commercial Bank, founded in Milan in 1894.

[37] Fondazione Marco Besso.

[38] *Translation:* For me Statistics was a science admirably suited to the spirit of the century, often censored with the mark of positivism, because, applying the logical process of natural method to moral sciences, it candidly proclaimed that it wanted to substitute patient analysis to sterile hypotheses, the inexorable reason of numbers to mystical and dubious intuitions. And while this science, with all its elaborate and rigorous methods, aimed at enclosing the domain of countries and peoples' conditions in the orbit of numbers too, it suggested not to forget that also numbers

The characteristic prose of Luzzatti is somewhat pompous and prolix, very different from his nephew's dry style, but very close in its substance, as it seems to me.

Verbale di Facoltà: Seduta del 10 novembre 1923 (vol. 9)

> (...) le Matematiche Attuariali (...) devono essere coltivate in un ambiente scientifico e in special modo a Roma, a vantaggio delle Amministrazioni dello Stato e degli Istituti di assicurazioni [36].[39]

4 Adele Levi Della Vida (1822–1915): Education and Emancipation

Another fundamental figure of Guido Castelnuovo's family is Adele Della Vida, Guido's grandmother and Luigi Luzzatti and Enrico Castelnuovo's mother-in-law. As we said before, she personally took care of the primary education of Guido and his sister.

Adele Della Vida was Samuele's daughter (one of the founders of "Assicurazioni generali," led by Marco Besso from 1877 to 1920) and Regina Pincherle.

Adele received a very accurate education and had the opportunity to approach in his parents' house many distinguished men in arts, letters, and politics, among which the patriot Daniele Manin, of whom his brother Cesare was collaborator and friend.

She got married at about 18yo to Moise Levi from Cuneo, brother of David, man of letters, poet, patriot, advocate of the emancipation of the Jews, and uncle of the positivist scientist Luigi Lombroso.

In Venice, her family spent difficult years, during which and after the siege of 1849. When she returned to Venice, the lagoon city was under the Austrian domain. Adele Levi Della Vida did not want that her children were educated at Austrian schools. Therefore, she took them every summer to Brescia, where, under the pretext of vacation, she made them take exams at Italian schools. It was in Brescia that Luigi Luzzatti met and fell in love with her daughter Amelia. Sometimes, on her trips to Brescia, she smuggled out of Venice some Italian patriot disguised as a servant.

> ...della pedagogista veneziana abbiamo una descrizione che se ben letta avrebbe fatto capire quale sarebbe stato il danno della propaganda antisemita: Adele Levi era cresciuta in ambiente aperto alle idee più belle e attraenti di Patria e di Libertà. Ebrea, ma di una

---

had their sophisms and that it was necessary to correct and direct them with the genius of progress, love, and freedom. Goethe, who, with his prodigious ingenuity had known how to fly in the most sublime fields of poetry and art, expressed a truth when he said: *It is affirmed that numbers rule the world; however, it is certain that numbers show how the world is governed.*

[39] *Translation:* Session of November 10, 1923 (vol. 9). Actuarial Mathematics (...) must be cultivated in a scientific environment, especially in Rome, for the benefit of State Administrations and Insurance Institutes.

di quelle famiglie, come i Nathan, i Rosselli, i Luzzati ecc., a cui l'Italia guarda con riconoscenza. Ebrea, ma la cui religione si confonde con il culto dell'Italia... [30].[40]

4.1 *Children's Education*

Adele Levi Della Vida was deeply concerned with education. She learned about Fröbelian kindergartens from the writings of Octavie Masson, and she traveled to Switzerland to observe them in practice.

According to Carla degli Esposti and Nicoletta Lanciano [17], Guido and Bice Castelnuovo, on the premature death of their mother, were entrusted to their grandmother Adele Levi Della Vida, who decided to open a kindergarten in Venice, to ensure a valid educational environment for her grandchildren. Unfortunately, I did not find any image of Adele Levi Della Vida. In the absence of anything better, I chose an image of an austere (perhaps too austere) teacher in a Fröbelian kindergarten where you can imagine Guido and his sister Bice sitting playing with a box of wooden shapes.

4.2 *Fröbel's Gifts*

Fröbel used to call *gifts* the training resources that he distributed to children and attached great importance to the didactic use of them. Perhaps, the importance of concrete materials in learning and teaching processes was transmitted by Guido Castelnuovo, who experienced them in kindergarten, to his daughter Emma.

When Adele moved to Florence in 1888, she set up a Kindergarten in Castello (northwestern district of the city), became inspector of elementary schools, and was among the founding members of "Scuola del Popolo," directed by Pietro Dazzi. She moved to Rome at the time when Ernesto Nathan was Mayor, established a school of home economics, continued to participate in popular education experiences, and inspired Maria Montessori's innovative educational experiments.

In the right picture of Fig. 2, you can see a box of Fröbel's gifts, similar to those with which Guido and Bice played when learning in their grandmother's kindergarten. In one of the first issues of *L' Educazione moderna*, an Italian magazine dedicated to the diffusion of Fröbelian doctrines, I found a proposed *laboratory* lesson which I like to imagine that the two children attended

[40] *Translation:* ...of the Venetian pedagogist we have a description that, if read properly, would have made it clear what would have been the damage of anti-Semitic propaganda: "Adele Levi had grown up in an environment open to the most beautiful and attractive ideas of Homeland and Freedom. Jewish, but from one of those families, such as the Nathans, the Rossellis, the Luzzatis, etc., which Italy looks to with gratitude. Jewish, but whose religion is confused with the cult of Italy."

Fig. 2 On the left, an image of children with their teacher in a kindergarten. On the right, a box of Fröbel's gifts

> Vedete o bambini questo bicchiere che contiene un liquido, limpido come l'acqua pura? E veramente questo liquido è acqua; ma non è propriamente pura; vi fu stemperata entro, una piccola quantità di calce che non si vede, perché è perfettamente sciolta. Ora state ad osservare: metto fra le labbra questo cannellino, e soffio leggermente per entro al liquido. Vedete come si intorbida; lasciamolo un po'riposare; ecco che torna limpido; ma sul fondo è deposta una polvere bianca che prima non c'era. Forse che il mio soffio ha avuta la virtù di raccogliere la calce che prima era sciolta e di farla precipitare in basso? No, o bambini, dalle mie labbra è uscita un'aria differente da quella che ho respirata, prima di accingermi a questa esperienza. Quest'aria è un miscuglio di due altre arie, diciamole gas, con termine più italiano; l'uno di quei gas l'avevo, a dir vero, respirato, perché è un componente dell'aria che ci sta d'intorno; l'altro no, esso è un composto di due differenti gas dei quali l'uno è pure componente del miscuglio aereo. Questo gas composto, uscito dal mio petto, dicesi *acido carbonico*. Io avevo inspirato *azoto* misto ad *ossigeno* ed ho espirato *azoto* misto ad acido carbonico. L'acido carbonico si è combinato con la calce che era sciolta nell'acqua ed ha fatto un composto che è quella polvere caduta in fondo. Quella polvere è marmo polverizzato. Vedete dunque o bambini che l'aria uscita dalla nostra bocca non è uguale a quella che vi entra; se volete accertarvene, invece del cannello fate uso d'un soffietto come quello che si adopera per le stufe, e siccome esso ridona l'aria che riceve, non vedrete intorbidarsi il liquido, non vedrete disposri la polvere bianca sul fondo del bicchiere [7][41]

[41] *Translation:* Do you see, children, this glass contains a liquid, as clear as pure water? And truly, this liquid is water; but it is not strictly pure; a small quantity of lime was dissolved within it, which cannot be seen, because it is perfectly dissolved in it. Now observe: I put this small tube between my lips, and I blow lightly into the liquid. You see how cloudy it gets; let it rest for a while; now it is clear again; but on the bottom, a white powder is deposited, that was not there before. Did my breath have the virtue of collecting the lime that was previously melted and making it fall down? No, children, a different air came out of my lips from the one I breathed, before beginning this experience. This air is a mixture of two other airs, let's say gas, with a more appropriate Italian term; I had actually breathed one of those gases, because it is a component of the air around us; the other is not, it is a compound of two different gases, one of which is also a component of the air mixture. This compound gas, which came out of my chest, is called *carbonic acid*. I had inhaled

Fig. 3 In his writings on teaching, Guido Castelnuovo often refers to the importance of simple non-trivial experiences in the teaching of science

And Fig. 3 shows how I imagine Guido's sister during the experience.[42]

I believe that the activities proposed by their grandmother Adele have marked Guido and Bice with an indelible learning experience, of which I think it is possible to find traces in Guido's reflections about school and teaching.

5 Castelnuovo and the Problem of Higher Education

In conclusion of this work, I want to highlight some aspects of the idea of school that emerges from Guido Castelnuovo's articles written between 1907 and 1912 and dedicated to the problem of teaching [11–13]. In my opinion, they are directly

nitrogen mixed with *oxygen* and exhaled *nitrogen* mixed with carbonic acid. The carbonic acid combined with the lime that was dissolved in the water and made a compound which is that dust that fell to the bottom. That dust is pulverized marble. So you see, children, that the air that comes out of our mouth is not the same as that that enters it; if you want to ascertain it, instead of the small pipe, use a bellows like the one used for stoves, and since it restores the air it receives, you will not see the liquid become cloudy, you will not see the white powder arranging itself on the bottom of the glass.

[42] The image has been downloaded from the Internet at the url: https://vivalascuola.studenti.it/3-esperimenti-scientifici-da-fare-con-i-bambini-249181.html.

linked to his personal experiences in kindergarten and to a critical rethinking of his grandmother's ideas on children education and those of his father on vocational training. In particular, it is likely that the experiences of his school days came back to his mind vividly when his children attended school and his grandmother participated in the lively debate on elementary and popular education in Rome. Of course, in addition to personal experiences, there are also other sources from which Castelnuovo developed his ideas on school, including the ideas on teaching of Felix Klein (see [18, 19, 29]) and his brother-in-law Federigo Enriques' dynamic vision of science and its learning [20]. An extensive analysis of these sources and of the synthesis made by Castelnuovo would deserve a much more in-depth work.

In this paper, I limit myself to delving into just two of the main themes considered by Castelnuovo in his work:

1. Which are the main student's attitudes to be developed in teaching and how to develop them.
2. The importance of linking disciplines in teaching.

I shall mainly translate some excerpts from [11–13] I hope that providing an (even imperfect) English translation would contribute to arousing interests in his contribution to thinking on learning and teaching mathematics and on the influences of his thinking.

5.1 The Attitudes to be Developed in Teaching, and How to Develop Them

According to Castelnuovo, the school should aim to balance the different attitudes of intelligence, without neglecting any to the advantage of others, whatever the future of the young person will be.

> Lo scopo precipuo che l'insegnante deve proporsi non è quello di dare ai giovani una indigesta ed effimera erudizione, bensì di educare armonicamente tutte le varie attitudini dell'intelligenza, risvegliando le assopite, e disciplinando le esuberanti. Le maggiori cure egli dovrà dedicare alla facoltà più nobile, la fantasia creatrice che risulta da un felice accordo dell'intuizione[43] con lo spirito di osservazione. Mancherà il tempo per estendere la cultura? E che importa? Le sole nozioni che la mente sappia conservare sono quelle che essa è adatta a ricevere, o quelle (oserei dire) che essa è in grado di procurarsi da sé.
>
> Preparare il terreno è la cosa essenziale. La natura è tutta piena di germogli fecondi. Se il terreno sarà fertile, non tarderanno a sbocciare i più mirabili fiori [11], pp. 15–16.[44]

[43] Which, for Castelnuovo, is simply: "the fruit of unconscious experiences or even imagined experiences."

[44] *Translation:* The main purpose that a teacher must pursue is not to give young people an indigestible and ephemeral erudition, but to harmoniously educate all the various attitudes of intelligence, awakening the dormant, and disciplining the exuberant. He will devote the greatest attention to the noblest of all faculties, the "creative fantasy," resulting from a happy agreement of intuition [(which is just "the fruit of unconscious experiences or even imagined experiences")] with

More precisely, what are the main attitudes or qualities that teaching must develop? Castelnuovo says:

> Nel campo intellettuale le qualità che meglio valgono a distinguere l'uomo elevato dalla mediocrità sono *la fantasia creatrice*, *lo spirito di osservazione*, che forniscono gli elementi ad ogni opera d'arte o di scienza, *le facoltà logiche* che, frenando gli slanci del pensiero, danno all'opera proporzioni giuste e coesione.
>
> Senza fantasia non vi è artista né scienziato. Ma di fantasia non possono difettare nemmeno l'ingegnere, il commerciante, l'uomo politico, quando essi non si rassegnino a seguire pedestremente i precetti dei loro predecessori, rinunziando a qualsiasi audace iniziativa[12], p. 25.[45]

But how to develop these qualities in teaching?

> Ogni opera grande di arte o di scienza, i poemi di Omero come le moderne teorie cosmologiche, fisiche o biologiche, quando siano commentati da uno spirito largo, possono risvegliare o disciplinare la divina fantasia [12], p. 25.[46]

And how to train the broad spirit[47] of a teacher?

> Dal maestro dovremmo quindi esigere, più ancora che una profonda e specialistica conoscenza di un campo ristretto, un larga visione delle scienze che colla propria hanno le maggiori affinità, e delle applicazioni a cui quella dà luogo [13] p. 57.[48]

Castelnuovo often insists on the importance, in the training of perspective teachers, of complementing an "intensive" disciplinary teaching (favoring in-depth studies) with an "extensive" one (favoring connections) [18, 19].

---

the spirit of observation. Will lack time to extend culture? What does it matter? The only notions that the mind knows how to keep are those that it is capable of receiving, or those (I dare say) that it is able of getting itself.

Preparing the ground is the essential thing. Nature is full of fruitful seeds. If the soil is rich, the most wonderful flowers will bloom soon.

45 *Translation:* In the intellectual field, the qualities that best distinguish the elevated man from mediocrity are *creative imagination*; *spirit of observation*, which provides the elements to every work of art or science; *logical abilities*, which, by holding back the impulses of thought, give the work correct proportions and cohesion.

Without imagination there is neither scientist nor artist. But even the engineer, the merchant, the politician cannot lack imagination when they do not accept following the precepts of their predecessors, renouncing any audacious initiative.

46 *Translation:* Any great work of art or science, Homer's poems, as well as modern cosmological, physical, or biological theories, when commented by a broad spirit, may awaken or discipline the divine fantasy.

47 "spirito largo".

48 *Translation:* We should therefore demand from the teacher, even more than a deep and specialized knowledge of a narrow field, a broad vision of the sciences which have the greatest affinity with their own and of the applications to which it gives rise.

5.2 The Importance of Linking Disciplines in Teaching

As we have seen, according to Castelnuovo, one of the fundamental attitudes to be developed through teaching is the *spirit of observation*. It is a transversal quality, in which teachers of different disciplines are invited to collaborate in their teaching.

> Lo studio dei capolavori artistici, gioverà pure a perfezionare una qualità a cui attribuisco un valore grandissimo: lo spirito di osservazione. In ciò il professore di materie letterarie o artistiche potrà aiutare i colleghi di scienze naturali e di fisica, ai quali particolarmente è affidato questo nobile intento.
>
> (...)
>
> lo spirito di osservazione viene acuito dalla fisica, la quale, ove sia insegnata in modo opportuno, dà insieme una squisita educazione dei sensi e della mente. Il più umile fenomeno, sfuggito mille volte alla nostra attenzione, viene fissato dal sagace osservatore., il quale lo associa con mille fatti che in apparenza non hanno con quello nessun rapporto, e che pur risultano espressioni d'unica legge grandiosa, valida nell'intero universo. Qual ginnastica intellettuale può reggere al paragone con questa, che alterna continuamente l'uso dei procedimenti deduttivi e induttivi, e ad ogni passo richiede l'intervento delle doti tecniche per controllare con l'esperimento le più ardite concezioni del pensiero?
>
> (...)
>
> Nell'educare la mente nell'acuire le qualità logiche, la fisica troverà una preziosa alleata nella matematica, purché questa si tolga quel superbo isolamento in cui, a torto, si è voluto rinchiuderla [12], pp 26–27.[49]

[49] *Translation:* The study of artistic masterpieces will also help to perfect a quality to which I attribute a great value: the spirit of observation. In this, the professor of literary or artistic subjects will be able to help his colleagues in natural sciences and physics, to whom this noble aim is particularly entrusted.

(...)

the spirit of observation is sharpened by physics, which, when taught in an appropriate way, gives at the same time an exquisite education of both senses and mind. The humblest phenomenon, which escaped our attention a thousand times, is fixed by the shrewd observer, who associates it with a thousand facts which apparently have no relation with it, and which are nevertheless expressions of a single grandiose law, valid in the whole universe. What intellectual gymnastics can stand up in comparison with this one, that continually alternates the use of deductive and inductive procedures, and at each step requires the intervention of technical skills to control the most daring conceptions of thought with experiment?

(...)

In educating the mind to sharpen the logical qualities, physics will find a precious ally in mathematics, provided that mathematics gets out of that superb isolation in which we, wrongly, wanted to lock it up.

and, moreover

> le lettere e le scienze, quando siano insieme associate, educano nel modo più armonico le varie facoltà dello spirito; mentre se gli uni insegnamenti troppo prevalgono sugli altri, la cultura diviene unilaterale e ristretta [12], p. 28.[50]

It is a question of overcoming, at least in school, one of the limits of modern science, the scientific specialism, about which Castelnuovo commented:

> Contro lo specialismo scientifico si va ora combattendo una vivace campagna, che io ritengo giusta e opportuna. Faccio solo qualche riserva sui termini di essa; riconosco infatti che, ove non si voglia rinunziare nell'indagine scientifica al concorso degli ingegni medi, è necessario consentire a questi di coltivare di coltivare un campo limitato, lasciando ai sommi l'ebbrezza di dominare un orizzonte più vasto. Ma dove ritengo funesto senza restrizioni lo specialismo è nell'insegnamento, e specialmente nell'insegnamento medio. La cultura generale che esso si propone di fornire non deve assomigliare ad un territorio selvaggio e montuoso, le cui vette illuminate dal sole sono separati da abissi profondi e inesplorati. Deve esser piuttosto un dominio già civilizzato, le cui province siano collegate da ponti e da strade. Non già i particolari più raffinati di una dottrina interessano il giovanetto che anela ad estendere il proprio sapere. La sua curiosità è spesso attratta verso quelle questioni elevate ed eterne, che mal si adattano alle artificiali divisioni dei nostri libri. O, se le sue attitudini lo portano verso le questioni concrete, egli si ribellerà contro l'eccessivo spirito astratto dei nostri corsi, e non comprenderà l'interesse di una teoria finché non ne avrà vista qualche pratica conseguenza. Disattento o passivo ascoltatore mentre l'insegnante si affatica a sviscerare con soverchia minuzia un capitolo del programma, il discepolo si anima di vita spirituale, quando rapporti imprevisti o inattese implicazioni vengono rivelate al suo intelletto [13], p. 77.[51]

Acknowledgments I like to acknowledge the help of Sergio Camiz in the revision of my English and the competent support of Carlo Urbani in searching the archives of Istituto di Scienze, Lettere ed Arti in Venice.

---

[50] *Translation:* Letters and sciences, when combined together, educate the various faculties of the spirit in the most harmonious way; while if one teaching too much prevails over the others, culture becomes one-sided and restricted.

[51] *Translation:* A lively campaign is now being fought against scientific specialism, which I believe to be right and suitable. I only have some concerns about the terms of it; in fact, I recognize that, if one does not want to renounce the contribution of average talents in scientific investigation, it is necessary to allow them to cultivate a limited field, leaving to the top talents the thrill of dominating a wider horizon. But where I find fatal specialism, without restriction, is in teaching, and especially in school teaching. The general culture it aims at providing must not resemble a wild and mountainous territory, whose sunlit peaks are separated by deep and unexplored chasms. Rather, it should be an already civilized domain, whose provinces are connected by bridges and roads. The finest details of a doctrine are not those which interest a young man who yearns to extend his knowledge. His curiosity is often drawn to those lofty and eternal questions, which are ill-suited to the artificial divisions of our books. Or, if his attitudes lead him to concrete questions, he will rebel against the excessively abstract spirit of our courses and will not understand the interest of a theory until he has seen some practical consequence. Inattentive or passive listener, while the teacher struggles to dissect a chapter of the program with excessive minutiae, the disciple becomes animated with spiritual life, when unexpected relationships or unexpected implications are revealed to his intellect.

References

1. Arbarello, E.: In memoria di Guido Castelnuovo, Conferenza alla giornata *Guido Castelnuovo: un ricordo a 150 anni dalla nascita*, organizzata dal Dipartimento di Matematica della "Sapienza, Università di Roma" il 5 Novembre (2015) http://www1.mat.uniroma1.it/people/arbarello/In%20honour%20of%20Guido%20Castelnuovo.pdf
2. Archivio Luzzati, consultabile presso l'Istituto Veneto di Scienze, Lettere ed Arti. Venezia.
3. Bano, D.: "La Scuola Superiore di Commercio" in *Storia di Venezia*, Istituto dell'Enciclopedia Italiana Treccani, Roma, 2002.
4. Berengo, M.: Luigi Luzzatti e la tradizione ebraica, in Pier Luigi Ballini e Paolo Pecorari, *Luigi Luzzatti e il suo tempo*, Istituto Veneto di Scienze, Lettere ed Arti, Venezia (1994)
5. Bordiga G.: "Enrico Castelnuovo", commemorazione in Atti dell'Istituto veneto di scienza, letteratura ed arti **75**(1), 27–47 (1915–16)
6. Capuzzo, E.: Luigi Luzzatti tra ebraismo e laicità. Clio **XLIII**(4), 693–701 (2007)
7. Cassani, P.: La scienza alla portata dei fanciulli. L'educazione di fanciulli, periodico mensile indirizzato alla diffusione delle teorie di Federico Frobel **1**, 4–9 (1883)
8. Castelnuovo, E.: I Moncalvo, Roma, Lucarini (1989)
9. Castelnuovo, G.: Discours de M. Castelnuovo. L'Enseignement mathématique **16**, 188–191 (1914)
10. Castelnuovo, G.: Calcolo delle Probabilità, Alighieri, Milano (1919)
11. Castelnuovo, G.: Il valore didattico della matematica e della fisica. Rivista di Scienza (Scientia), **1**, 329–337 (1907). Anche in Castelnuovo, G.: Opere Matematiche, Roma, Accademia Nazionale dei Lincei, p. 9–16 (2017)
12. Castelnuovo, G.: La scuola media e le attitudini che deve risvegliare nei giovani. L'istruzione media: giornale della Federazione nazionale Insegnanti di Scuole medie. **9**, 33–47 (1910). Anche in Castelnuovo, G.: Opere Matematiche, Roma, Accademia Nazionale dei Lincei, pp. 21–30 (2017)
13. Castelnuovo, G.: 'La scuola nei suoi rapporti con la vita e con la scienza moderna. In: Atti del III Congresso della Mathesis – Società Italiana di Matematica (Genova 21–24 Ottobre 1912), Cooperativa Tipografica Manuzio, Roma, 1913, pp. 15–21. Anche in Castelnuovo, G.: Opere Matematiche, Roma, Accademia Nazionale dei Lincei, pp. 76–81 (2017)
14. Castelnuovo, G.: Opere Matematiche, Roma, Accademia Nazionale dei Lincei (2017)
15. Croce, B.: Recensione a *Luigi Luzzatti, La libertà di coscienza e di scienza*. La critica **7**, 287–292 (1909)
16. Degli Esposti, C., e Lanciano, N., Castelnuovo, E.: L'asino d'oro, Roma (2016)
17. Degli Esposti, C., e Lanciano, N., Castelnuovo, E.: Dizionario Biografico degli Italiani, Istituto dell'Enciclopedia Italiana, Roma (2018)
18. De Marchis, M., Rogora, E.: Attualità delle riflessioni di Guido Castelnuovo sulla formazione dell'insegnante di Matematica. Periodico di Matematiche **9**, 71–79 (2017)
19. De Marchis, M., Menghini, M., Rogora, E.: The importance of Extensive Teaching in the education of perspective teachers of Mathematics. In: Szarková, D., Richtáriková, D., Prášilová, M. (eds.) Proceedings of APLIMAT 2020, Bratislava, pp. 344–353 (2020)
20. Enriques, F., Chisini, O.: Lezioni sulla Teoria Geometrica delle Equazioni e delle Funzioni algebriche, vol. I. Zanichelli, Bologna (1915)
21. Fradeletto, A.: Commemorazione di Enrico Castelnuovo. annuario 1921–22 della scuola per il commercio di Venezia. https://fc.cab.unipd.it/fedora/objects/o:50450/methods/bdef:Book/view?language=it#page/31/mode/1up
22. Gario, P.: Guido Castelnuovo: l'uomo e lo scienziato. Rendiconti di matematica e delle sue applicazioni **XXXVII**(3–4), 147–183 (2016)
23. Rosadi, G.: I poeti di Montecitorio, Luigi Luzzati. Il Marzocco, **7**, p. 1 (1913)
24. Levi, A.: Enrico Castelnuovo, l'autore dei Moncalvo. La Rassegna Mensile di Israel, terza serie **15**(8/9), 388–419 (1949)
25. Luzzatti, L.: Previdenza libera e previdenza legale. Hoepli, Milano (1882)

26. Luzzatti, L.: Memorie. Zanichelli, Bologna (1935)
27. Luzzatti, L.: Discorsi parlamentari (2 voll.). CRD della camera dei deputati, Roma, (2013)
28. Marucco, D.: Luigi Luzzatti e gli esordi della legislazione sociale. In Pier Luigi Ballini e Paolo Pecorari, Luigi Luzzatti e il suo tempo, Istituto Veneto di Scienze, Lettere ed Arti, Venezia (1994)
29. Menghini, M.: Guido Castelnuovo e l'insegnamento della Matematica. Rendiconti di Matematica **VII**(37), 185–197 (2016)
30. Morpurgo, P.: *Le scuole e gli ebrei.* https://www.edscuola.it/archivio/didattica/scuolebrei.html
31. Nachlassverzeichnis Friedrich Engel, Giessen
32. Negri Zamagni, V.: Luigi Luzzatti. In: Il Contributo italiano alla storia del Pensiero – Economia (2012)
33. Onoranze per il giubileo scientifico del prof. Guido Castelnuovo, Città di Castello, Tip. dell'Unione Arti Grafiche (1937)
34. Polizzi, G.: La polemica di Gentile con Federigo Enriques. In: Ciliberto, M. (ed.) Croce e Gentile, pp. 156–161. Istituto dell'Enciclopedia Italiana, Stamperia artistica Nazionale, Torino (2016)
35. Recchilongo, B.: Enrico Castelnuovo. In: Dizionario Biografico degli Italiani, vol. 21 (1978)
36. *Verbali delle sedute della Facoltà di Scienze*, Archivio Storico Sapienza, Università di Roma

The Genesis of the Italian School of Algebraic Geometry Through the Correspondence Between Luigi Cremona and Some of His Students

Nicla Palladino and Maria Alessandra Vaccaro

Abstract Luigi Cremona is considered the founder of the Italian school of algebraic geometry. He formed a group of students of great value, very active in scientific research. Examining the letters from Eugenio Bertini, Ettore Caporali, and Riccardo De Paolis to Cremona preserved in the archive of the Istituto Mazziniano in Genoa, we have reconstructed their biographies, careers, studies, and relationships with their teacher. They had the merit of cultivating the scientific innovations of the period and passing them on to the subsequent generations.

Keywords Italian school of algebraic geometry · Luigi Cremona correspondence · Bertini · Caporali · De Paolis

1 Introduction

In the first decades of the nineteenth century, the French and the German schools had built the foundations of modern geometry. The establishment of the Italian school of algebraic geometry was one of the aims of the broader post-Risorgimento plan of founding an Italian mathematical school related to the most advanced European studies.

Antonio Luigi Gaudenzio Giuseppe Cremona (Pavia 1830—Roma 1903) is considered one of the renovators of geometric studies in Italy. His fundamental scientific merit was probably the systematic study, since 1863, of "birational transformations" on the plane and the space named, after him, "Cremona transformations." See [1, 2]. In addition to the scientific ones, Cremona had the merit of having formed a group

N. Palladino (✉)
Dipartimento di Matematica e Informatica, University of Perugia, Perugia, Italy
e-mail: nicla.palladino@unipg.it

M. A. Vaccaro
Dipartimento di Matematica e Informatica, University of Palermo, Palermo, Italy
e-mail: marialessandra.vaccaro@unipa.it

G. Bini (ed.), *Algebraic Geometry between Tradition and Future*,
Springer INdAM Series 53, https://doi.org/10.1007/978-981-19-8281-1_14

of students of great value, very active in scientific research and also as teachers or directors in various Italian schools.

From 1860 to 1866, Cremona was Professor of Higher Geometry at the University of Bologna. In 1866, he moved to the Polytechnic University of Milan, founded in 1863 by Francesco Brioschi, teaching higher geometry and graphical statics. In 1873, he was called to Rome, recently become the capital of Italy, to organize and direct the Scuola di applicazione degli ingegneri. He was also appointed Professor of Higher Mathematics at the University of Rome. He was a member of the senate of the Kingdom of Italy and briefly minister for education.

Among the students who studied with him and scientifically followed his footsteps, at least for some phase of their career, we can mention:

a) Eugenio Bertini (1846–1933, graduated in Pisa) among the Bolognese students
b) Angelo Armenante (1844–1878, graduated in Naples), Ferdinando Aschieri (1844–1907, graduated in Pisa), Giulio Ascoli (1843–1896, graduated in Pisa), Giuseppe Jung (1845–1926, graduated in Naples), Carlo Saviotti (1845–1928), and Emil Weyr (1848–1894) among the students of the Polytechnic of Milan (where they came to perfect their studies)
c) Riccardo De Paolis (1854–1892), Ettore Caporali (1855–1886), Giuseppe Veronese (1854–1917), and Giovanni Battista Guccia (1855–1914) among the Roman ones

The students who had direct contacts with Cremona form the "first nucleus" of young Italian algebraic geometers. Actually, Corrado Segre (1863–1924), Guido Castelnuovo (1865–1952), Federigo Enriques (1871–1946), and Francesco Severi (1879–1961), who brought Italian algebraic geometry to full maturity, constitute the second generation of Italian algebraic geometers.

> To create a school, the value of the teacher is not enough nor is it enough that he knows how to project a research plan so great that it surpasses his own workforce. It is also necessary that he be able to communicate his passion and faith to the students and know how to require and direct their collaboration. Luigi Cremona had these qualities in an eminent degree. The students who were lucky enough to listen to him in full zeal of research tell us that his enthusiasm for the problems he presented transpired during the lesson and was transmitted to the audience making them participate in the enjoyment of the discovery. With willpower, which was his main gift, he influenced young people and attracted them to the field of study he favored. This strength and faith were such that even his direct disciples were able to convey them to us, the disciples of the second generation.[1]

[1] "Per dar vita ad una scuola non basta il valore del maestro, né basta che egli sappia tracciare un piano di ricerche così vasto da superare la propria forza di lavoro. Occorre altresì che egli riesca a comunicare la sua passione e la sua fede ai discepoli e sappia esigerne e dirigerne la collaborazione. Queste doti possedeva in grado eminente Luigi Cremona. Raccontano gli allievi che ebbero la fortuna di ascoltarlo quando egli era nel pieno fervore della ricerca, che l'entusiasmo per le questioni da lui esposte traspariva durante la lezione e si trasmetteva all'uditorio rendendolo partecipe del godimento della scoperta. Con la forza di volontà, che era una sua dote precipua, suggestionava i giovani e li attirava verso l'indirizzo da lui prediletto. Tali furono questa forza e questa fede che noi stessi della seconda generazione ne subimmo la influenza, trasmessaci dai discepoli diretti."

The above statement by Guido Castelnuovo in [3, pp. 615–616] well described Cremona's role in the creation of the Italian school of algebraic geometry.

In different ways and with different roles, and according to their possibilities, the aforementioned scholars contributed to raising the Italian mathematics by working with alacrity in scientific research. Following their career and their studies allows us to reconstruct the genesis of the Italian school of algebraic geometry; to do this, it was useful to make use of the unpublished correspondence they had with their professor, preserved in the archive of the Istituto Mazziniano in Genoa (Legato Itala Cozzolino Cremona). The archive contains a wide number of letters from almost all these scholars that give a complete idea of the role played by Cremona: Armenante (15 letters), Aschieri (5 letters), Ascoli (13 letters), Bertini (14 letters), Caporali (48 letters), De Paolis (1 letter), Guccia (44 letters), Jung (148 letters), and Weyr (1 letter). Furthermore, with regard to his direct students, 2 letters from Caporali, 6 letters from Saviotti, and 27 letters from Weyr to Cremona have been published in [4].

In this chapter, we directed our attention to Bertini, Caporali, and De Paolis not only because they are the first group of researchers who embraced Cremona's research, but above all because they can be credited with having represented with their studies a transition bridge to the next generation of algebraic geometers.

2 The First Student of Cremona: Eugenio Bertini

Bertini was in chronological order the first distinguished student of Cremona and, most likely, the one who had a role of major importance in the subsequent developments of algebraic geometry.

He was born in Forlì on 8 November 1846 from Vincenzo and Agata Bezzi and, after attending the technical school of the city, in 1863, at the age of 17, he entered the University of Bologna to study engineering. Following a mathematics course held by Cremona, he decided to devote himself to the studies of pure mathematics. In 1866, when the Third War of Independence broke out, Bertini interrupted the studies and he joined as a volunteer with Garibaldi. After starting again his studies, Cremona advised Bertini to move to the University of Pisa under the leadership of Enrico Betti and Ulisse Dini.

> With these few lines I recommend the young Eugenio Bertini from Forlì. He is a young man very dear to me for his excellent qualities: he is very clever and really wants to learn; if, as I do not doubt, he perseveres, he will be different from other scholars. Yesterday he passed the Higher Geometry exam which lasted more than an hour and made Professor Grassmann, who was present, amazed. But unfortunately Bertini knows little or nothing about analysis: and he comes to Pisa precisely to study algebra, calculus, etc. I kindly ask you to follow

> and advise him. Introduce him, also in my name, to Novi, to Dini, to those who can benefit him.[2]
>
> [Cremona to Betti. Bologna, 11 November 1866]

In 1867, Bertini graduated with honors in mathematics, and in 1868, as "aggregate student" of the Reale Scuola Normale Superiore, he obtained the teaching qualification by presenting a valuable thesis on Eulerian polyhedra, subsequently published in the first volume of Annali of the Reale Scuola Normale Superiore.

> As you already know, for Normale's thesis I continued to study those things on polyhedra I wrote to you about in a previous letter of mine, making various modifications. Pr. Betti and Dini were not dissatisfied and they told me that they would propose to publish them in the annali of mathematics.[3]
>
> [Bertini to Cremona. Tredozio (on the Apennines), 7 September 1868]

Cremona's charisma attracted Bertini to Milan, the city where the geometer had moved since 1867. During the academic year 1868–1869, he had the opportunity to follow a three-part course held by Brioschi, Felice Casorati, and Cremona on abelian integrals, analyzed from three different points of view: Jacobi's analytical method, Riemann's topological one, and Clebsch's and Gordan's algebraic geometric one. The course had a profound and lasting influence on the young mathematician, to the point that in 1869 he was able to give one of the first and simplest geometric proofs on the genus invariance of an algebraic curve with birational transformations.[4] This proof, included in three classical texts [5, pp. 683–684; 6, vol. 2, pp. 131–135; 7, p. 314], helped to increase Bertini's estimation.

From December 1869 to 1872, Bertini taught at the Parini high school in Milan and in 1872–1873 at the Ennio Quirino Visconti high school in Rome. In that same year, by the will of Cremona who also settled in Rome, he taught courses in descriptive and projective geometry.

---

[2] "Con queste poche righe ti raccomando il giovane Eugenio Bertini di Forlì. È un giovane a me assai caro per le egregie sue qualità: ha molto ingegno e molto desiderio di imparare; se, come non dubito, egli persevera, riuscirà qualche cosa di diverso dall'ordinario. Ha fatto l'altr'ieri un esame di geometria superiore che è durato più di un'ora e che ha fatto meravigliare il professore Grassmann ch'era presente. Ma disgraziatamente il Bertini sa poco o nulla di analisi: e viene a Pisa appunto per studiare algebra, calcolo, ecc. Ti prego caldissimamente di assisterlo e consigliarlo. Presentalo, anche a nome mio, al Novi, al Dini, a chi gli può giovare."

[3] "Come già le sarà noto, per la tesi di Normale seguitai quelle cose sui poliedri di cui le scrissi in una mia precedente, apportandovi varie modificazioni. Il Pr. Betti e il Dini non ne rimasero scontenti ed essi medesimi mi dissero che si sarebbero adoperati onde venissero pubblicate negli annali di matematica."

[4] E. Bertini, *Nuova dimostrazione del teorema: due curve punteggiate projettivamente sono dello stesso genere*, Giornale di Matematiche, 7, 1869, pp. 105–106.

> I was given the course of Descriptive Geometry and this too was approved by the superior council. I confess that this assignment gives me great pleasure and brings me a great deal of satisfaction. I know that for this too I am grateful to you for the words you wrote to Rome and that I doubt I deserve.[5]
>
> [Bertini to Cremona. Roma, 13 November 1872]

In 1875, Bertini won, with the support of Betti, the competition for the teaching position of Higher Geometry at the University of Pisa and after 3 years he became full professor. He remained in Pisa until 1880, when, by exchange with De Paolis, he moved to the University of Pavia, also attracted by the presence of two dear colleagues: Casorati and Eugenio Beltrami. In 1892, ceding to the friendly insistence of Luigi Bianchi and Ulisse Dini, Bertini returned to Pisa where the position of Higher Geometry had become vacant due to the untimely death of De Paolis and remained there for the rest of his life. In 1922, he retired at the age of 75 (succeeded by Carlo Rosati), but for 10 years more he continued to teach as professor emeritus a course of complements of projective geometry, whose lessons are published in a book he wrote at the age of 82 years. He died in Pisa on 24 February 1933.

Bertini is described by some of his direct students, such as Luigi Berzolari (1863–1949), Guido Fubini (1879–1943), and Gaetano Scorza (1876–1939), who affectionately named him "Papà Bertini," as a modest, selfless, good-natured person, unrelated to any form of exhibitionism, dedicated to his family and his work. "Passionate about teaching, he carefully prepared his lessons teaching with admirable clarity and precision" (Castelnuovo in [8, p. 748]). "For Bertini, research and teaching were simply two aspects of the same activity" (Berzolari in [9, p. 612]). "Often the subject of his lessons were the contributions he brought to science, so conversely many of his researches originated from didactic needs [...] many of Bertini's most beautiful works were precisely caused by the need to perfect this or that theory before presenting it into school, here filling a gap, there obviating possible objections" (Scorza in [10, pp. 110–111]).[6]

Educated at the Cremona school, Bertini soon departed from it, marking an interesting turning point in the history of Italian algebraic geometry. Between 1863 and 1865, Cremona had created the theory of birational transformations,[7] then named after him, first on the plane and then in space. He used them as a tool to reduce more complicated geometric entities into simpler ones and to convey the properties of the latter to them. Probably seduced by the many applications, it seems that Cremona had no interest in studying the group theoretical properties that remain unchanged under these transformations. On the contrary, in 1876, Bertini, through research on the plane involutions of pairs of points, decided to determine the

[5] "A me fu affidato l'incarico dell'insegnamento della Geometria descrittiva e anche questo fu approvato dal consiglio superiore. Le confesso che questo incarico mi fa molto piacere e mi reca moltissima soddisfazione. So che anche di ciò io debbo avere molta gratitudine a Lei che volle scrivere a Roma parole che io dubito di meritare."

[6] About Bertini's life, see also [11–13].

[7] With regard to the historical aspects of the Cremona transformations, see [14].

various typologies to which the plane involutions can be reduced through Cremona transformations.

> I also thank you for the inclusion of my work in the Annali [15]. Thinking in these days about the general question, I realized that some cases were missing. That is, when the fixed curve Γ does exist, it can be of order n with a (n-1)^{ple} point and it can break in parts: furthermore it can even be the case where there is a unique fixed dotted line. I have now completed the study of these three cases where there are no serious difficulties, and it seems to me that all three are deduced from harmonic homology with repeated quadratic transformations. I will write these additions in the proofs, since, as far as I have sent it to you, there are no major changes. But please do me the courtesy of reviewing the second drafts. I will try to be sure of what I do: but, to tell you the truth, experience has taught me to be very cautious and I would be much calmer if you wanted to look at these insertions (in which there are some delicate considerations).[8]
> [Bertini to Cremona. Pisa, 23 July 1876]

A few months later, Caporali also devoted himself to studying Bertini's first work on involutions. In fact, he wrote to Cremona:

> I have read a note by Prof. Bertini on the Jonquières transformations which are also involutory. It seems to me that the consideration of other cases of the same problem may be interesting. When I was working on my thesis I encountered another class of involutory transformations, and it is the following. Fixed on the plane a net of curves of order n and genus one which have in common so many fixed points as to absorb n-2 intersections (which is generally possible in several ways), all the curves that pass through an arbitrary point in the plane also pass through another point that can be taken as corresponding to the first. The simplest case (for n = 3) served me in my thesis.[9]
> [Caporali to Cremona. Roma, 11 September 1876]

Bertini's goal was clear, as he himself declared in the introduction to his Research [16, p. 244]: "Indicate all the possible involutory transformations of the plane, which

[8] "La ringrazio altresì della inserzione del mio lavoretto negli Annali [15]. Pensando in questi giorni alla questione generale mi sono accorto che mancava la trattazione di alcuni casi. Cioè, quando esiste la curva unita Γ, questa può essere d'ordine n con un punto $(n-1)$plo e può spezzarsi in parti: e può anche darsi che esista una sola retta punteggiata unita. Ho ormai condotto a termine lo studio di questi tre casi, ne' quali non s'incontrano gravi difficoltà, e mi pare che tutti tre si deducano dall'omologia armonica con ripetute trasformazioni quadratiche. Queste aggiunte le farò nelle bozze di stampa, giacché, per quello che le ho spedito, non vi sono modifiche di gran rilievo. Però voglia usarmi la cortesia di rivedere le seconde bozze. Cercherò di essere sicuro di ciò che faccio: ma, a dirle il vero, l'esperienza mi ha insegnato a diffidare assai e sarei assai più tranquillo se a queste aggiunte (nelle quali c'è qualche considerazione delicata) Ella volesse dare una occhiata."

[9] "Ho letto una memoria del Prof. Bertini sulle trasformazioni di Jonquières che sono anche involutorie. Mi pare che la considerazione di altri casi dello stesso problema, possa essere interessante. Quando facevo la mia tesi mi si era presentata un'altra classe di trasformazioni involutive, ed è la seguente. Fissata nel piano una rete di curve d'ordine n e di genere uno le quali abbiano in comune tanti punti fissi da assorbire n − 2 intersezioni (lo che è possibile in generale in più maniere), tutte le curve che passano per un punto preso ad arbitrio nel piano, passano anche per un altro punto che si può prendere come corrispondente al primo. Il caso più semplice (per n = 3) mi ha servito nella mia tesi."

are irreducible, that is, which cannot be inferred from each other by a series of quadratic transformations or, what is the same, by an one-to-one transformation."[10] He classified them into four irreducible types: the first two referable to involutory de Jonquières transformations; the third generated by the cubic plane curve passing through seven points, already considered by Geiser; and the fourth generated by sextics which have eight double points in common. To develop his studies, Bertini introduced some restrictive hypotheses that could have raised doubts about the incompleteness of this classification, a fear that was removed only later: "Other considerations would lead me to think that all the possible involutory transformations on the plane can be reduced to these cases. But I do not have a rigorous proof of this property. The difficulties arise above all from the consideration of those cases in which the fundamental curves are broken, and therefore several properties that exist in general are not valid."[11]

In [8, pp. 746–747], Castelnuovo affirmed that Bertini's important classification was received coldly by Cremona and that the latter, when Bertini personally communicated the result of his research to him, merely observed that he was already aware of the latest type of involution, thanks to an observation sent by the letter to Caporali. Thus, Bertini in [16, p. 273] inserted in a note the proof of Cremona introducing it in this way: "After having delivered this work to the editor of the Annali, I learned that prof. Cremona had already observed this property and communicated it by letter to Dr. Caporali."[12]

Castelnuovo in [8, p. 745] and Scorza in [10, p. 116] revealed that the good relations between Cremona and Bertini in a certain way deteriorated when the latter in 1875 left the chair of Descriptive Geometry in Rome to assume that of Higher Geometry in Pisa and that the two scholars reconciled only some years later. From the following letter, one infers how deeply disappointed Bertini was in 1878 by Cremona's behavior towards his last works:

> Professor Betti, returning from Rome, informed me that he had expressed to you my desire to ask for the chair of Higher Geometry and added that you thought it was more appropriate for me to ask for the chair of Projective Geometry. Now it is on this point that I would dare to write you some considerations.
>
> First of all, Betti assured me that you have definitively abandoned the intention of coming to Pisa. Consequently, together with the teaching of Projective Geometry, I should also take that of Higher Geometry. On the other hand, I do not hide to you that I care a lot about this last teaching, because from it I derive a lot of satisfaction and vigor for my

[10] "Indicare tutte le possibili trasformazioni involutorie del piano, che sono irriducibili, cioè non possono dedursi l'una dall'altra per una serie di trasformazioni quadratiche o, ciò che è lo stesso, per una trasformazione univoca."

[11] "Considerazioni di altra specie m'indurrebbero a pensare che a questi casi possano ridursi tutte le possibili trasformazioni involutorie del piano. Però una dimostrazione rigorosa di tale proprietà non mi è riuscita. Le difficoltà provengono soprattutto dalla considerazione di quei casi ne' quali le curve fondamentali si spezzano, e quindi cessano di essere vere parecchie proprietà che sussistono in generale."

[12] "Dopo aver consegnato il presente lavoro alla Direzione degli Annali, seppi che il prof. Cremona aveva già osservata questa proprietà e comunicatala per lettera al dott. Caporali."

studies. So I would be engaged in teaching at least as much as now, while it is one of my desires to be able to limit my occupations in order to make them as fruitful as possible.

But, leaving this consideration aside, I do not deny that, receiving the communication from Betti, I was truly disheartened. I thought that, after three years of teaching Higher Geometry and after having worked with the greatest possible effort, I would only be able to deserve the chair of Projective Geometry. I thought that a year ago I had already been judged worthy of this chair in the report of the commission for the Naples competition and that since then I have continued to work both for teaching and for science. I do not know if you have looked at my *Researches* but, if the love of a father does not deceive me, I believe that those researches (after the Naples competition) must be taken into account, regardless of the two little notes for the Lincei and the Giornale di Napoli[13] (even after the aforementioned competition) which are to be considered as a consequence of the same *Researches.* I am convinced that, if you want to examine this work carefully, you will find (I am sorry to enter into such particular facts, but I am not able to do without it) that the problem presented to me was not at all simple, that the appropriate way to solve it could not be predicted a priori, that the method I applied is (at least I think) new and applicable in similar research, and that finally, of the results I found, several are (or I am wrong) of interest. It is true that the problem was not entirely resolved, but you acknowledge that I have come very close to the solution; and I do not despair of completing it by resuming those researches in the future.[14]

[Bertini to Cremona. Pisa, 14 February 1878]

---

[13] E. Bertini, *Una nuova proprietà delle curve di ordine n con un punto (n-2)-plo*, Trasunti della Reale Accademia Nazionale dei Lincei, (3), 1, 1876–77, pp. 92–96 and *Sulle curve razionali per le quali si possono assegnare arbitrariamente i punti multipli*, Giornale di Matematiche, 15, 1877, pp. 329–335.

[14] "Il Pr. Betti, ritornando da Roma, mi ha comunicato di averle manifestato il desiderio che io aveva di chiedere la titolarità di Geometria superiore e mi ha soggiunto che Ella credeva invece più opportuno che io domandassi la titolarità di Geometria projettiva. Ora è su ciò che io oserei scriverle alcune considerazioni. Anzitutto il Betti mi ha assicurato che Ella ha abbandonato in modo definitivo il proposito di venire a Pisa. Per conseguenza insieme all'insegnamento di Geometria projettiva, dovrei assumere anche quello di Geometria superiore. D'altra parte non Le nascondo che a quest'ultimo insegnamento io tengo assai, perché da esso traggo molta soddisfazione e vigoria ne' miei studï. Sicché mi troverei impegnato nell'insegnamento per lo meno quanto ora, mentre è uno de' miei desideri di riuscire a limitare l'estensione delle mie occupazioni onde rendere queste più fruttuose che sia possibile. Ma, lasciando a parte una tale considerazione, non Le nego che, ricevendo la comunicazione del Betti, sono rimasto veramente sconfortato. Ho pensato che, dopo tre anni di straordinariato di Geometria superiore e dopo avere lavorato colla maggiore intensità possibile, io sarei riuscito soltanto a meritarmi la titolarità della Geometria proiettiva. Ho pensato che di questa titolarità io era già giudicato degno un anno fà nel rapporto della Commissione pel concorso di Napoli e che da quell'epoca ho continuato a lavorare sia per l'insegnamento, sia per la scienza. Io non so se Ella abbia data un'occhiata alle mie Ricerche: ma, se non m'illude l'amore di padre, giudico che quelle ricerche (posteriori al concorso di Napoli) debbano essere tenute in qualche conto, prescindendo dalle due noticine pei Lincei e pel Giornale di Napoli (pure posteriori al suddetto concorso) che sono da riguardare come conseguenza delle medesime Ricerche. Io sono persuaso che, se Ella vorrà usarmi la gentilezza di esaminare attentamente questo lavoro troverà (mi duole di entrare in cosifatti particolari, ma per il mio assunto non ne posso a meno) che il problema propostomi non era affatto semplice, che non poteva prevedersi a priori la via opportuna a risolverlo, che il metodo seguito è (almeno credo) nuovo e applicabile in ricerche analoghe e che infine dei risultati trovati parecchi sono (o m'inganno) interessanti. Vero che il problema non fu interamente risoluto, ma Ella riconoscerà che mi sono avvicinato assai alla soluzione; e non dispero di compierla riprendendo in avvenire quelle ricerche."

In 1879, Caporali published the note [17] which, discrediting the real feasibility of the reduction in a finite number of irreducible involutions proposed by Bertini, approached the problem from another point of view. By introducing the concept of *class* of an involution, that is, the number of pairs of corresponding points lying on an arbitrary line, he determined all the first-class involutory transformations and applied the general results obtained to two particular cases.

In 1880 and thereafter in 1883, Bertini returned to this subject by publishing the two papers [18] and [19] where, accepting the notion of class defined by Caporali, he dealt with the construction of all first- and second-class involutions. Later in 1889, making use of simple geometric considerations, in [20] he demonstrated that each of the four types of involutions he determined generated a double transformation, that is, a rational involution. Actually, this result had already been determined in 1878 by Max Noether[15] and later treated by Jacob Lüroth.[16]

Bertini also devoted himself to the theory of linear systems through the publication of three works[17] of 1882, 1889, and 1901. In the first of them, he proved two classical theorems of algebraic geometry, generalized later by Enriques and Severi, which constitute the research subject of an article by Steven Kleiman [21].

Since 1885, Bertini turned his interest to studies on the projective geometry of hyperspaces, thanks to the contributions of Veronese and above all of Segre with whom he became friends. Subsequently, he collected his lectures on hyperspaces held in Pisa in the volume *Introduzione alla geometria proiettiva degli iperspazi* (1907), where he presented the results of last years with order and clarity.

A second group of works of the years 1889–1891 was aimed at simplifying the proof of a classical theorem of Noether developed through a purely algebraic treatment and enriched with new and important properties. Applications of these results can be found in the monograph of 1894,[18] where, following the approach of Alexander von Brill and Noether and making use of the correspondence with Segre and Noether, Bertini presented the geometry of linear series on a curve in a clear and systematic way. We owe both our author and Segre the credit for having contributed to the dissemination in Italy of the theory of the aforementioned German mathematicians, as Castelnuovo stated in [8, p. 748]: "This monograph and a contemporary one by Corrado Segre on the same topic discussed with a different method, have disseminated among us a theory that, originated in Germany, took

[15] M. Noether, *Über die ein-zweideutigen Ebenentransformationen*, Erlangener Berichte, 10, 1878, p. 81.

[16] J. Lüroth, *Rationale Flächen und involutorische Transformationen*, Prorektoratsrede, Freiburg, 1889, pp. 1–25.

[17] E. Bertini, *Sui sistemi lineari*, Rendiconti del Reale Istituto Lombardo di Scienze e Lettere, (2), 15, 1882, pp. 24–29; *Sulle curve fondamentali dei sistemi lineari di curve piane*, Rendiconti del Circolo Matematico di Palermo, 3, 1889, pp. 5–21 and *Sui sistemi lineari di grado zero,* Rendiconti della Reale Accademia Nazionale dei Lincei, (5), 10, 1901, pp. 73–76.

[18] E. Bertini, *La geometria delle serie lineari sopra una curva piana secondo il metodo algebrico*, Annali di Matematica pura ed applicata, (2), 22, 1894, pp. 1–40.

on thereafter new and extensive developments in Italy. Those two works gave the impetus to the subsequent research of our geometric school."[19]

3 Ettore Caporali in His Correspondence with Cremona

Ettore Caporali took his degree with Luigi Cremona in July 1875 in Rome. He was born in Perugia on 17 August 1855, from Vincenzo and Tecla Campi, had attended secondary schools in his hometown, and then studied mathematics at the University of Rome, under Giuseppe Battaglini and Eugenio Beltrami as well as with Cremona.

His thesis *Sulla superficie del quinto ordine dotata di una curva doppia di quinto ordine*[20] was about the rational surface of the fifth order of the ordinary three-dimensional space with a double line of the fifth order that he studied by means of its one-to-one representation on the plane. He derived this representation from a spatial construction of the surface through a particular birational transformation on the space. The research, which extends the results of Cremona, shows how Caporali faced problems that were of interest to the greatest experts of algebraic geometry at the time.

The archive of the Istituto Mazziniano in Genoa contains 48 letters from Caporali to Cremona, from 1876 to 1886. The unpublished correspondence with his teacher highlights a very close relation between the two scholars, both from a scientific and a human point of view. Immediately after his graduation, Caporali became teacher at the Spedalieri high school in Catania; however, he lived the year he spent in Sicily with boredom, and he often wrote to his teacher about this, complaining of not being in the right disposition to develop his studies and research and asking him for advice on changing living places and jobs. In 1876, he obtained the position of teaching assistant at the Scuola d'applicazione per gli ingegneri in Rome.

In this period, the scientific activity of Caporali became more intensive; he often discussed with his teacher by letter about the research they were developing: for example, in a letter, we read that Cremona had sent him the drafts of his note *Über die Polar-Hexaeder bei den Flächen dritter Ordnung*[21] and Caporali answered with observations and his original ideas:

> I was impressed by the simplicity of the analytical definition of your hexahedra: it is also remarkable that they relate the 27 straight lines to Sylvester's pentahedron with Reye's theorems: especially because it is perhaps possible, by means of the properties of third-class developable surfaces, to give this relation a simpler form than the one directly provided by Reye's note.

[19] "Questa monografia ed un'altra contemporanea di Corrado Segre sullo stesso argomento trattato con metodo diverso, hanno divulgato tra noi una teoria che, sorta in Germania, prese poi in Italia nuovi ed ampi sviluppi. Quei due scritti hanno dato l'impulso alle successive ricerche della nostra scuola geometrica."

[20] Printed by decision of the examining commission in the Annali di Matematica pura ed applicata, serie 2°, tomo 7°.

[21] L. Cremona, *Ueber die Polar-Hexaeder bei den Flächen dritter Ordnung*, Mathematischen Annalen, Band XIII, 1878, pp. 301–304.

This topic made me think about Clebsch's Diagonalfläche: the 15 straight lines that determine it are arranged three by three in 15 planes; and since a pentahedron (the Sylvester pentahedron) is assumed as the starting point I believed your properties could provide a complete pentahedron theory. However, I realized shortly after that the hexahedron that should be deduced from the 15 lines is lost and with it almost all the properties. Since any edge of Sylvester's pentahedron absorbs three of the 60 Pascal lines, 30 lines remain that, perhaps, deserve to be studied.

Moving from one idea to another, I thought about the problem of assuming six fundamental points in the plane in order to have the representation of the Diagonalfläche: I resolved it analytically.[22]

[Caporali to Cremona. Massa Martana, 15 September 1877]

Caporali's ideas then converged into the work *Sull'esaedro completo*[23] which resumed Cremona's research of the aforementioned note, *Teoremi stereometrici dai quali si deducono le proprietà del l'esagrammo di Pascal* of Cremona[24] and *On Pascal's theorem* of Cayley.[25]

At that time, Caporali was trying to be hired at the University of Naples because some professorships had remained vacant after Battaglini's retirement. In 1878, being just 23 years old, he was then appointed extraordinary professor of Higher Geometry at the University of Naples. We can read the arguments of his course from a letter he wrote to Cremona:

I began by rapidly re-introducing projectivity in the forms of the 1st kind: I will quickly repeat the main projective properties of conics and then I will present the real program of the course, which consists of the 2nd part of the Reye,[26] proceeding as far as possible.[27]

[Caporali to Cremona. Napoli, 17 December 1878]

---

[22] "Mi ha colpito la semplicità della definizione analitica dei suoi esaedri: è poi notevole che essi mettono in relazione le 27 rette col pentaedro di Sylvester per mezzo dei teoremi del Reye: tanto più che è forse possibile, per mezzo delle proprietà delle sviluppabili di terza classe, di dare a questa relazione una forma più semplice di quella che è immediatamente fornita dalla memoria di Reye. Questo argomento mi aveva fatto pensare alla Diagonalfläche di Clebsch: le 15 rette che la determinano sono situate tre a tre in 15 piani; e siccome si assume come punto di partenza un pentaedro (di Sylvester) credevo che le di Lei proprietà potessero fornire una teoria del pentaedro completo. Però mi sono accorto poco dopo che l'esaedro che dovrebbe dedursi dalle 15 rette si perde e con esso quasi tutte le proprietà. Ma delle 60 rette di Pascal avviene che ogni spigolo del pentaedro di Sylvester ne assorbe tre, dimodoché ne rimangono 30: queste ultime meritano forse di essere studiate. Passando da un'idea all'altra ho pensato al problema di assumere sei punti fondamentali in un piano in modo da avere la rappresentazione della Diagonalfläche: l'ho risoluto analiticamente."

[23] E. Caporali, *Sull'esaedro completo*, Rendiconti della Reale Accademia delle Scienze fisiche e matematiche di Napoli, fascicolo 3°, marzo 1881.

[24] L. Cremona, *Teoremi stereometrici dai quali si deducono le proprietà dell'esagrammo di Pascal*, Memorie della Reale Accademia Nazionale dei Lincei, serie 3, vol. 1, 1876–1877, pp. 854–874.

[25] A. Cayley, *On Pascal's theorem*, Quarterly Journal of Pure and Applied Mathematics, vol. IX, 1868, pp. 348–353.

[26] T. Reye, *Die Geometrie der Lage*, Erste Abtheilung, Carl Rümpler, Hannover, 1866.

[27] "Ho cominciato col ripresentare rapidamente la proiettività nelle forme di 1a specie: ripeterò di volo le proprietà principali proiettive delle coniche e poi passerò al vero programma del corso, che consiste nella 2a parte del Reye, arrivando fin dove si potrà."

In October 1878, the Società Italiana delle Scienze awarded him a prize for his two notes: *Sui complessi e sulle congruenze di secondo grado*[28] and *Sopra i piani ed i punti singolari della superficie di Kummer.*[29] The commission was formed by Cremona, Beltrami, and Battaglini.[30]

Throughout his life, Caporali was very attached to his family, plagued by various health problems which resulted in economic problems. For this reason, he often asked for loans from people closest to him, including Cremona. In 1884, he became full professor, but he went through a rough period for the health (his own and of his relatives), the household economy, and also the scientific production, as we can clearly read:

> At the beginning of the summer I was indisposed for a slight reappearance of nervous heartbeat, to which, if you remember, I was more severely subject in the past. [...] Anyway, the school year just ended was almost lost for my studies except for some reading. But it gets worse. And it is, that the few aptitudes that I previously seemed to have in studies, I now find considerably diminished, which makes me becoming greatly discouraged, as well as the mortification of finding myself unworthy of a position that I had accepted almost with a certain boldness.[31]
>
> [Caporali to Cremona. Genzano di Roma, 24 August 1882]

In these words, the despair that will lead Caporali to death is already beginning to be manifest. Despite the difficulties, he continued to devote himself to the elaboration of some works, as can be seen from a letter published in [4]:

> The first note refers to a 6th degree complex which is the locus of the tangents of all cubics passing through 5 fixed points. It is a complex that can be represented on points in the space: and its representation is given by the surfaces of the 3rd order that pass through the 5 points and the diagonal points of the nonplanar pentagon that they form.
>
> The second note contains a theorem on the tangents to a plane curve drawn from its multiple points. When the tangents emanate from an arbitrary point, in general the first polar of this point is the simplest curve that passes through the contact points of those tangents. But when the point is multiple, there are in general curves of minor orders that define the

---

[28] E. Caporali, *Sui complessi e sulle congruenze di secondo grado*, Memorie della Reale Accademia Nazionale dei Lincei, s. III, vol. II, 1877–1878, pp. 749–769.

[29] E. Caporali, *Sopra i piani ed i punti singolari della superficie di Kummer*, Memorie della Reale Accademia Nazionale dei Lincei, s. III, vol. II, 1877–1878, pp. 791–810.

[30] As it appears in the *Rapporti* of the *Memorie di Matematica e di Fisica della Società Italiana delle Scienze* (serie III, vol. IV, pp. XXI–XXIV). The *Società Italiana delle Scienze* was founded by Antonio Maria Lorgna (1735–1796) in 1782 in Verona as the Società Italiana, comprising 40 scientists from various parts of Italy. For this reason, it was also called *Accademia dei XL*.

[31] "Al principio dell'estate io sono stato indisposto per un leggero riapparire di cardiopalmo nervoso, incontro al quale, se si rammenta, andavo più gravemente soggetto per l'addietro. [...] L'anno scolastico testé finito, è stato però quasi perduto pei miei studi se se ne eccettui qualche lettura. Ma c'è di peggio. Ed è che le poche attitudini che prima mi pareva d'avere agli studi, le trovo ora notevolmente diminuite, cosa che mi fa provare grande scoraggiamento, oltre alla mortificazione di trovarmi inferiore ad una posizione che avevo accettato quasi con una certa baldanza."

constraints on the contact points more simply. My theorem is very general and includes as a particular case that of Bertini, which you presented two or three years ago to the Lincei.[32]

As soon as I have prepared these two notes for the press, I will resume certain studies on systems of lines that I interrupted for military service but which I have now been encouraged to resume by reading Stahl's work[33] on 2nd order and 3rd class systems published in latest issue of Crelle's Journal.[34]

[Caporali to Cremona. Napoli, 29 May 1881]

In March 1881, Caporali was appointed honorary and resident member of the Reale Accademia delle Scienze fisiche e matematiche di Napoli. In 1882, the mathematician Seligmann Kantor, who had studied in Rome under Cremona in 1878, won the competition announced by the Mathematical Section of the Accademia of Naples on the subject: "Considering the birational transformation in two planes coinciding with each other, find the conditions so that by applying the same transformation several times, we return to the original figure."[35] Caporali wrote an elaborate report on Kantor's note, published on the Rendiconti dell'Accademia in December 1883.[36]

From the letters to Cremona, it comes out that in Naples, Caporali was also very active as a member of the faculty while trying to move to Rome taking advantage of Battaglini's wish to go back to Naples. He was also concerned with the preparation of the scientific cabinets of the Mathematics Institute with purchases from Germany and to ensure that Naples had a representative in the Consiglio Superiore della Istruzione Pubblica.

In his short life, Caporali received other awards for his scientific merits: on 31 December 1883, he became a corresponding member of the Reale Accademia

---

[32] E. Bertini, *Una nuova proprietà delle curve di ordine n con un punto (n-2)-plo,* Trasunti della Reale Accademia Nazionale dei Lincei, (3), 1, 1876–77, pp. 92–96.

[33] W. Stahl, *Das Strahlensystem dritter Ordnung zweiter Klasse,* Journal für die reine und angewandte Mathematik, 91, pp. 1–22.

[34] "La prima si riferisce ad un complesso di 6° grado che è il luogo delle tangenti di tutte le cubiche passanti per 5 punti fissi. È un complesso rappresentabile sui punti dello spazio: e la rappresentazione è data dalle superficie del 3° ordine che passano pei 5 punti e pei punti diagonali del pentagono gobbo che essi formano. La seconda nota contiene un teorema sulle tangenti condotte ad una curva piana da un suo punto multiplo. Quando le tangenti si conducono da un punto arbitrario, in generale la prima polare di questo punto è la curva più semplice che passa pei punti di contatto di quelle tangenti. Ma quando il punto è multiplo, vi sono in generale curve di ordine minore e che perciò definiscono più semplicemente i vincoli che legano i punti di contatto. Il mio teorema è molto generale e comprende come caso particolare quello del Bertini che Ella presentò due o tre anni fa ai Lincei. Appena avrò approntate queste due note per la stampa, riprenderò certi studi sui sistemi di rette che interruppi pel servizio militare ma che ora sono stato invogliato a riprendere per la lettura del lavoro di Stahl sui sistemi del 2° ordine e 3° classe pubblicato nell'ultimo fascicolo del giornale di Crelle."

[35] "Considerando la trasformazione birazionale in due piani tra loro coincidenti, trovare le condizioni affinché applicando più volte di seguito la stessa trasformazione, si ritorni alla figura da cui si parte."

[36] Competition report for the Premio accademico of 1882, published in the Rendiconti della Reale Accademia delle Scienze fisiche e matematiche di Napoli, issue 12°, December 1883.

Nazionale dei Lincei, and in May 1886, shortly before his death, ordinary and resident member of the Accademia Pontaniana of Naples.

On 2 July 1886, he committed suicide; for the circumstance, his colleague Dino Padelletti wrote these words in [22, pp. III–IV]: “He worried excessively about the diminished activity of his wits, which to him seemed to be irreparably decaying, turned his cruel hand against himself in a supreme moment of sadness and despair [...]. Important tasks entrusted to him by the unanimous vote of his colleagues showed the great trust that everyone had placed in that young man, in whom his judgment seemed to have preceded his age, and all whose actions were full of the very warm love of science and a feeling of irreproachable honesty.”[37]

Even Guccia, in his letters to Cremona, wrote the reasons for his gesture: “He told me that intelligence abandoned him, he was no longer capable of creating anything, he did not find the usual mental energy to develop any research, he saw himself reached and overtaken by others, he read about various topics but did not feel the strength to deepen some of them; he saw the sacred fire of Science extinguish in him, he was discouraged”[38] (Guccia to Cremona. Palermo, 6 July 1886, in [23, p. 103]).

Guccia wrote again to Cremona: “Discouragement had invaded our friend in last years, due to a slow, but persistent, brain disease that increasingly lowered his mental faculties”[39] (Guccia to Cremona. Palermo, 8 August 1886, in [23, p. 109]).

Furthermore, Guccia expressed the estimation that they both had for Caporali: “Your beloved student, the best of your students [...]. I often collected evidences of estimation for the young Italian geometer. His works were more than a promise. The Cremona school has lost one of its best elements, as a man of science and as a professor”[40] (Guccia to Cremona. Palermo, 6 July 1886, in [23, p. 105]).

During his life, Caporali published 12 notes [22, 32], but among his manuscripts, other works were found which, although incomplete, were considered appropriate to publish. In 1889, 3 years after Caporali’s death, Gino Loria published Caporali’s entire scientific work in the Giornale di Matematiche di Battaglini [24], on

---

[37] “Egli accuorandosi oltre misura della scemata attività del suo ingegno, che a lui sembrava decadenza irrimediabile, rivolse in un momento supremo di tristezza e di sconforto la mano crudele contro sé stesso [...]. Importanti ufficii affidatigli dal voto unanime dei colleghi mostravano la grande fiducia da tutti riposta in quel giovane, in cui il senno sembrava avesse precorsa la età, e tutte le cui azioni erano informate dall’amore caldissimo della scienza e dal sentimento di una onestà intemerata.”

[38] “Mi diceva che l’intelligenza lo abbandonava, non era più capace di crear nulla, non ritrovava la solita energia mentale per condurre a fine qualsiasi ricerca, si vedeva raggiunto e sorpassato da altri, curiosava intorno a diversi argomenti ma non si sentiva la forza di approfondirne alcuno; vedeva estinguersi in lui il fuoco sacro della Scienza, era scoraggiato.”

[39] “Scoraggiamento di cui era invaso il nostro amico negli ultimi anni, per via di una lenta, ma persistente, malattia celebrale che abbassava vieppiù le sue facoltà mentali.”

[40] “Suo amatissimo allievo, il migliore dei suoi allievi [...]. Raccolsi spesso all’estero delle autorevoli testimonianze di stima sul conto del giovane geometra italiano. I suoi lavori erano più che una promessa. La scuola del Cremona ha perduto uno dei migliori elementi, come uomo di scienza e come professore.”

the initiative of a group of colleagues, friends, and admirers, to remember the distinguished extinct scholar in the most lasting way. In [24], Loria sorted Caporali's works into the following topics: third-order curves, fourth-order curves, nth-order curves, third-order surfaces, one-to-one transformations on the plane, representation of surfaces on the plane, geometry of the line, theory of configurations, four-dimensional geometry, manifolds, and geometric applications of algebraic forms.

Among the most important students of Caporali, we include Pasquale del Pezzo (1859–1936), who graduated in mathematics in 1882 and, among other things, was rector of the University of Naples and mayor of the city.

In 1892, Corrado Segre published the manuscript [25] to complete the fragments collected in *Sulla teoria delle curve piane del quarto ordine* that had been included in [22]. Segre reported two letters that Caporali had addressed to him in 1885 (11 August and 13 September). The letters seem important as evidence of the acknowledgement (by one of the best students of Cremona) of the emergence of a new referent for the Italian school of algebraic geometry. In [25, pp. 172–173], Caporali wrote to Segre:

> What you write to me about my studies on fourth order curves is interesting and shows that you immediately understood the essence of those researches. Although not very advanced, they have a complicated history that is related to various causes extraneous to science that have prevented me for three years from applying myself to study with that regularity and perseverance that would allow me to have good results. [...] When you published your Memoria sulla geometria delle coniche, I immediately saw the advantage that could be gained from the systematic use of that mode of representation and which is confirmed to me by your letter: I did not resume, however, nor will I be able to resume the research immediately, although precisely now, having acquired all their generality, they are in the most interesting stage.[41]
>
> [Caporali to Segre. Torre del Greco, 13 September 1885].

This letter is particularly significant because Caporali, who in 1885 had already been full professor in Naples for some time, addressed to Segre, graduated only 2 years before and assistant of his thesis supervisor Enrico D'Ovidio (1843–1933), as a new referent. Furthermore, the letter contains Caporali's work projects and shows how the hyperspace methods developed by Segre and Veronese fit perfectly into the framework of the research programs of the Cremona school. In this regard, see also [26].

[41] "Ciò che Ella mi scrive intorno ai miei studi sulle curve del quarto ordine è interessante e dimostra che Ella ha immediatamente penetrato lo spirito di quelle ricerche. Per quanto poco avanzate, esse hanno una storia complicata e in relazione con diverse cause estranee alla scienza che m'impediscono da tre anni di attendere allo studio con quella regolarità e quella perseveranza che sole permettono di cavarne buoni frutti. [...] Quando Ella pubblicò la sua Memoria sulla geometria delle coniche, vidi immediatamente il partito che si poteva trarre dall'uso sistematico di quel modo di rappresentazione e che mi è confermato dalla sua... lettera: non ripresi però, né potrò subito riprendere le ricerche, benché precisamente ora, avendo acquistata tutta la loro generalità, siano nello stadio più interessante."

4 Riccardo De Paolis Through the Memory of Corrado Segre

The lives and careers of Riccardo De Paolis and Ettore Caporali followed parallel avenues, as suggested by De Paolis's commemoration by Corrado Segre in [27].

De Paolis was born in Rome on 9 January 1854 from Achille and Elena Chatelain; he studied in Rome immediately demonstrating a penchant for mathematics. In 1870, both Caporali and De Paolis enrolled at the University of Rome and became friends. During the first 2 years, they studied together solving many problems inspired by George Salmon and inventing new ones. Subsequently, they attended the Scuola di magistero per la matematica founded by Cremona in 1873, having as professors also Battaglini and Beltrami.

After graduating in July 1875,[42] De Paolis was appointed Professor of Mathematics in the high school of Caltanissetta while, as we have already said above, Caporali went to Catania. In 1876, both returned to Rome, De Paolis with the assignment of practical mathematics exercises at the University and then also as a substitute teacher for analytical geometry. In June 1878, he obtained his teaching qualification in analytical geometry and projective geometry. In these 2 years, he wrote and published some works on the double transformations on the plane that gave him notoriety in the scientific world. In the paper [28], he developed researches similar to those by Cremona for one-to-one transformations, providing a synthetic treatment. Segre drew attention to a substantial difference with the reduction of plane involutions through Cremona transformations made by Bertini in the same period. In fact, De Paolis did not deal with reducing plane (1, 2)-correspondences by applying Cremona transformations, but he analyzed their projective properties. This work was followed by two others on the study of two particular double transformations and their applications.[43]

In November 1878, winning a competition, he was appointed extraordinary professor of Algebra and Analytical Geometry at the University of Bologna and in January 1880, again by competition, he was called to Pavia as extraordinary professor of Higher Geometry. As already mentioned, the same year, he moved to Pisa in the same chair thanks to an exchange with Bertini and remained in Pisa for the rest of his life. In addition to higher geometry, De Paolis taught graphical statics from 1882–1883 to 1889, and projective and descriptive geometry from 1889–1890. In 1883, he was appointed, at the same time as Caporali, a corresponding member of the Accademia Nazionale dei Lincei and elected a member of the Consiglio direttivo of the Circolo Matematico di Palermo. De Paolis died in Rome on 24 June 1892 due to tuberculous peritonitis.

[42] De Paolis' degree thesis *Sopra un sistema omaloidico formato da superficie d'ordine n con un punto (n−1)plo* was published in the Giornale di Matematiche, 13, 1875, pp. 226–248, 282–297.

[43] R. De Paolis, *La trasformazione piana doppia di secondo ordine e la sua applicazione alla geometria non euclidea*, Memorie della Reale Accademia Nazionale dei Lincei, (3), 2, 1878, pp. 31–50 and *La trasformazione piana doppia di terzo ordine, primo genere, e la sua applicazione alle curve del quarto ordine*, Memorie della Reale Accademia Nazionale dei Lincei, (3), 2, 1878, pp. 851–878.

In the note [29], De Paolis continued and completed what Karl von Staudt,[44] Felix Klein,[45] and Jean Gaston Darboux[46] had done to establish the fundamental theorem of projective geometry and to arrive, without measurement of quantities, at the projective coordinates of the forms of the first kind. The work consists of two parts: In the first one, following Klein, Lüroth, and Darboux, starting from the concept of Euclidean metrics on the projective line, he proved the fundamental theorem of projective geometry. In the second part, using projective concepts, De Paolis showed that it is possible to put the straight line in direct correspondence with rational numbers and therefore, postulating its completion, with real numbers. Segre stated: "This work gives us a first example of a particular tendency of De Paolis: that of always going back to the foundations, to the principles of theories. Perhaps he did not have this tendency from the beginning. On the occasion of his graduation, Cremona, doubting that De Paolis neglected special cases or the most elementary things in order to deal only with general, elevated issues, made him understand that this was a mistake. The student did not forget teacher's lesson!"[47]

In 1884, De Paolis published the book *Elementi di Geometria*,[48] and in the preface, he wrote: "I had a double purpose; to abandon the ancient separation of plane Geometry from solid, to try to rigorously establish the fundamental truths of Geometry and the theories of equivalence, limits, measure." [49] The originality of this work, which according to Segre had benefited well from the fusion between plane and solid geometry, was due to the fact that De Paolis, when he wrote it, knew no other elements than those of Euclid and during the writing he examined very few texts.

In 1885, De Paolis turned back to the topic of double transformations [30], and, analogously to what he had done in the case of the plane, he studied the ones in the space with the same methods, although here a larger number of cases occur. In [31], he analyzed those involutions in which the straight lines joining the pairs of homologous points are ∞^2 instead of ∞^3 many.

Even De Paolis, like Bertini and Caporali, can be considered the link between the first and second generations of Italian algebraic geometers. In fact, in addition

[44] K. G. C. von Staudt, *Geometrie der Lage*, Nürnberg Bauer & Raspe, 1847.

[45] F. Klein, *Über die sogenannte Nicht-Euklidische Geometrie*, Mathematische Annalen, 6, 1873, pp. 112–145.

[46] M. G. Darboux, *Sur le théorème fondamental de la géométrie projective* (*Extrait d'une lettre à M. Klein*), Mathematische Annalen, 17, 1880, pp. 55–61.

[47] "Questo lavoro ci dà un primo esempio di una particolare tendenza del De Paolis: quella di risalire sempre ai fondamenti, ai principi delle teorie. Forse questa tendenza egli non aveva da principio. Nell'occasione della laurea il Cremona, dubitando che il De Paolis per occuparsi solo di questioni generali, elevate, trascurasse i casi speciali o le cose più elementari, gli fece intendere che questo era un errore. Il discepolo non dimenticò la lezione del maestro!"

[48] R. De Paolis, *Elementi di geometria*, Torino, E. Loescher, 1884.

[49] "Si propose un doppio scopo; abbandonare l'antica separazione della Geometria piana dalla solida, tentare di stabilire rigorosamente le verità fondamentali della Geometria e le teorie dell'equivalenza, dei limiti, della misura."

to having had scientific relationships and an intense correspondence with Segre, he had Enriques among his students.

His latest research mainly concerned a question of method: Segre revealed that although the analytical method was widely used in previous writings, and more precisely the algebra of invariants, there was a strong preference for essentially geometric methods, a typical preference of the Cremona school. Following these methods, De Paolis proposed to go further.

In the archive of the Istituto Mazziniano in Genoa, a single letter from De Paolis to Cremona was found and, because it is very relevant, we choose to transcribe it almost entirely:

> Together with this letter of mine, I sent you two little notes, one on equivalent figures, which I published last year, the other one on projective involutions where I proved a theorem that I have developed about some research I am doing and I want to talk to you about it, surely I will not bore you, because I know that you remember me always fondly, by the time your old student, and you have always followed my publications with the satisfaction of a teacher. Here is what it is: I think I told you in some other occasion about a fixed idea that has haunted me for several years, that is, the idea of freeing Geometry from the aid of Algebra, founding true pure Geometry. After many efforts and many useless attempts, justified however by the difficulty of the problem, I can finally claim that I have completely resolved the question. It would be impossible for me to tell you in few words the path I followed, the method I used is completely new, and besides solving the problem I had posed, it also provides new results and new research to do. The plan I have proposed is vast and I still need a lot of time to fully implement it. **I start with the definitions of surface, line and point, I do not assume any knowledge of Geometry, and after some general considerations, I study the general theory of correspondences between the points of any two linear, surface or solid fields, assuming as a condition the continuity of the correspondences themselves. In this part my work can be considered as a geometric theory of continuous functions of 1, 2 or 3 real variables, and in fact I prove some fundamental geometric theorems, which correspond perfectly to the fundamental theorems of Analysis by Weierstrass, Cantor, etc. I then show that each continuous correspondence [m, n], between the points of two surfaces, can be replaced with another [1, 1] between the points of two other surfaces, which I call the Riemann surfaces of the given correspondence, in this way I acquire a powerful geometric research tool that will then be very useful to me later and especially in the study of the curves which up to now have been called algebraic and which I still do not know how to call; I introduce in Geometry those considerations that Riemann has used so much in Analysis. Then I am naturally led to study the connection of the surfaces, etc. etc. After these general considerations, I move on to establish the projective Geometry, with the same rigor held by Pasch, but much more simply. In a certain sense, after having elaborated that part of Geometry that corresponds to the general theory of functions, I move on to the part that corresponds to the study of algebraic functions. I analyze the part concerning projective Geometry up to the 2nd degree forms, also considering the imaginary elements, by Staudt's method. I show thereafter, based on some results due to Thieme, that systems ∞^{n-1} of groups of n elements of a fundamental form of 1st kind can be built such that a group of the system is identified by any n-1 elements of its, I call these systems involutions of n degree and n-1 kind. After a complete study of these involutions and their linear systems, I prove, and herein lies the crux of the matter, that n of these involutions always have a group of n elements, real or imaginary, in common. I therefore study the intersections of these involutions and thus find the involutions of n degree and r $\leq$ n-1 kind, that are linear systems ∞^r of groups of n elements. I finish this part of the work by showing that two involutions of 1st kind, projective and coincident,**

always have m+n fixed elements, real or imaginary, in common. In order to arrive at this result, on which the whole theory of curves and surfaces can be based, I apply the properties of continuous correspondences, which I proved at the beginning. The theory of polar groups naturally arises from the theory of involutions. For now, with regard to curves and surfaces, I have only seen how they can be generated, with their linear systems, how it can be shown that two curves of order m, n always have mn common points, real or imaginary, etc. etc. Later I will go further and systematically evolve the whole theory with my method. But I will not stop there. I have also seen how the concept of multidimensional spaces can be introduced, and in a rigorous way, using a representation in our space. For multidimensional fields I also establish a theory of continuous correspondences, of connection, etc., and I introduce the concept of a multidimensional Riemann surface, a concept that I then use to determine the genus of the multidimensional surfaces. After this part, which corresponds to the study of continuous functions of n real variables, I proceed, with a method like that used for the two and three dimensions, to consider the elements that correspond to the algebraic surfaces.

You can see how the field I want to explore is large; but I have already found everything I need to easily deduce the remaining parts, in any case, the drafting of the work will still take me a long time, and I will not be able to finish it before the end of the year, although I work hardly on it. I had already found the main results when I learned that part of the subject I was studying, the part, which refers to establishing the Geometry of algebraic curves without using Algebra, had twice been the topic of a competition by the Berlin Academy, and that the last time Kötter received the prize. Now, Kötter's work has not been published so far, I read the report they made at the Academy and it seems to me that he did not go as far as I did, that he applied another, less rigorous method, and rigor in some problems is everything. In any case, in order not to lose priority, I wrote the main results I obtained in 6 notebooks, closed them, sealed them and sent them to the Accademia dei Lincei to record the date. I then decided to submit all the work to the competition for the royal prize, which expires with the current year, but before deciding I would like to have your advice. Does it seem to you that the topic is sufficiently important, and that, if I had done it well, I would not make a bad impression even if I do not win the prize?[50]

[De Paolis to Cremona. Pisa, 3 March 1887]

---

[50] "Insieme a questa mia le ho spedito due noticine, una, sulle figure equivalenti, che pubblicai l'anno scorso, l'altra sulle involuzioni proiettive, nella quale dimostro un teorema che ho trovato a proposito di alcune ricerche che sto facendo, e delle quali desidero parlarle, sicuro di non recarle noia, perché so che Ella si ricorda sempre affettuosamente di me, che sono ormai un suo antico allievo, e sempre ha tenuto dietro con soddisfazione di maestro alle pubblicazioni mie. Ecco di che cosa si tratta: Mi pare di averle altre volte parlato di una idea fissa che mi perseguita da più anni, dell'idea cioè di emancipare la Geometria dal sussidio dell'Algebra, di fondare la vera Geometria pura. Dopo molte fatiche e molti tentativi inutili, ma giustificati dalla difficoltà dell'argomento, posso finalmente asserire di avere completamente risoluto la quistione. Mi sarebbe impossibile dirle in poche parole la via che ho seguito, il metodo che ho adoperato è completamente nuovo, e non solo risolve il problema che mi ero proposto, ma fornisce anche nuovi risultati e nuove ricerche da fare. Il piano che mi sono proposto è vasto e ho bisogno ancora di molto tempo per svolgerlo completamente. Comincio dalle definizioni della superficie, della linea e del punto, non suppongo alcuna cognizione di Geometria, e dopo alcune considerazioni generali, studio la teoria generale delle corrispondenze tra i punti di due qualunque campi lineari, superficiali o solidi, ponendo per sola condizione la continuità delle corrispondenze stesse. In questa parte il mio lavoro si può considerare come una teoria geometrica delle funzioni continue di 1, 2 o 3 variabili reali, ed infatti dimostro alcuni teoremi geometrici fondamentali, che tengono perfettamente luogo di

On 1 July 1886, Ernst Kötter won the Steiner prize at the Berlin Academy of Sciences on the subject of founding a purely geometric theory of curves and surfaces

---

teoremi fondamentali dell'Analisi, trovati da Weierstrass, Cantor, ecc. Faccio poi vedere che ogni corrispondenza continua [m,n], tra i punti di due superficie, si può sostituire con un'altra [1,1] tra i punti di altre due superficie, che chiamo le superficie di Riemann della corrispondenza data, così acquisto un potente mezzo di ricerca geometrico, mezzo che mi sarà poi utilissimo in seguito e specialmente nello studio delle curve che fin qui si sono chiamate algebriche e che io ancora non so come chiamare; introduco nella Geometria quelle considerazioni che Riemann ha tanto utilizzato nell'Analisi. Poi sono naturalmente condotto a studiare la connessione delle superficie, ecc. ecc. Dopo queste considerazioni così generali, passo a stabilire la Geometria proiettiva, con lo stesso rigore tenuto da Pasch, ma molto più semplicemente. In un certo senso, dopo avere svolto quella parte della Geometria che corrisponde alla teoria generale delle funzioni, passo alla parte che corrisponde allo studio delle funzioni algebriche. La parte che riguarda la Geometria proiettiva la spingo fino alle forme di 2° grado, comprese, considerando anche gli elementi immaginarî, col metodo di Staudt. Faccio poi vedere, fondandomi sopra alcuni risultati dovuti a Thieme, che si possono costruire sistemi ∞^{n-1} di gruppi di n elementi di una forma fondamentale di 1ª specie, tali che un gruppo del sistema sia individuato da n-1 qualunque dei suoi elementi, questi sistemi li chiamo involuzioni di grado n e specie n-1. Dopo uno studio completo di queste involuzioni e dei loro sistemi lineari, dimostro, e qui sta il nodo della quistione, che n di queste involuzioni hanno sempre comune un gruppo di n elementi, reali o immaginari. Passo quindi a studiare le intersezioni di queste involuzioni e trovo così le involuzioni di grado n e specie $r \leq n-1$, che sono sistemi lineari ∞^r di gruppi di n elementi. Termino questa parte del lavoro dimostrando che due involuzioni, di 1ª specie, proiettive e sovrapposte hanno sempre $m + n$ elementi uniti, reali o immaginarî, comuni. Per arrivare a questo risultato, sul quale si può basare tutta la teoria delle curve e delle superficie, applico le proprietà delle corrispondenze continue, che ho dimostrato sul principio. Dalla teoria delle suddette involuzioni sorge naturalmente quella dei gruppi polari. Per ora, riguardo alle curve e superficie, ho veduto solamente come si possono generare, insieme ai loro sistemi lineari, come si può dimostrare che due curve di ordine m, n hanno sempre mn punti comuni, reali o immaginarî, ecc. ecc. In seguito mi spingerò più innanzi e svolgerò col mio metodo sistematicamente tutta la teoria. Ma non mi fermerò qui. Ho pure veduto come si può introdurre il concetto di spazî a più dimensioni, e in modo rigoroso, ricorrendo ad una effettiva rappresentazione nello spazio nostro. Per i campi a più dimensioni stabilisco pure una teoria delle corrispondenze continue, della connessione, ecc., e introduco il concetto di superficie di Riemann a più dimensioni, concetto che poi utilizzo per determinare i generi delle superficie a più dimensioni. Dopo questa parte, che corrisponde allo studio delle funzioni continue di n variabili reali, passo, con metodo analogo a quello tenuto per le due e tre dimensioni, a considerare gli enti che corrispondono alle superficie algebriche. Ella vede quanto è vasto il campo che voglio esplorare; però ho già trovato tutto quello che mi serve per dedurre senza difficoltà le parti rimanenti, in ogni modo la redazione del lavoro mi porterà via ancora molto tempo, e non potrò terminarlo prima della fine dell'anno, per quanto ci lavori assiduamente. Già avevo trovato i risultati principali, quando seppi che una parte del tema che stavo trattando, quella che si riferisce allo stabilire la Geometria delle curve algebriche senza ricorrere all'Algebra, era stata per due volte consecutive messa a concorso dall'Accademia di Berlino, e che l'ultima volta un certo Kötter aveva ricevuto il premio. Ora il lavoro di Kötter non è stato fin qui pubblicato, ho letto la relazione che ne hanno fatta all'Accademia e mi pare che non si sia spinto innanzi come me, e che abbia tenuto un altro metodo, meno rigoroso, ed il rigore in certe quistioni è tutto. In ogni modo per non perdere la priorità ho appuntato, in 6 quinterni, i principali risultati che ho ottenuto, li ho chiusi, sigillati e spediti all'Accademia dei Lincei, perché se ne prenda data. Ho poi pensato di presentare tutto il lavoro al concorso per il premio reale, che scade coll'anno corrente, prima però di decidermi avrei piacere di sentire un suo consiglio. Le pare che l'argomento sia sufficientemente importante, e che, qualora lo avessi bene svolto, non farei una brutta figura anche se non vincessi il premio?"

of higher order. The day after writing to Cremona, De Paolis wrote a similar letter to Segre:[51]

> I read the report on Kötter's work in the publications of the Berlin Academy; I have been looking forward to its publication, but I do not know if it has been published yet. [...] For many years the idea of freeing Geometry from the use of Algebra, of founding true *pure Geometry* has haunted me. When I heard the news[52] I had completely solved the problem, I could not publish my research immediately, because I had not yet done the secondary parts of all the work I had proposed to do; so I thought of waiting for the publication of Kötter's essay, whatever could happen, all the more so since, from the report made on this essay, it seems to me that Kötter did not go as far as I did and that he used another method. [...] But now my work has gone further and I am tired of waiting; therefore, I wrote the main results I obtained in 6 notebooks, and I sent them, in a sealed envelope, to the Accademia dei Lincei, to record the date. So I hope to arrive, in any case, in time not to lose priority, or at least to be able to prove that I had solved the problem independently of Kötter. [...] It seems to me that Kötter's result, which I also obtained independently, in any case, is one of the possible applications of my method.[53]
>
> [De Paolis to Segre. Pisa, 4 March 1887]

With these researches, in December 1887, De Paolis competed for the royal prize for mathematics at the Accademia Nazionale dei Lincei, presenting his essay *Fondamenti di una teoria, puramente geometrica, delle linee e delle superficie*. The paper was divided into three parts: general theory of correspondence between the points of several groups, general theory of projective correspondences in the basic forms with one dimension, and the one in two dimensions. The first part was published in 1890,[54] and the second and third parts[55] were published after his

[51] In order to avoid unnecessary repetition, the part selected in bold from the one sent to Cremona has been omitted from this letter.

[52] He is referring to the news of the result of the contest for the Steiner prize.

[53] "Lessi nei Rendiconti dell'Accademia di Berlino, la relazione del lavoro di Kötter; ho aspettato ansiosamente la sua pubblicazione, ma non so se ancora abbia veduto la luce. [...] Sono più anni che mi perseguita l'idea di emancipare la Geometria dal sussidio dell'Algebra, di fondare la vera *Geometria pura*. Quando seppi la notizia avevo completamente risolto la quistione, non potevo pubblicare subito le mie ricerche, perché ancora non avevo svolto le parti secondarie di tutto il lavoro che mi ero proposto di fare; perciò pensai di aspettare la pubblicazione della Memoria di Kötter, qualunque cosa potesse avvenire, tanto più che dalla relazione fatta su questa Memoria mi pare che Kötter non si sia spinto innanzi come me e che abbia tenuto un altro metodo. [...] Ora però il mio lavoro è spinto più innanzi e sono stanco di aspettare; perciò ho appuntato i principali risultati che ho ottenuto in 6 quinterni, e li ho spediti, in un plico sigillato, all'Accademia dei Lincei, perché se ne prenda data. Spero così di arrivare, in ogni caso, in tempo per non perdere la priorità, o almeno poter provare che avevo risoluto il problema indipendentemente dal Kötter. [...] Mi pare che il risultato di Kötter, che anche io del resto ho tenuto indipendentemente, in ogni modo sia una delle possibili applicazioni del mio metodo."

[54] R. De Paolis, *Teoria dei gruppi geometrici e delle corrispondenze che si possono stabilire tra i loro elementi*, Memorie della Società Italiana delle Scienze (detta dei XL), (3), 7, 1890.

[55] R. De Paolis, *Le corrispondenze proiettive nelle forme geometriche fondamentali*, Memorie della Accademia delle Scienze di Torino, (2), 42, 1892, pp. 495–584 and *Teoria generale delle corrispondenze proiettive e degli aggruppamenti proiettivi nelle forme fondamentali a due dimensioni*, Rendiconti della Reale Accademia Nazionale dei Lincei, (5), 3, 1894, pp. 225–227.

death in 1892 and in 1894, thanks to the contribution of Mario Pieri and Segre. The extracts from the second essay contain, according to the author's will, the dedication "If the work is worthy—of remembering a name—it must be the illustrious name—of Luigi Cremona—who in geometric science—I had as a teacher."[56]

5 Conclusions

The unpublished correspondence between Luigi Cremona and his students forms a rich source of evidences that highlight even better the dynamic that was developing in the period immediately after the unification of Italy, when the mathematicians of the Risorgimento succeeded in bringing Italian mathematics to the level of the European one. The numerous direct students of Cremona, including Bertini, Caporali, and De Paolis, formed a generation of scholars who had the merit of cultivating the scientific innovations of their teacher and passing them on to the subsequent generations, that is, those young mathematicians who were able to take up the baton by easily entering the framework of European research, having at their disposal cultural structures and tools (journals, libraries, scientific laboratories, etc.) by now largely sufficient for research developments.

From many points of view, Bertini, more than the others, served as a link between the Cremona work and that of the new school of algebraic geometry that was being formed. Caporali in Naples and De Paolis between Pavia and Pisa contributed to developing Cremona's ideas and researches and constantly considered him as a reference in their scientific life. Also, they both interacted with Segre, showing that he was already representing the new reference in the framework of the programs of research of the Cremona school.

The letters that the students of Cremona sent to their teacher form a new piece of that puzzle full of scientific and cultural ideas that has been reconstructing for years, making the influence that Cremona had in the scientific environment, the interactions, the collaborations that his students had with other European scientists, and the successes and results achieved nationally and internationally even more evident.

References

1. Conte, A., Ciliberto, C.: La seconda rivoluzione scientifica: matematica e logica. La scuola di geometria algebrica italiana. In: Storia della Scienza, Treccani (2004)
2. Palladino, N., Vaccaro, M.A.: L'ipocicloide tricuspide: il duplice approccio di Luigi Cremona ed Eugenio Beltrami. Bollettino di Storia delle Scienze Matematiche. **38**(1), 61–92 (2018). https://doi.org/10.19272/201809201003

[56] "Se l'opera è degna – di ricordare un nome – sia quello illustre – di Luigi Cremona – che nella scienza geometrica – ebbi maestro."

3. Castelnuovo, G.: Luigi Cremona nel centenario della nascita, Commemorazione. Rendiconti della Reale Accademia Nazionale dei Lincei. **XII**, 613–618 (1930)
4. Israël, G.: Correspondence of Luigi Cremona (1830–1903). Brepols (2017)
5. Clebsch, A., Lindemann, F.: Vorlesungen über Geometrie. Druck und Verlag, Leipzig (1876)
6. Enriques, F., Chisini, O.: Lezioni sulla teoria geometrica delle equazioni e delle funzioni algebriche, vol. 1, 2. Zanichelli, Bologna (1915, 1918)
7. Salmon, G.: A Treatise on the Higher Plane Curves, 2nd edn. Hodges, Foster and Co, Dublin (1873)
8. Castelnuovo, G.: Commemorazione del socio Eugenio Bertini, Atti Accademia Nazionale Lincei Cl. Sci. Fis. Mat. Natur. Rend. Lincei. **17**(6), 745–748 (1933)
9. Berzolari, L.: Commemorazione di Eugenio Bertini. Rendiconti del Reale Istituto Lombardo di Scienze e Lettere. **66**(2), 67–68 (1933) 609–635
10. Scorza, G.: Eugenio Bertini. Esercitazioni matematiche (Circolo matematico di Catania). **7**(2), 101–117 (1934)
11. Berzolari, L.: Eugenio Bertini. Bollettino Unione Matematica Italiana. **12**, 148–153 (1933)
12. Conti, A.: Onoranze ad Eugenio Bertini. Bollettino di mat. Firenze. **13**(2), 46–52 (1934)
13. Fubini, G.: Eugenio Bertini. Atti della Reale Accademia delle scienze di Torino, Classe di scienze fisiche matematiche e naturali. **68**, 447–453 (1933)
14. Vaccaro, M.A.: Dalle trasformazioni quadratiche alle trasformazioni birazionali. Un percorso attraverso la corrispondenza di Luigi Cremona. Bollettino di Storia delle Scienze Matematiche. **36**(1), 9–44 (2016)
15. Bertini, E.: Sopra una classe di trasformazioni univoche involutorie. Annali di Matematica. **8**(2), 11–23 (1877) Errata-Corrige, p. 146
16. Bertini, E.: Ricerche sulle trasformazioni univoche involutorie nel piano. Annali di Matematica. **8**(2), 244–286 (1877)
17. Caporali, E.: Sulle trasformazioni univoche piane involutorie. Rendiconti della Reale Accademia delle Scienze fisiche e matematiche di Napoli. **18**, 212–219 (1879)
18. Bertini, E.: Sulle trasformazioni univoche piane e in particolare sulle involutorie. Rendiconti del Reale Istituto Lombardo di Scienze e Lettere. **13**(2), 443–451 (1880)
19. Bertini, E.: Sopra alcune involuzioni piane. Rendiconti del Reale Istituto Lombardo di Scienze e Lettere. **16**(2), 89–101 (1883) 190–208
20. Bertini, E.: Deduzione delle trasformazioni piane doppie dai tipi fondamentali delle involutorie. Rendiconti del Reale Istituto Lombardo di Scienze e Lettere. **22**(2), 771–778 (1889)
21. Kleiman, S.: Bertini and his two fundamental theorems. Rendiconti Circolo Matematico di Palermo, 1–29 (1997)
22. Caporali, E.: Memorie di Geometria. Napoli Benedetto Pellerano (1888)
23. C. Cerroni, Il carteggio Cremona-Guccia., Mimesis edizioni, 2014
24. Loria, G.: L'opera scientifica di Ettore Caporali esaminata da G. Loria. In: Giornale di Matematiche, vol. XXVII, pp. 1–32 (1889)
25. Segre, C.: Alcune idee di Ettore Caporali intorno alle quartiche piane. Annali di Matematica. **XX**, 237–242 (1892)
26. Brigaglia, A.: Per una biografia scientifica di Corrado Segre, La Matematica nella Società e nella Cultura. Rivista dell'Unione Matematica Italiana. **6**(3), 415–474 (2013)
27. Segre, C.: Riccardo De Paolis; cenni biografici. Rendiconti del Circolo Matematico di Palermo. **6**, 208–224 (1892)
28. De Paolis, R.: Le trasformazioni piane doppie. Memorie della Reale Accademia Nazionale dei Lincei. **1**(3), 511–544 (1877)
29. De Paolis, R.: Sui fondamenti della geometria projettiva. Memorie della Reale Accademia Nazionale dei Lincei. **9**(3), 489–503 (1881)
30. De Paolis, R.: Le trasformazioni doppie dello spazio. Memorie della Reale Accademia Nazionale dei Lincei. **1**(4), 576–608 (1885)
31. De Paolis, R.: Alcune particolari trasformazioni involutorie dello spazio. Rendiconti della Reale Accademia Nazionale dei Lincei. **1**(4), 735–742 (1885) 754–758
32. Padelletti, D.: Ettore Caporali. Annali dell'Università di Napoli (1886) published also in [22]

Veronese, Cremona, and the Mystical Hexagram

Aldo Brigaglia

Abstract Apart from a brief historical introduction on what precedes them, this chapter limits itself to presenting the two almost simultaneous works (1877) of the young Veronese and the already famous Cremona in the belief that they constitute significant moments for both the scholars. Pascal mysticum hexagram, after the works of Steiner, Kirkman, Plücker, Cayley, and Salmon, provided a complex configuration made by 60 Pascal lines (one for each of the hexagons built on the same points of a conic), which converge 3 by 3 on Kirkman's 60 points (K) and Steiner's 20 points (S). The points K lie 3 by 3 on the Pascal lines p. Each line p passes through a point S; each of the 15 Plücker (pl) lines passes through 4 points S. The points K lie 3 by 3 in the Cayley lines (c); the lines c pass through 3 of the 15 Salmon's points (Sa); every line c passes through a point S; 4 lines c pass through each Sa point. In his paper, Veronese added three new results: It is possible to divide the complex of the mystical hexagram into 6 configurations, each with 10 Pascal lines and 10 Kirkman points, plus 10 Steiner points, shared with two other configurations; two different configurations always have in common 4 Steiner points and a Plücker line that joins them; it is possible to obtain infinite configurations that are isomorphic with the first (Conway and Ryba's multi-mystic). Cremona showed that the same results could be obtained from the configuration of 15 lines and 15 tritangent planes on a cubic surface (the Cremona–Richmond configuration). My chapter deals with the two different approaches to the two mathematicians.

Keywords Mystical hexagram · Cubic surfaces · Giuseppe Veronese · Luigi Cremona

Translated by Denis Gailor

A. Brigaglia (✉)
Dipartimento di Matematica e Informatica, Palermo, Italy

G. Bini (ed.), *Algebraic Geometry between Tradition and Future*,
Springer INdAM Series 53, https://doi.org/10.1007/978-981-19-8281-1_15

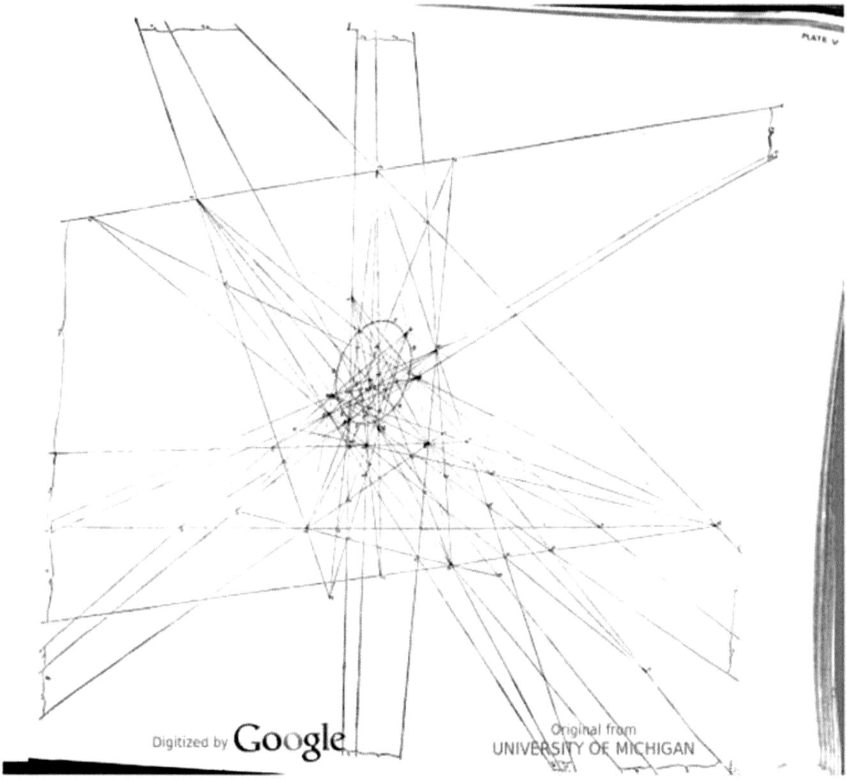

This figure is taken from [1], to my knowledge the only generous attempt to graphically reproduce the complexity of the mystical hexagram at a time when the technical means available were very limited.

> I have constructed on a very large scale a figure of the sixty Pascalian lines, and the forty-five Pascalian points, ...; but the figure is from its complexity and the inconvenient way in which the points are either crowded together or fly off to a great distance, almost unintelligible [2].

I chose to start with this quote from Cayley basically for two reasons: on the one hand, it is a preventive self-defense. I have included many figures in the text, certainly difficult to read, and I thought I found in Cayley a very authoritative defender. But there is more in the words of the English mathematician: in the vicissitudes of the mystical hexagram, a great complexity hides an eminently symmetrical structure, which in a sense is extremely simple. It is perhaps for this reason that such a large number of leading nineteenth-century mathematicians became interested in an apparently secondary problem which, moreover, can be solved with elementary methods. The list is remarkable: Steiner, Plücker, Staudt, Cayley, Salmon, Veronese, Cremona, ...

This is a largely incomplete work of mine. For example, it would be necessary to examine the developments deriving from the works of Veronese and Cremona (to which it is entirely dedicated), which are conspicuous, from various points of view: for example, those deriving from the hyperspatial interpretation of Cremona's hexahedral equation [3] and those connected with the Cremona-Richmond configuration [4, 5]. Apart from a brief historical introduction on what precedes them, this chapter of mine limits itself to presenting the two almost simultaneous works of the two mathematicians in the belief that they constitute significant moments both for the development of the young Venetian and for better understanding the Roman period of the elderly (?, 47 years!) and already very famous mathematician from Pavia, as well as the intrinsic development of some aspects of nineteenth-century mathematics.

1 Prologue

> Since I had to prepare a little work to give a lecture among my student friends at the Zurich Polytechnic in the Mathematische Seminar directed by the illustrious Profs. Fiedler and Frobenius, in June last year, I set out to solve the question of Hesse and Schröter, and since I believe I have not only completely solved it, but also have added other important theorems to it, I take heart to now present my little work to the professors at Rome University, with the hope that it will be well accepted.

With these words, Giuseppe Veronese presents his first scientific work to the professors at Rome University. [1]

Giuseppe Veronese (1854–1910) had studied at the Polytechnic in Zurich and, since the summer of 1875, had had as his teacher Wilhelm Fiedler, who corresponded with Cremona and was his inspiration for teaching projective-descriptive-graphical static geometry for engineering students. In 1866, Fiedler had also translated Salmon's text on conics into German, in which there is a Note on the Hexagram. It is therefore probable that it was his teacher who directed his attention to this problem.

Equipped with this work (and also with Fiedler's recommendation)[2] in the summer of 1876, he had already written to Cremona asking him if he could enroll directly in the fourth year of the course and proposing his *little work* as a thesis. Not only did Cremona accept, but, in the autumn of that year, the 22-year-old mathematician from Chioggia became assistant to the chair of Cremona himself.

[1] Here and afterwards, the English translation of the Italian citations is of the author. *Siccome io dovevo preparare un lavoretto per tenere una conferenza tra i miei amici studenti del Politecnico di Zurigo nel Mathematische Seminar diretto dagli illustri sigg. Prof. Fiedler e Frobenius, nel giugno dell'anno testé passato, così mi proposi allora di risolvere la questione di Hesse e Schröter, e poiché io credo di averla non solo completamente risoluta, ma ben anco avervi aggiunti altri teoremi importanti, così mi faccio animo di presentare ora questo mio piccolo lavoro ai sigg. Professori dell'Università Romana, con la speranza che sia ben accettato* [6, p. 651].

[2] Beltrami had also supported Veronese's requests. In a letter dated 4 October 1876, he wrote: He is a talented young man, especially in pure geometry, and I think he is perfectly eligible for the fourth year. (Biblioteca dell'Istituto Mazziniano di Genova, Fondo Itala Cremona Cozzolino, Sc. 49, 9992).

Veronese, who had just returned to Italy, continued his work on the mystical hexagram and kept his former teacher informed. On September 11, 1876,[3] the young Venetian mathematician expounds the state of his work to him: at this time, he has only taken the first step in the direction of the multi-mystic and therefore speaks of two equivalent systems. On October 24, Fiedler wrote to Cremona warmly recommending his pupil and praising his work. On December 24, the work is complete: the multi-mystic, with its infinite isomorphic configurations, is completely determined. On July 19, Veronese could send several copies of his work to his professors and colleagues. He gave a brief summary of Cremona's work, highlighting above all the realization of the multi-mystic through the hexahedron and the associated succession 1, 3, 1/3, ... This is an aspect that will be overlooked by almost all subsequent reviewers of Cremona's work.

In April of the following year, Battaglini presented the work to the Accademia dei Lincei, and it was published in the *Memoirs* in the same year. Cremona himself presented his own stereometric version in the same *Lincei* fascicle.

The work, 61 pages long, contains 64 theorems. It begins with a historical note in which the main results obtained so far are given. He goes on to give his own exposition of the preexisting material. I will refer mainly to this presentation in the following section.

2 Before Veronese (1639–1877)

Here, I will give a quick outline of the main results known in 1876. I will return to these results in more detail in examining Veronese's work.

Of course, the story begins with Pascal and is well known. In 1639, 16-year-old Blaise delineated his vision of the problems relating to conics [9] which he published the following year (a flying sheet, only surviving in two copies). In it there is stated what is known as *Pascal's theorem.* Since it is stated in a slightly different way from the usual one, I think it is better to express it in his words (Fig. 1): let 4 points M, S, K, V be given such that if MK and SV intersect in A and MV and SK in B, no triplet of aligned points is formed (let the 6 points M, S, K, V, A, B be in a general position). Let us consider any conic passing through K and V and let P, O, N, Q be the points at which MK, MV, SK, SV intersect the conic again: then MS, PQ, and ON constitute a sheaf (or equivalently PQ intersects ON at a point T on MS). We immediately see that M is the intersection of the opposite sides PK and OV of the hexagon $KPQVON$, while S is that of KN and QV and T that of PQ and ON, hence the usual statement of the theorem. I will use the notation $p\ (ABCDEF)$ to indicate the Pascal line relative to the hexagon $ABCDEF$.

[3] Veronese's and Cremona's letters to Fiedler can be found in [7], where there are also summaries of the ones to Cremona, the full texts of which can be found in [8].

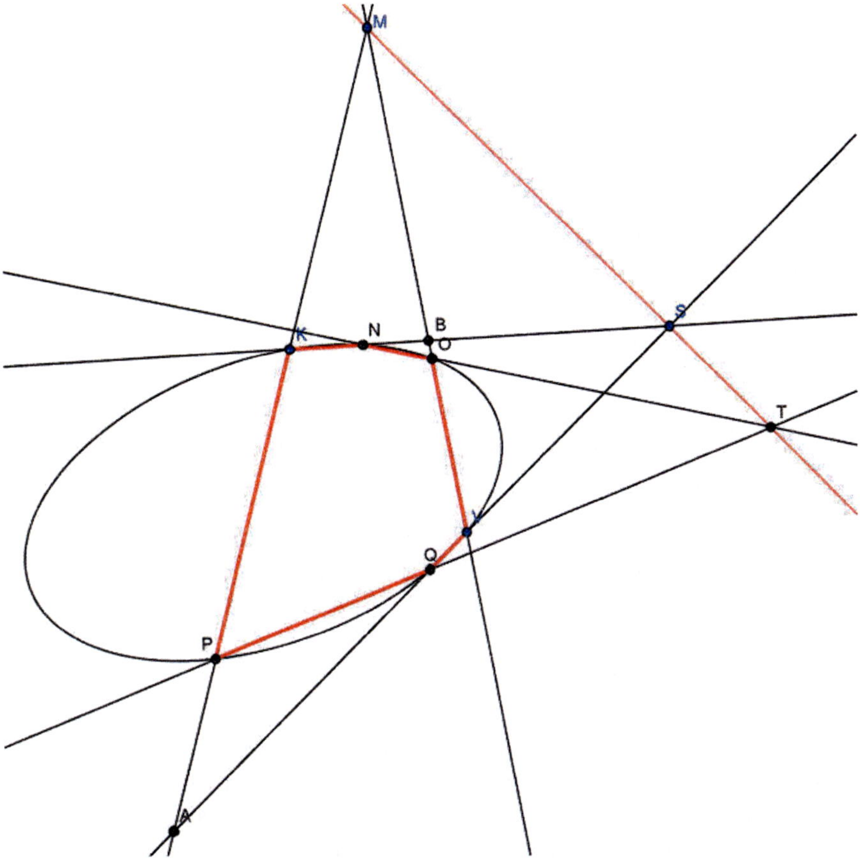

Fig. 1 Pascal's theorem (the letters are those referred to in the text)

In what follows I will indicate with (*AB*, *CD*, *EF*) the triangle formed by lines *AB*, *CD*, and *EF*.

We immediately note that two permutations of the letters *A*, *B*, *C*, *D*, *E*, *F* give rise to the same hexagon in two cases: if they are obtained from each other through a cyclic permutation (equivalent to choosing one vertex or another one as the starting point for reading the hexagon) or if they are obtained by exchanging the order (equivalent to choosing a clockwise or counterclockwise reading). We will state two of such equivalent permutations. In total, the 720 permutations of 6 letters are divided into 60 equivalence classes.[4] The name *mystical hexagram* is attributed by Leibniz to Pascal himself.

[4] Many nineteenth-century commentators attribute to Steiner the idea that Pascal's theorem refers to any of the 60 permutations. It does not seem to me in any way that such an idea was not already present in Pascal himself.

Before continuing, a few notations:

Given six points on a conic, we can determine 15 lines (chords) that join them 2 by 2. Let us consider the triangles as formed by three nonconsecutive chords 2 by 2 (which is equivalent to the fact that each letter appears once and only once). I will indicate such a triangle with (*AB*, *CD*, *EF*). Each chord belongs to three triangles. For example, the chord *AB* belongs to the following triangles: (*AB*, *CD*, *EF*), (*AB*, *CE*, *DF*), and (*AB*, *CF*, *DE*). A total of 15 triangles can be formed.

Two triangles will form a "pair" if they have no side in common. The set of pairs is in one-to-one correspondence with the set of hexagons constructed on the six points; for example, to the pair (*AB*, *CD*, *EF*), (*AC*, *BE*, *DF*), there corresponds the hexagon *ABEFDC* and vice versa.[5] There are therefore 60 pairs, and with each one we can associate a Pascal line (in our case, *p*(*ABEFDC*)). It should be noted that Pascal's theorem is equivalent to the statement that the two triangles of a pair are always perspective with a Pascal line axis. So, for example, (*AF*, *BC*, *DE*) is perspective with respect to (*AB*, *CD*, *EF*) (with line *p*(*ABCDEF*) as the axis), as can be seen in the figure and is easy to verify (Fig. 2).

Anyway, it was Steiner, three centuries after Desargues, who revived interest in extensions of the hexagram: he did so in 1828 through a "Question Proposée" in *Annales de Mathématiques* [10] of just one page in which he proposed four questions relating to Pascal line, their correlatives in relation to the Brianchon point, and their respective polarities. The four relating to Pascal line are as follows: 1) For each of the 60 hexagons inscribed in a conic and having the same vertices *A*, *B*, *C*, *D*, *E*, *F*, Pascal's theorem holds and so we have a line *r*. 2) The lines concur in threes in 20 points, which would later be called *Steiner points*. 3) These 20 points lie in fours on the same line, thus determining 5 lines. 4) The 5 lines concur at the same point. While the first two statements are correct, the third and fourth are incorrect. Conway and Ryba [11] observe: *Beware! His one-page paper contains as many false assertions as true ones*!

It was Plücker, the following year, who integrated and corrected Steiner's proposals [12]. The work is dedicated to the presentation of the principle of duality, the subject of a dispute over priority in those years, but now one of the most important tools of nascent projective geometry. The first section is dedicated to the extension of Pascal's theorem. Plücker initially uses an analytical method adapted to his favorite symbology. It would be interesting to examine his methods closely, but this would make me stray too far from the purpose of this chapter.

Plücker knows Steiner's proposal and reports it in full in his work, correcting point 3): the 20 points are aligned by 4, but in 15 lines (afterwards called *Plücker lines*) and not in 5 s. Point 4) is completely suppressed as it is erroneous. I will not give the demonstrations now. I will refer to those of Veronese in what follows.

---

[5] The rule for switching from the pairs of triangles to the hexagon is simple: just start from any side of the first and then continue alternating between the first and the second. Thus, *AB* is followed by *BE*, then *EF*, then *FD*, and then *DC*. Conversely, the triangles are obtained from the hexagon by alternating the sides. Thus, *AB* will be the first side of the first triangle and *BE* that of the second. There will, respectively, follow *EF* and *FD* and lastly *DC* and *CA*.

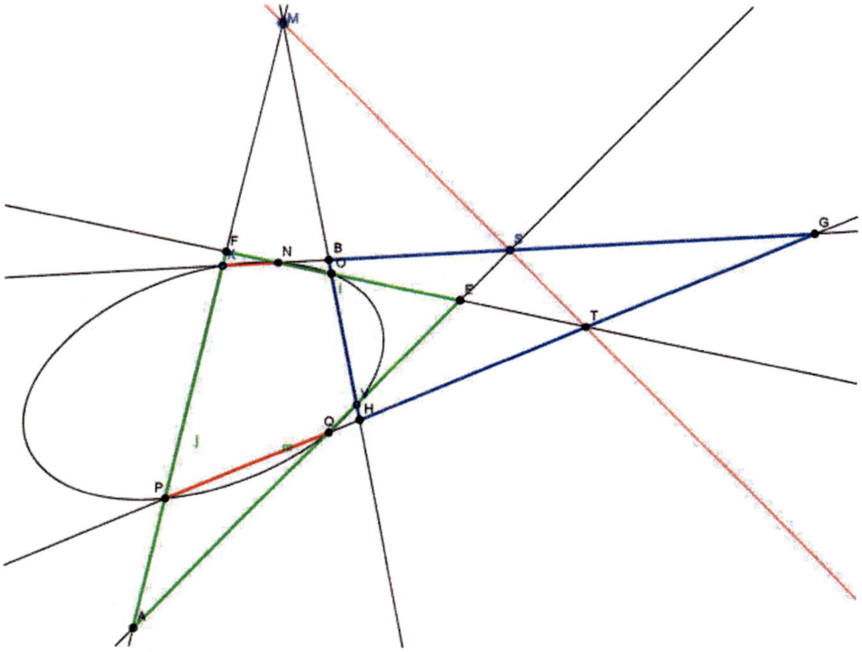

Fig. 2 A pair of triangles and Pascal's theorem

I will limit myself to a few hints regarding the important developments of the theory in the following years, even if they deserve a thorough study of the different combinatorial methods used for the different proofs. In 1832, in his very influential *Entwicklung* [13], Steiner returned to the subject giving[6] (once again also in the dual form) the results he had enounced in 1828, this time in the correct form: the lines (Plücker ones, but Plücker is not mentioned at all) have correctly become 15.

It seems to me that, perhaps due to the richness of Steiner's text, this Appendix has been almost completely unnoticed. However, a brilliant English "amateur," the Reverend Thomas Kirkman (1806–1896),[7] seems to have read it carefully. Kirkman was indeed an amateur, but an original and interesting mathematician. In 1846, he made his debut in the field of mathematics dealing with what are now called "Steiner systems," one of the key points of combinatorics. He was to remain famous for one of the founding problems of this branch of mathematics, the well-known one of the *fifteen schoolgirls* problem, which he addressed in 1850 in the *Lady's and Gentleman's Diary*. In the same year, Cayley's and his

[6] [13], Anhang, 54–58, pp. 311–313.

[7] I use the term "amateur" only in the sense of "not belonging to the academic world." On the biography of this interesting mathematician, I refer the reader to [14], also recalling that the author is Dikran Tahta, Stephen Hawking's teacher.

solutions were published. It is likely that in these years so intensely devoted to reflections on combinatorial mathematics, the reading of some of Cayley's works aroused his interest. In the 1840s, Cayley showed growing interest in the geometric interpretation of combinatorial problems. He was to refer in particular to the series of articles with the same title [Ca1–Ca4] which appeared between 1846 and 1851 in *Crelle's Journal.* The first of these works appeared in 1846 and constitutes a founding work for the study of projective configurations, tending to systematize the various scattered results on the subject. I will quote here the initial passage: *En prenant pour donné un système quelconque de points et de droites, on peut mener par deux points donnés des nouvelles droites, ou trouver des points nouveaux, savoir les points d'intersection de deux des droites données; et ainsi de suite. On obtient de cette manière un nouveau système de points et de droites, qui peut avoir la propriété que plusieurs des points sont situés dans une même droite, ou que plusieurs des droites passent par le même point; ce qui donne lieu à autant de théorèmes de géométrie de position. On a déjà étudié la théorie de plusieurs de ces systèmes; par exemple de celui de quatre points; de six points, situés deux à deux sur trois droites qui se rencontrent dans un même point; de six points trois à trois sur deux droites, ou plus généralement de six points sur une conique (ce dernier cas, celui de l'hexagramme mystique de Pascal, n'est pas encore épuisé; nous y reviendrons dans la suite)* [15, 16].

And the second part of the work is dedicated to the mystical hexagram. Cayley starts from what Steiner wrote in the previously cited Appendix to the *Entwicklung* and says that there the Swiss mathematician had proved (*prouvé*) the theorems on 20 points and 15 lines and then says that the first part is easily proved (and he will do so a little further on) while for the second, *je n 'ai pas réussi à trouver les combinaisons de quatre points qui doivent être situés en ligne droite, et il me paraît même qu'il est impossible de les trouver* [15, 16]. This is a wholly surprising statement! Indeed, first of all, Steiner has not proved at all what has been asserted (and if he had proved it, how could Cayley doubt the possibility of finding *les combinaisons de quatre points*); moreover, he still completely ignores Plücker's proof, which appeared, as we have seen, in the same magazine. What should make us reflect, I believe, is the inherent difficulty of this kind of problem. While it is not that difficult to prove propositions once the rules are discovered, it is difficult to establish these rules. Which permutations on the vertices of the hexagon give rise to concurrent lines at a point? And having established these permutations and then defined such a point through three different permutations, which (and how many) triplets give rise to aligned points? These are problems in which visual intuition can give few hints (unless it is directed towards "seeing" configurations of perspective triangles).

Cayley, however, immediately became aware of (or was informed about) Plücker's work, and the subsequent work (which appeared in 1847) begins with the acknowledgment that this *avait déjà été démontrée d 'une manière aussi simple qu'élégante par M. Plücker*. Cayley then gives his proof, entirely based on a careful study of the permutations connected with their geometric meaning and on a suitable notation.

The third article of the series (from 1848) also begins with Pascal's theorem: this time, it is an application to the alignment problems of the inflection points of a cubic.

Kirkman was probably influenced by reading Cayley's works and, in the summer of 1849, he published two articles on the subject in a newspaper (the *Manchester Courier*). A collaboration with two of the most famous English mathematicians of the time started around these problems, Cayley, precisely, and George Salmon. Certainly, in December, in an authoritative journal [17], Kirkman published not only his own results, but also some new ones provided by his authoritative colleagues. I will limit myself to a few mentions of some of the 32 theorems demonstrated in the text.

After presenting Steiner points and Plücker lines (which, strangely, given Cayley's correct attribution, he attributes to Steiner himself), Kirkman presents some new theorems (32 in all, including 1 from Cayley and 8 from Salmon). The main one due to him is the first: the 60 Pascal lines concur in threes on as many points (*Kirkman*).

Kirkman's theorem IV (which he owed, as he stressed, to Cayley) says that his 60 points are aligned 3 by 3 in 20 lines (later called *Cayley lines*). Theorem V (from Salmon) says that each Cayley line passes through a Steiner point. Theorem VI (again from Salmon) says that the Cayley lines concur in fours on 15 points (later called *Salmon points*).

In 1851 (but dated July 1849), in his fourth and last memoir of the series, Cayley published Kirkman's results with a proof of his theorem. In 1855, Salmon inserted a section of *Notes on Pascal's Theorem* in his widely used text on conics [18, pp. 317–321]. In this section, he summarized the aforementioned material and reorganized it systematically using the Desargues' theorem. In essence, what Kirkman wrote in his article was thus confirmed: *As the twenty points G of Steiner, each one the intersection of three of the Pascalian lines, lie four together on fifteen lines I* [Plücker lines], *so do the twenty lines X discovered by Mr. Cayley, each containing three of the sixty points H* [Kirkman points], *go four together through fifteen points Φ,—a remarkable correspondence* [17 , p. 191]. And in fact, this complex, but eminently symmetrical, configuration linked to a random and disordered choice of points deserved, I believe, the nickname of *mystic*.

3 Giuseppe Veronese's First Work: 1877

Among other works on the subject, I will limit myself to mentioning the 1868 work [19] by Hesse in which he demonstrates that 10 pairs of Steiner points are mutually harmonic with respect to the conic. For the rest, these are different and often interesting demonstrations rather than new results.[8] However, the work of

[8] As already noted, it was Hesse's conjecture that prompted Veronese to begin his work.

von Staudt [20], much appreciated by Veronese, should be mentioned. After brief historical notes, Veronese continues with 4 theorems on homological triangles with a purely combinatorial character which *as far as I know* [the author says, but I say too] *have not yet been considered* [6 p. 651].

Theorems 5 and 6 are none other than the well-known ones of Pascal and Brianchon.

Theorem 7 recovers Steiner points.

3.1 Steiner Points

I will start with some premises.

Each triangle will form eight pairs with as many triangles. For example, the triangle (*AB*, *CD*, *EF*) forms a pair with (*AC*, *BE*, *DF*), (*AC*, *BF*, *DE*), (*AD*, *BE*, *CF*), (*AD*, *BF*, *CE*), (*AE*, *BC*, *DF*), (*AE*, *BD*, *CF*), (*AF*, *BC*, *DE*), (*AF*, *BD*, *CE*):

We will say that three triangles form a triplet if by two, they form a pair. Each triangle belongs to 16 triplets and there are 80 triplets in all, as can easily be seen. In a triplet, there are three Pascal lines given by the three pairs that can be formed. The following theorem can be proved: the three axes of symmetry of a triplet concur on a point.

Veronese's proof is based on this property of triplets. For example, the triangles $GHI = (AB, CD, EF)$, $JKL = (DE, AF, BC)$, $MNO = (CF, BE, AD)$ form a triplet[9] (Fig. 3). As we have seen, the homology axis of the first two is *p*(*ABCDEF*), and lines *GJ*, *HK*, *IL* concur on point S'. It is easy to see that *M*, *N*, *O* belong, respectively, to the three lines.[10] Therefore, the triangles of the triplet have a single center of homology S'. Therefore, the axes of homology *p*(*ABCDEF*), *p*(*AFCBED*), *p*(*ABEFCD*) concur on a point, *S*, called *Steiner point*. Lines *GJM*, *HKN*, *ILO* are, respectively, *p*(*ABCFED*), *p*(*ADCBEF*), *p*(*ABEDCF*), i.e., S' is the Steiner point determined by the triplet (*AB*, *FC*, *ED*), (*BC*, *EF*, *AD*), (*AF*, *CD*, *BE*), which differs from the first by an exchange of the letters *D* and *F*. Steiner points *S* and S' are said to be conjugated. It can be noted that the three Pascal lines concurring on a Steiner point correspond to the hexagons obtained by fixing the three vertices of odd places and cyclically permuting those of even places. Using the notation of [11], I will denote *S* as (*ACE*; *BDF*) and hence S′ as (*ACE*; *BFD*).

Theorem 8 rediscovers a Hessian theorem from 1868 and gives a demonstration reprised by Staudt, a proof that uses refined reasoning on projective sheaves and its representation of imaginary elements. The theorem says the following:

Steiner points *S*(*ACE*; *BDF*) and S'(*ACE*; *BFD*) are harmonic conjugates with respect to the conic (i.e., the polar of *S* with respect to the conic passes through S').

Theorem 10 gives us the 60 Kirkman points.

---

[9] The triangles are presented in an orderly way, in the sense that the first, second, and third sides correspond to each other.

[10] Indeed, $M = (BE, CF)$ belongs to *p*(*ABEDCF*) and the same goes for *N* and *O*.

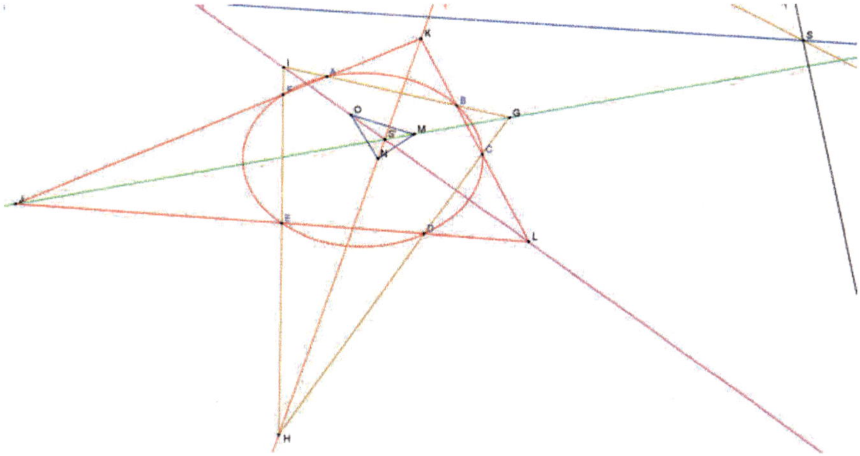

Fig. 3 A triplet of the first kind and the relative Steiner points

3.2 Kirkman Points

From a simple examination of the previously shown list, it can be established that the couple (*AB*, *CD*, *EF*) and (*DE*, *AF*, *BC*) forms a triplet, not only with the triangle (*CF*, *BE*, *AD*) but also with three other triangles: (*CE*, *BF*, *AD*), (*CF*, *AE*, *BD*), (*DF*, *BE*, *AC*).[11] These three triangles also form another triplet together. The five triangles will then form four triplets. The points determined by them, the intersection of three Pascal lines, are denoted as *Kirkman points*. From what has been said before, a pair can be completed to a triplet with three different triangles, and therefore on each Pascal line, there are three Kirkman points. There are therefore 60 Kirkman points, the same number as the Pascal lines. In conclusion, we can state that a pair of triangles (or a hexagon, which is the same thing) can be completed with a triplet in two ways: (a) with a triangle that does not form a couple with any of the others (triplet of the first kind): in this case, we have a Steiner point—the Steiner points therefore number 20, or (b) with each of the three triangles forming a triplet (triplets of the second kind): in this case, we have Kirkman points.

It is now appropriate to give, as Veronese does, an indication of the nomenclature adopted.

For the Kirkman points, the three lines will be determined by the three hexagons disjoined from one and the same. In our case, for example, we find that the triplet $(AB, CD, EF), (DF, BE, AC), (AD, BF, CE)$ is given by the three lines $p(ABEFDC), p(ACEBFD), p(ABFECD)$ and all three hexagons are disjoined from *AEDBCF*.

[11] The three triangles are obtained in the only way in which it is possible to use in one way the nine sides of the complete hexagon not used in the first two.

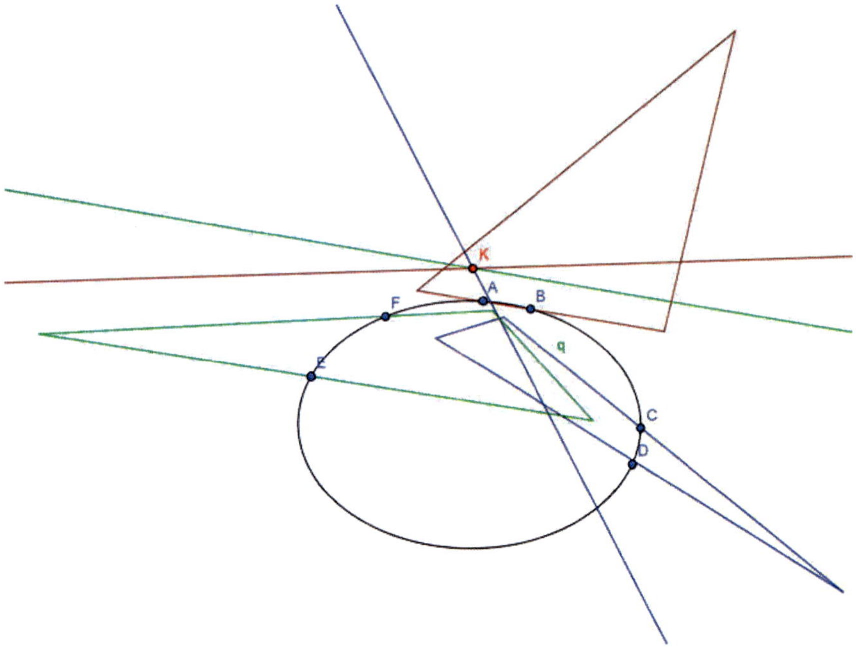

Fig. 4 A triplet of the second kind and the relative Kirkman point K

The corresponding point will therefore be indicated with *K*(*AEDBCF*). Line *p*(*AEDBCF*) and point *K*(*AEDBCF*) are said to be corresponding (Fig. 4).

Theorem 11 begins to prepare Veronese's extension, that is, it groups the Pascal lines and the corresponding Kirkman points in 6 configurations of 10 points and 10 lines. This is the first original result contained in the memoir by the young Venetian and the best known among his contemporaries, also because it disproved a well-known conjecture by Hesse, as we shall see.

3.3 *Veronese's Six Configurations*

Summarizing, starting from a couple, we can construct five triangles two by two forming a couple that gives rise to a Pascal line. Any three of the triangles will form a triplet of the second kind and thus give rise to a Kirkman point. Since in the five triangles there are ten pairs and ten triplets, we will have ten Pascal lines and ten Kirkman points.

At the risk of boring the reader, I would like to list these lines and these points and their relationships. It is to be remembered that a Kirkman point lies on a Pascal line if and only if the line and the point are defined by disjoined hexagons; the line and the point instead correspond if and only if they are defined by the same hexagon.

In our case, the hexagons that define (Fig. 5) the ten pairs are *ABEFDC*, *ABFECD*, *ABDCFE*, *ABCDEF*, *ACEBFD*, *ACFDBE*, *ACBEDF*, *ADBFCE*, *ADECBF*, *AEDBCF*; the ten triplets are determined by the same hexagons: for example *ACEBFD*, *ACFDBE*, *ADBFCE* constitute the triplet of hexagons disjoined from *ABCDEF* and so on. It can be seen that for each hexagon the three disjoined hexagons (i.e., the corresponding point *K*) are still among the ten. So, the set of ten lines and the ten corresponding points *K* is closed, in the sense that each line of the set contains three points of the set and three lines pass through each of its points. Since two of these sets are evidently disjoined, we will find that Pascal's configuration made up of 60 lines and 60 points split into 6 distinct ones. Veronese also demonstrates that for each configuration there is a different conic that makes it polar (i.e., such that corresponding points and lines are pole and polar with respect to the conic). With this, which remains the best known result contained in this work, Hesse's conjecture which had hypothesized the existence of a single polarity for the entire configuration of Pascal was denied.

Veronese's construction proceeds in the way I will now show (I exemplify the set of 10 lines to which there belongs *p*(*ABCDEF*)). Let us start precisely from *p*(*ABCDEF*) to *K*(*ABCDEF*). For each of the Pascal lines passing through *K*, we determine the corresponding point, hence *K*1(*ACEBFD*), *K*2(*ADBFCE*), *K*3(*AEBDFC*). The 6 lines that pass through these points (through each of the 3 points, there certainly passes *p*(*ABCDEF*)) together with the 4 previously listed, together with the corresponding points, form the configuration. Here are some images, with the suggestion of constructing them with GeoGebra (or other similar software), a way to understand the meaning and complexity of the construction.

In the figure, we have the starting line *r* = *p*(*ABCDEF*)) and the corresponding point, *K*, the intersection of the 3 lines *p*(*ACEBFD*), *p*(*ADBFCE*), *p*(*AEBDFC*) which have as their corresponding points, respectively, *K*1, *K*2, *K*3,, all on *r*. The lines that intersect on these 3 points are—omitting *r* – *p*(*AFCBDE*), *p*(*AEFCDB*); *p*(*ACDFEB*), *p*(*AFDEBC*); *p*(*ABFECD*), *p*(*ADECBF*). The corresponding points are, in order, *K*5, *K*4, *K*7, *K*6, *K*8, *K*9.

The scheme for determining the 6 configurations is therefore conceptually simple. The "recipe" given above allows us to obtain the configuration in which a given Pascal line is found.

In terms of permutations, we will find that the permutations involved, starting with *ABCDEF*, will be a) the 3 disjoined *ACEBFD*, *ADBFCE*, *AEBDFC*; b) the 2 disjoined from *ACEBDF*: *AEFCDB*, *AFCBDE*; c) the 2 disjoined from *ADBFCE*: *AFDEBC*, *ACDFEB*; and d) the 2 disjoined from *AEBDFC*: *ADECBF*, *ABFECD*. We immediately see that the set of 10 permutations is closed with respect to the "disjunction" operation. At this point, we can immediately determine which permutations (which hexagons) belong to the second configuration. It is a matter of taking any permutation not belonging to the first one and determining the relative disjoined ones in succession,

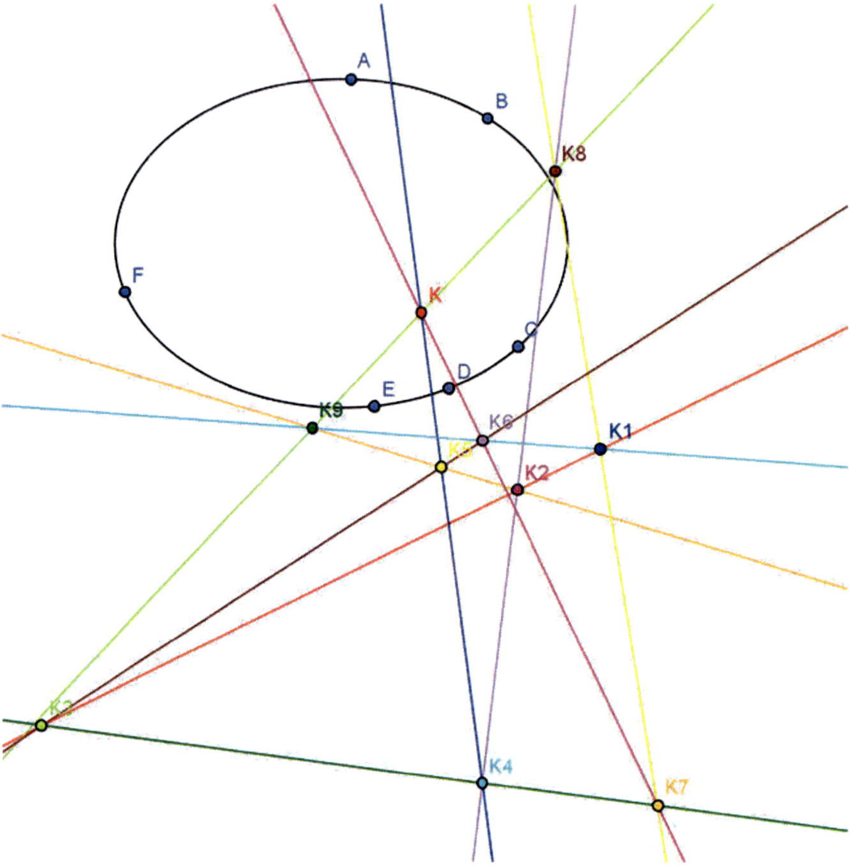

Fig. 5 One of Veronese's configurations

and to do this, it will be enough to replace each letter in the first one with its corresponding one. Thus, choosing the permutation as the starting point *ABEFCD*, we will have *AECBFD*, *AFBDEC*, *ACBFDE*, *ACDEFB*, *ADEBFC*, *ADFCBE*, *AEFDCB*, *AFCEBD*, *ABDCEF*.. The situation is analogous for the other 4.

Together with Veronese, I will indicate the 6 configurations indicating the generating element: *I*(*ABCDEF*), *II*(*ABEFCD*), *III*(*AFCBED*), *IV*(*ABEDCF*), *V*(*ADEFCB*), *VI*(*AFEBCD*).

So, we have six projective configurations made up of 10 lines and 10 points, such that each line contains exactly 3 points and 3 lines pass through each point (i.e., a configuration $(10)_3$, like that of Desargues).

Further, since the triangles $K9K6K5$ e $K8\,K7\,K4$ are evidently homologous (with center K and axis $p(ABCDEF)$), they are also (through a theorem of von Staudt) polar with respect to a conic Γ and therefore it follows (theorem XII) that the six configurations are such that each point is the pole of the corresponding line with

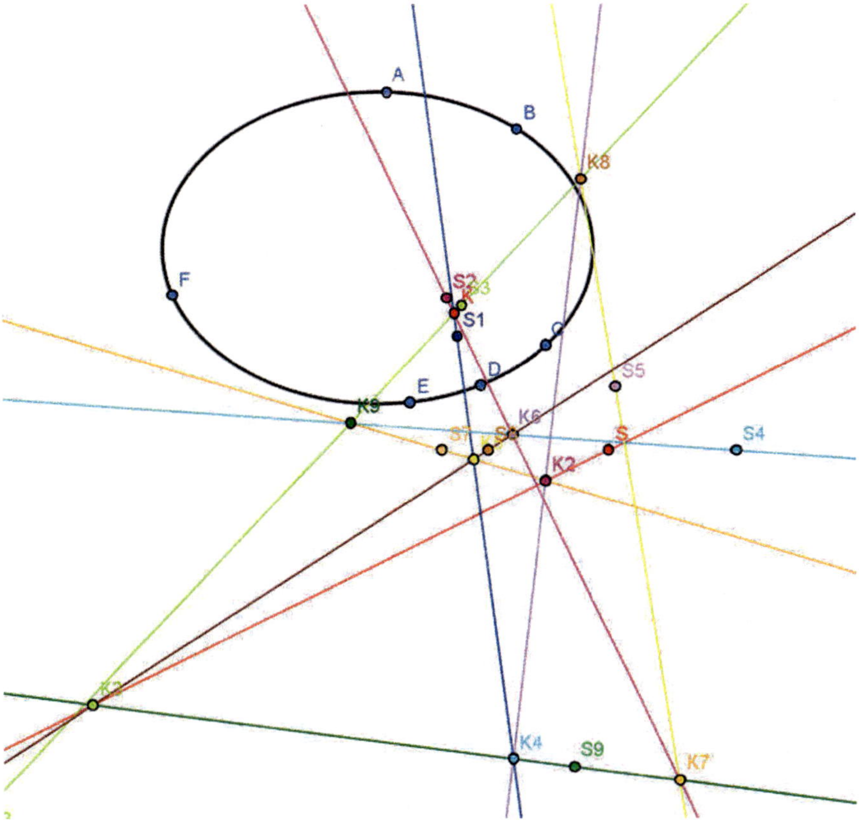

Fig. 6 The ten Steiner points of the first configuration

respect to the conic Γ (different for each of the 6 configurations, as will be shown below).

Each configuration then contains the Steiner points relating to each of its Pascal lines. Figure 6 shows the configuration with Steiner points *S*, *S*1, ...*S*9. It should be noted that 2 of the lines defining a Steiner point do not belong to the configuration. For example, *S* is the intersection of line *p*(*ABCDEF*) with *p*(*AFCBED*)and *p*(*ADCFEB*), which, as can be seen directly, do not belong to the configuration. It can be seen immediately—for instance by listing, like Veronese, the hexagons of each configuration—that the first belongs to the second configuration and the second to the third.

It therefore follows that each Steiner point belongs to 3 configurations and, vice versa, 3 different arbitrary configurations have a Steiner point in common; it is also shown that the conjugate of a Steiner point belongs to the three configurations complementary to those to which *S* belongs. Finally, 2 different configurations always have 4 Steiner points in common.

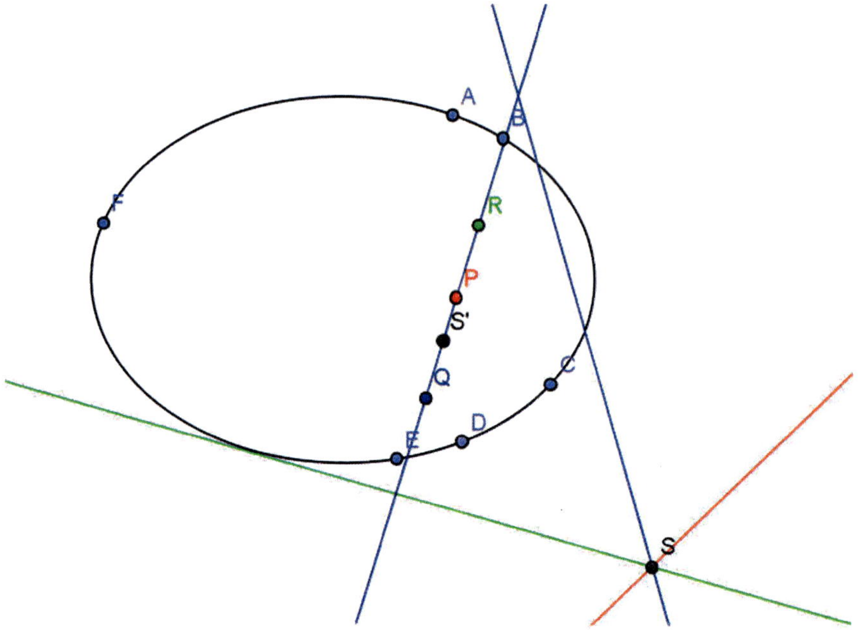

Fig. 7 A Cayley line of the first configuration

For convenience, I will list the Steiner points common to the 5 pairs obtained with the first configuration. In the following line:
$S, S4, S7, S8$; *I and III*: $S, S5, S6, S9$; *I and IV*: $S1, S2, S8, S9$; *I and V*: $S1, S3, S6, S7$; *I and VI*: $S2, S3, S4, S5$

Then, in an original way, Veronese finds the Cayley lines, obtaining the following result: given a Steiner point *S*, the intersection of 3 Pascal lines, *p*, *q*, *r*, and the 3 corresponding Kirkman points, *P*, *Q*, *R*, are aligned and the line that joins them (Cayley line) also goes through the conjugated Steiner point, S'. We will speak, with Veronese, of the Cayley line **corresponding** to the Steiner point. Figure 7 describes the situation for point *S*, with the symbols having obvious meanings.

Of course, of the Kirkman points aligned in a Cayley line, only one belongs to a given configuration while each Cayley line belongs to 3 different configurations, just like the corresponding Steiner point. Below (Fig. 8) there are marked the Steiner points $S1, \ldots S10$ of the first configuration and the corresponding Cayley lines, $c1$, $\ldots$ $c10$ (the lines correspond to the Steiner point of the same color and pass through the Kirkman point of the same color).

The correspondence between Steiner points and Cayley lines is complete: two configurations always have in common 4 Steiner points and the 4 corresponding Cayley lines; three configurations always have 1 point and the corresponding line in common (theorem XIV).

Furthermore, the 4 Steiner points common to 2 configurations are on a Plücker line (common). The 4 Cayley lines common to 2 configurations concur at a Salmon

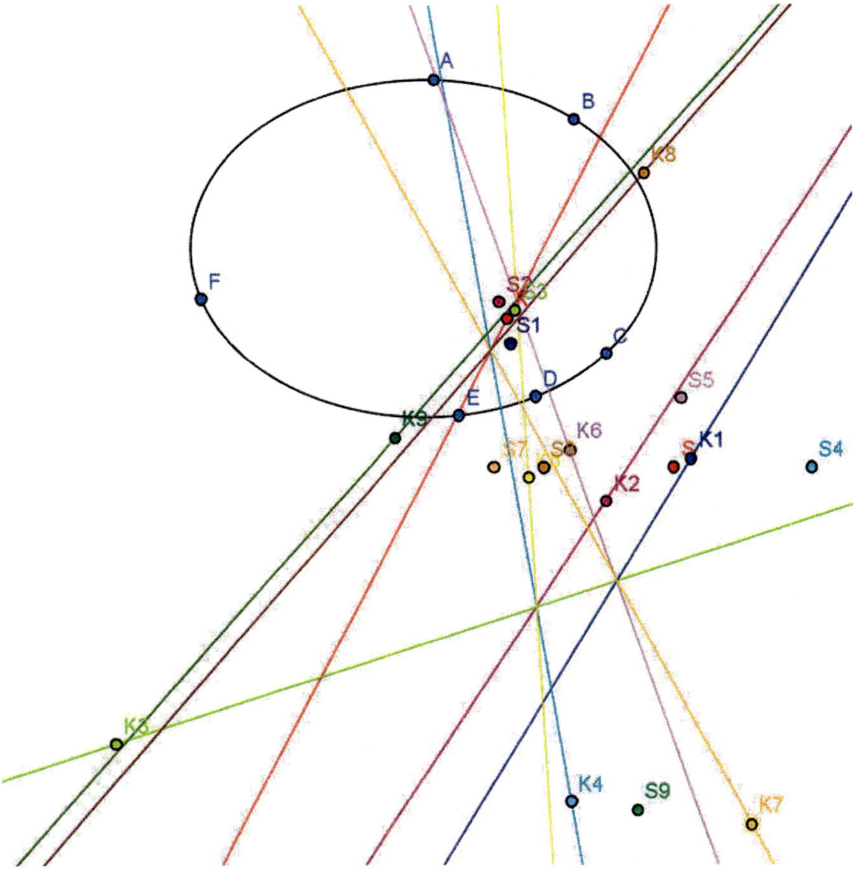

Fig. 8 Cayley lines and Steiner points of the first configuration

point. The Plücker lines and Salmon points number 15 in all, as was already known (theorem XV).

In Fig. 9, the Plücker line joining the four Steiner points common to configurations I and II is red, to I and III purple, to I and IV blue, to I and V green, and to I and VI magenta.

In the next figure, we see Veronese's procedure. We start again from the Pascal line r, in magenta, $p(ABCDEF)$. As we have seen, it contains a Steiner point $S = S(r)$, which is obtained as the intersection of r with lines $s = p(AFCBED), t = p(ADCFEB)$. Neither of these two lines belongs to the configuration $c(r)$. Let us consider the configuration $c(t)$. Obviously, S belongs to $c(r) \cap c(t)$. We see that Steiner points $S4$, $S7$, $S8$, which lie, respectively, on the lines of $c(r), p(ABFECD), p(ABDCEF), p(ABFECD)$, also lie, respectively,

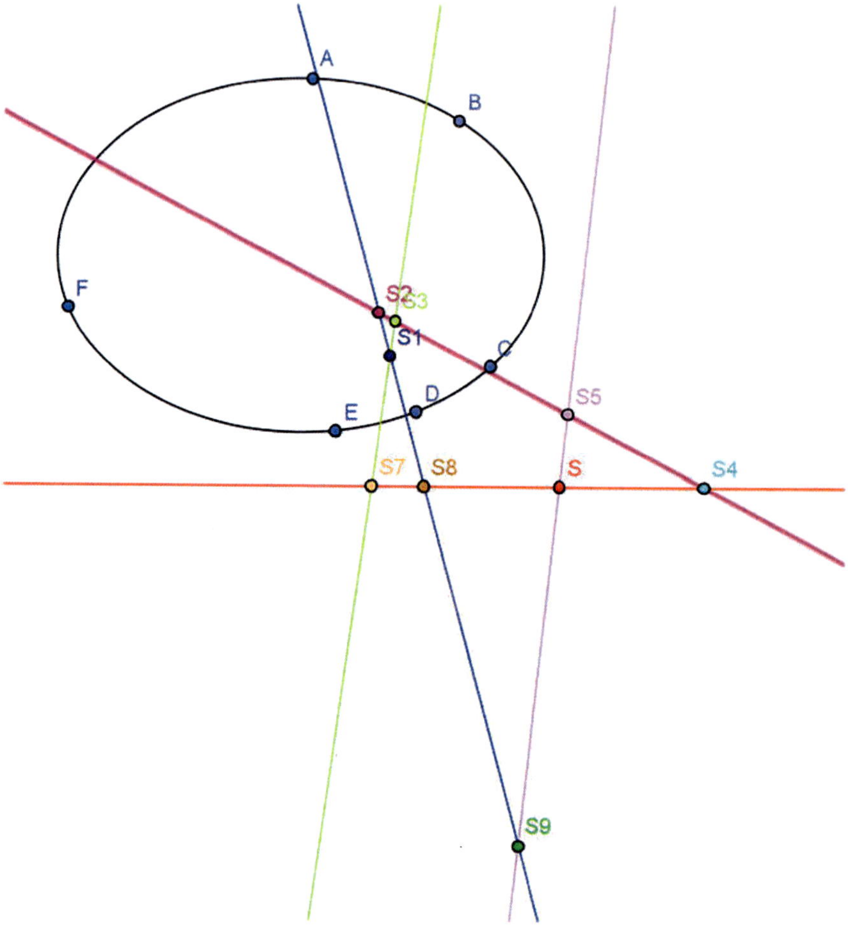

Fig. 9 The Plücker lines common to the first and the other configurations

on the lines of $c(t), p(ABDCFE), p(ABEFDC), p(ABCDFE)$. The 4 points are aligned in the Plücker line (Fig. 10).

If we now take the 4 Steiner points common to the first 2 configurations (as was done in the previous figure), we find the following (theorem XVI): the 4 Cayley lines corresponding to the 4 Steiner points concur at one point (Salmon point). In Fig. 11, the Steiner points are those obtained in the previous figure. The Cayley lines are the same color as the corresponding Steiner points. The Kirkman points that define the Cayley lines are all red. Salmon point V is highlighted. Some confusion is caused by the fact that some points are too close, but perhaps this can serve to show the complexity of identifying the alignments, all proved through a repeated use of Desargues' theorem.

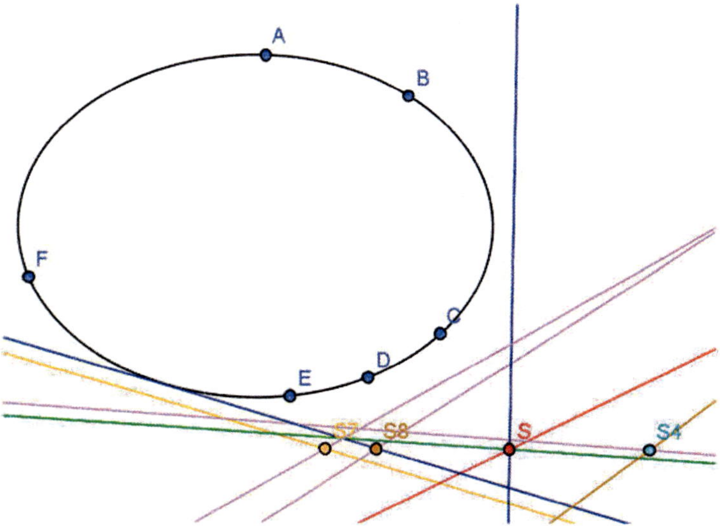

Fig. 10 Veronese's procedure for determining the Plücker line

The same consideration can be made for each pair of configurations. Figure 12 shows all the Steiner points (10), Cayley lines (10), Salmon points (5), and Plücker lines (5) present in configurations I and II. Corresponding points and lines are the same color. The Salmon points are V (intersection of the Cayley lines $c, c4, c7, c8$), O ($c, c5, c6, c9$), P ($c1, c2, c8, c9$; not visible in the figure), Q ($c1, c3, c6, c7$), R ($c2, c3, c4, c5$). Dually, the Plücker lines are *pl*2 (joining Steiner points $S, S4, S7, S8$), $pl3$ ($S, S5, S6, S9$), $pl4$ ($S1, S2, S8, S9$), $pl5$ ($S1, S3, S6, S7$), $pl6$ ($S2, S3, S4, S5$).

The reader may believe that this figure is just a messy tangle of points and lines, and he or she would be absolutely right. But it should make us reflect on the fact that this chaotic whole actually possesses a superior order and admirable symmetry.

I take from [21] a useful summary that links the "German" configurations with the "British" ones.

There are three correspondences between the two configurations:

1. One Kirkman point corresponds to each Pascal line.
2. One Steiner point corresponds to each Cayley line.
3. Each Plücker line corresponds to a Salmon point.

Three point/line incidence relationships:

1. On each Pascal line, there are 3 Kirkman points and one Steiner.
2. On each Cayley line, there are 3 Kirkman points, one Steiner point, and 3 Salmon points.
3. On each Plücker line, there are 4 Steiner points.

Three relations of line/point incidence:

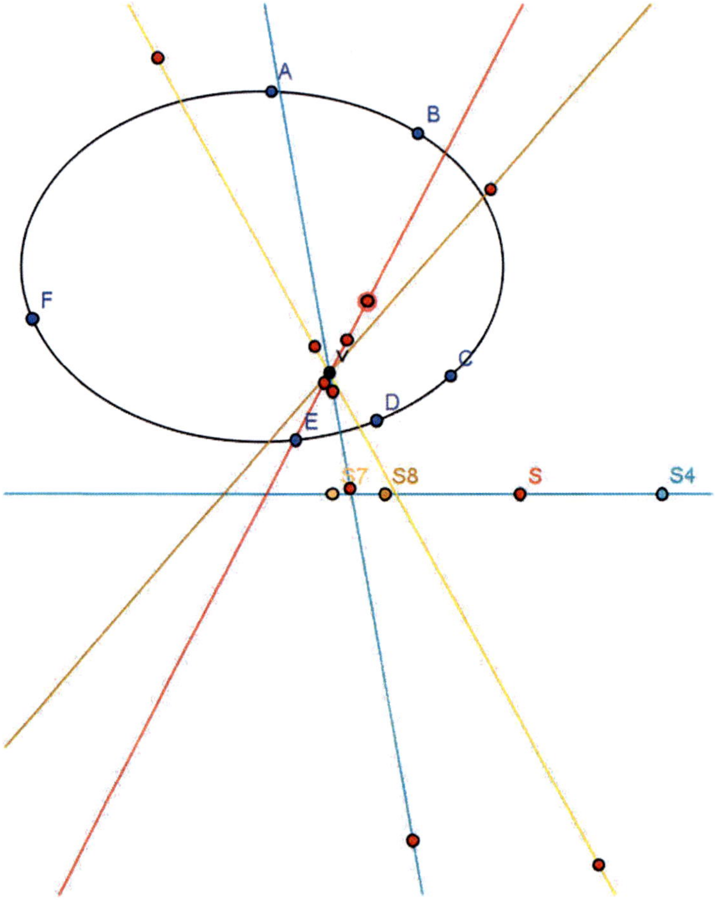

Fig. 11 Salmon point

1. For each Kirkman point, there are 3 Pascal and 1 Cayley lines.
2. For each Steiner point, there are 3 Pascal lines, 1 Cayley line, and 3 Plücker lines.
3. For each Salmon point, there are 4 Cayley lines.

I would like to clarify the correspondences. As we have seen, a Steiner point S is the intersection of 3 Pascal lines indicated by 3 permutations of the vertices of the hexagon. The same permutations define 3 Kirkman points that belong to a Cayley line c. We say that point S and line c correspond.

Finally, each Plücker line contains 4 Steiner points defined by the same permutations that define a Salmon point. Points *Sa* and Plücker lines thus obtained are referred to as corresponding.

To aligned Steiner points, there correspond concurrent Cayley lines.

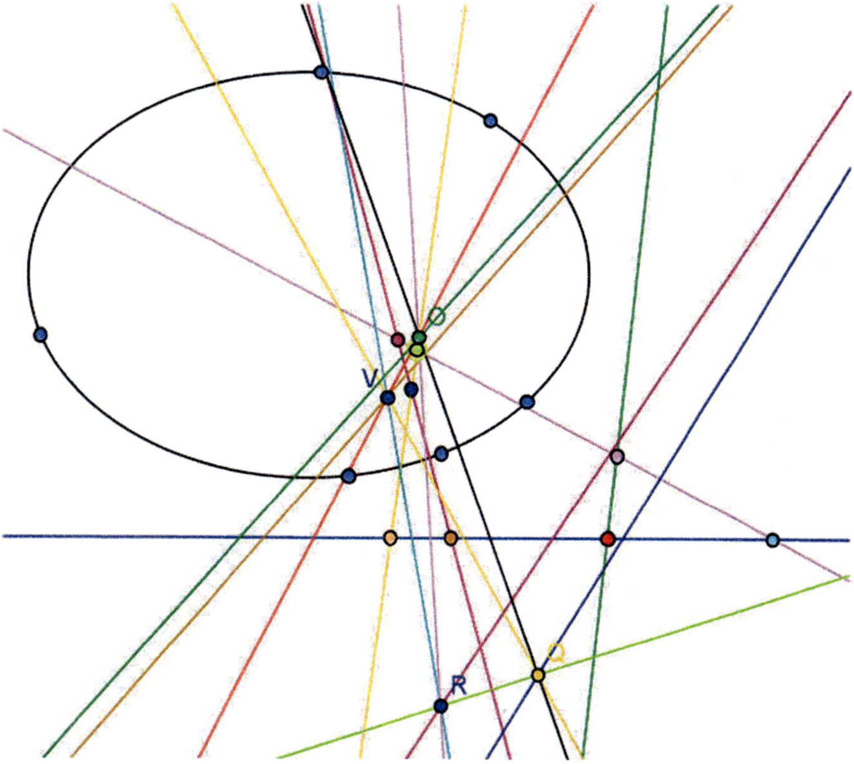

Fig. 12 Lines and points present in the first two configurations

This seems to me, as Enriques puts it, a *more hidden order of harmony where a marvelous beauty shines forth*. It is a harmony that in my opinion justifies the interest shown in this problem, which makes use of purely elementary methods provided by the greatest geometers of the time, and also justifies the enthusiasm with which Cremona welcomed the work of his young student, capable of not only giving order to the material so far accumulated, but also setting up further elements of profound harmony.

4 Veronese's Multi-Mystic

While working on that paper they discovered the results of this one, and were later surprised to find that a good number of them were already in an 1879 paper by Christine Ladd[12] which

[12] Christine Ladd (1847–1930) is best known for her contributions to psychology. In 1876, at the request of Sylvester himself, she had enrolled at Johns Hopkins University using, in presenting her

in turn attributed them to Veronese. They hope this paper will encourage other explorers [22 p. 45].

The complexity of the apparatus cannot escape us, but Veronese does not stop there; we have not even reached half of his work. I will limit myself to presenting those propositions that serve to realize what in [11] is called the ***multi-mystic***.

Indeed, the system contains new points and new lines. I will try to give a quick overview of it. Veronese introduces the new lines, called v, which connect two Kirkman points (for example, $K(ABCDEF)$ and $K(ABFDEC)$, aligned with point $P = (AE, BD)$). We will indicate this line as $v(AB;\ DE)$.[13] Through each Kirkman point, there pass 3 lines v: through $K(ABCDEF)$, there pass $v(AB; DE)$, $v(BC; EF)$, *and* $v(CD; FA)$. Hence, there are 90 lines v altogether (Veronese's theorem XXII).

In the figure, we have points $K = K(ABCDEF)$ and $H = K(ABFDEC)$, line $KH = v(AB; DE)$, which meets lines AE and BD at point G, and then points and $I = K(BCAEFD)$ and $J = K(CDBFAE)$ and lines $IK = v(BC; EF)$ and $JK = v(CD; FA)$ which pass, respectively, through points $N = (BF, CE)$ (not visible in the figure) and $M = (AC, DF)$ (Fig. 13).

To give a more complete picture, we can add the following:

1. On each of the 90 lines v lie 2 points K.
2. Through each of the 60 points K there pass 3 lines v.
3. Through each of the 45 points P there pass 4 lines p.
4. On each of the 60 lines p lie 3 points P.

The first 2 incidences are illustrated in the previous figure. Figure 14 illustrates 3 and 4. Point P is the point (AE, BD) through which there evidently pass lines $GH = p(AEFBDC)$, $RJ = p(AECBDF)$, $KL = p(AEFDBC)$, *and* $QS = p(AECDBF)$.

4.1 The First Step of the Multi-Mystic

Given a line $v(AB; DE)$, we can define its associated line, $v(AB; ED)$. Let us consider Veronese's configuration $V1(ABCDEF)$ (Fig. 5). For each point K, there are 3 lines v that pass through it. Their associates intersect at a point that we will call $K2$. Points $K2$ are collinear by 3. Points K2 are as many as points K, i.e., 10. The set of points

credentials, the name C. Ladd, which did not specify her gender. On the difficulties encountered in relation to the University's anti-female regulations, I refer to [16]. Here, it suffices to recall that in 1882, despite having regularly submitted the documents for the PhD, she was not allowed to obtain it. With a belated recognition of its error, the University conferred the qualification on her in 1926, when she had already turned 78! She was also a student of Sanders Peirce, who praised her studies on Boolean algebra.

[13] So, with this symbol, I mean the line joining the two Kirkman points obtained using the two permutations that have A, B as the first and second elements and D, E as the fourth and fifth and the remaining ones C, F in the two possible positions. Thus, $(BC; EF)$ refers to the line between $K(BCAEFD)$ and $K(BCDEFA)$. The lines $(AB; DE)$ and $(DE; AB)$ coincide, as is easily verified.

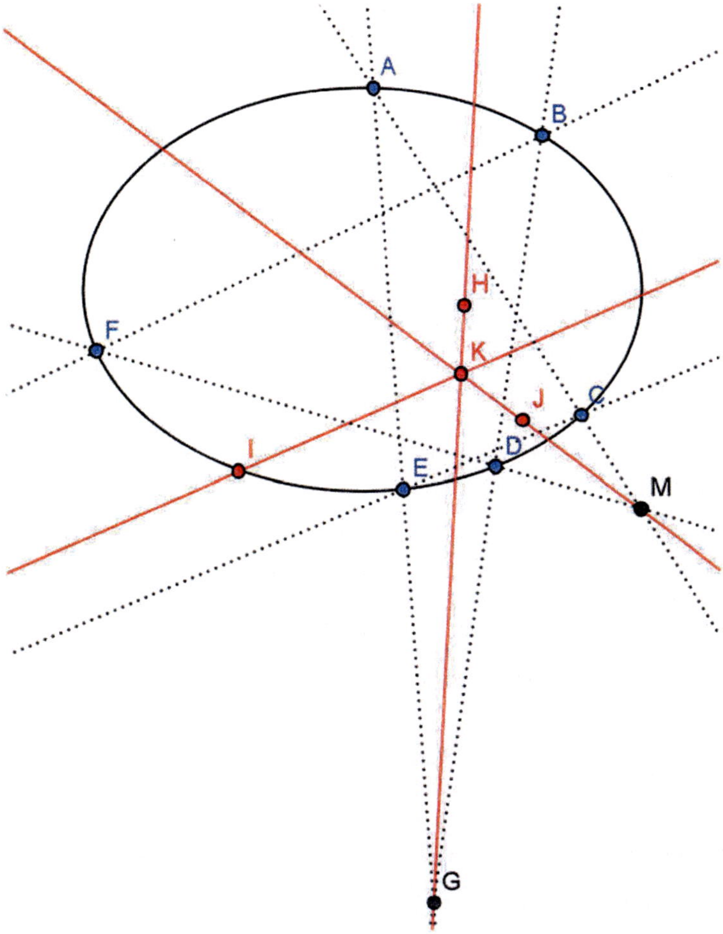

Fig. 13 Lines v

$K2$ and lines v that join them (which we will call lines $p2$) constitutes the second level of the multi-mystic.

Let us take a closer look at the process:

As we have already seen, through Kirkman point $K(ABCDEF)$, there pass lines v, $v(AB;DE) = KH$, $H = K(ABFDEC)$, $v(BC;EF) = KI$, $I = K(BCAEFD)$, $v(CD;FA) = KJ$, $J = K(CDBFAE)$. These lines are in red in the figure. The associated lines are the following: $v(AB;ED) = (K(ABCEDF), K(ABFEDC)) = NO$; $v(BC;FE) = (K(BCAFED), K(BCDFEA)) = PQ$; $v(CD;AF) = (K(CDBAFE), K(CDEAFD)) = RT$. In the figure, they are in purple. The three associated lines concur on a point, U, unfortunately very close to K. This is contained in Veronese's theorems XXII–XXIV, which also contain much more. In Fig. 15, I have kept the

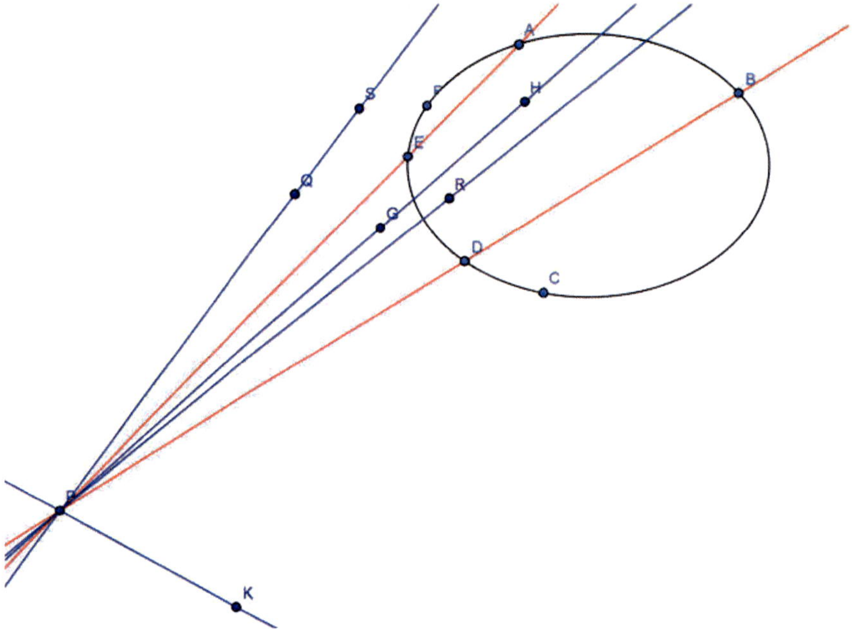

Fig. 14 Points P and lines p

Kirkman points relative to Veronese's first configuration to underline how the points that give rise to the second are, excluding K, points unrelated to this first one.

Point U and point K are in correspondence, and I will indicate them with the same symbol relating to the hexagon to which they correspond. Since K was denoted by $K(ABCDEF)$,I will indicate U with $K2 = K2(ABCDEF)$. We can do the same with points $K1, K2, \ldots, K9$ obtaining points $K2, 1, K2, 2, \ldots, K2, 9$. This represents the second level of Kirkman points $K2$. Given a Kirkman point, there are three concurrent Pascal lines in it. Points $K2$ corresponding to these 3 lines are aligned in a line v (theorem XXXI), which is therefore indexed as the corresponding Kirkman point. So, given that through $K(ABCDEF)$ there pass lines $p(ACEBFD)$, $p(ADBFCE)$, *and* $p(AEBDFC)$, points $K2(ACEBFD)$, $K2(ADBFCE)$, *and* $K2(AEBDFC)$ are aligned in a line that we will denote with $p2(ABCDEF)$. This procedure can be repeated for each Kirkman point, even simply applying the same transformations that generated it. For example, for point $K(ACEBFD)$, we find that through it there pass Pascal lines $p(AEFCDB)$, $p(ABCDEF)$, *and* $p(AFCBDE)$ and hence the relative points $K2$ will lie on line $p2(ACEBFD)$. At this point, it is very simple to reconstruct the configuration of the second level of the multi-mystic, indeed that of the 6 configurations into which it is split taking into account the incidences of the first level which remain identical. We just have to replace K with $K2$ and p with $p2$.

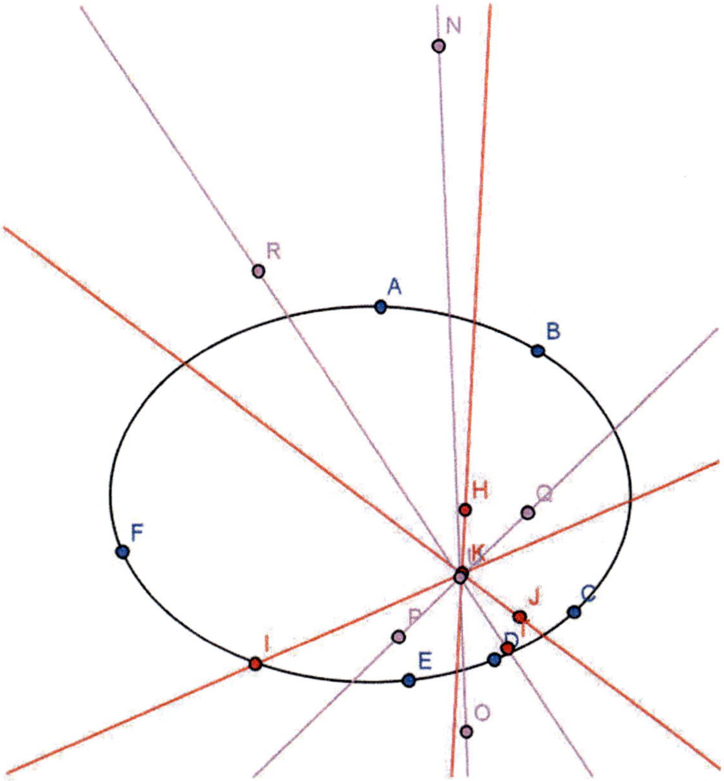

Fig. 15 Lines v

Thus, for the first configuration of the second level, we have red line $p2$, $p2(ABCDEF)$ and point $K2$, $K2(ABCDEF)$. For each of the Pascal lines passing through $K2$, we determine the corresponding point, and therefore $K2$, $1(ACEBFD)$, $K2$, $2(ADBFCE)$, *and* $K2$, $3(AEBDFC)$. The 6 lines (through each of the 3 points, there certainly passes $p2(ABCDEF)$) together with the 4 previously listed, together with the corresponding points, form the configuration.[14] Instead, the Steiner points, Cayley lines, Plücker lines, and Salmon points are the same as for the first level.

There follow some pictures[15] with the suggestion to construct them with GeoGebra (or other similar software). Making a comparison with Fig. 5, we can see in effect that the incidences are the same while the points and the lines are

[14] Perhaps, it is worth pointing out that here I have limited myself to a copy-paste of what was previously written about the first level. Perhaps, it is appropriate here to quote Poincaré saying that mathematics is "l'art de donner le même nom à des choses différents."

[15] I have slightly modified the arrangement of the vertices of the hexagon to make the figure more readable.

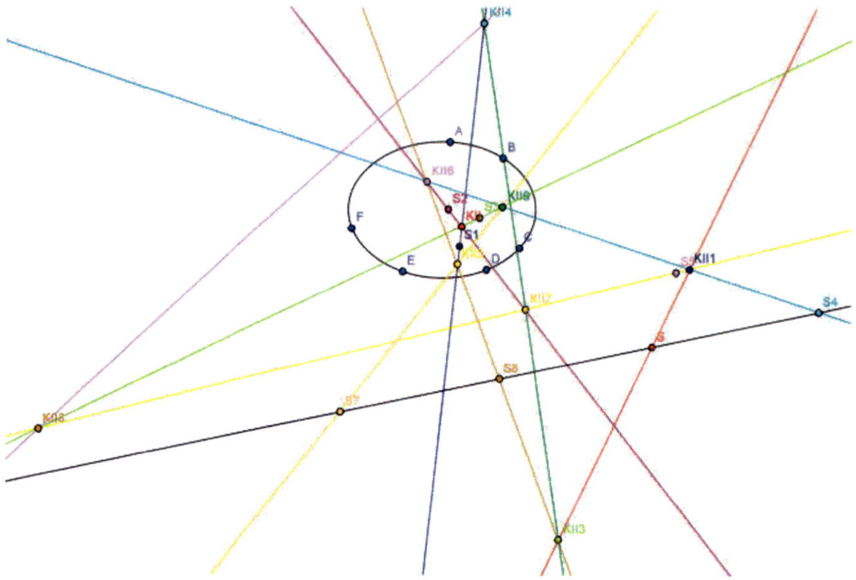

Fig. 16 The second level (partial)

completely different. We could say that the different levels are different models of the same structure (Fig. 16).

4.2 *The Second Step*

The passage from the second to the third level occurs in a perfectly dual way with respect to that from the first to the second. This time, the transition occurs through points V, which are the intersection points between two lines $p2$ connected by an exchange between the third and sixth letters. So, for example, $V(AB;DE)$ is the intersection between $p2(ABCDEF)$ and $p2$ $(ABFDEC)$. On each line $p2$, there are 3 points V. For instance, on line $p2(ABCDEF)$, we find $V(AB;DE)$, $V(BC;EF)$, *and* $V(CD;FA)$; the points associated with the 3, $V(AB;ED)$, $V(BC;FE)$, *and* $V(CD;AF)$, lie on a line, $p3(ABCDEF)$. Given a line $p2$, on it lie 3 points $K2$, whose corresponding lines concur at point $K3(ABCDEF)$. The rest continues in the exact same way, determining 60 lines $p3$ and 60 points $K3$ divided into 6 configurations, each of which determines a conic and the consequent polarity.

In Fig. 17, we have lines $p2(ABCDEF)$ in red and $p2$ $(ABFDEC)$ in green, which intersect at point $V = V(AB;DE)$ (not easy to see, being too near points $K2$, 1 and G, which will be defined below). On line $p2(ABCDEF)$, in addition to V, we find points $V1$ and $V2$, respectively, equal to $V(BC;EF)$ and $V(CD;FA)$, associated

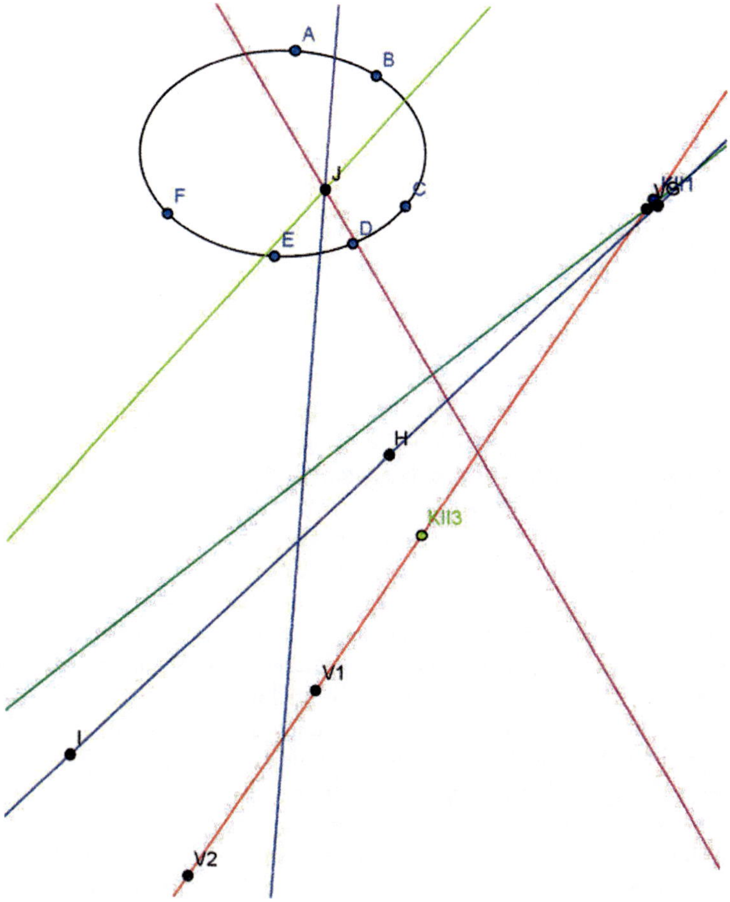

Fig. 17 Points *V* in the second level

with which are $G = (AB; ED)$, $H = (BC; FE)$, *and* $I = (CD; AF)$, lying on the line named $p3(ABCDEF)$, in blue. Further, in line $p2(ABCDEF)$, we find points $K2$, 1, $K2$, 2 (out of the picture), $K2$, 3, whose corresponding lines $p2$ meet at point $J = K3(ABCDEF)$.

The method can be repeated over and over again, achieving the multi-mystic using what Conway and Rybe define *a conceptually elegant inductive procedure to define an infinite sequence of structures* $\mathbf{M}(i)$ *for* $i > 1$.[16]

Veronese's article is still very rich in results, in particular, not only on the demonstrations of the results I have mentioned, but also on the involution relations

[16] By chance, I wrote this on April 18, 2021, a few days after the first anniversary of the death of John Conway (April 11, 2020), the recollection of whom, as a mathematician who was in many ways extraordinary and completely unusual, remains indelible.

existing for many of the geometric objects present. I will stop here, as I believe that I have given a sufficient idea of the innovations present in the work of the young mathematician from Chioggia.

5 Cremona's Interpretation of the Mystical Hexagram in the Theory of Cubic Surfaces

5.1 Cremona's Study on a Singular Cubic Surface

On April 8, 1877, before his graduation, the work was presented by Battaglini to the Accademia dei Lincei. Immediately after the presentation and the reading of a summary by Battaglini, Cremona spoke: *Having Mr. Veronese asked me to read his memoir, I thought to verify the results contained therein in a way other than that which A. had followed. While he always adhered to plane geometry, I had recourse to three-dimensional space and properly a third-order surface with a double point, and thus obtained figures that projected from the double point immediately provide that of Mr. Veronese.* [17]

Cremona's long talk continues with an exposition of his results, which can be considered a summary of the work [23] afterwards published in the *Memorie Lincee*. Cremona had immediately grasped the analogy of what Veronese described in the plane with configurations present in the cubic surfaces, of which he had deep knowledge.[18] I will try, within the limits of my scarce knowledge, to give an idea of the way of proceeding of the mathematician from Pavia.

Cremona knows that through the singular point O of a cubic surface Σ, there pass 6 of the lines on the surface, lines that are on a quadric cone with vertex O. Obviously, these lines (a, b, c, d, e, f) are projected from 0 into any plane at 6 points (A, B, C, D, E, F). The other 15 lines of the surface lie in planes ab, ac, $\ldots ef$, and they are indicated with the same name as the plane. These lines lie in threes on a tritangent plane, which I will indicate with a symbol that also identifies the triangle on it. The list of these tritangent planes (I use the same table as in Cremona) is as

[17] *Avendomi il sig. Veronese pregato di leggere la sua Memoria, io feci pensiero di verificare i risultati in essa contenuti per una* via *diversa da quella che l'A. aveva seguita. Mentr'egli si è attenuto sempre alla geometria piana, io ebbi ricorso allo spazio a tre dimensioni e propriamente ad una superficie di terzo ordine dotata di un punto doppio, ed ottenni così delle figure che proiettate dal punto doppio somministrano immediatamente quella del sig. Veronese* [Transunti della R. Accademia dei Lincei (3),1, seduta dell'8 aprile 1877, p. 146].

[18] The reader is reminded that not many years earlier, in 1866, Cremona had won the Steiner prize with a profound memoir [24] precisely on cubic surfaces. On them, and also on the memoir cited, see [25].

follows:

$$(ab, cd, ef)\ (ab, ce, df)\ (ab, cf, de)$$

$$(ac, bd, ef)\ (ac, be, df)\ (ac, bf, de)$$

$$(ad, bc, ef)\ (ad, be, cf)\ (ad, bf, ce)$$

$$(ae, bc, df)\ (ae, bd, cf)\ (ae, bf, cd)$$

$$(af, bc, de)\ (af, bd, ce)\ (af, be, cd)$$

This can also be said in the sense that the surface contains 15 triangles whose vertices are all different from 0 and each of the lines xy is the common side of three of these triangles.

Two triangles are said to form a pair when they have no common sides.[19] It is easy (you can count them directly in the table) to see that there are exactly 60 pairs. The tritangent planes of the pair intersect in a line (not belonging to the surface) called *Pascal line*.

An example can clarify the reasons for this nomenclature. The straight-line intersection of a pair intersects the surface in three points. For example, $\pi 1 = (ab, cd, ef)$ and $\pi 2 = (ac, be, df)$ intersect in a line. Lines *ab and df* are coplanar (in the plane (*ab*, *ce*, *df*)—it is sufficient to look at the table). Hence, they intersect at a point on the surface which will be one of the points sought. The same goes for *cd* and *be* and *ef* and *ac*. All this in projection becomes the Pascal line on which the intersections of the opposite sides of the hexagon are aligned.

Let us again consider the couple $\pi 1, \pi 2$. Their projection from *O* consists of the triangles *JKL* and *OMN* illustrated in the figure, which are evidently homological with axis *p*(*ABEFDC*), the image of the straight intersection between the two tritangent planes. Points $G = (AB, DF)$, $H = (BE, CD)$, and $I = (AC, EF)$ are the projections of the points at which Pascal line meets the surface (Fig. 18).

It is shown that a couple $\pi 1$ and $\pi 2$ can be completed in two ways: 1) Trihedron of the first kind: with a plane $\pi 3$ that forms a couple with both, but with none of the others. For example: the pair $\pi 1$ and $\pi 2$ can be completed with the plane $\pi 3 = (af, bd, ce)$. The edges are Pascal lines, and the vertex is a Steiner point, *S*(*abefdc*). Another possibility with the same sides is the triangle (*ab*, *df*, *ce*), (*ac*, *ef*, *bd*), (*af*, *cd*, *be*),which gives rise to the conjugated trihedron. 2) Trihedron of the second kind: it is also possible to choose three planes formed with the nine lines not lying on the first two planes, lines that also form a couple together. In our case,

[19] Of course, this is equivalent to saying that the corresponding tritangent planes intersect in a line not lying on the surface. Dolgachev calls the pair of tritangent planes thus obtained a *Cremona pair*, attributing this nomenclature to Reye.

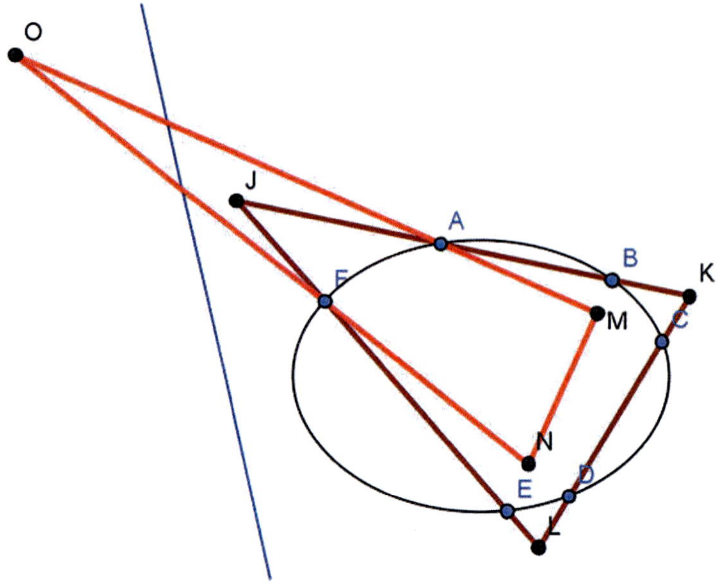

Fig. 18 The projection of triangles π1 and π2

they are (*ad, bf, ce*), (*ae, bd, cf*), *and* (*af, bc, ed*). We thus obtain another trihedron whose edges are Pascal lines and the vertex a Kirkman point. By projecting, one obtains Pascal's configuration (60 Pascal lines, 3 by 3 concurrent points in one of the 20 points Steiner and 3 by 3 in one of the 60 Kirkman points). The 5 planes considered (the initial 2 plus the 3 of the last trihedron) constitute the Cremona pentahedron, a configuration of 10 Pascal lines and 10 Kirkman points. Projected, they give the corresponding configuration of Veronese. It is not difficult to see that Pascal's configuration is divided into 6 pentahedra which have neither Pascal lines nor Kirkman points in common.

The edges of this trihedron are the corresponding Pascal lines, while the vertex is the *Kirkman point*. I present (Fig. 19) its projection. It can easily be seen that, in projection, the constructions on the cubic surface correspond to those in the plane[20] and, above all, that they make use of the same combinatorics.

I will not dwell on Cremona's further stereometric constructions. The fact is that it is necessary to move on to the mystical hexagram and its link with cubic surfaces.

If we consider the five triangles constituted by a pair and by the corresponding trihedron as in Fig. 20, we have the possibility of forming ten pairs and therefore ten trihedra which are all of the second kind. Each pair corresponds to a Pascal

[20] For example, it is enough to replace the word "triplet" of triangles with "trihedron" to pass identically from one construction to another. It will in fact be noted that the figures constructed for the projections correspond exactly to those created directly in the plane.

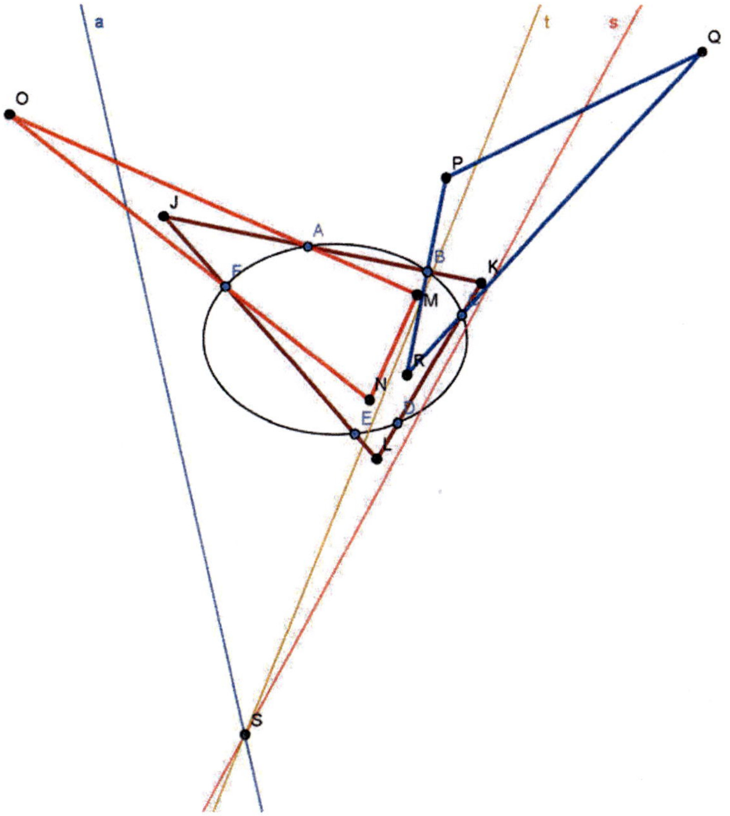

Fig. 19 The projection of a trihedron of the second kind

line (intersection of the tritangent planes) and to each trihedron there corresponds a Kirkman point (vertex) and three Pascal lines (edges). The five tritangent planes considered here constitute a pentahedron made up of ten trihedra. In total, we therefore have six pentahedra (*Cremona Pentahedra*) [5].

The 6 pentahedra are (the tritangent planes that are part of two different pentahedra are the same color):

$$I: ABCDEF, ACEBFD, ACFDBE, AECFBD, ABDCFE, AEDBCF,$$
$$ABFECD, ADECBF, ABEFDC, ACBEDF$$

$(ab, cd, ef), (af, bc, de), (ac, be, fd), (ad, bf, ec), (db, cf, ae),$

$$II: ABEFCD, AECBDF, AEDFBC, ACEDBF, ABFEDC, ACFBED,$$
$$ABDCEF, AFCEBD, ABCDFE, AEBCFD$$

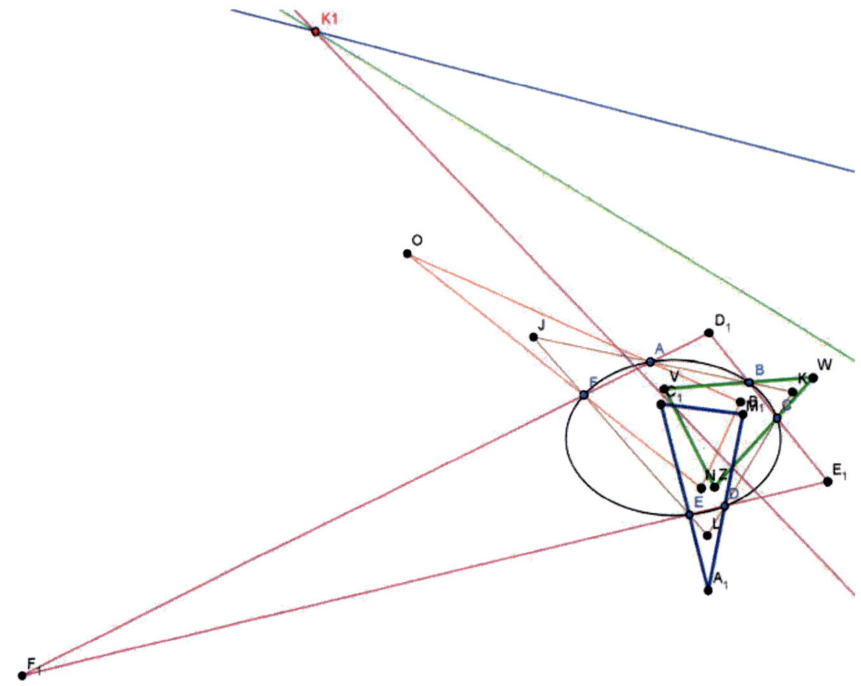

Fig. 20 The projection of a pentahedron

(ab, cd, ef), (cf, be, ad), (ae, df, bc), (bd, ce, af), (bf, de, ac)

$$III: AFCBED, AECDFB, AEFBDC, ACEFDB, ADBEFC, ACBDEF, ADFCEB, ABCEDF, ADCFBE, AEDCBF$$

(af, de, bc), (be, cf, ad), (ae, bf, cd), (df, ce, ab), (ac, bd, fe)

$$IV: ABEDCF, AECBFD, AEFDBC, ACEFBD, ABDEFC, ACDBEF, ABFCED, ADCEBF, ABCFDE, AEBCDF$$

(ab, cf, de), (cd, be, af), (ae, df, bc), (bf, ce, ad), (bd, ef, ac)

$$V: ADEFCB, ACEBDF, ACDFBE, AECDBF, ABFCDE, AEFBCD, AFCEDB, AFECBD, ABEDFC, ACBEFD$$

(ad, bc, ef), (cf, de, ab), (ac, df, be), (bd, ce, af), (bf, cd, ae)

$$VI: AFEBCD, ACEDFB, ACFBDE, AECFDB,$$
$$ADBCFE, AEBDCF, ADFECB, ABECDF, ADEFBC, ACDEBF$$

$(af, cd, be), (bc, ef, ad), (ac, bf, de), (df, ce, ab), (bd, cf, ae)$.

In summary: there are 21 lines on the surface, 6 of which pass through the singular point. The other 15 are 3 by 3 coplanar in the 15 tritangent planes, which in turn pass 3 by 3 through each of the lines.

The intersections between tritangent planes number 105, counting the repeated ones. 45 (15x3) are lines on the surface. The other 60 are in lines not belonging to the surface (Pascal lines). The planes are identified by the triangle of the 3 lines in which they cut the surface; the planes that intersect outside the surface correspond to triangles without common sides (pairs). Given a pair, the points at which its Pascal line intersects the surface are easily determined. The line constitutes the axis of homology between the two triangles.

Each tritangent plane π intersects the quadric cone on which the 6 lines through O lie in a conic intersected by the lines at points A, B, C, D, E, F. For example, $\pi 1$ contains 3 lines of the cubic (in black in the figure), forming the triangle (*ab*, *cd*, *ef*) with which I indicate the entire plane. $\Pi 1$ contains 8 Pascal lines (one for each of the 8 planes that form a pair with $\pi 1$). Four of these lines belong to a pentahedron and 4 to the other that has π in common. In the figure, they are indicated in red and blue, respectively. The lines meet at 28 points. The $6 + 6$ points of intersection between lines of the same color are Kirkman points (respectively belonging to the 2 pentahedra). Four of the intersections between lines of different colors are Steiner points (common to the 2 pentahedra and aligned in the Plücker line, in green). In the other 12 lines, *AB*, *CD*, *EF* also converge and they are the points at which Pascal's lines "pierce" the surface (Fig. 21).

As we have seen, Cremona achieved a significant and, in many ways, unexpected result. In a cubic surface with a singular point O, each tritangent plane intersects the other tritangent planes either in three lines of the surface (six planes, two for each line) or in eight lines which are the Pascal lines relative to the eight hexagons that can be inscribed in the conic in which the plane intersects the characteristic cone, completing the three chords intersected by the three lines. Similar constructions can also be given for the other relevant intersections via the mystical hexagram.

The four Steiner points found in the $\pi 1$ plane have only two Pascal lines of the plane concurrent in each of them. The third line will therefore not belong to $\pi 1$. Thus, we will have four Pascal lines outside the plane. It is not difficult (just a little boring) to demonstrate that these lines lie in a single plane. Cremona calls this plane "Plücker plane" and denotes it as the tritangent plane from which it derives, in our case (*ab*, *cd*, *ef*). We therefore have 15 Plücker planes that intersect the corresponding tritangent plane in a Plücker line.

Let us now consider the five (Plücker) lines found in the pentahedron (intersections of a tritangent plane with the corresponding Plücker plane). It can be seen

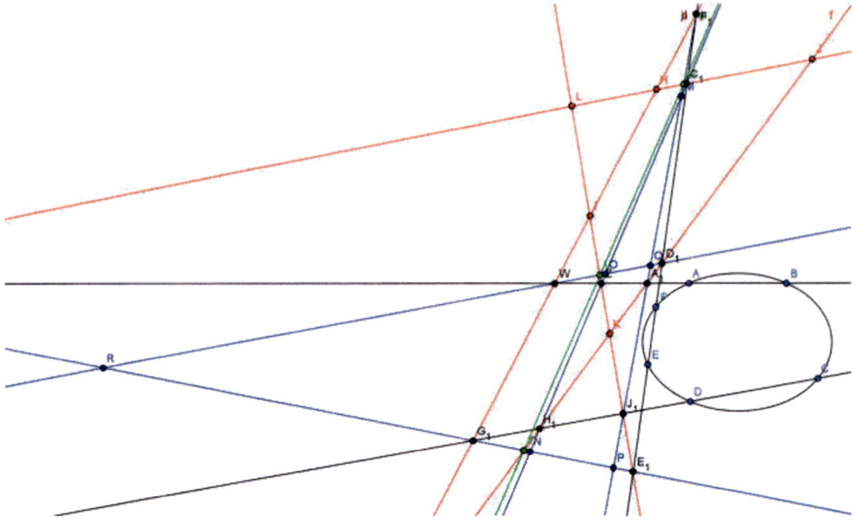

Fig. 21 The "geography" of a tritangent plane

that these lines concur in twos and are therefore on a plane.[21] The six planes thus obtained form a complete hexahedron. Its 15 edges are Plücker lines, while the 20 vertices are Steiner points. Each face of the hexahedron will contain five Plücker lines and ten Steiner points (Fig. 22).

I omit to follow the fascinating considerations regarding the determination of the Cayley lines and the Salmon points and their respective position with respect to the planes and lines already determined.

5.2 *Cremona and the Multi-Mystic*

To pass from the mystical to the multi-mystic, Cremona makes a significant change of point of view. The presence of the double point is not really necessary. In reality, it only serves to project the configuration of space on the plane, that is, to give a specific correspondence between the objects of one and the other; but this correspondence lies only in the invariance of formal relations and nothing else. This seems to me to be a significant step towards a more abstract vision of the substantial identity between the two configurations. I therefore leave the floor to Cremona himself:

> In the things explained so far, the hypothesis of a third-order surface Φ with a conical point O was only necessary because we wanted to have a projection center to deduce the

[21] For example, the Plücker line of $\pi 1$ intersects that of the plane $\pi 2$ at the Steiner point (*ACE*; *BDF*), as can be easily verified from the table.

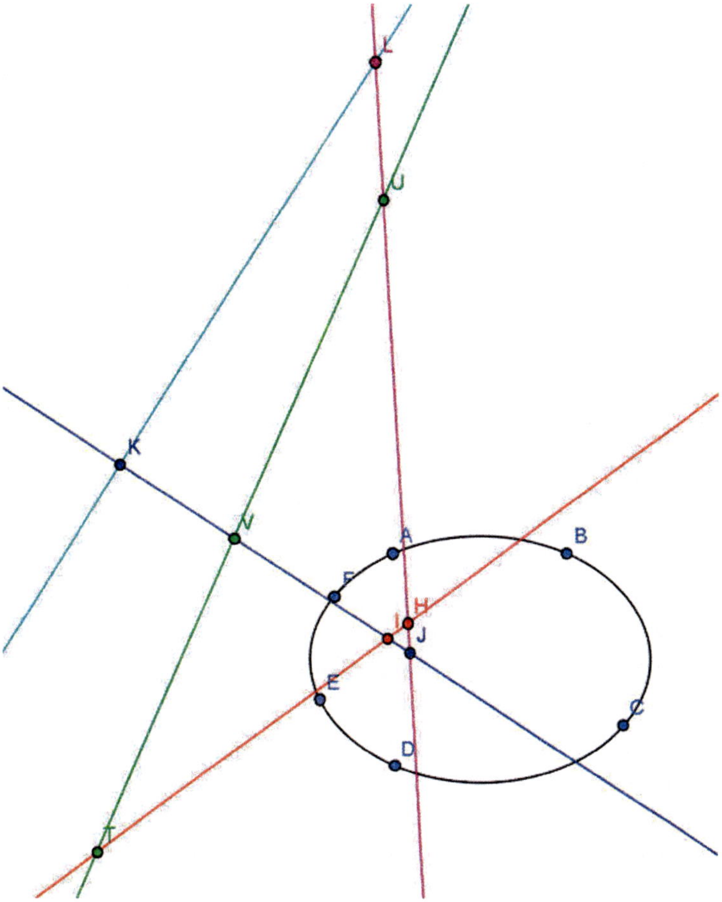

Fig. 22 The "geography" of the hexahedron's face linked to the first pentahedron (two points are outside the image)

planimetric theorems of the hexagrammum mysticum from the stereometric propositions. If we disregard this projection, the properties demonstrated in the first part of this memoir only presuppose the existence of a system of fifteen lines located three by three at fifteen planes. The fifteen planes are grouped into six pentahedra whose vertices and edges I have called Kirkman points and Pascalian lines, and are also grouped in twenty combined trihedra two by two, whose vertices are Steiner points: points located four by four in fifteen lines, called Steiner lines.[22] Steiner's lines and points are edges and vertices of a hexahedron, which in a way constitutes the core of the entire figure, with whose six faces the six pentahedra are coordinated. [23]

[22] In the foregoing, I have called these lines Plücker lines.

[23] *Nelle cose esposte sinora, l'ipotesi di una superficie Φ di terz'ordine dotata di un punto conico O non è stata necessaria se non in quanto s'è voluto avere un centro di projezione per dedurre*

Cremona then finds the 15 lines lying in threes in 15 planes in a general cubic using the known properties of double sixes. These are pairs of systems of six lines on the surface such that the lines of each system are skewed with respect to each other, while each is skewed with respect to one and only one of those of the other system. The lines in the two systems that are skewed with respect to each other are called conjugated. If we remove the 12 lines of a double six from the set of 27 lines of the general cubic, we obtain 15 lines that lie in threes on a tritangent plane. Since there are 36 double sixes, we obtain 36 systems isomorphic to the mystic. Then he reconstructs the whole mystical hexagram in Veronese's version.

More precisely, we have the following: we denote the 6 + 6 lines of the double six with $a_1 \dots a_6$; $b_1 \dots b_6$. Let the plane Pij be the plane (ai, bj) i $\neq$ j and ij the straight intersection between Pij and Pji. The 15 lines ij are grouped into 15 triplets such that all numbers 1 ... 6 are used once and only once. These 15 triangles lie on 15 defining planes. This determines the 15 tritangent planes and the 15 3-by-3 lines on them. The table of the tritangent planes is now identical to that of the previous case (it is only necessary to replace the letters a, b, ... f with the numbers 1, 2, ...6), and it can be organized in 6 pentahedra:

I: (12, 34, 56), (16,23, 45), (13, 25, 46), (14, 26, 35), (24, 36, 15),
II: (12, 34, 56), (36, 25, 14), (15, 46, 23), (24, 35, 16), (26, 45, 13)
III: (16, 45, 23), (25, 36, 14), (15, 26, 34), (45, 35, 12), (13, 24, 56)
IV: (12, 36, 45), (34, 25, 16), (15, 46, 23), (26, 35, 14), (24, 56, 13)
V: (14, 23, 56), (36, 45, 12), (13, 46, 25), (24, 35, 16), (26, 34, 15)
VI: (16, 34, 25), (23, 56, 14), (13, 26, 45), (45, 25, 12), (24, 36, 15).

The central point that allows Cremona to move on to Veronese's multi-mystic consists of the observation that the hexahedron (*the core of the entire figure*) can be considered independently of the surface.

Cremona observes that, instead of starting from the surface, the whole theory could be reconstructed starting from an arbitrary hexahedron.

We choose 4 of the hexahedron planes as coordinated planes: $x = 0$; $y = 0$; $z = 0$; $w = 0$. For one of the other 2 planes, we can choose the equation $x + y + z + w \equiv t = 0$; the equation of the sixth one will remain indeterminate: $ax + by + cz + dw \equiv u = 0$. Let us consider 30 planes passing 2 by 2 through the edges:

dalle proposizioni stereometriche i teoremi planimetrici dell'hexagrammum mysticum. Che se si prescinde da tale projezione, le proprietà dimostrate nella prima parte di questa Memoria presuppongono unicamente l'esistenza di un sistema di quindici rette situate tre a tre in quindici piani. I quindici piani si aggruppano in sei pentaedri i cui vertici e i cui spigoli ho chiamati punti di Kirkman e rette di Pascal; e si aggruppano inoltre in venti triedri conjugati due a due, i cui vertici sono i punti di Steiner: punti situati quattro a quattro in quindici rette, dette rette di Steiner. Le rette ed i punti di Steiner sono spigoli e vertici di un esaedro, che costituisce in certo modo il nucleo dell'intera figura ed alle cui sei facce sono coordinati i sei pentaedri [23 , p. 866].

$$(r-a)\,x \pm (r-b)\,y = 0;\ (r-a)\,x \pm (r-c)\,z = 0 : (r-a)\,x \pm (r-d)\,w = 0;$$
$$(r-a)\,x \pm rt = 0;$$

$$(r-a)\,x \pm u = 0;\ (r-b)\,y \pm (r-c)\,z = 0;\ (r-b)\,y \pm (r-d)\,w = 0;\ (r-b)\,y \pm rt = 0;$$
$$(r-b)\,y \pm u = 0 : (r-c)\,z \pm (r-d)\,w = 0;\ (r-c)\,z \pm rt = 0;$$
$$(r-c)\,z \pm u = 0;\ (r-d)\,w \pm rt = 0;$$

$$(r-d)\,w \pm u = 0;\ rt \pm u = 0.$$

Here, r is an indeterminate parameter and the $+$ and $-$ signs, respectively, indicate two harmonic planes with respect to the two faces on the edge.

Setting $(r-a)x = X$; $(r-b)y = Y$; $(r-c)z = Z$; $(r-d)w = W$; $rt = T$; $u = U$, we obtain $X + Y + Z + W + T + U = 0$.

The six planes $X = Y = Z = W = T = U = 0$ therefore constitute the hexahedron whose 15 edges are called the "Plücker lines" of the system and the 20 vertices the "Steiner points."

The 6 faces of the hexahedron will therefore be

$$X = 0;\ Y = 0;\ Z = 0;\ W = 0;\ T = 0;\ U = 0.$$

The 15 Plücker lines will be the edges, and therefore:

$$X = 0 = Y;\ X = 0 = Z;\ \ldots$$

Steiner's 20 points will be the vertices of the hexahedron, and therefore:

$$(X = Y = Z = 0)\ \text{con il coniugato}\ (W = T = U = 0)$$

$$(X = Y = W = 0)\ \text{con il coniugato}\ (Z = T = U = 0)$$

...

Six pentahedra are then constructed:

1. $X + kY = 0;\ X + kZ = 0;\ X + kW = 0;\ X + kT = 0;\ X + kU = 0$
2. $Y + kX = 0;\ Y + kZ = 0;\ Y + kW = 0;\ Y + kT = 0;\ Y + kU = 0$
3. $Z + kX = 0;\ Z + kY = 0;\ Z + kW = 0;\ Z + kT = 0;\ Z + kU = 0$
4. $W + kX = 0;\ W + kY = 0;\ W + kZ = 0;\ W + kT = 0;\ W + kU = 0$
5. $T + kX = 0;\ T + kY = 0;\ T + kZ = 0;\ T + kW = 0;\ T + kU = 0$
6. $U + kX = 0;\ U + kY = 0;\ U + kZ = 0;\ U + kW = 0;\ U + kT = 0$

Cremona therefore notes that, by calling the edges and vertices of these pentahedra "Pascal lines" and "Kirkman points," these pentahedra have, *whatever the value of k, ... properties similar to those of the pentahedra* defined in the previous case.

In particular, the set of their edges and their vertices is analogous to the system of the sixty Pascal lines and the sixty Kirkman points.[24]

Let us take a closer look at this statement: three edges of the pentahedron (e.g., $X + kY = X + kZ$; $X + kY = X + kW = 0$) obviously converge in the vertex $X + kY = Y + kZ = X + kZ = 0$. On each edge, there are three vertices, the completions of the two planes that define the edge to the three that define a vertex in the three possible ways.

Cremona now describes the remaining properties which, in the chosen coordinate system, are almost obvious.

For example, one edge of the pentahedron (let us say $X + kY = 0 = X + kZ$) contains a vertex of the hexahedron (Steiner point), in our case $X = Y = Z = 0$, through which there pass three edges of the pentahedron (Pascal lines), $X + kY = 0 = X + kZ$; $Y + kX = 0 = Y + kZ$; $Z + kX = 0 = Z + kY$.

The 15 planes $X + Y = 0$; $X + Z = 0$; ... correspond to the tritangent planes and the harmonic ones combined with them $X - Y = 0$; $X - Z = 0$; ... to the Plücker planes.

It is now quite evident that each Plücker line (edge of the hexahedron) contains 4 Steiner points (vertices): e.g., the edge $X = 0 = Y$ contains the vertices

$$X = Y = Z = 0;\ X = Y = W = 0;\ X = Y = T = 0;\ X = Y = U = 0.$$

Three Plücker planes ($X - Y = 0$; $Y - Z = 0$; $X - Z = 0$) intersect in a Cayley line ($X = Y = Z$) which passes through 3 Kirkman points:

$$(W + kX = 0,\ W + kY = 0,\ W + kZ = 0)\,,$$

$$(T + kX = 0,\ T + kY = 0,\ T + kZ = 0)\,,$$

$$(U + kX = 0,\ U + kY = 0,\ U + kZ = 0)\,.$$

The four Cayley lines $X = Y = Z$; $X = Y = W$; $X = Z = W$; $Y = Z = W$ intersect at the (Salmon) point: $X = Y = Z = W$.

Cremona is now ready to face the construction of the multi-mystic following Veronese's footsteps in the new context. I will give a brief example of the method followed:

We start from a Plücker line, $X = Y = 0$. It goes through four Steiner points: $X = Y = Z = 0$; $X = Y = W = 0$; $X = Y = T = 0$; $X = Y = U = 0$. Their conjugates are:

$$W = T = U = 0;\ Z = T = U = 0;\ Z = W = U = 0;\ Z = W = T = 0.$$

[24] *l'insieme dei loro spigoli e de' loro vertici è analogo al sistema delle sessanta rette di Pascal e de' sessanta punti di Kirkman* [23 , p. 871].

Three Pascal lines pass through each of these four points, 12 in total. These 12 lines concur in twos on six points (the V points):

$$Z = W = -kT = -kU;\ Z = T = -kW = -kU;\ Z = U = -kW - kT$$

$$W = T = -kZ = -kU;\ W = U = -kZ = -T;\ T = U = -kZ - kW.$$

It is easy to calculate that the V points number 90. They lie in threes in 30 lines. For example, points

$$Z = W = -kT = -kX;\ Z = W = -kT = -kY;\ Z = W = -kU$$

lie in line $Z = W = \ -kT$. These lines are the edges of six pentahedra obtained from the previous ones by replacing k with k^{-1}. For example, the previous line is the edge corresponding to $-kZ = \ -kW = T$, which is one of the edges of the fifth pentahedron of the previous series.

This is the first step in the multi-mystic series.

For the second step, we start from a Plücker plane, e.g., $X - Y = 0$, which contains 12 Kirkman points (vertices of pentahedra) and lies in twos on the six lines (lines v), intersections of the $X = Y$, and six planes:

$$(k-1)(X+Y)+2(Z+W)=0;\ (k-1)(X+Y)+2(Z+T)=0;\ (k-1)(X+Y)+2(Z+U)=0$$

$$(k-1)(X+Y)+2(W+T)=0;\ (k-1)(X+Y)+2(W+U)=0;\ (k-1)(X+Y)+2(T+U)=0$$

The v lines therefore number 90 (6 for each of Plücker's 15 planes). They concur in threes on 60 points, the vertices of a new pentahedron, e.g., the three v lines

$$(k+1)(Y+Z)+2(W+T)=0,\ Y=Z;$$

$$(k+1)(Z+X)+2(W+T),\ Z=X;$$

$$(k+1)(X+Y)+2(W+T),\ X=Y$$

intersect at point $W + T = (1 - k)X = (1 - k)Y = (1 - k)Z$. From $X + Y + Z + W + T + U = 0$, we obtain $-U = (4 - k)X = (4 - k)Y = (4 - k)Z$, which is a vertex of the pentahedron

$$U + (4-k)X = 0;\ U + (4-k)Y = 0;\ U + (4-k)Z = 0;\ U + (4-k)W = 0;$$
$$U + (4-k)T = 0$$

obtained from the sixth one of the previous series by substituting k with $4 - k$.

Therefore, starting from a pentahedron, we have two different ways to obtain a system with the characteristics given in relation to edges and vertices (Pascal lines and Kirkman points): (a) changing k to k^{-1} and (b) changing k to $4 - k$. Alternating the two mutations, we have Veronese's multi-mystic, obtaining, for example, starting from $k = 1$, the succession[25] 1, 3, 1/3, 11/3, 3/11, 41/11, ... As Cremona says: *To obtain precisely the figures ... considered by the young geometer ... it is enough to start from the pentahedron k = 1, deduce from it the conjugate in the involution k + k '= 4, then from this the conjugate in the involution k' k "=1, then from the latter the conjugate in the involution k "+ k "'= 4, and so on, alternating the two involution indefinitely.* [26]

Cremona had already shown previously that, once the 15 tritangent planes have been set,

$X + Y = 0$, ..., the generically smooth surface of the equation.

$(X + Y)(X + Z)(Y + Z) + (W + T)(W + U)(T + U) = 0$ is determined. We therefore have a representation of the mystical hexagram through a smooth surface, without a singular point from which to project the configuration onto a plane.

The reconstruction of the surface proceeds as follows: each tritangent plane contains three lines belonging to the surface. For example, the plane $X + Y = 0$ is cut by the planes $W + T = 0$; $W + U = 0$; $T + U = 0$ in three lines clearly belonging to the surface. We thus have the 15 lines that belong to the surface and to the tritangent planes. The remaining 12 lines belonging to the surface form a double six, the construction of which is indicated by Cremona.

At this point, the circle closes: starting from a nodal cubic surface, Cremona has constructed a configuration isomorphic with Pascal's through a projection that allows you to "see" the isomorphism. This representation includes both the "classic" results of Steiner, Plücker, Kirkman, Cayley, and Salmon and the breakdown into the six configurations found by Veronese. Cremona therefore takes an important step in the direction of strong abstraction, a typical point in mathematics and in particular in that of the late nineteenth century. I like to hypothesize that he asked himself the meaning of this "structural" analogy between two questions a priori as distant from each other as Pascal's theorem and the geometry of a cubic surface (one of the "furtive caresses" of which André Weil speaks?). Cremona finds an answer in the fact that the 15 planes tritangent to a cubic correspond to as many lines of the surface, spread out three by three on them.[27] Practically, everything follows from this. It is a matter of giving new names to old objects. So, if we call the 60 intersections between tritangent planes not belonging to the surface "Pascal lines," group them into six pentahedra and call their vertices "Kirkman points"; we have, as we have seen, a formal structure perfectly corresponding to that of the mystical hexagram—60 mystical hexagrams where, as Dolgachev notes, *there is no conic in which the hexagon is inscribed!*

---

[25] A similar succession is obtained in [26] (where it is called a *Veronese sequence*).

[26] *Per ottenere precisamente le figure ... considerate dal giovane geometra ... basta partire dai pentaedri k = 1, dedurre da esso il conjugato nell'involuzione k + k' = 4, poi da questo il coniugato nell'involuzione k'k" = 1, indi da quest'ultimo il conjugato nell'involuzione k" + k"' = 4, e così* via *di seguito, alternando le due involuzioni indefinitamente* [23 , p. 874].

[27] As is well known, today this configuration $(15)_3$ is called the Cremona-Richmond one [4, 5]. The name of Richmond derives from the hyperspace reinterpretation made by the American mathematician of Cremona's construction in [27].

A further step towards abstraction is taken by Cremona when he notices that there is not even a need to have the cubic surface as a starting point, and it is enough to start from a complete hexahedron, interpreting as "tritangent planes" an appropriate choice of planes passing through its 15 edges and so on (as we have seen).

We therefore have a new vision of the problem that unifies different topics in a deep and in some respects unexpected way. We can say that Cremona is already on Poincaré's wavelength when he says: *Les mathématiciens n'étudient pas des objets, mais des relations entre les objets; il leur est donc indifférent de remplacer ces objets par d'autres, pourvu que les relations ne changent pas. La matière ne leur importe pas, la forme seule les intéresse* [28 , p. 32].

I will return to these topics in the conclusions. It seems right to me that this section should end with a mention of the work [3], closely connected with this one, the last one he dedicated to his beloved cubic surfaces and one which is still cited today.

During the elaboration of his work, Cremona probably already dealt with a question deriving from his studies on the hexagram, but entirely concerning the study of cubic surfaces, and decided to make it the subject of his speech at the Congress of Naturalists that would be held in September in Munich. He wrote as follows to his friend Hirst on August 24, 1877: *I am now busy with the problem of determining the pentahedron of a third-order surface of which 27 lines are given. Basically, I have already obtained the geometric solution, deducing it from certain recent theorems of Reyes on polar pentahedra and hexahedra ... with the results I obtained with the stereometric treatment of the hexagrammum mysticum (About the memoir of Mr. Veronese)*[28]; and on November 7th (after the presentation on September 19th): *I made a small communication to the Mathematics Section about the solution of the problem of reducing the equation of a general surface of the third order to the form*

$$\sum_{r=1}^{r=6} x_r^3 = 0 \text{ under the condition } \sum x_r = 0$$

It is a result I arrived at while I was in the Alps... continuing the research contained in the small work printed by Lincei.[29]

Cremona's demonstration is linear: the 20 Steiner points, the vertices of the hexahedron, are vertices of 20 trihedra conjugated by two: for example, the trihedra.

---

[28] Letter from Cremona to Hirst of August 24, the Italian original in [29 , p. 181], reads: *Sono ora occupato intorno al problema di determinare il pentaedro d'una superfice del 3° ordine della quale siano date le 27 rette. In sostanza ho già ottenuta la soluzione geometrica, deducendola da certi teoremi recenti di Reye sui pentaedri ed esaedri polari ... coi risultati da me ottenuti colla trattazione stereometrica dell'hexagrammum mysticum (A proposito della memoria del Sig. Veronese).*

[29] Letter from Cremona to Hirst of November 7, 1877, *Io feci una piccola comunicazione alla Sezione Matematica* circa *la soluzione del problema di ridurre l'equaz.^{e} di una superficie generale del 3° ordine alla forma*

$(x_1 = x_2 = 0; x_2 = x_3 = 0; x_3 = x_1 = 0)$, $(x_4 = x_5 = 0; x_5 = x_6 = 0; x_6 = x_4 = 0)$ give rise, as was well known, to the Cayley-Salmon equations of the surface:

$$(x_1 + x_2)(x_2 + x_3)(x_3 + x_1) + (x_4 + x_5)(x_5 + x_6)(x_6 + x_4) = 0.$$

Bearing in mind the identity

$$(x_1 + x_2 + x_3 + x_4 + x_5 + x_6)\left[(x_1 + x_2 + x_3)^2 + (x_4 + x_5 + x_6)^2 - (x_1 + x_2 + x_3)(x_4 + x_5 + x_6)\right]$$

$$= x_1{}^3 + x_2^3 + x_3^3 + x_4^3 + x_5^3 + x_6^3 + 3(x_1 + x_2)(x_2 + x_3)(x_3 + x_1) + 3(x_4 + x_5)(x_5 + x_6)(x_6 + x_1)$$

we find that, setting $x_1 + x_2 + x_3 + x_4 + x_5 + x_6 = 0$, the equation of the general cubic surface can be put in the form $\sum_{r=1}^{r=6} x_r^3 = 0$,[30] precisely as Cremona had found during his alpine sojourn, *continuing the researches contained in the small work printed by Lincei.*

6 Conclusions

The arrival in Rome of the young Veronese coincides with a period of serious crisis in Cremona. The yoke of Rome bureaucratic duties oppresses him, and he wants to escape this *slavery of Babylon* (to Hirst in July 27 [29 , p. 179]). As he wrote to his wife: *You will remember that when I was offered the current office in Rome I did not accept it without hesitation ... I expected to have to undertake very serious tasks, but I would never have dreamed of such a huge and complete sacrifice. All my time and all my strength taken up by administrative work, I could no longer do anything for science, in which my only ambition lies.* [31] The opportunity had come precisely in that summer of 1877, when, on July 9, he talked about it with Enrico Betti, who offered him an attractive prospect, moving to Pisa: *Having science and the professorship as my sole occupation, having you as colleagues, Dini, Bertini* etc.

$$\sum_{r=1}^{r=6} x_r^3 = 0 \textit{ sotto la condizione } \sum x_r = 0$$

È un risultato al quale pervenni mentre mi trovavo nelle Alpi ... in continuazione delle ricerche contenute nel piccolo lavoro stampato dai Lincei [29 , p. 183].

[30] This is the famous and much-cited hexahedral equation of Cremona. For a modern treatment of the subject, see [25].

[31] *Ti ricorderai che quando mi fu offerto l'attuale ufficio in Roma io non accettai senza esitazione ... Prevedevo di dovermi sobbarcare a occupazioni assai gravi, ma non mi sarei mai sognato un sacrificio così enorme e completo. Assorbito tutto il mio tempo, tutte le mie forze da lavori amministrativi, non potei più far nulla per la scienza, nella quale è riposta l'unica mia ambizione* [Letter to his wife of July 29 1877, Lascito Itala Cremona Cozzolino–Archivio dell'Istituto mazziniano di Genova, 051–11,831].

and living in a quiet city like Pisa ... this is my place! It seemed like a done deal. In particular, Betti's academic and political strength ensures the minister's support. At the end of August, he is now convinced that it is done. At San Martino di Castrozza, preparing for the Munich talk, he can indulge in optimistic lyricism: *In this cool and healthy place where the eye admires the gigantic peaks of the Dolomites and rests on the green of meadows and vast woods, where no echoes of Roman troubles come, I lead a truly happy life. The time divided between walks and excursions, together with my family, and scientific work, flows simply, serenely—very differently from what happened to me in Rome.* ***And the present is all the sweeter to me because I hope for the future.*** [32]

The work continues inspired by his young alumnus, but much more his previous studies. Presented at the Munich congress, where he will meet Klein, Lie, Brill, Noether, Sturm, Reye, Gordan, Lüroth, and Geiser, ... it seems to confirm the hopes of the still young Cremona (not yet 47 years old) in a future entirely dedicated to research.

It is well known that things turned out differently: *Arriving in Rome ... I underwent formidable assaults from Blaserna, Battaglini, Cannizzaro and from several teachers of the school.* [33] Eventually, Cremona capitulated. He was to stay in Rome. He had written to his wife: *Now or never. I am now at an age where there is no more time to waste. If I delayed a few more years, returning to my studies would be impossible for me and there would be nothing left of me but a bad bureaucratic tool.* [34]

Now or never. Cremona was well aware that in the moment of profound transformation that mathematics was going through in the 1870s, contact with the most advanced points of mathematical research could not be lost. Even leaving aside the great innovations made in those years by the works of Weierstrass, Cantor, Dedekind, Klein, and Lie, ... it should be remembered that in those years Max Noether, in collaboration with Alexander Brill and in continuation of the works carried out with Rudolf Clebsch (probably Cremona's closest friend and collaborator in the second half of the 1860s), had laid the foundations for a revolution in the methods of algebraic geometry. Cremona's failure to respond to

---

[32] *In questo luogo fresco e salubre dove l'occhio ammira le gigantesche guglie dolomitiche e si riposa sul verde de' prati e de' boschi estesissimi, dove non giunge alcun'eco de' fastidi romani, io conduco una vita veramente felice. Il tempo diviso tra le passeggiate e le escursioni, insieme con la mia famiglia, e il lavoro scientifico, scorre semplicemente, serenamente - in modo assai diverso da quello che mi accadeva in Roma. E il presente mi riesce tanto più dolce perché spero nell'avvenire* (Letter to Hirst, August 24 1877, in [29 , p. 181]).

[33] *Arrivando a Roma ... ebbi assalti formidabili da Blaserna, Battaglini, Cannizzaro e da parecchi professori della scuola* (Letter of Cremona to Enrico Betti, October 8 1877, in [30 , p. 79]).

[34] *Ora o mai. Sono ormai ad una età in cui non c'è più tempo da perdere. Se tardassi ancora qualche anno il ritorno agli studj mi sarebbe impossibile e non resterebbe di me che un cattivo arnese burocratico* [Letter to his wife, July 29, 1877 already quoted].

these stimulations is one of the main aspects of that **never** that he feared. Thus, Castelnuovo observes:

> Introducing a new group of transformations into geometry, not artificially imposed, but imposed by nature itself means, in the first place, offering the means of transporting known properties of simple entities to more complex entities ... But secondly it means giving rise to the study of those geometric properties that are not altered by the transformations themselves. Of these two parts of the program that the discovery of Cremona made it possible to formulate he carried out the first; ... The second part of the program ... inspired the main researches that were carried out in the field of algebraic geometry in the last fifty years ... **It is difficult to say if he foresaw its subsequent developments, but I can affirm, also from pleasant personal memories, that he followed its progress with the greatest interest, up to the last days of his life.** [35]

The "missed opportunities" of Cremona in his last 25 years of life were numerous, and it is not appropriate to go over them here.

Cremona felt the great opportunity that was being offered to him at the moment of his greatest creativity and fame slipping away from him. Perhaps, it is no coincidence that in 1877 when he saw his attempted move to Pisa fail, Cremona manifested intolerance towards Bertini's detachment from his preferred methods.[36] Perhaps, he had hoped that things had gone as with Veronese and the mystical hexagram: that is, demonstrating that he was able, not only to follow the work of the students, but also to highlight the deep links in it with his methods and his favorite topics. This is the frame of reference in which I place what is his last significant contribution to algebraic geometry.

Before ending this section, I would like to emphasize that, if in regard to active research Cremona's "now or never" is in a way prophetic, certainly the other prediction, that of being reduced to a *bad bureaucratic tool*, is not. The 30 years spent in Rome were years in which he made an essential contribution to the creation of a "Rome capital," a cultural center (through the National Library, which he directed), a political center (also through his work in the Senate, of which he was also vice president), a scientific one (through the Accademia dei Lincei, to whose affirmation he made a very valid contribution alongside Quintino Sella), a university (through his 30-year direction of the Application School), and a mathematical one (through the creation of a mathematical school represented by Bertini, Veronese, De Paolis, Caporali, but also by engineers such as Saviotti, ...).

---

[35] *Introdurre nella geometria un nuovo gruppo di trasformazioni, non imposte artificialmente, ma imposte dalla natura stessa vuol dire, in primo luogo, offrire il mezzo di trasportare proprietà note di enti semplici ad enti più complessi ... Ma vuol dire in secondo luogo dar origine allo studio di quelle proprietà geometriche che non vengono alterate dalle trasformazioni stesse. Di queste due parti del programma che la scoperta del Cremona permetteva di formulare egli svolse la prima; ... La seconda parte del programma ... ha ispirato le principali ricerche che nel campo della geometria algebrica furono condotte nell'ultimo cinquantennio ... È difficile dire se egli ne prevedesse i successivi sviluppi, ma posso affermare, anche per graditi ricordi personali che egli ne seguì i progressi col più grande interesse, fino agli ultimi giorni della sua vita* [31].

[36] On the "coldness" with which Cremona received Bertini's important results of 1877, I refer to [32].

In the 1880s, there were numerous developments of the different ideas contained in the two works of the teacher and the alumnus, which were so different, but so intimately linked. But here, I can only promise myself to trace out these developments (*deo favente*) following further studies.

References

1. Linton, A., Linton, E.: Pascal's Mystic Hexagram, its History and Graphical Representation. Sagwan Press, Philadelphia (1921)
2. Cayley, A.: A notation for points and lines in Pascal's theorem. Quarterly Mathematical Journal. **9**, 268–274 (1868)
3. Cremona, L.: Über die Polar-Hexaeder bei den Flächen dritter ordnung. Math. Ann. **13**, 301–304 (1878)
4. Coxeter, H.: Self-dual configurations and regular graphs. Bull. Am. Math. Soc. **56**, 413–455 (1950)
5. Dolgachev, I.: Abstract configurations in algebraic geometry. In: Collino, A., Conte, A., Marchisio, M. (eds.) The Fano Conference, pp. 423–462. Dipartimento di Matematica, Torino (2004)
6. Veronese, G.: Nuovi Teoremi sull'Hexagrammum Mysticum. Memorie della Reale Accademia dei Lincei. **3**(1), 649–703 (1877)
7. Confalonieri, S., Schmidt, P., Volkerts, K.: Der Briefwechsel von Wilhelm Fiedler mit Alfred Clebsch, Felix Klein und italienischen Mathematikern. Universi (2019)
8. Israel, G. (ed.): The Correspondence of Luigi Cremona. Brepols, Turnhout (2017)
9. Pascal, B., Essai sur les Coniques. (1640)
10. Steiner, J.: Question proposées. Théorèmes sur l'hexagramum mysticum. Annales de Mathématiques pures et appliquées. **18**, 339–340 (1827/1828)
11. Conway, J., Ryba, A.: The Pascal mysticum demystified. Math. Intell. **34**, 4–8 (2012)
12. Plücker, J.: Über ein neues Princip der Geometrie und den Gebrauch allgemeiner Symbole und unbestimmter Coefficienten. Journal für die reine und angewandte Mathematik. **5**, 268–286 (1829)
13. Steiner, J.: Systematische Entwickelung der Abhängigkeit geometrischer Gestalten. Engelmann, Berlin (1832)
14. Tahta, D.: The Fifteen Schoolgirls. Black Apollo Press, Cambridge (2006)
15. Cayley, A.: Sur quelques Théorèmes de la Géométrie de position. Journal für die reine und angewandte Mathematik. **31**, 213–227 (1846) 34, 1847, pp. 270–275; 38, 1848, pp. 97–104; 41, 1851, pp. 66–72
16. Vaughn, K. Profile. Christine Ladd-Franklin. https://web.archive.org/web/20180925065213/http://www.feministvoices.com/christine-ladd-franklin/
17. Kirkman, T.: On the complete hexagon inscribed in a conic section. The Cambridge and Dublin Mathematical Journal. **5**, 185–200 (1850)
18. Salmon, G.: A Treatise on Conic Sections, 3rd edn. C. J. Clay and Sons, London (1855)
19. Hesse, O.: Über die Reciprocität der Pascal - Steinerschen und die Kirkman - Cayley - Salmonschen Sätze von der Hexagrammum mysticum. Journal für die reine und angewandte Mathematik. **68**, 193–207 (1968)
20. von Staudt, C.: Über die Steiner'sche Gegenpunkte. Journal für die reine und angewandte Mathematik. **62**, 142–150 (1863)
21. Ladd, C.: The Pascal hexagram. Am. J. Math. **2**, 1–12 (1879)
22. Conway, J., Ryba, A.: Extending the Pascal Mysticum. Math. Intell. **35**, 44–51 (2013)

23. Cremona, L.: Teoremi stereometrici dai quali si deducono le proprietà dell'esagrammo di Pascal. Atti della R. Accademia dei Lincei, Memorie della classe di scienze fisiche, matematiche e naturali. **3**(1), 854–874 (1876/1877)
24. Cremona, L.: Mémoire de Géométrie pure sur les Surfaces du troisième ordre. Journal des Mathématiques pures et appliquées. **68**, 1–133 (1868)
25. Dolgachev, I.: Luigi Cremona and Cubic Surfaces. Cornell University Library, Ithaca. http://www.math.lsa.umich.edu/~idolga/cremona.pdf
26. Chipalkatti, J., Ryba, A.: Absolute projectivities in Pascal's Multimysticum. Int. J. Geom. **9**, 114–136 (2020)
27. Richmond, H.: The figure formed by six points in the space of four dimensions. Q. J. Math. **31**, 125–160 (1900)
28. Poincaré, H.: La Science et l'Hypotèse. Flammarion, Paris (1902)
29. Nurzia, L. (ed.): La corrispondenza di Luigi Cremona, vol. IV. Università degli studi di Roma "La Sapienza", Palermo (1999)
30. Menghini, M. (ed.): La corrispondenza di Luigi Cremona, vol. III. Università degli studi di Roma "La Sapienza", Palermo (1996)
31. Castelnuovo, G.: Luigi Cremona. Rendiconti della Reale Accademia dei Lincei. **6**(12), 613–618 (1930)
32. Kleiman, S.: Bertini and his Two Fundamental Theorems. *Rendiconti del Circolo Matematico di Palermo – Supplemento*. Cornell University Library, Ithaca (1997)